Rays, Waves, and Scattering

PRINCETON SERIES IN APPLIED MATHEMATICS

Ingrid Daubechies (Duke University); Weinan E (Princeton University); Jan Karel Lenstra (Centrum Wiskunde & Informatica Fellow, Amsterdam); Endre Süli (University of Oxford)

The Princeton Series in Applied Mathematics publishes high quality advanced texts and monographs in all areas of applied mathematics. Books include those of a theoretical and general nature as well as those dealing with the mathematics of specific applications areas and real-world situations.

For a list of books in the series, see page 589.

Rays, Waves, and Scattering

Topics in Classical Mathematical Physics

John A. Adam

PRINCETON UNIVERSITY PRESS

PRINCETON AND OXFORD

Published by Princeton University Press, 41 William Street, Princeton, New Jersey 08540

In the United Kingdom: Princeton University Press, 6 Oxford Street, Woodstock, Oxfordshire OX20 1TR

press.princeton.edu

Jacket photograph courtesy of the author

ISBN 978-0-691-14837-3

Library of Congress Control Number 2017934760

British Library Cataloging-in-Publication Data is available

This book has been composed in Times New Roman

Printed on acid-free paper ∞

Typeset by Nova Techset Pvt Ltd, Bangalore, India

Printed in the United States of America

10 9 8 7 6 5 4 3 2 1

Which is the way to the place where light is scattered and the east wind is spread across the earth?

Job 38:24

Contents

Preface   xvii

Acknowledgments   xxiii

Chapter 1   Introduction   1

1.1   The Rainbow Directory   3
         1.1.1   The Multifaceted Rainbow   3
1.2   A Mathematical Taste of Things to Come   5
         1.2.1   Rays   5
         1.2.2   Waves   6
         1.2.3   Scattering (Classical)   7
         1.2.4   Scattering (Semiclassical)   9
         1.2.5   Caustics and Diffraction Catastrophes   11

PART I.   RAYS   15

Chapter 2   Introduction to the "Physics" of Rays   17

2.1   *What Is a Ray?*   17
         2.1.1   Some Mathematical Definitions   18
         2.1.2   Geometric Wavefronts   19
         2.1.3   Fermat's Principle   21
         2.1.4   The Intensity Law   21
         2.1.5   Heuristic Derivation of Snell's Laws   23
         2.1.6   Generalization   24
2.2   Geometric and Other Proofs of Snell's Laws of Reflection
         and Refraction   25
         2.2.1   The Law of Reflection   25
         2.2.2   The Law of Refraction   26
         2.2.3   A Wave-Theoretic Proof   28
         2.2.4   An Algebraic Proof   29

Chapter 3   Introduction to the Mathematics of Rays   33

3.1   Background   33
3.2   The Method of Characteristics   34

3.3	Introduction to Hamilton-Jacobi Theory	37
	3.3.1 Hamilton's Principle	39
	3.3.2 Rays and Characteristics	39
	3.3.3 The Optical Path Length Revisited	43
3.4	Ray Differential Geometry and the Eikonal Equation Again	46
	3.4.1 The Mirage Theorem for Horizontally Stratified Media	49
	3.4.2 A Return to Spherically Symmetric Media: $n(r)$ Continuous	51
3.5	Dispersion Relations: A Wave-Ray Connection	54
	3.5.1 Fourier Transforms and Dispersion Relations	55
	3.5.2 The Bottom Line	56
	3.5.3 Applications to Atmospheric Waves	61
3.6	General Solution of the Linear Wave Equation: Some Asymptotics	64
	3.6.1 Stationary Phase	64
	3.6.2 Asymptotics for Oscillatory Sources: Wavenumber Surfaces	65
3.7	Rays and Waves in a Slowly Varying Environment	70
	3.7.1 Some Consequences	71
	3.7.2 Wavepackets and the Group Speed Revisited	75
Chapter 4 Ray Optics: The Classical Rainbow		**76**
4.1	Physical Features and Historical Details: A Summary	76
4.2	Ray Theory of the Rainbow: Elementary Mathematical Considerations	78
	4.2.1 Some Numerical Values	84
	4.2.2 Polarization of the Rainbow	85
	4.2.3 The Divergence Problem	87
4.3	Related Topics in Meteorological Optics	89
	4.3.1 The Glory	89
	4.3.2 Coronas (Simplified)	92
	4.3.3 Rayleigh Scattering—a Dimensional Analysis Argument	93
Chapter 5 An Improvement over Ray Optics: Airy's Rainbow		**95**
5.1	The Airy Approximation	95
	5.1.1 Some Ray Prerequisites	95
	5.1.2 The Airy Wavefront	100
	5.1.3 How Are Colors Distributed in the Airy Rainbow?	104
	5.1.4 The Airy Wavefront: A Derivation for Arbitrary p	105
Chapter 6 Diffraction Catastrophes		**113**
6.1	Basic Geometry of the Fold and Cusp Catastrophes	114
	6.1.1 The Fold	114
	6.1.2 The Cusp	115
6.2	A Better Approximation	122
	6.2.1 The Fresnel Integrals	124

6.3 The Fold Diffraction Catastrophe 126
  6.3.1 The Rainbow as a Fold Catastrophe 128
6.4 Caustics: The Airy Integral in the Complex Plane 130
  6.4.1 The Nature of $Ai(X)$ 133

Chapter 7 Introduction to the WKB(J) Approximation: All Things Airy 137
7.1 Overview 137
  7.1.1 Elimination of the First Derivative Term 139
  7.1.2 The Liouville Transformation 141
  7.1.3 The One-Dimensional Schrödinger Equation 143
  7.1.4 Physical Interpretation of the WKB(J) Approximation 144
  7.1.5 The WKB(J) Connection Formulas 145
  7.1.6 Application to a Potential Well 148
7.2 Technical Details 149
7.3 Matching Across a Turning Point 152
7.4 A Little More about Airy Functions 153
  7.4.1 Relation to Bessel Functions 154
  7.4.2 The Airy Integral and Related Topics 156
  7.4.3 Related Integrals 159

Chapter 8 Island Rays 162
8.1 Straight and Parallel Depth Contours 163
  8.1.1 Plane Wave Incident on a Ridge 164
  8.1.2 Wave Trapping on a Ridge 166
8.2 Circular Depth Contours 167
8.3 Constant Phase Lines 169
  8.3.1 Case 1 169
  8.3.2 Case 2 170
  8.3.3 Case 3 170
8.4 Waves and Currents 170

Chapter 9 Seismic Rays 173
9.1 Seismic Ray Equations 173
9.2 Ray Propagation in a Spherical Earth 175
  9.2.1 A Horizontally Stratified Earth 178
  9.2.2 The Wiechert-Herglotz Inversion 179
  9.2.3 Further Properties of X in the Horizontally Stratified
    Case 181

PART II WAVES 187

Chapter 10 Elastic Waves 189
10.1 Basic Notation 190
10.2 Plane Wave Solutions 193

10.3	Surface waves	195
10.4	Love Waves	198

Chapter 11 Surface Gravity Waves 200

11.1	The Basic Fluid Equations	201
11.2	The Dispersion Relation	203
	11.2.1 Deep Water Waves	203
	11.2.2 Shallow Water Waves	204
	11.2.3 Instability	205
	11.2.4 Group Speed Again	210
	11.2.5 Wavepackets	212
11.3	Ship Waves	214
	11.3.1 How Does Dispersion Affect the Wave Pattern Produced by a Moving Object?	214
	11.3.2 Whitham's Ship Wave Analysis	218
	11.3.3 A Geometric Approach to Ship Waves and Wakes	221
	11.3.4 Ship Waves in Shallow Water	227
11.4	A Discrete Approach	229
	11.4.1 Long Waves	229
	11.4.2 Short Waves	230
11.5	Further Analysis for Surface Gravity Waves	231

Chapter 12 Ocean Acoustics 237

12.1	Ocean Acoustic Waveguides	237
	12.1.1 The Governing Equation	237
	12.1.2 Low Velocity Central Layer	239
	12.1.3 Leaky Modes	240
12.2	One-Dimensional Waves in an Inhomogeneous Medium	241
	12.2.1 An Eigenfunction Expansion	242
	12.2.2 Poles	245
12.3	Model for a Stratified Fluid: Cylindrical Geometry	247
12.4	The Sech-Squared Potential Well	250
	12.4.1 Positive Energy States	250
	12.4.2 Bound States	253

Chapter 13 Tsunamis 255

13.1	Mathematical Model of Tsunami Propagation (Transient Waves)	255
13.2	The Boundary-Value Problem	257
13.3	Special Case I: Tsunami Generation by a Displacement of the Free Surface	258
	13.3.1 A Digression: Surface Waves on Deep Water (Again)	263
	13.3.2 How Fast Does the Wave Energy Propagate?	265
	13.3.3 Kinematics Again	267
13.4	Leading Waves Due to a Transient Disturbance	268
13.5	Special Case 2: Tsunami Generation by a Displacement of the Seafloor	270

Chapter 14 Atmospheric Waves 273

14.1   Governing Linearized Equations 274
14.2   A Mathematical Model of Lee/Mountain Waves over
        an Isolated Mountain Ridge 285
        14.2.1 Basic Equations and Solutions 286
        14.2.2 An Isolated Ridge 288
        14.2.3 Trapped Lee Waves 290
14.3   Billow Clouds, Wind Shear, and Howard's Semicircle Theorem 292
14.4   The Taylor-Goldstein Equation 296

PART III   CLASSICAL SCATTERING 299

Chapter 15 The Classical Connection 301

15.1   Lagrangians, Action, and Hamiltonians 301
15.2   The Classical Wave Equation 304
15.3   Classical Scattering: Scattering Angles and Cross Sections 308
        15.3.1 Overview 308
        15.3.2 The Classical Inverse Scattering Problem 313

Chapter 16 Gravitational Scattering 316

16.1   Planetary Orbits: Scattering by a Gravitational Field 317
        16.1.1 Repulsive Case: $k > 0$ 318
        16.1.2 Attractive Case: $k < 0$ 319
        16.1.3 The Orbits 319
16.2   The Hamilton-Jacobi Equation for a Central Potential 325
        16.2.1 The Kepler Problem Revisited 326
        16.2.2 Generalizations 327
        16.2.3 Hard Sphere Scattering 328
        16.2.4 Rutherford Scattering 329

Chapter 17 Scattering of Surface Gravity Waves by Islands, Reefs,
    and Barriers 332

17.1   Trapped Waves 333
17.2   The Scattering Matrix $S(\alpha)$ 334
17.3   Trapped Modes: Imaginary Poles of $S(\alpha)$ 337
17.4   Properties of $S(\alpha)$ for $\alpha \in \mathbb{R}$ 338
17.5   Submerged Circular Islands 340
17.6   Edge Waves on a Sloping Beach 342
        17.6.1 One-Dimensional Edge Waves on a Constant Slope 345
        17.6.2 Wave Amplication by a Sloping Beach 345

Chapter 18 Acoustic Scattering   348

18.1   Scattering by a Cylinder   350
18.2   Time-Averaged Energy Flux: A Little Bit of Physics   352
18.3   The Impenetrable Sphere   354
        18.3.1 Introduction: Spherically Symmetric Geometry   354
        18.3.2 The Scattering Amplitude Revisited   356
        18.3.3 The Optical Theorem   358
        18.3.4 The Sommerfeld Radiation Condition   358
18.4   Rigid Sphere: Small ka Approximation   359
18.5   Acoustic Radiation from a Rigid Pulsating Sphere   361
18.6   The Sound of Mountain Streams   364
        18.6.1 Bubble Collapse   367
        18.6.2 Playing with Mathematical Bubbles   369

Chapter 19 Electromagnetic Scattering: The Mie Solution   371

19.1   Maxwell's Equations of Electromagnetic Theory   378
19.2   The Vector Helmholtz Equation for Electromagnetic Waves   379
19.3   The Lorentz-Mie solution   383
        19.3.1 Construction of the Solution   386
        19.3.2 The Rayleigh Scattering Limit: A Condensed Derivation   392
        19.3.3 The Radiation Field Generated by a Hertzian Dipole   394

Chapter 20 Diffraction of Plane Electromagnetic Waves by a Cylinder   397

20.1   Electric Polarization   398
20.2   More about Classical Diffraction   406
        20.2.1 Huygen's Principle   406
        20.2.2 The Kirchhoff-Huygens Diffraction Integral   406
        20.2.3 Derivation of the Generalized Airy Diffraction Pattern   409

PART IV   SEMICLASSICAL SCATTERING   413

Chapter 21 The Classical-to-Semiclassical Connection   415

21.1   Introduction: Classical and Semiclassical Domains   415
21.2   Introduction: The Semiclassical Formulation   416
        21.2.1 The Total Scattering Cross Section   418
        21.2.2 Classical Wave Connections   419
21.3   The Scalar Wave Equation   420
        21.3.1 Separation of Variables   420
        21.3.2 Bauer's Expansion Again   422
21.4   The Radial Equation: Further Details   423
21.5   Some Examples   426
        21.5.1 Scattering by a One-Dimensional Potential Barrier   426
        21.5.2 The Radially Symmetric Problem: Phase Shifts
                and the Potential Well   428

Chapter 22 The WKB(J) Approximation Revisited   434

22.1   The Connection Formulas revisited: An Alternative Approach   435
22.2   Tunneling: A Physical Discussion   437
22.3   A Triangular Barrier   438
22.4   More Nuts and Bolts   440
     22.4.1   The Phase Shift   445
     22.4.2   Some Comments on Convergence   445
     22.4.3   The Transition to Classical Scattering   446
22.5   Coulomb Scattering: The Asymptotic Solution   448
     22.5.1   Parabolic Cylindrical Coordinates (ξ, η, ϕ) 449
     22.5.2   Asymptotic Form of ${}_1 F_1(-i\mu, 1; ik\xi)$ 450
     22.5.3   The Spherical Coordinate System Revisited   451
22.6   Coulomb Scattering: The WKB(J) Approximation   453
     22.6.1   Coulomb Phases   453
     22.6.2   Formal WKB(J) Solutions for the TIRSE   454
     22.6.3   The Langer Transformation: Further Justification   456

Chapter 23 A Sturm-Liouville Equation: The Time-Independent
One-Dimensional Schrödinger Equation   459

23.1   Various Theorems   460
23.2   Bound States   463
     23.2.1   Bound-State Theorems   463
     23.2.2   Complex Eigenvalues: Identities for $\mathrm{Im}(\lambda_n)$ and $\mathrm{Re}(\lambda_n)$ 467
     23.2.3   Further Theorems   468
23.3   Weyl's Theorem: Limit Point and Limit Circle   471

PART V   SPECIAL TOPICS IN SCATTERING THEORY   475

Chapter 24 The S-Matrix and Its Analysis   477

24.1   A Square Well Potential   477
     24.1.1   The Bound States   480
     24.1.2   Square Well Resonance: A Heuristic Derivation
             of the Breit-Wigner Formula   480
     24.1.3   The Watson Transform and Regge Poles   481
24.2   More Details for the TIRSE   487
24.3   Levinson's Theorem   489

Chapter 25 The Jost Solutions: Technical Details   491

25.1   Once More the TIRSE   491
25.2   The Regular Solution Again   494
25.3   Poles of the S-Matrix   498
     25.3.1   Wavepacket Approach   501

Chapter 26 One-Dimensional Jost Solutions: The S-Matrix Revisited   504

26.1   Transmission and Reflection Coefficients   504
      26.1.1   Poles of the Transmission Coefficient: Zeros of $c_{12}(k)$ 506
26.2   The Jost Formulation on $[0, \infty)$: The Radial Equation Revisited   507
      26.2.1   Jost Boundary Conditions at $r = 0$ 507
      26.2.2   Jost Boundary Conditions as $r \to \infty$ 508
      26.2.3   The Jost Function and the S-Matrix   508
      26.2.4   Scattering from a Constant Spherical Inhomogeneity   509

Chapter 27 Morphology-Dependent Resonances: The Effective Potential   512

27.1   Some Familiar Territory   512
      27.1.1   A Toy Model for $l \neq 0$ Resonances: A Particle Analogy   516
      27.1.2   Resonances   521

Chapter 28 Back Where We Started   523

28.1   A Bridge over Colored Water   523
28.2   Ray Optics Revisited: Luneberg Inversion and Gravitational Lensing   531
      28.2.1   Abel's Integral Equation and the Luneberg Lens   531
      28.2.2   Connection with Classical Scattering and Gravitational Lensing   534

Appendix A   Order Notation: The "Big O," "Little o," and "$\sim$" Symbols   537

Appendix B   Ray Theory: Exact Solutions   539

B.1   Profile 1   540
B.2   Profile 2   541
B.3   Profile 3   542
B.4   Profile 4   543
B.5   Profile 5   543
B.6   Profile 6   544
B.7   Profile 7   545
B.8   Profile 8   546
B.9   Profile 9   546
B.10   Profile 10   547

Appendix C   Radially Inhomogeneous Spherically Symmetric Scattering: The Governing Equations   550

C.1   The Tranverse Magnetic Mode   550
C.2   The Tranverse Electric Mode   551

Appendix D Electromagnetic Scattering from a Radially
   Inhomogeneous Sphere   553

  D.1   A classical/Quantum connection for Transverse Electric and
          Magnetic Modes   553
  D.2   A Liouville Transformation   556

Appendix E Helmholtz's Theorem   559

  E.1   Proof of Helmholtz's Theorem   559
  E.2   Lamé's Theorem   560

Appendix F Semiclassical Scattering: A Précis (and a Few More Details)   562

Bibliography   567

Index   585

Preface

> But examine everything carefully; hold fast to that which is good.
>
> [156]

In his Preface to the first edition of *Wind Waves* [165], Blair Kinsman wrote:

> As you might expect ... this is a very personal book. It is not *the truth* about waves. It is an effort to communicate to you what I have so far managed to understand about waves ... One point I want to make clear. I have myself contributed only a few minor footnotes to the substance of the subject. This book represents a kind of intellectual janitorial service. I have tried to tidy up the subject. Don't ever confuse the men who create with those who arrange. Janitors are useful, but the janitorial activity is different in kind.
>
> Because I am trying to put in order for you things I did not create myself, there are any number of ways confusion may enter. I may have managed to garble a perfectly good idea, or I may have managed to misunderstand it entirely. I may have been taken in by a wrong idea and offer it to you as true. I may have been taken in by a wrong idea and then have compounded the agony by further fogging it. You should not accept passively anything in this book. Test everything by your own reason and experience. And you can always go back to the original sources. I cannot give you my own understanding of waves, and, if I could, you wouldn't want it. My best service to you is as a punching bag on which to exercise yourself. Then perhaps you can arrive at your own understanding—hopefully, a better one than mine. There is only one favor I would ask of you. Don't jump. I haven't written this book lightly.

The book you hold is the type of book that I would love to have read as a graduate student, and therefore it is an intensely personal one. I have long been enthralled by the unifying nature of mathematics in its ability to describe many of the patterns we see all around us. By separating the manuscript into several parts, I have attempted to provide an anthology of mathematical techniques utilized (at several complementary levels) in studying a wide range of phenomena falling under the umbrella of rays, waves, and scattering. The scope of this book is certainly very broad, as evidenced by the table of contents (even though there is nothing about relativistic scattering here). I have tried to include as many topics as possible under this tripartite umbrella. In one sense the underlying philosophy is similar to that of my first book, *Mathematics in Nature: Modeling Patterns in the Natural World* [122], in which I wrote that "I wanted to limit the topics covered to those objects that could be seen with the naked eye by anyone who takes their eyes outside; there are many books written on the mathematical principles behind phenomena that take place at the microscopic and submicroscopic levels, and also from planetary to galactic scales. But leaves, trees, spider webs, bubbles, waves, clouds, rainbows. These are elements of the stuff we can easily see."

Over the years my academic background has been influenced by a plethora of books and papers in the mathematical sciences. As the reader will note from even a cursory glance at the references, I have used many excellent resources in the preparation of this manuscript, including, naturally, many of the classics. Even relatively inexperienced readers will recognize some of these. To amend a well-known phrase, no book is an island, and this volume is no exception. There are many superb books and journal articles that discuss the topics I have included, and although it is not necessary to reinvent the wheel, I have adapted and modified some of those presentations to suit my objectives here. In short, I have made much use of resources I have come to love along the way. They are by now well-thumbed reminders of the thrill I experienced (and still do) when I read them. Many of the topics and examples I encountered therein are therefore perfectly suited for inclusion here. Since those early years even more favorites have lined my bookshelves (sometimes to the chagrin of my family). Consequently, yet another purpose of the book is to introduce readers (where necessary) to these classic texts and to some of the relevant literature. For any given topic, the reader is strongly encouraged, in the spirit of Kinsman's comments, to go to the original sources for more details (and also for secondary citations). Indeed, if reading this book encourages readers to acquire or access the sources cited here, I will be delighted. As noted above, I have obtained most of my library in this way over the course of my career. Occasionally in the book I have felt it useful to include 'asides' that are related (directly or indirectly) to the material in particular sections. Frequently these asides can be read independently of the surrounding material, and they have been placed in gray boxes throughout the text.

I'd like to share a little of my academic history here. As an undergraduate in theoretical physics, I was completely entranced by a final year course in quantum mechanics and a related project on (atomic) scattering theory. What fascinated me was not so much the physics but the mathematics, specifically, the plethora of mathematical methods, special functions, and how exact solutions to the time-independent Schrödinger equation could be expressed in terms of them. Later, but still while an undergraduate, I happened to meet a new member of the mathematics faculty whose research specialty was scattering theory. I explained to her my enthusiasm for the scattering topics I was studying, and she told me that her research was much more theoretical than that! I was puzzled by this remark, in part because at that point I had not been exposed to any of the underlying concepts of Hilbert space, and the importance of this framework in studying quantum mechanics "properly" (not to mention its beauty).

A little later as a graduate student, I developed an interest in wave problems on semi-infinite domains, and realized almost by accident that freely propagating and trapped waves could be associated with branch cuts and poles, respectively, of the corresponding Green's function in the complex "λ-parameter" plane. This is of course intimately connected with the properties of the underlying ordinary differential operators, the spectra of which (especially in my research field of astrophysical magnetohydrodynamics) were often much more complicated than that of the one-dimensional time-independent Schrödinger equation [154]. This was indeed a far cry from the atomic spectral theory I had encountered in my early physics courses. In retrospect my almost primal fascination with the continuous/discrete dichotomy was enhanced not so much by exposure to functional analysis as by acquaintance with another beautiful subject: functions of a complex variable.

As I grew in mathematical maturity (with perhaps a very shallow saturation level), I realized how broad and deep the topic of scattering theory really is. Whether it concerns scattering of particles from atomic, molecular, or nuclear targets (and the concomitant potential well and barrier problems); electromagnetic and/or acoustic waves from geometrically idealized objects; the gravitational scattering of planets, asteroids, and comets by the sun (as in the simplest of two-body problems); the refraction and scattering of surface gravity waves by submerged islands, shoals and reefs, acoustic-gravity waves by mountain ranges, or seismic waves by the earth's interior, all can be described using similar (or at least related) mathematical ideas and techniques. Interestingly, much use is made in high-energy physics of terms directly associated with phenomena in meteorological optics: rainbow scattering and glory scattering. And because I consider these phenomena in nature to be among the most beautiful and esoteric sights to behold, the rainbow forms a type of template for this book. This phenomenon, like many others, can be studied at several complementary levels of mathematical description, and it serves in that sense as a unifying thread throughout much of the book. *And every equation, no matter what the context, tells a story.* We just have to learn to read it (even if it takes a lifetime of learning). I should note that while there are extremely valuable books on highly mathematical aspects of scattering theory [166], [167], my emphasis here—with the exception of Chapters 23, 25, and 26—is almost entirely devoid of "applied analysis" in the American sense of the phrase. Indeed, it is more appropriately described as "classical mathematical physics/applied mathematics" in the British sense, which is naturally part of the reason for the title of this book. Nevertheless, it is not my intent to compartmentalize these subjects. Indeed, it is quite the reverse: I wish instead to present a big picture, or panorama, insofar as I can see it.

As will be apparent even from the table of contents, in academic terms I am very much a Jack of all trades, a generalist rather than a specialist. (That quaint English phrase continues "and master of none"; it seems it was used as early as 1592 by Robert Greene in his booklet *Greene's Groats-Worth of Wit*, in which he dismissively refers to William Shakespeare!) It should be clear from these comments that this is not a book for the specialist researcher. It is more akin to *An Introduction to World Literature* than *War and Peace*. But it *is* a book for anyone fascinated by the mathematical connections existing in the union of ray, wave, and scattering problems, and indeed for those who seek an introduction to some of the methods of applied mathematics and mathematical physics. It paints a big picture in a way that a monograph is not intended to do. I would like the reader to be able to gain what I might have done, had I a copy of this book as a student: an early appreciation for the way that mathematics serves as the fabric for describing the immense tapestry of wave phenomena that exists all about us. Whether we are thinking about waves in the electromagnetic spectrum, acoustic waves, gravity waves, and so forth, all can exhibit similar behavior under appropriate circumstances: reflection, refraction, diffraction, interference, and the like, often manifesting quite spectacular displays in nature. Mathematics is aptly described as the science of patterns, and it is this unifying nature of mathematics that enables practitioners to recognize and utilize similarities and connections between very different fields of science. A reader interested in one field (e.g., ray or wave optics) may wish to see how corresponding mathematical ideas carry over to (say) seismology or the behavior of waves in the vicinity of reefs or islands. Someone noticing both ripples upstream and gravity waves downstream of a rock may be pleasantly surprised to find another physical analogue in the context of wave clouds adjacent to isolated mountains or hills.

Another objective of mine is to provide in one place a compilation of material that is not so readily available online or in a typical university library. As noted earlier, some of the arguments in this book have already been penned by different authors, past and present, and I have tried to adaptively emulate their approaches where possible. Indeed, books by many eminent mathematical scientists have been written on every chapter topic included here. However, I certainly would not wish this book to be a mere compendium of methods tied by a rope of three strands: rays, waves, and scattering. I hope that I have achieved a fairly seamless account of many facets of the subject while at the same time contributing to students' understanding of topics that can be quite confusing. If I have contributed more confusion than clarification, I beg the reader's forgiveness!

In a book of this size, a recurring problem can also be the following: suppose that in Chapter 1, say, I have introduced a specific equation, say (E), and need to refer to it again in Chapters 7, 12, 15, and 23; should I then expect the reader to thumb back to Chapter 1 every time? Or do I save the reader some time and frustration by restating the equation (while still referring to the location of its first incarnation)? In most cases I have chosen the latter option, because while not wishing to keep reinventing the wheel, I wish to save the reader continual backtracking (though sometimes it is unavoidable). All this means, however, that on occasion I have modified the notation to reflect the context in which it is used. A case in point is the celebrated Airy integral—the argument is not always the same: usually it will be X, but sometimes m or s. I trust that this small lack of consistency will not give the reader heartburn. And in the same spirit, sometimes arguments made in one chapter may be repeated in a slightly different form in another.

As with *Mathematics in Nature* [122] this is not a textbook per se; the style is relatively informal, and in place of end-of-chapter problem sets I have embedded exercises and examples throughout the text. Nevertheless, it could be used as a supplementary (or even standalone) text by instructors wishing to select topics from the wide array included in the table of contents. But for what courses might this book be used? I suggest that the subject matter is relevant to several types of senior/graduate course: (i) waves and/or scattering theory; (ii) mathematical methods of physics; and (iii) methods of mathematical physics. It may appear mere pedantry to separate methods courses into items (ii) and (iii), but I believe there is a subtle distinction here. I employ the phrase "mathematical methods of physics" as a generic term for classes in which fairly advanced mathematical techniques are taught as a general toolbox for use in the physical sciences. Such tools would surely include the method of separation of variables, common special functions and their associated transforms, vector field theory, functions of a complex variable, and variational methods. In contrast, "methods of mathematical physics" can be in my view more content-specific, focusing on techniques that are commonly required in wave theory and scattering problems, such as perturbation methods for solving ordinary differential equations (e.g., the almost universally known WKB(J) approximation named for its independent co-inventors Wentzel, Kramers, Brillouin, and Jeffreys—the nonstandard choice of acronym is explained in Chapter 1), the asymptotic evaluation of integrals (stationary phase and steepest descent) and the Sommerfeld-Watson transform. Of course, the distinction I have made is somewhat artificial; others might use different topics as examples or even consider that there is no distinction at all.

I hope the book will be valuable because it fills a gap between, on the one hand broad-based but less mathematical books directed at undergraduates and Courant-Hilbert types of book [202] (and also specialist monographs) on the other. The latter is more appealing (and generally only accessible) to advanced practitioners of the subject. Because monographs are just that—monographs—this book, whose audience might include advanced undergraduates, graduate students, and instructors, could be thought of as an anthology, handbook, or treasury. Some chapters go into considerably more mathematical depth or physical detail than others do. In this sense it has a monograph-like flavor that may interest those in one field of wave motion or scattering who wish to learn more in depth about topics outside their area of expertise.

I suggest then that *Rays, Waves, and Scattering* is effectively, to use an engineering analogy, a standalone volume near the center of mass of a triangular lamina with vertices corresponding to anthology, monograph, and textbook. I find this appealing, given the manuscript's triune partition, so that as an anthology of ray/wave/scattering phenomena, it could serve as a valuable supplement (or indeed complement) to courses in wave theory or mathematical methods for students of applied mathematics, theoretical physics, or engineering. It introduces the reader to a variety of mathematical techniques invoked in the studying wave motion in acoustic, fluid dynamic, elastic, and electromagnetic contexts, properties of the time-independent Schrödinger equation (e.g., as a special case of Sturm-Liouville problems), scattering theory, and much more. Thus the readership could include beginning graduate students and final-year undergraduates who have had a standard exposure to mathematical courses, such as advanced calculus, ordinary and partial differential equations, and functions of a complex variable. Some exposure to classical mechanics, electromagnetic theory, and elementary quantum mechanics would be valuable, but I hope the introductory material will render the book accessible to those who do not possess this background.

At the end of the Acknowledgments in [122] I wrote:

Finally, a gentle plea of mine to the reader is as follows: *be observant*—there are many optical and fluid dynamical phenomena (particularly the latter) taking place in the everyday world around you—in the sky, in clouds, rivers, lakes, oceans, puddles, faucets, sinks, coffee cups and bathtubs. I hope that as a result of reading this book you will be better able to understand such phenomena, both mathematically and physically … it is a fascinating interdisciplinary area of applied mathematics, which richly rewards those who invest some time and effort to study it.

Indeed,

Rest is not idleness, and to lie sometimes on the grass under trees on a summer's day, listening to the murmur of the water, or watching the clouds float across the sky, is by no means a waste of time. [John Lubbock, *The Use of Life*, 1894].

And here I would like to add the following. Don't be afraid to pursue diligently the mathematics behind these and other phenomena, no matter where they lead. The thrill and joy of personal discovery, regardless of how well known the results are to others, is something that cannot be easily described; it must be experienced. There is very little (if any) new mathematics or physics in this book (though I hope that the synthesis is new), but—and here's the point—it was all new to me at various points in time. And as I have encountered different aspects of these elegant concepts over the years it has been wonderful (at times) to wrestle with them and try to reformulate them in my own mind. For readers who, like me, are "academic plodders," I sincerely hope that this book proves to be enjoyable and helpful. And for those who are not such plodders, I trust that you will nonetheless find valuable material here (even if only in the references!).

Needless to say, any errors in this book are my own. Given the enormous number of mathematical symbols I have typed (with probably close to 3500 equations!), the probability of there being no errors from me or from the production process is vanishly small, so I would appreciate being informed of them.

Acknowledgments

I am very grateful to many individuals along the way in the pursuit of this enterprise. The book has evolved in a staccato-like fashion over a period of nearly seven years, though its roots go back several decades (as should be apparent from the Preface). The current chair of my department, Dr. Hideaki Kaneko, and his predecessor, Dr. J. Mark Dorrepaal, have been most understanding and generous in arranging my teaching schedule, insofar as has been possible, around this major project, and I am very grateful for their continued support and encouragement. I am grateful to Old Dominion University for the opportunity to spend the 2015 spring semester on research leave, where much of the material came into clearer focus. My original idea for the book was very ambitious in the breadth of topics I envisaged being able to include, but as reality kicked in I became somewhat discouraged by the magnitude of the task ahead. I was reminded of a pithy comment from an older and far wiser person when, as a graduate student in the United Kingdom, I was bemoaning the fact that I might never complete my dissertation because there was so much material I wished to include. The comment was essentially that "in your first year of your Ph.D. research you plan to change the world. In the second year you get discouraged because you realise that will not occur. And in the third year you write up what you *have* done."

During the summer of 2014 I was given the opportunity to teach a graduate course—Advanced Applied Mathematics I—in an rather unusual and informal way. At that point, I had "hunted-and-pecked" out about half of this material; the rest was in the form of handwritten notes, so I charged the students to critique and where necessary clarify my notes as they saw fit, and then type them up for my detailed review. It was designed to be a mutually advantageous class (shouldn't all courses be so?), and I trust they would agree with me that it was. They encountered many of the mathematical techniques from prior courses, but applied in some interesting physical contexts that were for most part entirely new to them, and I had acquired a conscientious cadre of highly efficient mathematical stenographers! But of course, they were far more than that. By now though, they may not be able to recognize their individual contributions, because I have added to or otherwise modified them (and in some cases removed material that seemed in retrospect to be redundant). In alphabetical order they are Boampong Asare, Jacob Bishop, Carol Buchan, Charles Harris, Nancy Nelson, Michelle Pizzo, Lindsey Santos Koos, Brian Swiger, Anthony Williams and Justin Wolfe. Furthermore, Michelle, ably assisted by Anthony, kept track of who was working with whom on what, assigned deadlines, kept me informed of progress on a weekly basis, and established a robust communication network. I thank them all for their input. Again, I remain responsible for any errors that may be present.

I am grateful to my former Ph.D. students Umaporn "Lucky" Nuntaplook and Michael Pohrivchak; to Lucky for permission to use some joint work in Chapter 26, and to Michael for his permission to summarize one of his dissertation chapters in Appendix B (and also for the complex contour diagram used on the back cover). My colleague, Dr. Przemyslaw Bogacki, with his very considerable technical expertise, has been enormously helpful to this techno-dinosaur, and I am deeply appreciative for his assistance with several issues. I have typed the majority of this material (a labor of love) using Scientific Workplace (http://www.mackichan.com/). The technical support from mackichan.com has been, without exception, first rate and I am so thankful that they have been able to help me understand how to incorporate various software packages into my manuscript. And also for saving me from a meltdown when, after saving my document, a message "file truncated" appeared. It appeared that I had lost about 80 percent of five years' worth of writing, but fortunately, despite my technical incompetence, it was recoverable. My particular thanks go to George Pearson at mackichan.com.

I am very grateful to the American Mathematical Society for their permission to include in both section 1.2.5 and Chapter 28 part of my review of *The Rainbow Bridge: Rainbows in Art, Myth, and Science* by Raymond L. Lee, Jr., and Alistair B. Fraser [169] in the *Notices*. The full review can be found in [148]. My friend and co-author Philip Laven graciously supplied me with details of his presentation on supernumerary bows and the modified Young theory, details of which can be found at the end of Chapter 5.

My editor Vickie Kearn has always been a paragon of support. She encouraged me to pursue my "magnum opus" even when it seemed like it would never be finished. Thank you, Vickie, for recognizing that all I needed was a little electronic shove from Princeton University Press. I am also extremely grateful to members of Vickie's team: Brigitte Pelner, Lauren Bucca (both of whom were a delight to work with and were very patient with me), and the talented Dimitri Karetnikov (who made silk purses out of the "sow's ear" figures I sent him!). I am indebted to Princeton University Press for permission to use material from my books [122], [203], and [212] in Chapters 4, 11, and 19. My copyeditor Cyd Westmoreland is, in a word, superb; she is clarity personified and made what was for me a necessary but unenviable task (responding to her comments) almost fun! She has an eagle eye for inconsistencies. It has been a great pleasure working with her, so much so that I am tempted to get a T-shirt with "Cyd rocks!" written on the front (subject to copyediting of course).

Last, and as they say, most definitely not least, I extend my love and gratitude to Susan, my wife of 45 years, without whose support and encouragement this book might never have been completed (though all those trips we made to visit grandchildren did extend the completion time a little, but very happily so).

Rays, Waves, and Scattering

Chapter One

Introduction

Probably no mathematical structure is richer, in terms of the variety of physical situations to which it can be applied, than the equations and techniques that constitute wave theory. Eigenvalues and eigenfunctions, Hilbert spaces and abstract quantum mechanics, numerical Fourier analysis, the wave equations of Helmholtz (optics, sound, radio), Schrödinger (electrons in matter) ... variational methods, scattering theory, asymptotic evaluation of integrals (ship waves, tidal waves, radio waves around the earth, diffraction of light)—examples such as these jostle together to prove the proposition.

M. V. Berry [1]

There is a theory which states that if ever anyone discovers exactly what the Universe is for and why it is here, it will instantly disappear and be replaced by something even more bizarre and inexplicable. There is another theory which states that this has already happened.

Douglas Adams [155]

Douglas Adams's famous *Hitchhiker* trilogy consists of *five* books; coincidentally this book addresses the three topics of rays, waves, and scattering in five parts: (i) Rays, (ii) Waves, (iii) Classical Scattering, (iv) Semiclassical Scattering, and (v) Special Topics in Scattering Theory (followed by six appendices, some of which deal with more specialized topics). I have tried to present a coherent account of each of these topics by separating them insofar as it is possible, but in a very real sense they are inseparable. We are in effect viewing each phenomenon (e.g. a rainbow) from several different structural directions or different mathematical levels of description. It is tempting to regard descriptions and explanations as synonymous, but of course they are not. In fact, regarding the rainbow in particular

Aristotle and later scientists in antiquity "constructed theories that primarily *describe* natural phenomena in mathematical or geometric terms, with little or no concern for physical mechanisms that might *explain* them." This contrast goes to the heart of the difference between "Aristotelian" and mathematical modeling. [169]

The first two chapters of Part I (Rays) introduce the topic from a physical and a mathematical point of view, respectively. Thereafter the subject of rays is viewed in an atmosphere–sea–earth sequence, merging with Part II (Waves) via the reverse sequence earth–sea–atmosphere. Part III (Classical Scattering) examines the relationship between the elegance of the classical Lagrangian and Hamiltonian formulations of mechanics and optics, and having set the scene, so to speak, moves on to develop Kepler's laws of planetary motion (arising from what I refer to as gravitational scattering). This is followed by a revisitation of the topics of surface gravity waves, acoustics, electromagnetic scattering (including the Mie solution), and diffraction. Part IV (Semiclassical Scattering) provides a transition from Part III and addresses more of the nuts and bolts of the underlying mathematics. In so doing it allows us to reexamine the WKB(J) approximation and apply it to some simple one-dimensional potentials. Readers may wonder at my notation for this approximation—why WKB(J)? It is to pay homage to the legacy of Sir Harold Jeffreys in this and many other areas of applied mathematics and mathematical physics. As an eighteen-year-old student I bought a copy of the celebrated *Methods of Mathematical Physics* [98], co-written with his wife, Lady Bertha Jeffreys. I vowed that I would try to understand as much of it as possible (and I have tried). It is a magnificent book. But writing WKBJ might be a bridge too far, as they say—the majority of citations in the literature appear to use the WKB form, so I have adapted the acronym accordingly. And it appears that I am not the only one to think Jeffreys deserves more credit than he has been given: see the first quotation in Chapter 22.

Writing the final chapter in this part was a fascinating undertaking; it is a collection of salient properties of Sturm-Liouville systems with particular reference to the time-independent Schrödinger equation. It touches on several aspects of the theory of differential equations that have profound implications for the topics in this book and is distilled from a variety of sources (some of which are now quite difficult to find). I hope that the reader will benefit from having these all in one place. Part V (Special Topics in Scattering Theory) is an eclectic anthology of material that has intrigued me over the years, and some of the topics are reflections of joint research with recent graduate students. Six appendices round out the material, the second of which is a condensed version of a chapter in [68] (see also [236]). Appendix D is also based on my contribution to a joint article written with former students [149].

I should also say something about the quotations I have chosen. At the beginning of most chapters I have placed a quotation (and occasionally more than one) from a book, scientific paper, or internet article that I consider to be a brief but pertinent introduction to the subject matter of that chapter—an appetite-whetter if you will. I have even cited material from Wikipedia (gasp!). In some of the chapters I have merely waffled on for a bit instead of using an introductory quotation. Regarding short excerpts from scientific articles on the internet, in the few cases for which I have been unable to find the name of the author (after extensive searching), I trust that the link provided will suffice for the interested reader to pursue the topic (and author) further.

In the Preface I mentioned that this book is an intensely personal one, and how the beautiful phenomenon of a rainbow serves as an implicit template or directory for much of the rest of the book. It does so because much of the motivation for the design of the book springs from it, as evidenced below. And as to *why* I have chosen to structure things in this way, you will need to read the first paragraph of the last chapter ("Back where we started") to find out!

1.1 THE RAINBOW DIRECTORY

Optical phenomena visible to everyone have been central to the development of, and abundantly illustrate, important concepts in science and mathematics. The phenomena considered from this viewpoint are rainbows, sparkling reflections on water, mirages, green flashes, earthlight on the moon, glories, daylight, crystals and the squint moon. And the concepts involved include refraction, caustics (focal singularities of ray optics), wave interference, numerical experiments, mathematical asymptotics, dispersion, complex angular momentum (Regge poles), polarisation singularities, Hamilton's conical intersections of eigenvalues ('Dirac points'), geometric phases and visual illusions. [151]

The theory of the rainbow has been formulated at many levels of sophistication. In the geometrical-optics theory of Descartes, a rainbow occurs when the angle of the light rays emerging from a water droplet after a number of internal reflections reaches an extremum. In Airy's wave-optics theory, the distortion of the wave front of the incident light produced by the internal reflections describes the production of the supernumerary bows and predicts a shift of a few tenths of a degree in the angular position of the rainbow from its geometrical-optics location. In Mie theory, the rainbow appears as a strong enhancement in the electric field scattered by the water droplet. Although the Mie electric field is the exact solution to the light-scattering problem, it takes the form of an infinite series of partial-wave contributions that is slowly convergent and whose terms have a mathematically complicated form. In the complex angular momentum theory, the sum over partial waves is replaced by an integral, and the rainbow appears as a confluence of saddle-point contributions in the portion of the integral that describes light rays that have undergone m internal reflections within the water droplet. [168]

1.1.1 The Multifaceted Rainbow

What follows is a partial list of context-useful descriptions of a rainbow; they are certainly not mutually exclusive categories. References are made to (italicized) topics that will be expanded in later chapters, so the reader should not be unduly concerned about words or phrases in this Chapter that have yet to be defined (e.g., as in the above quotations). Such topics will be unfolded in due course, so be patient, dear reader. But for those who wish to know *now* whether "the Butler did it," and if not, who did, the answers *are* in the back of the book, and it is recommended that you turn immediately to the short last chapter, titled "Back Where We Started" prior to returning here.

In part, a rainbow is:

(1) A concentration of light rays corresponding to an extremum of the scattering angle $D(i)$ as a function of the angle of incidence i. In particular it is a minimum for the primary bow, and the exiting ray at this minimum value is called the *Descartes or rainbow ray*; as noted below, the notation $D(i)$ is replaced by $\theta(b)$ in most of the scattering literature, where b is the *impact parameter*. The angle θ is much used in connection with the equations of scattering in the chapters that follow, but

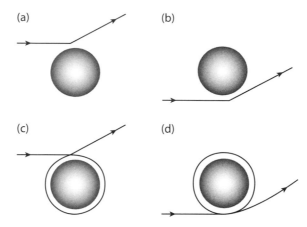

Figure 1.1 (a) $\Theta = \theta$; (b) $\Theta = -\theta$; (c) $\Theta = \theta - 2\pi$; (d) $\Theta = -\theta - 2\pi$.

the former notation will be retained when discussing the optical rainbow (I trust this will be at worst only a minor annoyance for my colleagues in meteorological optics);

(2) An integral superposition of waves over a (locally) cubic wavefront (*the Airy approximation*). This approximation is valid only for large size parameters (i.e., the drop size is much larger than the wavelength of the incident light) and for small deviations from the angle of the rainbow ray; and relatedly

(3) An interference phenomenon within the same light wave (the origin of the *supernumerary bows*);

(4) A *caustic*, separating a 2-ray region from a 0-ray (or shadow) region; and relatedly

(5) A *fold diffraction catastrophe*.

(6) Related to the behavior of the *scattering amplitude* for the third term in what is referred to as a *Debye series expansion*. The dominant contribution to the third Debye term in the rainbow region is a *uniform asymptotic expansion*. On the bright side of the rainbow, this result matches smoothly with the *WKB(J) approximation* in the 2-ray region. On the dark side, it does so again, this time with the damped complex *saddle-point contribution*. Put differently, diffraction into the shadow side of a rainbow occurs by *tunneling*. And relatedly

(7) A coalescence of two *real saddle points*;

(8) A result of ray/wave interactions/tunneling with an *effective potential* comprising a *square well* and a *centrifugal barrier*.

DEFLECTION ANGLE

The *classical deflection angle* is denoted by $\Theta(b)$, where b is the impact parameter. The scattering angle is $\theta(b)$, and the two angles are related by

$$\Theta(b) + 2n\pi = \pm\theta, \tag{1.1}$$

where n is a non-negative integer chosen such that $\theta \in [0, \pi]$. Illustrative examples of the relationship (1.1) for the *same value* of θ are shown in Figure 1.1. For a repulsive potential (as in Figure 1.1a), $\Theta = \theta$, but in principle for an attractive potential Θ can be arbitrarily negative, because the particle (or ray) may orbit the scattering center many times before emerging from it. Thus there may be several different trajectories that lead to the same scattering angle θ (this will be discussed further in Chapter 16).

1.2 A MATHEMATICAL TASTE OF THINGS TO COME

1.2.1 Rays

The Ray-Theoretic Rainbow

A rainbow occurs when the the scattering angle $D(i)$, as a function of the angle of incidence i, passes through an extremum. The folding back of the corresponding scattered or deviated ray takes place at this extremal scattering angle (the rainbow angle $D_{\min} = \theta_R$; note that sometimes in the popular literature the rainbow angle is loosely interpreted as the *supplement* of the deviation, $\pi - D_{\min}$). Two rays scattered in the same direction with different angles of incidence on the illuminated side of the rainbow ($D > \theta_R$) fuse together at the rainbow angle and disappear as the "dark side" ($D < \theta_R$) is approached. This is one of the simplest physical examples of a *fold catastrophe* in the sense of Thom, as will be discussed later. It will also be shown later that rainbows of different orders can be associated with so-called *Debye terms* of different orders; the primary and secondary bows correspond to $p = 2$ and $p = 3$, respectively, where $p > 1$ is the number of ray paths inside the drop, so the number of *internal reflections* is $k = p - 1$.

Details

For $p - 1$ such internal reflections the total deviation is

$$D_{p-1}(i) = (p - 1)\pi + 2(i - pr), \qquad (1.2)$$

where by Snell's law, $r = r(i)$. This approach can be thought of as the elementary classical description.

For the primary rainbow ($p = 2$),

$$D_1(i) = \pi - 4r(i) + 2i, \qquad (1.3)$$

and for the secondary rainbow ($p = 3$),

$$D_2 = 2i - 6r(i) \qquad (1.4)$$

(modulo 2π). For $p = 2$ the angle through which a ray is deviated is

$$D_1(i) = \pi + 2i - 4\arcsin\left(\frac{\sin i}{n}\right), \qquad (1.5)$$

where n is now the relative refractive index of the medium. In general

$$D_{p-1}(i) = (p-1)\pi + 2i - 2p \arcsin\left(\frac{\sin i}{n}\right), \tag{1.6}$$

or equivalently,

$$D_k(i) = k\pi + 2i - 2(k+1)\arcsin\left(\frac{\sin i}{n}\right). \tag{1.7}$$

For $p = 2$ the minimum angle of deviation occurs when

$$i = i_c = \arccos\left(\frac{n^2 - 1}{3}\right)^{1/2}, \tag{1.8}$$

and more generally when

$$i = i_c = \arccos\left(\frac{n^2 - 1}{p^2 - 1}\right)^{1/2}, \tag{1.9}$$

(this clearly places constraints on both n and p if bows are to occur). For the primary rainbow $i_c \approx 59°$, using an approximate value for water of $n = 4/3$; from this it follows that $D_1(i_c) = D_{\min} = \theta_R \approx 138°$. For the secondary bow $i_c \approx 72°$; and $D_2(i_c) = \theta_R \approx 129°$. For $p = 2$, in terms of n alone, $D_1(i_c) = \theta_R$ (the rainbow angle) is defined by

$$D_1(i_c) = \theta_R = 2\arccos\left[\frac{1}{n^2}\left(\frac{4 - n^2}{3}\right)^{3/2}\right]. \tag{1.10}$$

1.2.2 Waves

Surprising as it may seem to anyone who thinks in terms of rays alone, there is also a wave-theoretic approach to the rainbow problem. The essential mathematical problem for scalar waves can be thought of either in classical terms (e.g., the scattering of sound waves) or in wave-mechanical terms (e.g., the nonrelativistic scattering of particles by a square potential well (or barrier) of radius a and depth (or height) V_0). In either case we can consider a scalar plane wave in spherical geometry impinging in the direction $\theta = 0$ on a penetrable ("transparent") sphere of radius a. The wave function $\psi(r)$ satisfies the scalar Helmholtz equation

$$\nabla^2\psi + n^2k^2\psi = 0, \ r < a; \tag{1.11}$$
$$\nabla^2\psi + k^2\psi = 0, \ r \geq a, \tag{1.12}$$

where $n > 1$ is the refractive index of the sphere (a similar problem can be posed for gas bubbles in a liquid, for which $n < 1$). The boundary conditions are that $\psi(r)$ and $\psi'(r)$ are continuous at the surface.

1.2.3 Scattering (Classical)

In what follows θ is the angle of observation, measured from the forward to the scattered direction, and thus defines the *scattering plane*. At large distances from the sphere ($r \gg a$) the wave field ψ can be decomposed into an incident wave + scattered field:

$$\psi \sim e^{ikr\cos\theta} + \frac{f(k,\theta)}{r} e^{ikr}, \tag{1.13}$$

where the scattering amplitude $f(k,\theta)$ (which can be made dimensionless with respect to the radius a) is defined as

$$f(k,\theta) = \frac{1}{2ik} \sum_{l=0}^{\infty} (2l+1)(S_l(k) - 1) P_l(\cos\theta). \tag{1.14}$$

This is the Faxen-Holtzmark formula, and it will be encountered in several different guises later in the book (e.g., Section 21.5). $S_l(k)$ is an element of the scattering matrix (or function or operator, depending on context)) for a given l, and P_l is a Legendre polynomial of degree l. But what *is* the S-matrix? In optical terms, it is the partial-wave scattering amplitude with diffraction omitted—of course, this just kicks the can down the road—what is the partial-wave scattering amplitude? What (really) is diffraction? The reader's forbearance is requested; for now float in a sea of relatively undefined terms and just soak. Fundamentally, the S-matrix is the portion of the partial-wave scattering amplitudes that corresponds to a direct interaction between the incoming wave and the scattering particle. In very simplistic terms it converts an ingoing wave to an outgoing one, and for a spherical square well or barrier we shall find that

$$S_l = -\frac{h_l^{(2)}(\beta)}{h_l^{(1)}(\beta)} \left\{ \frac{\ln' h_l^{(2)}(\beta) - n \ln' j_l(\alpha)}{\ln' h_l^{(1)}(\beta) - n \ln' j_l(\alpha)} \right\}, \tag{1.15}$$

where, following the notation of [26], $\ln'$ represents the logarithmic derivative operator, and j_l and h_l are spherical Bessel and Hankel functions, respectively. $\beta = 2\pi a/\lambda \equiv ka$ is the dimensionless external wavenumber, and $\alpha = n\beta$ is the corresponding internal wavenumber. S_l may be equivalently expressed in terms of cylindrical Bessel and Hankel functions. The lth partial wave in the *Mie solution* is associated with an impact parameter

$$b_l = \left(l + \frac{1}{2}\right) k^{-1}, \tag{1.16}$$

that is, only rays hitting the sphere ($b_l \lesssim a$) are significantly scattered, and the number of terms that must be retained in the Mie series to get an accurate result is of order β. As implied earlier, for visible light scattered by water droplets in the atmosphere, $\beta \sim$ several thousand. This is why, to quote Arnold Sommerfeld [170]:

> The electromagnetic study of light diffraction on an object is a very complicated problem even in the case of the sphere, the simplest possible one. The field outside a sphere can be represented by series of spherical harmonics and Bessel functions of half-integer indices. These series have been discussed by G. Mie for colloidal particles of arbitrary compositions. But even there a mathematical difficulty develops which quite generally is a drawback of this method of series development: for fairly large particles ($\beta = ka$, a = radius, $k = 2\pi/\lambda$) the series converge so slowly that they become practically useless. Except for this difficulty, we could, in this way, obtain a complete solution of the problem of the rainbow.

This problem can be remedied by using the *Poisson summation formula* (related to the *Watson transform*) to rewrite $f(k, \theta)$ in terms of the integral

$$f(\beta, \theta) = \frac{i}{\beta} \sum_{m=-\infty}^{\infty} (-1)^m \int_0^\infty [1 - S(\lambda, \beta)] P_{\lambda - \frac{1}{2}}(\cos \theta) e^{2im\pi\lambda} \lambda \, d\lambda. \quad (1.17)$$

For fixed β, $S(\lambda, \beta)$ is a meromorphic function of the complex variable $\lambda = l + 1/2$, which should not be confused with the wavelength (the context should always make this distinction clear). In particular in what follows it is the *poles* of this function that are of interest. In terms of cylindrical Bessel and Hankel functions, they are defined by the condition

$$\ln' H_\lambda^{(1)}(\beta) = n \ln' J_\lambda(\alpha), \quad (1.18)$$

and are called *Regge poles* in the scattering theory literature (see, e.g., [51]). Typically they are associated with surface waves for the *impenetrable* sphere problem; for the *transparent* sphere two types of Regge poles arise—one type leading to rapidly convergent residue series (diffracted or creeping rays), and the other type associated with resonances via the internal structure of the potential. Many of these are clustered close to the real axis, spoiling the rapid convergence of the residue series. Mathematically, the resonances are complex eigenfrequencies associated with the poles λ_j of the scattering function $S(\lambda, k)$ in the first quadrant of the complex λ-plane; they are known as Regge poles (for real β). The imaginary parts of the poles are directly related to resonance widths (and therefore lifetimes). As the index j decreases, $\text{Re} \, \lambda_j$ increases and $\text{Im} \, \lambda_j$ decreases very rapidly (reflecting the exponential behavior of the barrier transmissivity). As β increases, the poles λ_j trace out Regge trajectories, and $\text{Im} \, \lambda_j$ tends exponentially to zero. When $\text{Re} \, \lambda_j$ passes close to a "physical" value, $\lambda = l + 1/2$, it is associated with a resonance in the lth partial wave; the larger the value of β, the sharper the resonance becomes for a given node number j.

In [83] it is shown that

$$S(\lambda, \beta) = \frac{H_{\lambda}^{(2)}(\beta)}{H_{\lambda}^{(1)}(\beta)} \left(R_{22}(\lambda, \beta) + T_{21}(\lambda, \beta) T_{12}(\lambda, \beta) \frac{H_{\lambda}^{(1)}(\alpha)}{H_{\lambda}^{(2)}(\alpha)} \sum_{p=1}^{\infty} [\rho(\lambda, \beta)]^{p-1} \right),$$

(1.19)

where

$$\rho(\lambda, \beta) = R_{11}(\lambda, \beta) \frac{H_{\lambda}^{(1)}(\alpha)}{H_{\lambda}^{(2)}(\alpha)}.$$

(1.20)

This is the *Debye expansion*, arrived at by expanding the expression $[1 - \rho(\lambda, \beta)]^{-1}$ as an infinite geometric series. R_{22}, R_{11}, T_{21}, and T_{12} are, respectively, the external/internal reflection and internal/external transmission coefficients for the problem. This procedure transforms the interaction of wave + sphere into a series of surface interactions. In so doing it unfolds the stationary points of the integrand, so that a given integral in the Poisson summation contains at most one stationary point. This permits a ready identification of the many terms in accordance with ray theory. The first term inside the parentheses represents direct reflection from the surface. The pth term represents transmission into the sphere (via the term T_{21}) subsequently bouncing back and forth between $r = a$ and $r = 0$ a total of p times with $p - 1$ internal reflections at the surface (this time via the R_{11} term in ρ). The middle factor in the second term, T_{12}, corresponds to transmission to the outside medium. In general, therefore, the pth term of the Debye expansion represents the effect of $p + 1$ surface interactions.

1.2.4 Scattering (Semiclassical)

On the way to the scattering representation, so to speak, there is a *semiclassical* description. In a primitive sense, the semiclassical approach is the 'geometric mean' between classical and quantum mechanical descriptions of phenomena in which interference and diffraction effects enter the picture. The latter do so via the transition from geometrical optics to wave optics. This is a characteristic feature of the 'primitive' semiclassical formulation. Indeed, the infinite intensities predicted by geometrical optics at focal points, lines and caustics in general are "breeding grounds" for diffraction effects, as are light/shadow boundaries for which geometrical optics predicts finite discontinuities in intensity. Such effects are most significant when the wavelength is comparable with (or larger than) the typical length scale for variation of the physical property of interest (e.g. size of the scattering object). Thus a scattering object with a "sharp" boundary (relative to one wavelength) can give rise to *diffractive scattering* phenomena.

There are 'critical' angular regions where the primitive semiclassical approximation breaks down, and diffraction effects cannot be ignored, although the angular ranges in which such critical effects become significant get narrower as the wavelength decreases. Early work in this field contained *transitional asymptotic approximations* to the scattering amplitude in these 'critical' angular domains, but they have very narrow domains of validity, and do not match smoothly with neighboring 'non-critical' angular domains. It is therefore of considerable importance to seek *uniform asymptotic approximations* that by definition do not suffer from these failings. Fortunately, the problem of plane wave scattering by a homogeneous sphere exhibits all of the critical scattering effects (and it can be solved exactly, in principle), and is therefore an ideal laboratory in which to test both the efficacy and accuracy of the various approximations. Furthermore, it has relevance to both quantum mechanics (as a square well or barrier problem) and optics (Mie scattering); indeed, it also serves as a model for the scattering of acoustic and elastic waves, and was studied in the early twentieth century as a model for the diffraction of radio waves around the surface of the earth. [88]

It transpires that the integral for $f(\beta, \theta)$ can be expressed as an infinite sum:

$$f(\beta, \theta) = f_0(\beta, \theta) + \sum_{p=1}^{\infty} f_p(\beta, \theta), \tag{1.21}$$

where

$$f_0(\beta, \theta) = \frac{i}{\beta} \sum_{m=-\infty}^{\infty} (-1)^m \int_0^{\infty} \left[1 - \frac{H_\lambda^{(2)}(\beta)}{H_\lambda^{(1)}(\beta)} R_{22} \right] P_{\lambda-\frac{1}{2}}(\cos\theta) \exp(2im\pi\lambda)\lambda d\lambda. \tag{1.22}$$

The expression for $f_p(\beta, \theta)$ involves a similar type of integral for $p \geq 1$. The application of the modified Watson transform to the third term ($p = 2$) in the Debye expansion of the scattering amplitude shows that it is this term which is associated with the phenomena of the primary rainbow. More generally, for a Debye term of given order p, a rainbow is characterized in the λ-plane by the occurrence of two real saddle points $\bar{\lambda}$ and $\bar{\lambda}'$ between 0 and β in some domain of scattering angles θ, corresponding to the two scattered rays on the light side. As $\theta \to \theta_R^+$ the two saddle points move toward each other along the real axis, merging together at $\theta = \theta_R$. As θ moves into the dark side ($\theta < \theta_R$), the two saddle points become complex, moving away from the real axis in complex conjugate directions. Therefore, from a mathematical point of view, *a rainbow can be defined as a coalescence of two saddle points in the complex angular momentum plane* (see Figure 1.2).

Thus the rainbow light/shadow transition region is associated physically with the confluence of a pair of geometrical rays and their transformation into complex rays; mathematically this corresponds to a pair of real saddle points merging into a complex saddle point. Then the problem is to find the asymptotic expansion of an integral having two saddle points that move toward or away from each other. The generalization of the standard saddle-point technique to include such problems was made by Chester et al. [164] and using their method, Nussenzveig [84] was able to find a uniform asymptotic expansion of the scattering amplitude that was valid throughout the rainbow region and that matched smoothly onto results for neighboring regions. The lowest order approximation in this expansion turns out to be the celebrated *Airy approximation*, which, despite several attempts to improve on it, was the best approximate treatment prior to the analyses of Nussenzveig and coworkers. However, Airy's theory had a limited range of applicability as a result

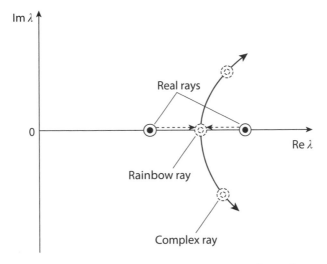

Figure 1.2 The coalescence of two real saddle points in the complex λ-plane as the rainbow angle is approached.

of its underlying assumptions; by contrast, the uniform expansion is valid over a much larger range.

1.2.5 Caustics and Diffraction Catastrophes

An alternative way of describing the rainbow phenomenon is by way of catastrophe theory, the rainbow being one of the simplest examples in catastrophe optics. Before summarizing the mathematical details of the basic rainbow diffraction catastrophe, it will be useful to introduce some of the language used in catastrophe-theoretic arguments. As we know from geometrical optics the primary bow deviation angle D_1 (in particular) has a minimum corresponding to the rainbow angle (or Descartes ray) $D_{min} = \theta_R$ when considered as a function of the angle of incidence i. Clearly the point $(i, D_1(i))$ corresponding to this minimum is a singular point (approximately $(59°, 138°)$) insofar as it separates a two-ray region $(D_1 > D_{min})$ from a zero-ray region $(D_1 < D_{min})$ in the geometrical-optics level of description. This is a singularity or *caustic point*. The rays form a directional caustic at this point, and this is a *fold catastrophe* (symbol: A_2), the simplest example of a catastrophe. It is the only stable singularity with *codimension* one (the dimensionality of the control space (one) minus the dimensionality of the singularity itself, which is zero). In space the caustic surface is asymptotic to a cone with semi-angle 42°.

Optics is concerned to a great extent with families of rays filling regions of space; the *singularities* of such ray families are *caustics*. For optical purposes this level of description is important for classifying caustics using the concept of *structural stability*: this enables one to classify those caustics whose topology survives perturbation. Structural stability means that if a singularity S_1 is produced by a generating function ϕ_1 (see below for an explanation of these terms), and ϕ_1 is perturbed to

ϕ_2, the correspondingly changed S_2 is related to S_1 by a *diffeomorphism* of the control set C (that is by a smooth reversible set of control parameters; in other words a smooth deformation). In the present context this means, in physical terms, that distortions of incoming wavefronts by deviations of the raindrop shapes from their ideal spherical form does *not* prevent the formation of rainbows, though there may be some changes in the features. Another way of expressing this concept is to describe the system as well-posed in the limited sense that small changes in the input generate correspondingly small changes in the output. For the so-called elementary catastrophes, structural stability is a generic (or typical) property of caustics. Each structurally stable caustic has a characteristic diffraction pattern, the wave function of which has an integral representation in terms of the standard polynomial describing the corresponding catastrophe. From a mathematical point of view these diffraction catastrophes are especially interesting, because they constitute a new hierarchy of functions, distinct from the special functions of analysis. A review of this subject has been made by Berry and Upstill [102] wherein may be found an introduction to the formalism and methods of catastrophe theory as developed particularly by Thom [171] but also by Arnold [173]. The books by Gilmore [105] (Chapter 13 of which concerns caustics and diffraction catastrophes) and Poston and Stewart [172] are noteworthy in that they also provide many applications.

Diffraction can be discussed in terms of the scalar Helmholtz equation in some spatial region $\mathcal{R}$,

$$\nabla^2 \psi(\mathcal{R}) + k^2 n^2(\mathcal{R}) \psi(\mathcal{R}) = 0, \tag{1.23}$$

for the complex scalar wavefunction $\psi(\mathcal{R})$, k being the free-space wavenumber and n the refractive index. The concern in catastrophe optics is to study the asymptotic behavior of wave fields near caustics in the short-wave limit $k \to \infty$ (semiclassical theory). In a standard manner, $\psi(\mathcal{R})$ is expressed as

$$\psi(\mathcal{R}) = a(\mathcal{R}) e^{ik\chi(\mathcal{R})}, \tag{1.24}$$

where the modulus a and the phase $k\chi$ are both real quantities. To the lowest order of approximation χ satisfies the *Hamilton-Jacobi equation*, and ψ can be determined asymptotically in terms of a phase-action exponent (surfaces of constant action are the wavefronts of geometrical optics). The integral representation for ψ is

$$\psi(\mathcal{R}) = e^{-in\pi/4} \left(\frac{k}{2\pi} \right)^{n/2} \int \cdots \int b(s; \mathcal{R}) \exp[ik\phi(s; \mathcal{R})] d^n s, \tag{1.25}$$

where n is the number of state (or behavior) variables s, and b is a weight function. In general there is a relationship between this representation and the simple ray approximation [102]. According to the principle of stationary phase, the main contributions to the above integral for given $\mathcal{R}$ come from the stationary points (i.e., those points s_i for which the gradient map $\partial\phi/\partial s_i$ vanishes) caustics are *singularities* of this map, where two or more stationary points coalesce. Because $k \to \infty$, the integrand is a rapidly oscillating function of s, so other than near the points s_i, destructive interference occurs and the corresponding contributions are negligible. The stationary points are well separated, provided $\mathcal{R}$ is not near a caustic; the simplest form of stationary phase can then be applied and yields a series of terms of the form

$$\psi(\mathcal{R}) \approx \sum_{\mu} a_{\mu} \exp[i g_{\mu}(k, \mathcal{R})], \qquad (1.26)$$

where the details of the g_{μ} need not concern us here. Near a caustic, however, two or more of the stationary points are close (in some appropriate sense), and their contributions cannot be separated without a reformulation of the stationary phase principle to accommodate this [164]. The problem is that the "ray" contributions can no longer be considered separately; when the stationary points approach closer than a distance $O(k^{-1/2})$, the contributions are not separated by a region in which destructive interference occurs (See Appendix A for the "order" nomenclature used this book). When such points coalesce, $\phi(s; \mathcal{R})$ is stationary to higher than first order, and quadratic terms as well as linear terms in $s - s_{\mu}$ vanish. This implies the existence of a set of displacements ds_i, away from the extrema s_{μ}, for which the gradient map $\partial\phi/\partial s_i$ still vanishes, that is, for which

$$\sum_{i} \frac{\partial^2 \phi}{\partial s_i \partial s_j} ds_i = 0. \qquad (1.27)$$

The condition for this homogeneous system of equations to have a solution (i.e., for the set of control parameters X to lie on a caustic) is that the Hessian

$$H(\phi) \equiv \det \left(\frac{\partial^2 \phi}{\partial s_i \partial s_j} \right) = 0, \qquad (1.28)$$

at points $s_{\mu}(X)$ where $\partial\phi/\partial s_i = 0$ (details can be found in [102]). The caustic defined by $H = 0$ determines the bifurcation set for which at least two stationary points coalesce (in the present circumstance this is just the rainbow angle). In view of this discussion there are two other ways of expressing this: (i) rays coalesce on caustics, and (ii) caustics correspond to singularities of gradient maps. To remedy this problem the function ϕ is replaced by a simpler "normal form" Φ with the same stationary-point structure; the resulting diffraction integral is evaluated exactly. This is where the property of structural stability is so important, because if the caustic is structurally stable it must be equivalent to one of the catastrophes (in the diffeomorphic sense described above). The result is a generic diffraction integral that will occur in many different contexts. The basic diffraction catastrophe integrals (one for each catastrophe) may be reduced to the form

$$\Psi(X) = \frac{1}{(2\pi)^{n/2}} \int \cdots \int \exp[i \Phi(s; X)] d^n s, \qquad (1.29)$$

where s represents the state variables and X the control parameters (for the case of the rainbow there is only one of each, so $n = 1$). These integrals stably represent the wave patterns near caustics. The *corank* of the catastrophe is equal to n: it is the minimum number of state variables necessary for Φ to reproduce the stationary-point structure of ϕ; the *codimension* is the dimensionality of the control space minus the dimensionality of the singularity itself. It is interesting to note that in ray catastrophe optics, the state variables s are removed by differentiation (the vanishing of the gradient map); in wave catastrophe optics they are removed by integration (via the diffraction functions). For future reference we state the functions $\Phi(s; X)$ for both the fold (A_2) and the cusp (A_3) catastrophes; the list for the remaining five elementary catastrophes can be found in the references above. For the fold

$$\Phi(s; X) = \frac{1}{3}s^3 + Xs, \tag{1.30}$$

and for the cusp

$$\Phi(s; X) = \frac{1}{4}s^4 + \frac{1}{2}X_2 s^2 + X_1 s.$$

By substituting the cubic term (1.30) into the integral (1.29), it follows that

$$\Psi(X) = \frac{1}{(2\pi)^{1/2}} \int_{-\infty}^{\infty} \exp\left[i\left(\frac{s^3}{3} + Xs\right)\right] ds = (2\pi)^{1/2} \text{Ai}(X),$$

where the integral here denoted by $\text{Ai}(X)$ is one form of the Airy integral. It will be encountered in several different guises as we proceed through the book. For $X < 0$ (corresponding to $\theta > \theta_R$) there are two rays (stationary points of the integrand) whose interference causes oscillations in $\Psi(X)$; for $X > 0$ there is one (complex) ray that decays to zero monotomically (and faster than exponentially). This describes diffraction near a fold caustic. In 1838 the Astronomer Royal Sir George Biddle Airy introduced this function to study diffraction along the asymptote of a caustic (although he did not express it in these terms) and provided a fundamental description of the supernumerary bows [23]. (There is also a sequel to this paper, published ten years later [268], which contains a transcript of a fascinating letter from the British mathematician Augustus De Morgan (1806–1871)). As noted above, this integral (in one form or another) will be a recurring topic in several later chapters. The corresponding integral for the cusp catastrophe is frequently referred to as the *Pearcey integral* [157].

PART I
Rays

Chapter Two

Introduction to the "Physics" of Rays

> It is common physical knowledge that wavefields (acoustic, electromagnetic, etc.) rather than rays are a physical reality. Nonetheless, the traditions to endow rays with certain physical properties, traced back to Descartes' times, have been deeply enrooted in natural science. Rays are discussed as if they were real objects. This handling of rays is justified when rays can be localized in space.
>
> [174]

2.1 *WHAT IS A RAY?*

In optics, a ray is a mathematical idealization of an infinitesimally narrow beam of light. In other words, it doesn't exist in the physical world, only in a mathematical realm. They are mathematical models that, like many models, can be extremely valuable despite their shortcomings. Specifically, they are geometrical objects as opposed to physical ones, and this gives rise to the subject of *geometrical optics*. As we shall see, rays may (usually) be regarded as normals to wavefronts (or to surfaces of constant phase). Clearly at this point the more an attempt is made to describe rays, the more things that have to be described! We shall therefore proceed as if rays were real physical entities. In this spirit, rays have positions, directions, and speeds, and they "carry" energy, while the "density" of rays is a measure of power per unit area. They can propagate in homogeneous or inhomogeneous media, but in the former they do so in straight lines, these being mathematical models of the path in which light travels in such media. In the latter, if the properties of the medium vary continuously, the rays paths of light are in general curved; if the properties of the medium are discontinuous the rays will undergo reflection (and possibly refraction). It is particularly important to note that ray paths are reversible. In this and some other chapters of this book, we will encounter curved interfaces, generally spherical, between media, so it is important to establish the behavior of rays at such interfaces. There are two features implicit in the approximate solution for the radiation field offered by the method of geometrical optics: (i) the incident field behaves locally as a plane wave, and (ii) each point on the interface behaves as if it were part of an infinite planar interface.

Given that light rays are the best physical models of straight lines [3], it is no surprise that the developments of geometry and optics from Euclid (born ca. 300 BCE) onward have been closely related. Euclid was familiar with reflection from smooth (highly polished) surfaces, having deduced the properties of parabolic, spherical, and elliptical reflectors. Furthermore, Heron of Alexandria (ca.)10–ca.(70 CE) essentially established the law of reflection before it was rediscovered by (amongst others) Willebrord Snellius (1580–1626; born Willebrord Snel van Royen) The correct spelling of his name is "Snel", by the way [4]; but with apologies to Professor Bohren I shall continue to use the almost universal form "Snell". The law of refraction was first accurately described by Ibn Sahl at the Baghdad court in 984.

> Note that *caustic surfaces* (or simply "caustics") *do* exist. They are envelopes of families of rays [174] and are readily noticed (as anyone wearing spectacles on a rainy night or tea/coffee drinkers will readily attest). More details will be found in Chapter 6.

2.1.1 Some Mathematical Definitions

These will be encountered and restated in various guises in later chapters, so for now the definitions are stated in simplest terms [177].

The refractive index n

The refractive index is simply defined as the ratio of the speed of light c in vacuo to that in the medium (v), be it water, glass, etc.:

$$n = \frac{c}{v}. \tag{2.1}$$

It will be constant for a homogeneous medium or a function of position for an inhomogeneous one (both types will be considered in this book). The *relative* refractive index is the ratio of refractive indices in two different media. There also exist non-optical environments for which geometrical optics is useful, as we shall see, so it will come as no surprise that it is possible to define a type of refractive index there also.

Geometric path length

Suppose that the ray path C is a smooth curve joining two points $\mathbf{r}_1$ and $\mathbf{r}_2$, where the ray is described parametrically by $\mathbf{r}(\tau)$, where $\dot{\mathbf{r}}(\tau) = d\mathbf{r}/d\tau$, then the geometric path length between the points is

$$s(\mathbf{r}_1; \mathbf{r}_2) = \int_C ds = \int_{\mathbf{r}_1}^{\mathbf{r}_2} ds = \int_{\mathbf{r}_1}^{\mathbf{r}_2} \frac{ds}{d\tau} d\tau = \int_{\mathbf{r}_1}^{\mathbf{r}_2} \left| \dot{\mathbf{r}}^2(\tau) \right|^{1/2} d\tau. \tag{2.2}$$

Optical path length ψ

The optical path length is defined by

$$\psi\,(\mathbf{r}_1;\mathbf{r}_2) = \int_C n(\mathbf{r})ds = \int_{\mathbf{r}_1}^{\mathbf{r}_2} n(\mathbf{r})ds, \tag{2.3}$$

so that for constant c it may be written as

$$\psi\,(\mathbf{r}_1;\mathbf{r}_2) = c\int_{\mathbf{r}_1}^{\mathbf{r}_2} \frac{ds}{v(\mathbf{r})} \equiv cT, \tag{2.4}$$

where T is the travel time for the light between the two points. In a continuously varying medium the ray path is a smooth curve that bends toward the region of higher refractive index at each point on the curve (this will be the subject of Chapter 3, in which the mirage theorem is introduced). A running optical path length from a fixed initial point $\mathbf{r}_0$ may be defined as

$$\psi\,(\mathbf{r}_0;\mathbf{r}) = \int_{\mathbf{r}_0}^{\mathbf{r}} n(\mathbf{x})ds. \tag{2.5}$$

A surface $\psi\,(\mathbf{r}_0;\mathbf{r}) = $ constant can be interpreted as a *geometric wavefront* emanating from a source at $\mathbf{r}_0$.

2.1.2 Geometric Wavefronts

A geometric wavefront is the locus of constant optical path from a point source. If the light (or other type of wave) leaves $\mathbf{r}_0$ at time t_0, the wavefront at a later time time t is simply

$$\psi\,(\mathbf{r}_0;\mathbf{r}) = c\,(t - t_0)\,. \tag{2.6}$$

As will be noted later, $\psi\,(\mathbf{r}_0;\mathbf{r})$ satisfies the *eikonal equation*

$$|\nabla\psi\,(\mathbf{r}_0;\mathbf{r})|^2 = n^2(\mathbf{r}). \tag{2.7}$$

The optical path length is the eikonal function, and because it has dimensions of length it is related to a dimensionless phase function θ (or sometimes, depending on context, S, to be introduced later) by $\theta = k\psi$, where $k = 2\pi/\lambda$ is a wavenumber. For a given k the geometric wavefront is equivalent to a surface of constant phase in wave optics, and indeed, the eikonal equation can be derived from the wave equation in the geometrical optics limit, that is, the limit of vanishingly small wavelength λ.

The Eikonal Equation and Pythagoras's Theorem

It is interesting to note that one form of the eikonal equation was known (in a sense) to Pythagoras, encapsulated in what has been termed the "secret Pythagorean theorem" or SPT [123]. Pythagoras's theorem, as everyone reading this book should know, states that in a right triangle the square on the hypotenuse is equal to the sum of the squares on the other two sides, so from Figure 2.1,

$$(OA)^2 + (OB)^2 = (AB)^2\,. \tag{2.8}$$

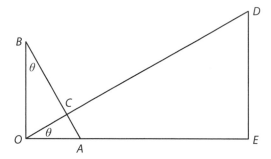

Figure 2.1 The secret Pythagorean theorem.

In contrast, the SPT states that

$$(OA)^{-2} + (OB)^{-2} = (OC)^{-2}, \tag{2.9}$$

or in other words, the sum of the squares of the reciprocals of the legs of a right triangle equals the square of the reciprocal of the altitude. The proof follows directly from Figure 2.1; we construct line segment OD collinear with OC and such that $OC \times OD = 1$, and form the altitude DE to the segment OE collinear with OA. Then

$$DE = OD \sin \theta = OD \left(\frac{OC}{OB} \right) = (OB)^{-1}. \tag{2.10}$$

Similarly,

$$OE = OD \cos \theta = OD \left(\frac{OC}{OA} \right) = (OA)^{-1}. \tag{2.11}$$

Referring to triangle ODE, Pythagoras's theorem states that

$$(OE)^2 + (DE)^2 = (OD)^2, \tag{2.12}$$

and hence, by virtue of the results (2.10) and (2.11) the SPT for triangle BOA is given by (2.9).

So what does this have to do with the eikonal equation (particularly in geophysical applications [123])? In a uniform medium suppose that O is a point on a wavefront at a given moment of time, and suppose further that at a time δt later the wavefront is represented by the line BCA, and the distance traveled by the ray (being orthogonal to the wavefront) is $OC = \delta s$. If $OA = \delta x$ and $OB = \delta y$, then multiplying throughout by δt, the SPT tells us that

$$\left(\frac{\delta t}{\delta x} \right)^2 + \left(\frac{\delta t}{\delta y} \right)^2 = \left(\frac{\delta t}{\delta s} \right)^2. \tag{2.13}$$

But this can be interpreted as an approximate relation between the various components of "slowness"; indeed, the usual limiting process implies that we may write (extending the result to three dimensions)

$$\left(\frac{\partial t}{\partial x}\right)^2 + \left(\frac{\partial t}{\partial y}\right)^2 + \left(\frac{\partial t}{\partial z}\right)^2 = \left(\frac{\partial t}{\partial s}\right)^2 = \left(\frac{1}{v}\right)^2, \qquad (2.14)$$

or

$$|\nabla t|^2 = \left(\frac{1}{v}\right)^2. \qquad (2.15)$$

This relates an eikonal (in this case travel time $t(x, y, z)$) to the *slowness* (or reciprocal speed) v^{-1}. Of course, in general $c = c(x, y, z)$, and the wavefronts and corresponding ray paths are curved, but the SPT will enable us (following [123]) to make an interesting connection later with seismic rays in Chapter 9. In particular if $c(x, y, z)$ is known and the initial source or wavefront location is specified, the travel time can be computed as a function of position in the medium. Frequently of course it is the inverse problem, that is, determination of $c(x, y, z)$ that is desired.

2.1.3 Fermat's Principle

Specifically including reflection and refraction, Fermat's principle may be stated as follows. *Reflected and refracted rays from a source point S to a point P are such that the optical path length between S and P with one point on the interface between two media is stationary with respect to infinitesimal variations in path.* More generally it tells us that the optical path length between two given points is an extremum (which may not be global). Thus by Fermat's principle, for infinitesimal variations in path, $\psi(\mathbf{r}_1; \mathbf{r}_2)$ is stationary, that is,

$$\delta \int_{\mathbf{r}_1}^{\mathbf{r}_2} n(\mathbf{r}) ds = 0. \qquad (2.16)$$

To determine the power flow in a radiation field, the ray paths need to be accompanied by information about the field amplitude and phase at each point. In what follows, reference is made to Figure 2.2.

2.1.4 The Intensity Law

The energy E in the field moves along the rays with speed $v = c/n$. It is proportional to the square of the field amplitude (or more generally, the square of its modulus), and the intensity is the energy crossing unit area in unit time, that is, $|E|^2 v$. By conservation of energy within a tube of rays, incident electric fields E_{i1} and E_{i2} are related via the area elements dA_1 and dA_2 as follows:

$$|E_{i1}|^2 dA_1 = |E_{i2}|^2 dA_2, \qquad (2.17)$$

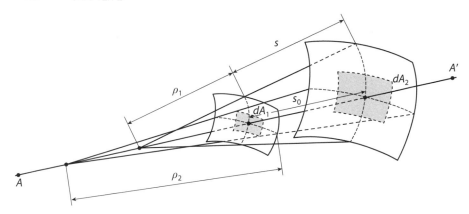

Figure 2.2 Local wavefront geometry for intensity variation.

or in terms of the local radii of curvature (ρ_1, ρ_2) of the wavefronts (see Figure 2.2) where s is the arc length [5],

$$\frac{dA_1}{dA_2} = \frac{\rho_1\rho_2}{(\rho_1 + s)(\rho_2 + s)} > 0. \tag{2.18}$$

Thus the phase and amplitude of the incident field are related by the fundamental result

$$E_{i2} = E_{i1} \left[\frac{\rho_1\rho_2}{(\rho_1 + s)(\rho_2 + s)} \right]^{1/2} \exp(iks). \tag{2.19}$$

There are four basic types of field in geometrical optics. If $\rho_1 = \rho_2$, the field propagates as a spherical wave; if one of these radii is infinite, it does so as a cylindrical wave, and if both are infinite, the field propagates as a plane wave. The most general case, for which both radii are finite and unequal, is said to be *astigmatic*. Note that because equation (2.19) is singular if ρ_1 or ρ_2 is equal to $-s$, geometrical optics predicts infinite intensity there. These regions are referred to as caustics; they may be surfaces, curves, or points, and clearly the geometrical optical solution is invalid there. At a caustic surface, the (generic) quantity $\rho/(\rho + s)$ changes sign, which introduces a phase shift of $\exp(\pm i\pi/2)$. The correct sign is +, as will be demonstrated below. Then

$$\left(\frac{\rho}{\rho + s} \right)^{1/2} = \left| \frac{\rho}{\rho + s} \right|^{1/2}; \quad \frac{\rho}{\rho + s} > 0, \tag{2.20}$$

$$= \left| \frac{\rho}{\rho + s} \right|^{1/2} e^{i\pi/2}; \quad \frac{\rho}{\rho + s} < 0. \tag{2.21}$$

In the event that rays encounter a (possibly curved) boundary, the field will in general consist of both reflected and refracted (or transmitted) portions. The resulting fields can be written by modifying equation (2.19) to one premultiplied by, respectively, a reflection and a transmission matrix. These are 2×2 matrices expressed

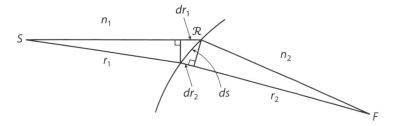

Figure 2.3 Boundary geometry for Snell's law of refraction.

in terms of a ray-based coordinate system. This will not be pursued here, but for excellent reviews, see [3], [5], and [183].

2.1.5 Heuristic Derivation of Snell's Laws

What follows is a nonrigorous account of the basic ideas involved in this derivation; it can readily be made rigorous by geometric and/or algebraic arguments, as will be seen below. Consider two closely adjacent rays emanating from a source S incident on a smooth reflecting or refracting surface $\mathcal{R}$ and intersecting again at a point F. The surface $\mathcal{R}$ separates uniform media with refractive indices n_1 and n_2. Depending on whether we consider the refracted or reflected ray, F will be on the other side or the same side as S, respectively (Figure 2.3). If r_1 and r_2 are the two-center bipolar coordinates of any point with respect to S and F, then by Fermat's principle, the optical path length for each ray is the same:

$$n_1 r_1 + n_2 r_2 = n_1 \left(r_1 + dr_1 \right) + n_2 \left(r_2 - dr_2 \right),$$

or

$$n_1 dr_1 = n_2 dr_2. \tag{2.22}$$

If the points S and F are on the same side as $\mathcal{R}$, (i.e, they undergo reflection), the image is said to be virtual, and $n_1 = n_2$. Thus for reflection

$$dr_1 = dr_2,$$

or

$$r_1 d\theta_1 = r_2 d\theta_2. \tag{2.23}$$

The laws of refraction and reflection are obtained by dividing these equations respectively by the incremental surface element ds at any point on $\mathcal{R}$. Thus from (2.22), the resulting expression

$$n_1 \frac{dr_1}{ds} = n_2 \frac{dr_2}{ds} \tag{2.24}$$

establishes *the law of refraction*. When $n_1 = n_2$ the *law of reflection* is obtained (in accordance with the concavity of the reflecting surface) [6], [7].

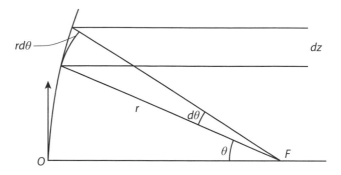

Figure 2.4 Geometry for example.

Example: Suppose that we wish to design a reflecting surface that focuses all the rays from S at infinity. Find the shape of the reflecting surface that will accomplish this.

Solution: (See Figure 2.4) equation (2.23) implies that $rd\theta = dz$; also $z = r\sin\theta$, whence

$$\ln r = \int (\csc\theta - \cot\theta)\, d\theta = \ln\left|\frac{\csc\theta - \cot\theta}{\sin\theta}\right| + \ln K$$

$$= \ln\left|\frac{1}{1+\cos\theta}\right| + \ln K = \ln\left|\frac{\sec^2(\theta/2)}{2}\right| + \ln K$$

$$= \ln\left|c\sec^2(\theta/2)\right|.$$

Therefore

$$r = \frac{2c}{1+\cos\theta},$$

which is the equation of a parabola with focal length c.

2.1.6 Generalization

The above ideas concerning refraction can be extended to an indefinitely large number of surfaces, ultimately progressing, in the appropriate limit, to an integral formulation. Thus if ds_i is the infinitesimal arc length element for the ith surface, the extended version of Fermat's principle for piecewise uniform media is

$$\sum_{i=1}^{N} n_i ds_i = \delta \sum_{i=1}^{N} n_i s_i = 0. \tag{2.25}$$

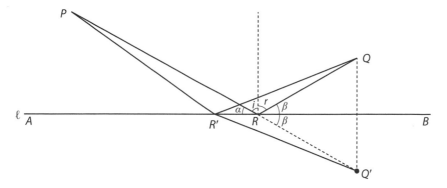

Figure 2.5 Geometry for the law of reflection (Heron's problem).

Extended formally to a continuously varying medium, we can define the optical length of a curve C joining two distinct points in the medium by writing equation (2.3) as

$$\psi\,(C) = \int_{C} n\,(s)\,ds, \qquad (2.26)$$

and by Fermat's principle this is stationary, that is, a form of (2.16) is recovered. Again, what this means in essence is that for small first-order variations of C (with respect to fixed endpoints), the variation of ψ is of higher order. Most commonly, ψ is a minimum for this stationary path. We shall return to this discussion later. In the next section we discuss various proofs of Snell's laws of reflection and refraction; further details may be found in [8], [9], [10], [11].

2.2 GEOMETRIC AND OTHER PROOFS OF SNELL'S LAWS OF REFLECTION AND REFRACTION

2.2.1 The Law of Reflection

This is also known as *Heron's problem* for the law of reflection. Suppose that P and Q are two given points on the same side of a line l (Figure 2.5). Find a point R on l such that the sum of the distances from P to R and from R to Q is a minimum. Consider the point Q' (the mirror image of Q in the line l) together with any second point R' on the line l. Clearly $|RQ| = |RQ'|$, and by the triangle inequality,

$$\left|PR'\right| + \left|R'Q'\right| \ge \left|PQ'\right| = |PR| + \left|RQ'\right| = |PR| + |RQ|. \qquad (2.27)$$

Since R' is arbitrary, it follows that the point R minimizes the sum of these distances. Snell's law of reflection follows from the fact that angle α ($\angle ARP$) is equal to angle β ($\angle B'RQ'$) and hence to angle BRQ. In terms of the common optical notation, the complementary angles to α and β are the angles of incidence (i) and reflection (r), respectively, are equal. This is Snell's law of reflection. Note

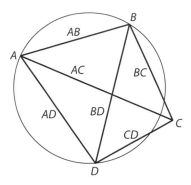

Figure 2.6 Geometry for Ptolemy's inequality.

also that if the medium above l is homogeneous, the speed of light (c) is constant there, and the ray path PRQ also minimizes the travel time T from P to Q, namely,

$$T = \frac{|PR|}{c} + \frac{|RQ|}{c}. \tag{2.28}$$

2.2.2 The Law of Refraction

We will examine several different proofs of the equivalence of Fermat's principle of least time and Snell's law of refraction, but first it will be necessary to examine another important result from geometry.

Ptolemy's Inequality

If A, B, C, and D are any distinct points in a plane, then

$$|AB||CD| + |BC||AD| \geq |AC||BD|, \tag{2.29}$$

with equality if and only if A, B, C, and D lie on a circle in that order. Ptolemy's inequality concerns the noncyclic quadrilateral $ABCD$ shown in Figure 2.6. When the quadrilateral is cyclic, the equality holds. The points need not all be in the same plane, and in fact a vector proof is indicated below for a Euclidean space of arbitrary dimension. The inequality states that the sum of the products of opposite sides is never less than the product of the diagonals, so in a slightly modified form:

$$|AB||CD| + |BC||DA| \geq |AC||BD|. \tag{2.30}$$

There are many different proofs of this inequality. The vector proof is as follows. Let any four points be denoted by the vectors $\mathbf{a}, \mathbf{b}, \mathbf{c}, \mathbf{d}$ (in two

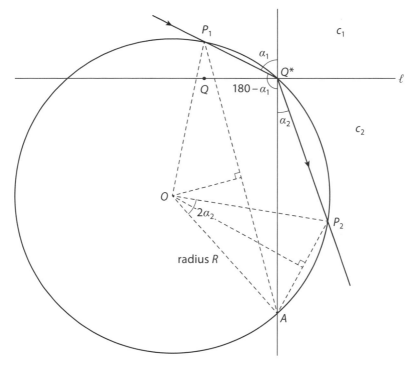

Figure 2.7 Use of Ptolemy's inequality for Fermat's principle (speeds c_1, c_2).

dimensions, one can instead use complex numbers a, b, c, d). Then

$$(\mathbf{a} - \mathbf{b}) \cdot (\mathbf{c} - \mathbf{d}) + (\mathbf{a} - \mathbf{d}) \cdot (\mathbf{b} - \mathbf{c}) \tag{2.31}$$

$$= \mathbf{a} \cdot \mathbf{c} - \mathbf{a} \cdot \mathbf{d} - \mathbf{b} \cdot \mathbf{c} + \mathbf{b} \cdot \mathbf{d} + \mathbf{a} \cdot \mathbf{b} - \mathbf{a} \cdot \mathbf{c} - \mathbf{d} \cdot \mathbf{b} + \mathbf{d} \cdot \mathbf{c} \tag{2.32}$$

$$= \mathbf{a} \cdot \mathbf{b} - \mathbf{a} \cdot \mathbf{d} - \mathbf{b} \cdot \mathbf{c} + \mathbf{d} \cdot \mathbf{c} \tag{2.33}$$

$$= (\mathbf{a} - \mathbf{c}) \cdot (\mathbf{b} - \mathbf{d}). \tag{2.34}$$

But from the triangle inequality,

$$|(\mathbf{a} - \mathbf{b}) \cdot (\mathbf{c} - \mathbf{d})| + |(\mathbf{a} - \mathbf{d}) \cdot (\mathbf{b} - \mathbf{c})| \geq |(\mathbf{a} - \mathbf{c}) \cdot (\mathbf{b} - \mathbf{d})|, \tag{2.35}$$

and the theorem is established.

Since any three distinct points P_1, Q^*, and P_2 lie on a circle, we now apply the theorem to Figure 2.7 to establish the equivalence of Fermat's principle of least time and Snell's law of refraction. In fact, we wish to show that for any other point Q on the boundary line l,

$$\frac{|P_1 Q|}{c_1} + \frac{|Q P_2|}{c_2} > \frac{|P Q^*|}{c_1} + \frac{|Q^* P_2|}{c_2}. \tag{2.36}$$

Recall the theorem from elementary geometry that the measure of an angle inscribed in a circle is equal to half the measure of its intercepted arc. Given the point A on the circle of radius R, it follows that

$$|AP_1| = 2R \sin(\pi - \alpha_1) = 2R \sin \alpha_1, \text{ and} \tag{2.37}$$

$$|AP_2| = 2R \sin \alpha_2. \tag{2.38}$$

Hence

$$\frac{|AP_1|}{|AP_2|} = \frac{\sin \alpha_1}{\sin \alpha_2} \equiv \frac{c_1}{c_2} \tag{2.39}$$

on using Snell's law of refraction. This means that we can write, for any positive constant k,

$$|AP_1| = \frac{k}{c_2}; \quad |AP_2| = \frac{k}{c_1}. \tag{2.40}$$

Applying Ptolemy's *equality* to the points on the circle,

$$|P_1 Q^*| \, |AP_2| + |P_2 Q^*| \, |AP_1| = |P_1 P_2| \, |AQ^*|. \tag{2.41}$$

For any point Q on l (but not on the circle) the strict *inequality* holds:

$$|P_1 Q| \, |AP_2| + |P_2 Q| \, |AP_1| > |P_1 P_2| \, AQ. \tag{2.42}$$

From the results (2.40), (2.41), and (2.42) we have therefore

$$k\left(\frac{|P_1 Q^*|}{c_1} + \frac{|P_2 Q^*|}{c_2}\right) = |P_1 P_2| \, |AQ^*|, \tag{2.43}$$

and

$$k\left(\frac{|P_1 Q|}{c_1} + \frac{|P_2 Q|}{c_2}\right) > |P_1 P_2| \, |AQ|. \tag{2.44}$$

Since $|AQ| > |AQ^*|$, we have proved that a ray path satisfying Snell's law of refraction has a minimum travel time.

2.2.3 A Wave-Theoretic Proof

For this proof we must formally state a form of Huygen's principle: every point of a wavefront S_t itself becomes a secondary source, and in time Δt a family of wavefronts is obtained from all these secondary sources. The wavefront $S_{t+\Delta t}$ is the envelope of this family (i.e., the surface tangent to all secondary wavefronts). The wavefront $AA'A_1$ moves with speed c_1, and suppose that at time t the point A reaches the point D on l; D now becomes a secondary source for the second medium, propagating with speed $c_2 < c_1$ as drawn in Figure 2.8. The ray passing through the point B_1 reaches the point D_1 on l at the later time

$$t_1 = t + \frac{|B_1 D_1|}{c_1} = t + \frac{|DD_1| \sin \alpha_1}{c_1}. \tag{2.45}$$

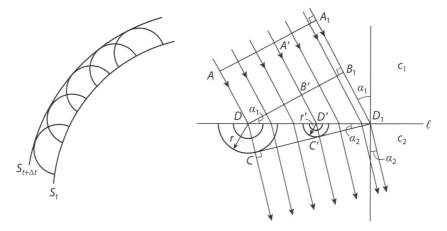

Figure 2.8 Using Huygen's principle to establish the law of refraction.

For intermediate (but otherwise arbitrary) points B' and D' on segments DB_1 and DD_1 the corresponding time is

$$t' = t + \frac{|DD'| \sin \alpha_1}{c_1}. \tag{2.46}$$

However, by time t_1 the spherical wave due to the secondary source D will have radius

$$r_1 = c_2 (t_1 - t) = |DD_1| \left(\frac{c_2}{c_1} \right) \sin \alpha_1. \tag{2.47}$$

Correspondingly the wave emanating from D' will have radius

$$r' = c_2 \left(t_1 - t' \right) = |DD'| \left(\frac{c_2}{c_1} \right) \sin \alpha_1. \tag{2.48}$$

Obviously the angles $\angle DD_1C$ and $\angle D'D_1C'$ are equal so the tangent planes (or lines as drawn in two dimensions) to these spheres coincide. But since D' is an arbitrary point on the plane wavefront DD_1, all secondary waves are tangent to the line CD_1 at time t_1, and this tangent line is at an angle α_2 to the line l. Since

$$\sin \alpha_2 = \frac{r_1}{|DD_1|} = \frac{r'}{|D'D_1|} = \left(\frac{c_2}{c_1} \right) \sin \alpha_1, \tag{2.49}$$

Snell's law of refraction has been established once more.

2.2.4 An Algebraic Proof

Suppose that the path PRQ is that of minimum travel time, T (Figure 2.9), where

$$T = \frac{|PR|}{c_1} + \frac{|RQ|}{c_2}. \tag{2.50}$$

On this path i is the angle of incidence of the ray in medium 1 (with speed of light c_1) and r is the angle of refraction of the ray into medium 2 (with speed of light c_2,

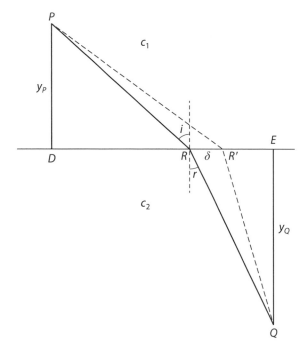

Figure 2.9 An algebraic proof for the law of refraction.

where as drawn $c_1 > c_2$). To prove that on this path $c_2 \sin i = c_1 \sin r$, the method of contradiction will be used. Suppose first that

$$c_2 \sin i < c_1 \sin r, \tag{2.51}$$

and that points P and Q lying on this path PRQ are chosen such that $|PR| = |RQ| = d$. Also we set $|DR| = x_1$ and $|RE| = x_2$. Following [11], we note the assumed inequality (2.51) implies that

$$\frac{x_1}{x_2} = \frac{(x_1/d)}{(x_2/d)} = \frac{\sin i}{\sin r} < \frac{c_1}{c_2} \Rightarrow \frac{x_2}{c_2} - \frac{x_1}{c_1} > 0. \tag{2.52}$$

Next we choose a positive $\delta < x_2$ such that

$$\delta < 2 \frac{(x_2/c_2) - (x_1/c_1)}{(1/c_1) + (1/c_2)}, \tag{2.53}$$

and pick point R' to the right of R so that $|RR'| = \delta$. Define

$$\Delta = \left(\frac{|PR'|}{c_1} + \frac{|R'Q|}{c_2} \right) - \left(\frac{|PR|}{c_1} + \frac{|RQ|}{c_2} \right). \tag{2.54}$$

Δ is just the difference in travel times between the path $PR'Q$ and PRQ. If $\Delta < 0$, the travel time for the latter path exceeds that of the former, and we have a contradiction, since the latter path is assumed to be minimal. Now

$$\Delta = \frac{\sqrt{(x_1 + \delta)^2 + y_P^2}}{c_1} + \frac{\sqrt{(x_2 - \delta)^2 + y_Q^2}}{c_2} - \left(\frac{d}{c_1} + \frac{d}{c_2}\right) \tag{2.55}$$

$$= \frac{\sqrt{\delta^2 + 2\delta x_1 + d^2} - d}{c_1} + \frac{\sqrt{\delta^2 - 2\delta x_2 + d^2} - d}{c_2}. \tag{2.56}$$

Lemma: For any $a > 0$ and any b such at $a + b > 0$,

$$\sqrt{a + b} - \sqrt{a} < \frac{b}{\sqrt{a}}. \tag{2.57}$$

Proof:

$$\left(\sqrt{a + b} - \sqrt{a}\right)\left(\sqrt{a + b} + \sqrt{a}\right) = b. \tag{2.58}$$

Therefore

$$\sqrt{a + b} - \sqrt{a} = \frac{b}{\sqrt{a + b} + \sqrt{a}} < \frac{b}{\sqrt{a}}. \tag{2.59}$$

Applying this to (2.56) with $a = d^2$ and $b = \delta^2 + 2\delta x_1$ or $b = \delta^2 - 2\delta x_2$ leads to the inequalities

$$\sqrt{\delta^2 + 2\delta x_1 + d^2} - d < \frac{\delta^2 + 2\delta x_1}{d}, \quad \text{and} \tag{2.60}$$

$$\sqrt{\delta^2 - 2\delta x_2 + d^2} - d < \frac{\delta^2 - 2\delta x_2}{d}. \tag{2.61}$$

Using these in (2.56), we have

$$\Delta < \frac{1}{c_1}\left(\frac{\delta^2 + 2\delta x_1}{d}\right) + \frac{1}{c_2}\left(\frac{\delta^2 - 2\delta x_2}{d}\right) \tag{2.62}$$

$$= \frac{\delta}{d}\left[\delta\left(\frac{1}{c_1} + \frac{1}{c_2}\right) + 2\left(\frac{x_1}{c_1} - \frac{x_2}{c_2}\right)\right]. \tag{2.63}$$

Also

$$\delta\left(\frac{1}{c_1} + \frac{1}{c_2}\right) + 2\left(\frac{x_1}{c_1} - \frac{x_2}{c_2}\right) < 0 \tag{2.64}$$

if and only if

$$\delta < 2\frac{(x_2/c_2) - (x_1/c_1)}{(1/c_1) + (1/c_2)}, \tag{2.65}$$

which is exactly the condition (2.53). Hence $\Delta < 0$ and we have arrived at a contradiction, and so

$$c_2 \sin i \not< c_1 \sin r. \tag{2.66}$$

By replacing the set of quantities $\{c_1, c_2, i, r, A, B\}$ by $\{c_2, c_1, r, i, B, A\}$ in order, respectively, the same argument establishes that

$$c_2 \sin i \not> c_1 \sin r, \tag{2.67}$$

and so

$$c_2 \sin i = c_1 \sin r, \tag{2.68}$$

that is, Fermat's principle of least time implies Snell's law of refraction. See [11].

Chapter Three

Introduction to the Mathematics of Rays

The theory of geometrical optics can be formulated by means of two seemingly different types of equations, which are (i) the ray equations, i.e. a set of ordinary non-linear differential equations, or (ii) the Hamilton-Jacobi equation, which is a first-order partial differential equation. We know, of course, that both formulations are equivalent: the ray equations are the characteristic equations of the Hamilton-Jacobi equation. This remark leads to the following beautiful geometrical interpretation. The family of rays of geometrical optics is perpendicular to the wavefronts $S = $ constant, if S denotes the appropriate solution of the Hamilton-Jacobi equation.

[178]

3.1 BACKGROUND

The Helmholtz equation (or reduced wave equation) for a monochromatic wave with time-harmonic ($e^{-i\omega t}$) behavior is

$$\nabla^2 u\left(\mathbf{r}\right) + k_0^2 n^2\left(\mathbf{r}\right) u\left(\mathbf{r}\right) = 0, \tag{3.1}$$

where the free-space wavenumber $k_0 = \omega/c$, and for an inhomogeneous medium, $n(\mathbf{r})$ is the refractive index. According to the geometrical optics approximation, the resulting wave field can be represented as an asymptotic series in negative powers of k_0 such that [183]

$$u\left(\mathbf{r}\right) = \sum_{n=0}^{\infty} \frac{A_n\left(\mathbf{r}\right)}{\left(ik_0\right)^n} e^{ik_0\psi\left(\mathbf{r}\right)}. \tag{3.2}$$

The quantity $\psi\left(\mathbf{r}\right)$ is known as the *eikonal* function [175], [176].

An alternative but equivalent approach based on the existence of a characteristic scale L for the problem is to expand $u(\mathbf{r})$ in terms of the (small) parameter $\varepsilon = (k_0 L)^{-1}$. This can often be a more useful version [175]. Then the difference between the sum u (which is in general divergent) and its truncated version u_N ($0 \leq n \leq N$) vanishes as $\varepsilon \to 0$; specifically, using the order notation of appendix A,

$$|u - u_N|_{\varepsilon \to 0} = O(\varepsilon^{N+1}); \tag{3.3}$$

so this is the short-wavelength (or high-frequency) asymptotic limit. However, as noted, generally for $\varepsilon \neq 0$,

$$\lim_{N \to \infty} |u - u_N| = \infty. \tag{3.4}$$

By substituting this series into the Helmholtz equation and formally equating coefficients of respective powers of k_0, a sequence of equations is obtained for the eikonal term ψ and for the amplitudes A_i (these are the transport equations; they will not concern us at this point). As will be shown below, the resulting nonlinear partial differential equations are simpler than (3.1), being first-order and reducible to ordinary differential equations: the *characteristic equations*, the basic concept of which is discussed in the next section.

3.2 THE METHOD OF CHARACTERISTICS

Suppose that a first-order partial differential equation for the dependent variable $u = u(x_i)$, $i = 1, 2, \ldots, n$ has the form

$$F(x_i, u, p_i) = 0, \tag{3.5}$$

where $p_i = u_{x_i} \equiv \partial u / \partial x_i$, $i = 1, 2, \ldots, n$. The general solution of this equation can always be found from the general solution of a related system of ordinary differential equations known as the characteristic equations. As already noted, these equations are intimately connected with such seemingly unconnected topics as geometrical optics, the calculus of variations, and Hamiltonian mechanics.

A common subset of the type (3.5) is the *quasilinear* form; equation (3.5) is said to be quasilinear if F is linear in the p_i, that is, if there exist coefficients a_i and b such that

$$F(x_i, u, p_i) = \sum_{i=1}^{n} a_i(x_i, u) p_i + b(x_i, u) = 0. \tag{3.6}$$

This of course includes the subset of linear equations, where

$$F(x_i, u, p_i) = \sum_{i=1}^{n} A_i(x_i) p_i + B(x_i)u + C(x_i) = 0. \tag{3.7}$$

In light of the subject matter in this book, we concentrate on quasilinear equations in two independent variables x, y taking the form

$$a(x, y, u)u_x + b(x, y, u)u_y = c(x, y, u). \tag{3.8}$$

A solution $u = u(x, y)$ may be regarded as the equation of the surface $f(x, y, u) = u(x, y) - u = 0$ (sometimes referred to as an integral surface). There is a nice geometrical interpretation of equation (3.8) given that it can be written as an inner product of two vectors in $\mathbb{R}^3$ as

$$\langle a, b, c \rangle \cdot \langle u_x, u_y, -1 \rangle = 0. \tag{3.9}$$

But $\langle u_x, u_y, -1 \rangle$ is just ∇f, the gradient of $f(x, y, u)$, so the vector $\langle a, b, c \rangle$ lies in the tangent plane of the surface $u = u(x, y)$ at any point (x, y, u). More generally (3.8) is the requirement for any integral surface $u = u(x, y)$ through the point $P(x, y, u)$ to be tangent there to the vector $\langle a(x, y, u), b(x, y, u), c(x, y, u) \rangle$. As we shall see, the characteristics are curves in the directions determined by those vectors such that

$$\frac{dx}{a(x, y, u)} = \frac{dy}{b(x, y, u)} = \frac{du}{c(x, y, u)}. \tag{3.10}$$

Lemma: A characteristic is the curve of intersection of two integral surfaces.

Proof: Suppose that the integral surfaces are defined, respectively, as $u = u_1(x, y)$ and $u = u_2(x, y)$. Then if $u_{1x} = \partial u_1 / \partial x$ etc., it follows that

$$u_{1x}dx + u_{1y}dy - du = 0, \tag{3.11}$$
$$u_{2x}dx + u_{2y}dy - du = 0. \tag{3.12}$$

Hence

$$(u_{1x} - u_{2x}) \, dx + \left(u_{1y} - u_{2y}\right) dy = 0, \tag{3.13}$$

so on multiplying equation (3.11) by u_{2y} and equation (3.12) by u_{1y} and subtracting one from the other, we obtain the result

$$\frac{dx}{u_{1y} - u_{2y}} = \frac{dy}{u_{2x} - u_{1x}} = \frac{du}{u_{2x}u_{1y} - u_{2y}u_{1x}}. \tag{3.14}$$

But from equation (3.8) we know that

$$au_{1x} + bu_{1y} = c = au_{2x} + bu_{2y}, \tag{3.15}$$

so in a similar fashion we can derive the result

$$\frac{a}{u_{1y} - u_{2y}} = \frac{b}{u_{2x} - u_{1x}} = \frac{c}{u_{2x}u_{1y} - u_{2y}u_{1x}}. \tag{3.16}$$

Therefore we have established that the result (3.10) is valid on the intersection of two integral surfaces. Note that if derivatives with respect to some parameter τ are denoted by $\dot{x} \equiv dx/d\tau$ etc., then (3.10) can be written as

$$\frac{\dot{x}}{a} = \frac{\dot{y}}{b} = \frac{\dot{u}}{c} = \eta \text{ (say)}. \tag{3.17}$$

Exercise: Show that the integral surface for the equation

$$(y - u) u_x + (u - x) u_y = x - y \tag{3.18}$$

that goes through the curve $u = 0$, $xy = 1$ is given explicitly by

$$u = \frac{1 - xy}{x + y}. \tag{3.19}$$

The Characteristic Strip Equations

Here we examine the conditions under which there exists an integral surface through a given regular arc Γ. Suppose we restrict ourselves to equation (3.5) for a solution $u(x, y)$, that is, such that

$$F(x, y, u, u_x, u_y) = 0, \tag{3.20}$$

so

$$F_x + F_u u_x + F_{u_x} u_{xx} + F_{u_y} u_{xy} = 0, \tag{3.21}$$

and

$$F_y + F_u u_y + F_{u_y} u_{yy} + F_{u_x} u_{yx} = 0. \tag{3.22}$$

Furthermore, as above, if τ is a parameter defining the curve Γ, then we also have

$$\dot{u} = u_x \dot{x} + u_y \dot{y}, \tag{3.23}$$

from which we arrive at

$$\dot{u}_x = u_{xx} \dot{x} + u_{xy} \dot{y}, \tag{3.24}$$

and

$$\dot{u}_y = u_{yy} \dot{y} + u_{yx} \dot{x}. \tag{3.25}$$

By eliminating u_{xx} from equations (3.21) and (3.24), it follows that

$$\left(\dot{x} F_{u_y} - \dot{y} F_{u_x} \right) u_{xy} = -(F_x + F_u u_x) \dot{x} - F_{u_x} \dot{u}_x. \tag{3.26}$$

A similar elimination of u_{yy} yields

$$\left(\dot{x} F_{u_y} - \dot{y} F_{u_x} \right) u_{yx} = -(F_y + F_u u_y) \dot{y} - F_{u_y} \dot{u}_y. \tag{3.27}$$

If along every point of the arc Γ the expression $\dot{x} F_{u_y} - \dot{y} F_{u_x}$ vanishes, there can be no integral surface unless the right-hand sides of these last two equations also vanish. As with a quasilinear system, such conditions define the characteristics, so here the characteristic strips are those for which the set $\{u, u_x, u_y\}$ renders the right-hand sides of equations (3.26) and (3.27) to be zero. Hence the characteristic strip equations are

$$\frac{\dot{x}}{F_{u_x}} = \frac{\dot{y}}{F_{u_y}} = \frac{\dot{u}}{u_x F_{u_x} + u_u F_{u_y}} = -\frac{\dot{u}_x}{F_x + F_u u_x} = -\frac{\dot{u}_y}{F_y + F_u u_y}. \tag{3.28}$$

Exercise: Prove this result.

The set (3.28) can easily be transformed into the equivalent form with $p = u_x$ and $q = u_y$ and indeed generalized to the general case $F(x_i, u, p_i) = 0$. The following lemma also may readily be established (and extended beyond $i = 1, 2$).

Lemma: Along every characteristic strip for $F(x, y, u, p, q) = 0$, the function $F(x, y, u, p, q)$ is independent of τ.
Exercise: Use (3.28) to prove this.

3.3 INTRODUCTION TO HAMILTON-JACOBI THEORY

In view of the above material we know that it is possible to represent systems of ordinary differential equations arising in mechanics as the characteristic system for a first-order partial differential equation: this is the celebrated *Hamilton-Jacobi equation* discussed below. We summarize the essential features involved in such a representation. In a mechanical system with n degrees of freedom, the motion of the system can be described in terms of certain *generalized coordinates* $x_1, x_2, ..., x_n$ as functions of time t. For a conservative system, external forces are derived from from a potential $V(x_i), i = 1, 2, ..., n$. In such a case the kinetic energy T is defined in an obvious way by

$$T = \frac{1}{2} \sum_{i=1}^{n} m_i \dot{x}_i^2, \tag{3.29}$$

and the *Lagrangian* $\mathcal{L}$ by

$$\mathcal{L} = \mathcal{L}(x_i, \dot{x}_i, t) = T - V. \tag{3.30}$$

The Euler-Lagrange Equation

The classical problem of the calculus of variations, as it is called, is to find the function $y = y(x)$ (from among a class of functions $Y(x)$) that (in many cases) minimizes a definite integral of the form

$$I(Y, Y') = \int_{x_1}^{x_2} f(x, Y(x), Y'(x)) \, dx. \tag{3.31}$$

More accurately, it is to find the $y(x)$ that renders the integral I *stationary*; in most cases, certainly all of interest to us, this will indeed correspond to a minimum, and we will operate under that assumption in what follows. We assume furthermore that the integrand f is a sufficiently differentiable function of its arguments x, Y, and Y' that we can carry out the analysis below in a mathematically "legal" manner. Geometrically, we are seeking that path $y(x)$ in the plane that connects the endpoints $(x_1, y(x_1))$ and $(x_2, y(x_2))$ while at the same time minimizing I. To do this, we construct a $Y(x)$ that differs from the minimal path $y(x)$ by a class of functions as follows:

$$Y(x) = y(x) + \varepsilon \eta(x); \quad \eta(x_1) = 0 = \eta(x_2). \tag{3.32}$$

Clearly, equation (3.31) can be written as

$$I(Y, Y'; \varepsilon) = \int_{x_1}^{x_2} f(x, y + \varepsilon \eta, y' + \varepsilon \eta') \, dx. \tag{3.33}$$

Since $I(Y, Y'; 0)$ is the required stationary value, it follows that

$$I'(Y, Y'; \varepsilon)|_{\varepsilon=0} = \left. \frac{\partial I}{\partial \varepsilon} \right|_{\varepsilon=0} = 0. \tag{3.34}$$

We can formally justify differentiation (with respect to ε) under the integral sign because the integral is with respect to x. Therefore, writing only the ε dependence explicitly in what follows and using the chain rule and integration by parts,

$$I'(\varepsilon) = \int_{x_1}^{x_2} \frac{\partial}{\partial \varepsilon} [f(x, Y(\varepsilon), Y'(\varepsilon))] \, dx$$

$$= \int_{x_1}^{x_2} \left\{ \frac{\partial f}{\partial Y} \eta + \frac{\partial f}{\partial Y'} \eta' \right\} dx$$

$$= \int_{x_1}^{x_2} \left(\frac{\partial f}{\partial Y} \right) \eta(x) dx + \left(\frac{\partial f}{\partial Y'} \right) \eta(x) \Bigg|_{x_1}^{x_2}$$

$$- \int_{x_1}^{x_2} \eta(x) \frac{d}{dx} \left(\frac{\partial f}{\partial Y'} \right) dx. \tag{3.35}$$

The second term on the right vanishes by virtue of the boundary conditions on η, so $I'(\varepsilon)$ can be written in a more compact form as

$$I'(\varepsilon) = \int_{x_1}^{x_2} \eta(x) \left\{ \frac{\partial f}{\partial Y} - \frac{d}{dx} \left(\frac{\partial f}{\partial Y'} \right) \right\} dx. \tag{3.36}$$

Therefore, from condition (3.34), it follows that

$$I'(0) = \int_{x_1}^{x_2} \eta(x) \left\{ \frac{\partial f}{\partial y} - \frac{d}{dx} \left(\frac{\partial f}{\partial y'} \right) \right\} dx = 0. \tag{3.37}$$

Now the function $\eta(x)$ is arbitrary, apart from appropriate smoothness and vanishing endpoint conditions. From the lemma below, we see that (3.37) implies the fundamental result of the calculus of variations, namely, the *Euler-Lagrange equation*:

$$\frac{\partial f}{\partial y} - \frac{d}{dx} \left(\frac{\partial f}{\partial y'} \right) = 0. \tag{3.38}$$

Lemma: If $p(x)$ is a function, continuous on an interval $[x_1, x_2]$, and if, for every continously differentiable function $\eta(x)$ such that $\eta(x_1) = 0 = \eta(x_2)$,

$$I = \int_{x_1}^{x_2} \eta(x) p(x) dx = 0, \tag{3.39}$$

then $p(x) \equiv 0$ on $[x_1, x_2]$.

Proof: We choose some number c in $[x_1, x_2]$ such that $p(c) \neq 0$; without loss of generality we may assume that $p(c) > 0$ (if otherwise, we can use $-p(c)$). By virtue of the continuity of p, there is obviously a subinterval (a, b) in which $p(x) > 0$, with $p(a) = p(b) = 0$, where $x_1 < a \leq x \leq b < x_2$. Clearly we may choose an $\eta(x) > 0$ in (a, b) (the following choice will do)

$$\eta(x) = (x - a)^2 (x - b)^2.$$

But this means that $I > 0$, in contradiction to the condition stated. The result follows.

3.3.1 Hamilton's Principle

Hamilton's principle states that an equilibrium is attained for motions $x_i(t)$ that render the integral

$$I = \int_{t_0}^{t_1} \mathcal{L}(x_i, \dot{x}_i, t) \, dt \tag{3.40}$$

stationary among all possible trajectories that join (for all i) a fixed pair of terminal points $x_i(t_0) = X_0$, $x_i(t_1) = X_1$. A necessary condition for I to be stationary is that the Euler-Lagrange equations are satisfied, namely,

$$\frac{d}{dt}\left(\frac{\partial \mathcal{L}}{\partial \dot{x}_i}\right) - \frac{\partial \mathcal{L}}{\partial x_i} = 0, \tag{3.41}$$

for all i. It is these equations that will be identified as the system of characteristic equations. We can define the so-called generalized momenta p_i as

$$p_i = \frac{\partial \mathcal{L}}{\partial \dot{x}_i} \tag{3.42}$$

for each fixed choice of x_i and t. The Hamiltonian $\mathcal{H}$ is defined as

$$\mathcal{H} = \mathcal{H}(x_i, \dot{x}_i, t) \equiv \sum_{i=1}^{n} p_i \dot{x}_i - \mathcal{L}, \tag{3.43}$$

[179], [180].

As a very simple example of this, consider a particle of mass m falling under the influence of a constant gravitational field:

$$m\ddot{x} = -mg. \tag{3.44}$$

Then

$$\mathcal{L} = \frac{1}{2}m\dot{x}^2 - mgx, \tag{3.45}$$

and

$$\mathcal{H} = p\dot{x} - \mathcal{L} = \frac{\partial \mathcal{L}}{\partial \dot{x}}\dot{x} - \mathcal{L} = \frac{1}{2}m\dot{x}^2 + mgx. \tag{3.46}$$

Equation (3.46) is of course the total energy of this one-particle system and provides a ready-made physical interpretation of the Hamiltonian (if the system is conservative). We will encounter this type of formulation in several other locations in the book.

3.3.2 Rays and Characteristics

The zero-order or eikonal equation is, as noted in Chapter 2,

$$|\nabla \psi|^2 = n^2(\mathbf{r}), \tag{3.47}$$

and the *characteristics* of this equation are the *rays* for the system. In fact (3.47) is a special case of Hamilton-Jacobi form

$$\mathcal{H}(p_i, q_i) = 0; \quad p_i = \frac{\partial \psi}{\partial q_i}, \tag{3.48}$$

where the $q_i (i = 1, 2, \ldots, n)$ are generalized coordinates, and the p_i are the associated generalized momenta. Noting that

$$d\mathcal{H}(p_i, q_i) = \sum_{i=1}^{n} \left[\frac{\partial \mathcal{H}}{\partial p_i} dp_i + \frac{\partial \mathcal{H}}{\partial q_i} dq_i \right] = 0, \tag{3.49}$$

the resulting set of differential equations is given by

$$\frac{dq_i}{\partial \mathcal{H} / \partial p_i} = -\frac{dp_i}{\partial \mathcal{H} / \partial q_i} = \frac{d\psi}{\sum_{i=1}^{n} p_i (\partial \mathcal{H} / \partial p_i)} = d\tau, \text{ say}, \tag{3.50}$$

or equivalently, the characteristic equations

$$\frac{dq_i}{d\tau} = \frac{\partial \mathcal{H}}{\partial p_i}; \tag{3.51}$$

$$\frac{dp_i}{d\tau} = -\frac{\partial \mathcal{H}}{\partial q_i}; \tag{3.52}$$

supplemented by

$$\frac{d\psi}{d\tau} = \sum_{i=1}^{n} p_i \frac{\partial \mathcal{H}}{\partial p_i}. \tag{3.53}$$

These last three are the characteristic strip equations discussed earlier; their solution is the characteristic strip defined by $q_i = q_i(\tau)$, $p_i = p_i(\tau)$, and $\psi = \psi(\tau)$. Specializing to the space $\mathbb{R}^3$, let the q_i ($i = 1, 2, 3$) be Cartesian coordinates and write $\mathcal{H} = \mathcal{H}(\mathbf{p}, \mathbf{r})$. Then supplemented by the definition of $\mathbf{p} = \nabla \psi$, the equations (3.51)–(3.53) take the shorthand form (with some vector notational abuse; e.g., here $\partial \mathcal{H} / \partial \mathbf{r}$ means the gradient vector $\nabla_{\mathbf{r}}$, i.e., the vector differential operator with components $\partial / \partial q_i$):

$$\frac{d\mathbf{r}}{d\tau} = \frac{\partial \mathcal{H}}{\partial \mathbf{p}}; \tag{3.54}$$

$$\frac{d\mathbf{p}}{d\tau} = -\frac{\partial \mathcal{H}}{\partial \mathbf{r}}; \tag{3.55}$$

$$\frac{d\psi}{d\tau} = \mathbf{p} \frac{\partial \mathcal{H}}{\partial \mathbf{p}}. \tag{3.56}$$

If $\psi^i = \psi^i(\tau^i)$ is the initial value of the phase (or eikonal) at $\tau = \tau^i$, then along a ray defined by $\mathbf{r} = \mathbf{r}(\tau)$, $\mathbf{p} = \mathbf{p}(\tau)$ it follows from (3.56) that

$$\psi = \psi^i + \int_{\tau^i}^{\tau} \mathbf{p} \frac{\partial \mathcal{H}}{\partial \mathbf{p}} d\varpi, \tag{3.57}$$

(ϖ being a dummy variable for τ).

A commonly used representation of the ray equations is the Hamiltonian form, expressed in terms of a "ray momentum" (continuing the physical analogy) $\mathbf{p} = \nabla \psi$ and a parameter τ measured along the ray in terms of its arc length s, where

$$d\tau = \frac{ds}{|\partial \mathcal{H}/\partial \mathbf{p}|}. \tag{3.58}$$

These equations are

$$\frac{d\mathbf{r}}{d\tau} = \mathbf{p}, \quad \text{and} \quad \frac{d\mathbf{p}}{d\tau} = \frac{1}{2}\nabla n^2, \tag{3.59}$$

and they follow from the choice

$$\mathcal{H} = \frac{1}{2}\left[\mathbf{p}^2 - n^2(\mathbf{r})\right], \tag{3.60}$$

(using the eikonal equation) and the set (3.51)–(3.52). Because

$$\frac{d\mathbf{r}}{d\tau} = \nabla \psi \tag{3.61}$$

it follows that the rays are orthogonal to the surfaces of constant phase (i.e., the wavefronts).

Suppose that the initial field u^i is defined on some surface S such that $\mathbf{r} = \mathbf{r}^i(\xi, \eta)$ in terms of curvilinear coordinates (ξ, η) by

$$u^i = u^i(\xi, \eta) = u_0^i(\xi, \eta) \exp\left[ik\psi^i(\xi, \eta)\right], \tag{3.62}$$

and let an initial condition for the ray trajectory $\mathbf{r} = \mathbf{R}(\xi, \eta, \tau)$ be $\mathbf{r}^i(\xi, \eta) = \mathbf{R}(\xi, \eta, 0)$. In addition to the initial source point of the ray, the outgoing direction must be identified; this is equivalent to to defining the initial value of the momentum $\mathbf{p}$. Since $\mathbf{p} = \nabla \psi$ (i.e., $d\psi = \mathbf{p} \cdot d\mathbf{r}$), the components of the initial momentum $\mathbf{p}(\tau^i) = \mathbf{p}^i(\xi, \eta)$ may be found from the derivatives of the initial phase $\psi(r^i) = \psi^i(\xi, \eta)$ as follows to obtain

$$\frac{\partial \psi^i}{\partial \xi} = \mathbf{p}^i \cdot \frac{\partial \mathbf{r}^i}{\partial \xi}, \quad \text{and} \tag{3.63}$$

$$\frac{\partial \psi^i}{\partial \eta} = \mathbf{p}^i \cdot \frac{\partial \mathbf{r}^i}{\partial \eta}. \tag{3.64}$$

Furthermore, we can use the eikonal equation to set

$$\left|\mathbf{p}^i\right|^2 = n^2(\mathbf{r}^i). \tag{3.65}$$

Then in conjunction with the initial data (3.63)–(3.65) the solution of the Hamiltonian ray equations (3.59) can be expressed in terms of the ray coordinates (ξ, η, τ) for a family of rays as

$$\mathbf{r} = \mathbf{R}(\xi, \eta, \tau), \quad \mathbf{p} = \mathbf{P}(\xi, \eta, \tau).$$

Provided that the Jacobian $J = \partial(x, y, z)/\partial(\xi, \eta, \tau)$ is nonzero in the domain of interest, the first of these expressions may be solved uniquely for the coordinates (ξ, η, τ) as $\xi = \xi(\mathbf{r})$, $\eta = \eta(\mathbf{r})$, and $\tau = \tau(\mathbf{r})$. Then the initial conditions for the phase on the surface S define a family of (spatial) phase fronts that is orthogonal to the family of rays $r = R(\xi, \eta, \tau)$. Thus it is possible to express the phase ψ as an integral along the ray from (3.57) such that

$$\psi = \psi^i + \int_{\tau^i}^{\tau} p^2 d\eta = \psi^i + \int_{\tau^i}^{\tau} n^2 [\mathbf{r}(\varpi)] d\varpi. \tag{3.66}$$

Then if for a given ψ^i, we are able to identify $\tau = \tau_\phi(\xi, \eta, \psi^i)$, from equation (3.66) [175], [176] it follows that the vector equation for the phase front is

$$\mathbf{r} = \mathbf{R}\left(\xi, \eta, \tau_\phi(\xi, \eta, \psi^i)\right) \equiv \mathbf{r}_\phi\left(\xi, \eta, \psi^i\right).$$

Connection with Fermat's Principle

The solution of the Hamilton-Jacobi equation (3.48) is equivalent to finding a function that renders stationary the functional

$$\int \mathbf{p} \cdot \frac{\partial \mathcal{H}}{\partial \mathbf{p}} d\tau,$$

that is, the first variation is zero:

$$\delta \int_{\tau_a}^{\tau_b} \mathbf{p} \cdot \frac{\partial \mathcal{H}}{\partial \mathbf{p}} d\tau = 0.$$

Recalling the famous analogy made by Hamilton between mechanical and geometrical-optic systems, in view of this condition the former corresponds to the *principle of least action* and the latter, once again, to Fermat's principle of least time. In particular, when the Hamiltonian is defined by (3.60), Fermat's principle states that for rays connecting the endpoints r_a and r_b, the functional

$$\int_{\mathbf{r}_a}^{\mathbf{r}_b} n^2(\mathbf{r}) d\tau = \int_{\mathbf{r}_a}^{\mathbf{r}_b} n(\mathbf{r}) ds$$

is stationary on paths satisfying equations (3.59). This is just another way of arriving at equation (2.3)

It is possible to reformulate the Lagrangian equations of motion (3.41) in terms of the Hamiltonian. If we differentiate the equations (3.43) with respect to x_i and p_i, we have

$$\frac{\partial \mathcal{H}}{\partial x_i} + \frac{\partial \mathcal{L}}{\partial x_i} + \sum_{j=1}^{n} \frac{\partial \mathcal{L}}{\partial \dot{x}_j} \frac{\partial \dot{x}_j}{\partial x_i} = \sum_{j=1}^{n} p_j \frac{\partial \dot{x}_j}{\partial x_i}, \tag{3.67}$$

(note the change of subscript in the summations) and

$$\frac{\partial \mathcal{H}}{\partial p_i} + \sum_{j=1}^{n} \frac{\partial \mathcal{L}}{\partial \dot{x}_j} \frac{\partial \dot{x}_j}{\partial p_i} = \sum_{j=1}^{n} p_j \frac{\partial \dot{x}_j}{\partial p_i} + \dot{x}_i. \tag{3.68}$$

From the definition of p_i in equation (3.42) it can be seen that the summation terms cancel in both of the above equations. Furthermore in view of (3.41), these equations take following the beautifully simple forms for all $i = 1, 2, \ldots, n$:

$$\frac{dp_i}{dt} = -\frac{\partial \mathcal{H}}{\partial x_i},$$

and

$$\frac{dx_i}{dt} = \frac{\partial \mathcal{H}}{\partial p_i}.$$

These $2n$ first-order ordinary equations for $p_i(t)$ and $x_i(t)$ are known as Hamilton's equations and replace the n second-order Lagrange equations given by (3.41). Clearly there is a striking similarity between this formulation and the system of characteristic equations for (3.5), where originally $p_i = \partial u/\partial x_i$ (i.e., for $F(x_i, u(x_i), p_i) = 0$). In fact by introducing an auxiliary unknown function $u(t, x_1, \ldots, n)$ and setting $p = \partial u/\partial t$, it follows analogously that we can write down a partial differential equation of the form

$$p + \mathcal{H}(x_i, p_i, t) = 0. \tag{3.69}$$

This is yet another representation of the Hamilton-Jacobi equation [180], [202].

3.3.3 The Optical Path Length Revisited

Suppose that a curve $\mathcal{C}$ is defined parametrically by $x = x(u)$, $y = y(u)$, $z = z(u)$ for some as yet undefined parameter u. For this to represent the path of a geometrical ray in three dimensions, it must satisfy Fermat's principle, so that the *optical path length* ψ can be written as

$$\psi = \int_{u_1}^{u_2} n(x, y, z) \left\{ \left(\frac{dx}{du} \right)^2 + \left(\frac{dy}{du} \right)^2 + \left(\frac{dz}{du} \right)^2 \right\}^{1/2} du \equiv \int_{u_1}^{u_2} W(u) du. \tag{3.70}$$

Then if $\dot{x} = d(x)/du$, $\delta \dot{x} = d(\delta x)/du$, etc., we can write the variation for two near neighboring curves as

$$\delta \psi = \int_{u_1}^{u_2} \sum_i \left(\frac{\partial W}{\partial \dot{x}} \delta \dot{x} + \frac{\partial W}{\partial x} \delta x \right) du = 0, \tag{3.71}$$

where the summation is now over $i = x, y, z$. Integrating by parts we obtain

$$\delta \psi = \left[\sum_i \frac{\partial W}{\partial \dot{x}} \delta x \right]_{u_1}^{u_2} - \int_{u_1}^{u_2} \sum_i \left(\frac{d}{du} \left(\frac{\partial W}{\partial \dot{x}} \right) - \frac{\partial W}{\partial x} \right) \delta x \, du = 0. \tag{3.72}$$

If the curves have common endpoints, the integrated terms vanish, and for arbitrary displacements δx we have recovered three Euler-Lagrange equations for a ray, namely,

$$\frac{d}{du} \left(\frac{\partial W}{\partial \dot{p}} \right) - \frac{\partial W}{\partial p} = 0, \tag{3.73}$$

where p is an abbreviation for each of x, y, z. Inserting the original expression for W in this result, the following equation is obtained:

$$\frac{d}{du}\left\{ \frac{n\dot{x}}{\left(\dot{x}^2 + \dot{y}^2 + \dot{z}^2\right)^{1/2}} \right\} - \frac{\partial n}{\partial x}\left(\dot{x}^2 + \dot{y}^2 + \dot{z}^2\right)^{1/2} = 0, \qquad (3.74)$$

and again, there are corresponding equations for y and z. At this point it is useful to make either of two specific choices for u:

(1) $u = s$, the arc length along C. Then it follows that

$$\frac{d}{ds}\left(n\frac{dx}{ds} \right) - \frac{\partial n}{\partial x} = 0,$$

and combining this with the remaining two equations yields the vector form

$$\frac{d}{ds}\left(n\frac{d\mathbf{r}}{ds} \right) = \nabla n, \qquad (3.75)$$

$\mathbf{r} = \langle x, y, z \rangle$ being the position vector of a point on the ray (referred to a specified origin). This is a particularly useful and succinct form for the ray equations. Note that with this choice for u the terms in equations (3.73) above,

$$\frac{\dot{x}}{\left(\dot{x}^2 + \dot{y}^2 + \dot{z}^2\right)^{1/2}}, \frac{\dot{y}}{\left(\dot{x}^2 + \dot{y}^2 + \dot{z}^2\right)^{1/2}}, \quad \text{and} \quad \frac{\dot{z}}{\left(\dot{x}^2 + \dot{y}^2 + \dot{z}^2\right)^{1/2}},$$

are the direction cosines of the ray at a given point on C, namely, α, β, and γ, respectively. These will be referred to below as α_i, $i = 1, 2, 3$, respectively.

(2) Another representation arises from the choice of

$$u = \int \frac{ds}{n}. \qquad (3.76)$$

The operator d/du now becomes nd/ds, so the vector equation (3.75) takes the form

$$\frac{d^2\mathbf{r}}{du^2} = \nabla\left(\frac{1}{2}n^2 \right). \qquad (3.77)$$

This equation has a direct mechanical analogue; it resembles the equation of motion of a particle of unit mass in a potential $-n^2/2$. From this perspective the rays are the orbits or trajectories of the point (assuming n is a continuous function), and the "velocity" of the particle, $d\mathbf{r}/du$, is $n(\mathbf{r})$ along its trajectory [12]. In any medium supporting rays, wavefronts can be defined as the set of normals to the rays at a given point in time. A form of *Huygen's construction* can be used to determine some properties of the wavefronts. Thus suppose that $\psi(\mathbf{r}) = 0$ is the equation of a wavefront that reaches the point $\mathbf{r} = \langle x, y, z \rangle$ at time $t = 0$. By Huygen's construction, if c is the speed of light (or indeed, sound) in a uniform medium, the position of the wavefront at a later (or earlier time) t is given by $\psi(\mathbf{r}) = ct$. In a nonuniform medium, the wavefront speed is a function of position, $c(\mathbf{r})$, say, and the refractive index of the medium is $n(\mathbf{r}) = c/v(\mathbf{r})$. The normals to the wavefront surface $\psi(\mathbf{r})$ are the rays, hence

$$\nabla\psi = \sigma\langle \alpha, \beta, \gamma \rangle = \sigma\left\langle \frac{dx}{ds}, \frac{dy}{ds}, \frac{dz}{ds} \right\rangle = \sigma\frac{d\mathbf{r}}{ds}, \qquad (3.78)$$

where σ is a constant of proportionality. Then in the wave frame of reference, defining α, β, γ as α_i, $i = 1, 2, 3$, respectively,

$$d\psi = cdt = \nabla\psi \cdot d\mathbf{r} = \frac{\partial\psi}{\partial x_i}dx_i = \alpha_i\frac{\partial\psi}{\partial x_i}ds = \sigma ds = \sigma vdt, \qquad (3.79)$$

where $ds = vdt$ is an increment in the ray path. Therefore $\sigma = c/v \equiv n$, the refractive index, and from (3.78) it follows that

$$\frac{\partial\psi}{\partial x_i} = n\alpha_i, \qquad i = 1, 2, 3. \qquad (3.80)$$

Then

$$\sum_i \left(\frac{\partial\psi}{\partial x_i}\right)^2 = |\nabla\psi|^2 = n^2. \qquad (3.81)$$

This equation is a form of the by-now familiar *eikonal equation*. But are the "rays" associated with the surface ψ identical with the rays of Fermat's principle? To answer this in the affirmative, we proceed as follows by concentrating (for simplicity) on the x-component of equation (3.80):

$$\frac{d}{ds}(n\alpha) = \frac{d}{ds}\left(\frac{\partial\psi}{\partial x}\right) = \frac{\partial^2\psi}{\partial x^2}\alpha + \frac{\partial}{\partial y}\left(\frac{\partial\psi}{\partial x}\right)\beta + \frac{\partial}{\partial z}\left(\frac{\partial\psi}{\partial x}\right)\gamma$$

$$= \frac{1}{n}\left[\frac{\partial^2\psi}{\partial x^2}\frac{\partial\psi}{\partial x} + \frac{\partial}{\partial y}\left(\frac{\partial\psi}{\partial x}\right)\frac{\partial\psi}{\partial y} + \frac{\partial}{\partial z}\left(\frac{\partial\psi}{\partial x}\right)\frac{\partial\psi}{\partial z}\right]$$

$$= \frac{1}{2n}\frac{\partial}{\partial x}\left[\sum_i\left(\frac{\partial\psi}{\partial x_i}\right)^2\right] = \frac{1}{2n}\frac{\partial n^2}{\partial x} = \frac{\partial n}{\partial x}, \text{ etc.} \qquad (3.82)$$

Since $\alpha = \alpha_1 = dx/ds$, we have established the original system of ray equations based on Fermat's principle. Furthermore, (3.79) indicates that the increment in ψ in passing from it to the surface $\psi + d\psi$ is the optical length nds of the ray. Coming full circle, so to speak, the rays are orthogonal to the surfaces of constant ψ, and by virtue of equation (3.81) it follows that they satisfy

$$\nabla\psi = n\frac{d\mathbf{r}}{ds} \equiv n\mathbf{s}, \qquad (3.83)$$

where $\mathbf{s}$ is the unit tangent vector to the ray. Hence

$$\nabla \times (n\mathbf{s}) = \mathbf{0}. \qquad (3.84)$$

Applying Stokes' theorem to any closed curve results in

$$\oint n\mathbf{s} \cdot d\mathbf{l} = 0, \qquad (3.85)$$

and this implies that the optical path between any two points

$$\int n\mathbf{s} \cdot d\mathbf{l} \qquad (3.86)$$

is independent of the integration path [13]. Along a ray, $\mathbf{dl} = \mathbf{dr}$, and since $|\mathbf{dr}| = ds$, this invariant is the optical length ψ given in general by

$$\psi = \int_C n\,ds. \tag{3.87}$$

3.4 RAY DIFFERENTIAL GEOMETRY AND THE EIKONAL EQUATION AGAIN

We have seen that geometric light rays are the orthogonal trajectories to the geometrical wavefronts $S(\mathbf{r}) = \Psi(\mathbf{r}) = \text{constant}$, where the phase function Ψ is related to the eikonal ψ by $\Psi = k_0\psi$, where ψ satisfies the eikonal equation

$$\nabla\,[\psi(\mathbf{r})]^2 = n^2(\mathbf{r}), \tag{3.88}$$

The function $n(\mathbf{r})$ is defined as the ratio of the speed of light in a vacuum to that in the medium (i.e., it is the refractive index of the medium),

$$n\,(\mathbf{r}) = \frac{c_0}{c(\mathbf{r})} \geq 1,$$

so that the free space wavenumber is related to the wavenumber k by $k = nk_0$. Equation (3.88) is the by-now familiar eikonal equation. Since $n = \nabla S$ we can define a unit vector

$$\mathbf{t} = \frac{\nabla\psi}{|\nabla\psi|}. \tag{3.89}$$

Suppose that $r(s)$ denotes the position vector of a point P on a ray; then $t = r'(s)$, and since $n = |\nabla\mathcal{S}|$,

$$\mathbf{t} = \frac{\nabla\psi}{n(\mathbf{r})}. \tag{3.90}$$

Hence

$$n(\mathbf{r})\frac{d\mathbf{r}}{ds} = \nabla\psi. \tag{3.91}$$

To examine the implications of this important result, consider two neigboring wavefronts $\psi = k_1$ and $\psi + \delta\psi = k_2$. In the limit as $\delta\psi \to 0$, it follows that

$$\frac{d\psi}{ds} = \nabla\psi\cdot\frac{d\mathbf{r}}{ds} = n(\mathbf{r})\left(\frac{d\mathbf{r}}{ds}\right)^2 = n(\mathbf{r})\,(\mathbf{t}\cdot\mathbf{t}) = n(\mathbf{r}). \tag{3.92}$$

Since $\psi(\mathbf{r})$ is sometimes called the optical path, the quantity

$$\int_C n(\mathbf{r})ds$$

along a curve C (defining the ray) is called the optical path length of the curve (see Chapter 2 for definitions). Thus the optical length between two points P_1 and P_2 on the ray is $\psi(\mathbf{r}_{P_1}) - \psi(\mathbf{r}_{P_2})$, and we have established the equivalence of the

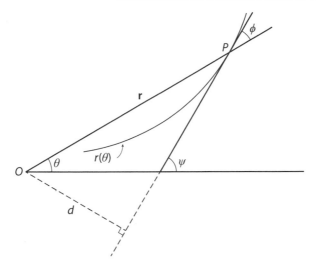

Figure 3.1 Geometry for Bouguer's law.

quantities ψ and S. The differential equations for the rays may be determined from equations (3.91) and (3.92) to be

$$\frac{d}{ds}\left(n(\mathbf{r})\frac{d\mathbf{r}}{ds}\right) = \nabla n(\mathbf{r}). \tag{3.93}$$

In particular, for a homogeneous medium n is constant, and equation (3.93) simplifies greatly, with solution $\mathbf{r} = \mathbf{a}s + \mathbf{b}$, where $\mathbf{a}$ and $\mathbf{b}$ are constant vectors. This is of course the vector equation of a straight line in the direction of $\mathbf{a}$, and passing through the point $\mathbf{r} = \mathbf{b}$. In systems for which n is piecewise constant (e.g., as would arise in an elementary description of the rainbow formation), the rays are described by a continuous combination of line segments. A more interesting special case, again related (among other things) to meteorological optics, is that of spherical symmetry, that is, $n = n(r)$, where $r = |\mathbf{r}|$. We prove a fundamental result, sometimes known as *Bouguer's law*:

$$\mathbf{r} \times n(\mathbf{r})\mathbf{t} = \mathbf{A}, \tag{3.94}$$

where $\mathbf{A}$ is a constant vector.

Proof: Consider the variation of the quantity $\mathbf{r} \times n\mathbf{t}$ with respect to arc length, where $\mathbf{t}$ is the unit tangent vector to the ray trajectory:

$$\frac{d}{ds}(\mathbf{r} \times n\mathbf{t}) = \frac{d\mathbf{r}}{ds} \times n\mathbf{t} + \mathbf{r} \times \frac{d}{ds}(n\mathbf{t}) = 0 + \mathbf{r} \times \frac{d}{ds}\left(n\frac{d\mathbf{r}}{ds}\right)$$

$$= \mathbf{r} \times \nabla n(r) = \mathbf{r} \times \left(\frac{\mathbf{r}}{r}\frac{dn}{dr}\right) = \mathbf{0}, \tag{3.95}$$

where we recall that $\mathbf{t} = \mathbf{r}'(s)$ and make use of equation (3.93). The result follows immediately, and it implies that all the rays are curves lying in a plane through the origin. Furthermore, with reference to Figure 3.1, where ϕ is the angle between the radius vector and the tangent to the ray, it follows from (3.94) that [13]

$$rn\,(r)\sin\phi\,(=nd)=|\mathbf{A}|\equiv K. \tag{3.96}$$

This result is the ray analogue of conservation of angular momentum for motion in a plane.

Question: How does the refractive index vary with radial distance if the ray trajectory is a circle centered at O?

It may be shown using elementary differential geometry that

$$\sin\phi=\frac{r(\theta)}{\left[r^2\,(\theta)+(dr/d\theta)^2\right]^{1/2}}, \tag{3.97}$$

so solving for $dr/d\theta$ we obtain

$$\frac{dr}{d\theta}=\pm\frac{r}{K}\left[n^2(r)r^2-K^2\right]^{1/2}, \tag{3.98}$$

from which

$$\theta(r)=\theta_0\pm K\int_{r_0}^{r}\frac{d\xi}{\xi\left[n^2\,(\xi)\,\xi^2-K^2\right]^{1/2}}, \tag{3.99}$$

where $K^2=n^2(r_0)r_0^2$ and $\theta(r_0)=\theta_0$, an "initial condition" to be identified later. We shall encounter this equation and its Cartesian cousin in Chapter 8 in connection with refraction of surface gravity waves as a result of variable topography on the seafloor.

Leaving spherical symmetry for the moment to return to the general case, use is made of the curvature vector

$$\kappa=\frac{d\mathbf{t}}{ds}=\frac{\boldsymbol{\nu}}{\rho}, \tag{3.100}$$

where ρ is the radius of curvature of the ray path at any point, and $\boldsymbol{\nu}$ is the principal unit normal vector at that point, directed inward. From equations (3.93) and (3.100) it follows that

$$\nabla n=\frac{dn}{ds}\mathbf{t}+n\kappa, \tag{3.101}$$

indicating that the refractive index gradient lies in the osculating plane (spanned by the vectors $\mathbf{t}$ and $\boldsymbol{\nu}$) of the ray path. But there is more: from (3.101),

$$n\,|\kappa|^2=\kappa\cdot\nabla n-\frac{dn}{ds}\mathbf{t}\cdot\frac{\boldsymbol{\nu}}{\rho}=\kappa\cdot\nabla n, \tag{3.102}$$

since this latter term is zero. Recalling that $|\kappa|=\rho^{-1}$, this result can be written as

$$\frac{1}{\rho^2}=\frac{\boldsymbol{\nu}}{\rho}\cdot\frac{\nabla n}{n},\quad\text{or}\quad\boldsymbol{\nu}\cdot\nabla\,(\ln n)=|\kappa|>0. \tag{3.103}$$

Thus, as we proceed along the direction of $\boldsymbol{\nu}$ from any point on the ray, $\ln n$ and hence n increases (i.e., the ray bends toward the region of higher refractive index). For reasons discussed below, we refer to this as the *mirage theorem*, and examine a special case of it below (after noting another derivation of equation (3.77)). A space curve can be described in terms of arc length s or some general parameter t; a natural question to ask is how these representations might be related. Specifically, let C be a piecewise-smooth curve given by the vector function $r(t) = \langle f(t), g(t), h(t) \rangle$, $a \le t \le b$, with the arc length function defined by

$$s(t) = \int_a^t |\mathbf{r}'(u)| \, du = \int_a^t \left([f'(u)]^2 + [g'(u)]^2 + [h'(u)]^2 \right)^{1/2} du. \quad (3.104)$$

The fundamental theorem of calculus implies that

$$\frac{ds}{dt} = |\mathbf{r}'(t)|. \quad (3.105)$$

Recall from equation (3.93) that

$$\frac{d}{ds} \left(n(\mathbf{r}) \frac{d\mathbf{r}}{ds} \right) = \nabla n(\mathbf{r}). \quad (3.106)$$

If the parameter t is normalized by the physical requirement that $|\mathbf{r}'(t)| = n$, the above equation can be written as

$$n \frac{d}{ds} \left(n \frac{d\mathbf{r}}{ds} \right) = n \nabla n = \nabla \left(\frac{n^2}{2} \right), \quad \text{or} \quad \frac{d^2 \mathbf{r}}{dt^2} = \nabla \left(\frac{n^2}{2} \right). \quad (3.107)$$

This is reminiscent of Newton's second law of motion [12] and is in fact one form of a mechanical-optical analogy.

Problem: Use the polar form of this equation and the fact that the ray trajectory lies in a plane to derive the result (3.99).

3.4.1 The Mirage Theorem for Horizontally Stratified Media

Does light or sound travel in straight lines, line segments, or curves? It all depends! Curved light paths can give rise to mirages. Suppose that the speed of light (or sound) depends on position in the medium in which the continuous refractive index varies vertically (i.e. $n = n(y)$), and so can be characterized by the position along the ray trajectory parametrized by arc length s. Let the propagation time from, say, $(0,0)$ to $(x_1, 0)$ be T. Then since

$$\frac{ds}{dt} = c(s), \quad (3.108)$$

we can write

$$T = \int_0^T dt = \int_0^{s(x_1)} \frac{ds}{c(s)}. \quad (3.109)$$

Thus if the speed varies in a given manner, $c = c(y)$, say, where y is the altitude above the surface of a locally flat earth, then we can write c in terms of the refractive index $n(y)$, such that

$$c(y) = \frac{c_0}{n(y)}, \quad (3.110)$$

where c_0 is (for light) the speed of light in vacuo (about 186,000 mi/s or 300,000 km/s). Obviously sound waves do not exist in a vacuum (so the sounds of explosions in deep space on "Star Trek" are filthy lies), and another reference speed has to be used. Thus the refractive index of a medium, if constant, can be defined as the ratio of the speed of light in vacuo to the speed in the medium. Typically for water it is about 4/3 and for glass about 3/2, though n is slightly wavelength dependent (without which we would have only "whitebows" instead of rainbows). Then the corresponding integral for the travel time T can be written

$$T = \frac{1}{c_0} \int_0^{x_1} n(y) \left[1 + \left(\frac{dy}{dx} \right)^2 \right]^{1/2} dx \equiv \int_0^{x_1} f\{y(x), y'(x)\}\, dx, \qquad (3.111)$$

the integrand being explicitly independent of x. The resulting path traced out by the ray is actually a consequence of *Fermat's Principle of Least Time*. Indeed, in 1662 Fermat espoused a more general principle: "Nature operates by the simplest and most expeditious ways and means." From the Euler-Lagrange equation (3.38)

$$\frac{\partial f}{\partial y} - \frac{d}{dx} \left(\frac{\partial f}{\partial y'} \right) = 0. \qquad (3.112)$$

In view of fact that the integrand does not depend explicitly on x, it can be shown (**Exercise**) that the Euler-Lagrange equation reduces to a first-order partial differential equation, specifically,

$$f - y' \left(\frac{\partial f}{\partial y'} \right) = K, \qquad (3.113)$$

where K is a constant. This equation has the solution

$$\frac{n(y)}{\left(1 + y'^2 \right)^{1/2}} = K, \qquad (3.114)$$

or as a simple sketch shows,

$$n(y) \sin \theta = K, \qquad (3.115)$$

where θ is the angle between the vertical direction (y) and the ray path. Note that $K > 0$ because it is the value of the refractive index when the slope of the ray path is zero (i.e., it is horizontal). Equation (3.115) is Snell's law of refraction for a continuously varying refractive index in a plane-stratified medium. Solving equation (3.114) as an initial-value problem for rays passing through the point (x_0, y_0), it follows that the ray trajectory satisfies

$$x - x_0 = \pm K \int_{y_0}^{y} \frac{d\xi}{\left[n^2(\xi) - K^2 \right]^{1/2}}, \qquad (3.116)$$

where $K > 0$ without loss of generality. So finally we can prove the mirage theorem.

The Mirage Theorem:
The ray path is concave toward regions of higher refractive index.

Proof: From equation (3.114) it follows that

$$y'^2 + 1 = \frac{n^2(y)}{K^2}. \tag{3.117}$$

Differentiating this with respect to y, it follows that

$$y'(x)\left[y''(x) - \frac{n(y)n'(y)}{K^2}\right] = 0, \tag{3.118}$$

so for $y' \neq 0$ the sign of y'' is the sign of n'. Thus if $n' > 0$ (n increases vertically), $y'' > 0$ and the curve is concave upward; if $n' < 0$ (n decreases vertically), $y'' < 0$ and the curve is concave downward. These are necessary and sufficient conditions.

Exercise: Establish the results (3.113) and (3.114).

3.4.2 A Return to Spherically Symmetric Media: $n(r)$ Continuous

From equation (3.98) we have the condition for an extremum to occur when $dr/d\theta = 0$ or $r = r_0$, where r_0 is the closest approach of the ray to the origin. Thus over the ray trajectory, $r \geq r_0$. For $\theta_0 < \theta < \pi$ we choose the $+$ sign in (3.98) (so that r increases with θ, and vice versa) and the $-$ sign for $\theta < \theta_0$. If we require that $n(r) \to 1^+$ as $r \to \infty$, we can establish some interesting relationships between the various quantities of physical interest. Of course, many practical problems require that n is discontinuous at some $r = a < \infty$ such that $n(r) = 1$ for $r > a$, and we shall investigate this problem in more detail later. Because the problem possesses cylindrical symmetry by virtue of the implied plane wave of amplitude $u(z) = \exp(ikz)$ arriving from (say) $z = -\infty$, we note that $z \to -\infty$ is equivalent in this plane polar problem to $\theta \to \pi$ as $r \to \infty$. Thus it follows that

$$\theta_0 = \pi - K \int_{r_0}^{\infty} \frac{dr}{r\left[n^2(r)r^2 - K^2\right]^{1/2}}. \tag{3.119}$$

We define the total deflection angle θ_D of the ray (see Figure 3.2) as the asymptotic angle of the exiting ray for $\theta < \theta_0$:

$$\theta_D = \theta_0 - K \int_{r_0}^{\infty} \frac{dr}{r\left[n^2(r)r^2 - K^2\right]^{1/2}}. \tag{3.120}$$

Obviously the standard ray-theoretic approach for the "rainbow problem" involves a sphere of finite radius, as described in Chapter 1. A version of the result (3.119) for spherical scatterers of radius $\tilde{a}$ will be discussed in Appendix B for several refractive index profiles $n(r)$.

Further Implications

Two things should be noted here. First, although Figure 3.2 is drawn for the case of $n'(r) < 0$, it is also the case for $n'(r) > 0$ that for $\theta < \theta_0$, $\theta \to |\theta_D|^-$ as $r \to \infty$. Second, $\theta_D = 2\theta_0 - \pi$. Two other important quantities in ray and wave theory are the phase $\Psi = k_0\psi$ and the optical path length (Ψ), such that $n[\mathbf{r}(s)] = \psi'(s)$, where s is the arc length along the ray. If we temporarily

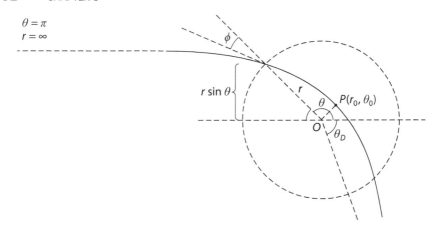

Figure 3.2 Ray trajectory for (3.119).

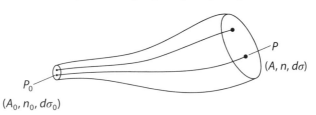

Figure 3.3 A narrow tube of rays.

reintroduce the independent variable t now as time and utilize the definition of n as the ratio of the speed of light in vacuo (c_0) to its speed c in the medium, that is,

$$n = \frac{c_0}{c} = \frac{c_0}{ds/dt},$$
(3.121)

then $nds = c_0 dt$ is the differential element of distance traveled in vacuo in time dt, expressed in terms of the distance traveled in the medium. If we assign $s = 0$ at some point P_0 on the ray, then

$$\psi(s) = \psi_0(P_0) + \int_0^s n[\mathbf{r}(\eta)]\, d\eta.$$
(3.122)

At each point on a ray, the geometrical optics field can be written as

$$u(s) = A(s)e^{ik\psi(s)},$$
(3.123)

where the amplitude $A(s)$ can be determined using the principle of conservation of energy. The energy flux (energy per unit area per unit time) must be the same through any cross section $d\sigma$ of a narrow tube of rays (Figure 3.3). Thus equating the two fluxes for the effective volume elements in the figure, we have

$$A^2(s)n(s)d\sigma = A_0^2(s)n_0 d\sigma_0, \quad \text{or}$$

$$A(s) = A_0 \left[\frac{n_0}{n(s)} \frac{d\sigma_0}{d\sigma} \right]^{1/2}.$$
(3.124)

If $P_0 = (r_0, \theta_0)$ it is sometimes more convenient to rewrite (3.122) as

$$\psi(r) = \psi(r_0) - \int_{r_0}^{r} \frac{\xi n^2(\xi) d\xi}{\left[n^2(\xi)\xi^2 - K^2\right]^{1/2}} \quad \text{for} \quad \theta_0 < \theta < \pi, \text{ and} \quad (3.125)$$

$$\psi(r) = \psi(r_0) + \int_{r_0}^{r} \frac{\xi n^2(\xi) d\xi}{\left[n^2(\xi)\xi^2 - K^2\right]^{1/2}} \quad \text{for} \quad \theta < \theta_0. \quad (3.126)$$

Physically, these equations correspond to the ray penetrating less far or farther into the medium, respectively. Since the point at infinity $z = -\infty$ corresponds to (∞, π) in polar coordinates, it follows that $z = re^{i\theta} \to -r$ as $\theta \to \pi$, so $u(z) = e^{ikz} \to e^{-ikr}$ (the phase must tend to $-r$ as $\theta \to \pi$), and hence equation (3.125) implies that

$$\psi_0 \equiv \psi(r_0) = \lim_{r \to \infty} \left\{ -r + \int_{r_0}^{r} \frac{\xi n^2(\xi) d\xi}{\left[n^2(\xi)\xi^2 - K^2\right]^{1/2}} \right\}. \quad (3.127)$$

If we wish to evaluate $A(s)$ using $z = -\infty$ as the initial point at which (say) $A_0 = 1$, $n_0 = 1$, then for impact parameters in the interval $[K, K + dK]$ we have $d\sigma_0 = 2\pi K dK$ and

$$d\sigma = (2\pi r \sin\theta) (rd\theta) (\cos\phi) = \frac{2\pi r^2 \sin\theta}{\left[1 + r^2 (d\theta/dr)^2\right]^{1/2}} d\theta. \quad (3.128)$$

Hence from equations (3.98) and (3.128) there follows the result

$$A(s) \equiv A(r, \theta) = \left[\frac{r}{K} \sin\theta \frac{d\theta}{dK}\right]^{-1/2} \left[n^2 r^2 - K^2\right]^{-1/4}. \quad (3.129)$$

Problem: Establish the result 3.129.

Combining equations (3.123), (3.126), and (3.129), we find that (for $\theta < \theta_0$) the geometrical optics field is given by

$$u = \left[\frac{r}{K} \sin\theta \frac{d\theta}{dK}\right]^{-1/2} \left[n^2 r^2 - K^2\right]^{-1/4} \exp\left\{ ik \left(\psi_0 + \int_{r_0}^{r} \frac{\xi n^2(\xi) d\xi}{\left[n^2(\xi)\xi^2 - K^2\right]^{1/2}}\right)\right\}. \quad (3.130)$$

To set this result in a broader context, recall that we are examining solutions of the Helmholtz equation for a spherically symmetric medium:

$$\left[\nabla^2 + k^2 n^2(r)\right] u(\mathbf{r}) = 0, \quad (3.131)$$

where $n(r) \to 1$ as $r \to \infty$. If the inhomogeneous region has compact support, (or more crudely, is appreciable only in a region of finite size), as will be the case in most of the topics discussed here, then the scattered wave has the asymptotic form

$$u_s \sim f(\theta)\frac{e^{ikr}}{r}, \quad (3.132)$$

$f(\theta)$ being the scattering amplitude. By considering the form of (3.130) as $r \to \infty$, and noting from (3.127) that

$$\exp\left\{ik \int_{r_0}^{\infty} \frac{\xi n^2(\xi)d\xi}{\left[n^2(\xi)\xi^2 - K^2\right]^{1/2}}\right\} \sim e^{ik(\psi_0+r)}, \tag{3.133}$$

we find that

$$u \sim \left[\frac{1}{K}\sin\theta \frac{d\theta}{dK}\right]^{-1/2} e^{2ik\psi_0} \frac{e^{ikr}}{r}. \tag{3.134}$$

Thus we can make the formal identification

$$f(\theta) = \left[\frac{1}{K}\sin\theta \frac{d\theta}{dK}\right]^{-1/2} e^{2ik\psi_0}. \tag{3.135}$$

This expression has singularities when (i) $d\theta/dK = 0$ (rainbow scattering); (ii) $\sin\theta = 0$ (glory scattering), and (iii) as $K \to \infty$. These features will be discussed in more detail in Chapter 15 (see also [81]).

3.5 DISPERSION RELATIONS: A WAVE-RAY CONNECTION

Before investigating ray trajectories in more detail for specific spatially varying media, it will be helpful to establish some general principles regarding ray and wave propagation in slowly varying systems. We will find this to be useful not only in this chapter but in later ones. Let us consider the simplest possible second-order wave equation that has enough structure to develop the concepts we need here and elsewhere in the book. The one-dimensional wave equation on the real line is familiar to all, namely,

$$\frac{\partial^2 \phi}{\partial t^2} - c^2 \frac{\partial^2 \phi}{\partial x^2} = 0, \quad -\infty < x < \infty, \tag{3.136}$$

where c is a constant wave speed. This equation describes the spatial and temporal evolution of the (here generic) quantity $\phi(x,t)$ in a homogeneous conservative system. The wave operator can be factored as follows:

$$\left(\frac{\partial}{\partial t} - c\frac{\partial}{\partial x}\right)\left(\frac{\partial}{\partial t} + c\frac{\partial}{\partial x}\right)\phi = 0. \tag{3.137}$$

Now let

$$\phi(x,t) = \widetilde{\phi}(\xi, \eta),$$

where

$$\eta = x - ct, \quad \xi = x + ct, \tag{3.138}$$

so that

$$\frac{\partial \phi}{\partial x} = \frac{\partial \widetilde{\phi}}{\partial \xi}\frac{\partial \xi}{\partial x} + \frac{\partial \widetilde{\phi}}{\partial \eta}\frac{\partial \eta}{\partial x}, \text{ etc.} \tag{3.139}$$

The governing equation can therefore be reduced to the simpler form

$$\frac{\partial^2 \widetilde{\phi}}{\partial \xi \partial \eta} = 0 \Rightarrow \widetilde{\phi} = f(\eta) + g(\xi), \tag{3.140}$$

that is,

$$\phi(x, t) = f(x - ct) + g(x + ct). \tag{3.141}$$

This is the well-known *D'Alembert solution* for the infinite real line. Before proceeding with a particular form of this solution, it will be useful to introduce the definitions of dispersion relations via integral superposition of plane waves (i.e., using Fourier transforms).

3.5.1 Fourier Transforms and Dispersion Relations

Fourier Transforms

Consider the following definition of the spatial Fourier transform $\mathcal{F}_x(\phi(x, t)) \equiv \Phi(k, t)$ of a function $\phi(x, t)$ defined on the real line $(-\infty, \infty)$ in both time and one spatial dimension (x), which exists provided that ϕ and its derivatives tend to zero as $|x|$ and $|t| \to \infty$:

$$\mathcal{F}_x(\phi(x, t)) = \frac{1}{(2\pi)^{1/2}} \int_{-\infty}^{\infty} \phi(x, t) e^{-ikx} dx \equiv \Phi(k, t). \tag{3.142}$$

Therefore

$$\mathcal{F}_x\left(\frac{\partial \phi}{\partial x}\right) = \frac{1}{(2\pi)^{1/2}} \int_{-\infty}^{\infty} \frac{\partial \phi}{\partial x} e^{-ikx} dx,$$

$$= \frac{1}{(2\pi)^{1/2}} \left\{ [\phi e^{-ikx}]_{-\infty}^{\infty} + ik \int_{-\infty}^{\infty} \phi e^{-ikx} dx \right\}$$

$$= 0 + ik\mathcal{F}_x(\phi) = ik\Phi(k, t), \tag{3.143}$$

provided

$$\lim_{|x| \to \infty} \phi(x, t) = 0. \tag{3.144}$$

Similarly,

$$\mathcal{F}_x\left(\frac{\partial^2 \phi}{\partial x^2}\right) = (ik)^2 \Phi(k, t) = -k^2 \Phi(k, t), \tag{3.145}$$

provided

$$\lim_{|x| \to \infty} \frac{\partial \phi}{\partial x} = 0. \tag{3.146}$$

In a similar manner, it can be shown that

$$\mathcal{F}_x\left(\frac{\partial^2 \phi}{\partial t^2}\right) = \frac{d^2 \mathcal{F}_x(\phi)}{dt^2} = \frac{d^2 \Phi}{dt^2}. \tag{3.147}$$

If instead we Fourier transform in *time t*, then we correspondingly define

$$\mathcal{F}_t\left(\phi(x,t)\right) = \frac{1}{(2\pi)^{1/2}} \int_{-\infty}^{\infty} \phi(x,t)e^{i\omega t}\,dt \equiv \Phi(x,\omega). \tag{3.148}$$

If ϕ and $\partial\phi/\partial t$ also tend to zero as $|t| \to \infty$, then

$$\mathcal{F}_t\left(\frac{\partial\phi}{\partial t}\right) = -i\omega\mathcal{F}_t(\phi) = -i\omega\Phi(x,\omega), \tag{3.149}$$

and

$$\mathcal{F}_t\left(\frac{\partial^2\phi}{\partial t^2}\right) = (-i\omega)^2\,\mathcal{F}_t(\phi) = -\omega^2\Phi(x,\omega). \tag{3.150}$$

Inverse Fourier Transforms

Let

$$\mathcal{F}_x^{-1}(\Phi(k,t)) = \frac{1}{(2\pi)^{1/2}} \int_{-\infty}^{\infty} \Phi(k,t)e^{ikx}\,dk \equiv \phi(x,t), \tag{3.151}$$

and

$$\mathcal{F}_t^{-1}\left(\Phi(x,\omega)\right) = \frac{1}{(2\pi)^{1/2}} \int_{-\infty}^{\infty} \Phi(x,\omega)e^{-i\omega t}\,d\omega \equiv \phi(x,t). \tag{3.152}$$

We can combine the Fourier transforms (and their inverses) to write, for example,

$$\Phi(k,\omega) = \frac{1}{2\pi} \int_{-\infty}^{\infty}\int_{-\infty}^{\infty} \phi(x,t)e^{i(\omega t - kx)}\,dx\,dt, \tag{3.153}$$

and

$$\phi(x,t) = \frac{1}{2\pi} \int_{-\infty}^{\infty}\int_{-\infty}^{\infty} \Phi(k,\omega)e^{i(kx-\omega t)}\,dk\,d\omega. \tag{3.154}$$

3.5.2 The Bottom Line

If we Fourier transform the one-dimensional wave equation (3.136), for example, we obtain

$$\left(-\omega^2 + c^2k^2\right)\Phi(k,\omega) = 0, \tag{3.155}$$

where the polynomial expression

$$D(\omega,k) = -\omega^2 + c^2k^2 = 0 \tag{3.156}$$

is called the *dispersion relation*, in that it relates the frequency ($\nu = \omega/2\pi$) of the "wave" (or Fourier component) to the wavelength ($\lambda = 2\pi/k$). The quantity k is called the *wavenumber* To interpret this wave terminology note that the definition of the two-dimensional inverse transform

$$\phi(x,t) = \frac{1}{2\pi} \int_{-\infty}^{\infty}\int_{-\infty}^{\infty} \Phi(k,\omega)e^{i(kx-\omega t)}\,dk\,d\omega, \tag{3.157}$$

is essentially an *integral superposition* of complex exponentials (often called *plane waves*) whose real and imaginary parts are sinusoidal in character. So the quick and dirty way to obtain a dispersion relation from a partial differential equation is to consider a plane wave of the form

$$\phi(x, t) = a e^{i(kx - \omega t)}, \tag{3.158}$$

for constant amplitude a and directly substitute it into the governing partial differential equation to obtain the dispersion relation. This assumes of course that the coefficients of the various derivatives are constant (and that the Fourier transform exists). In particular, if L is a polynomial operator such that

$$L\left(\frac{\partial \phi}{\partial t}, \frac{\partial \phi}{\partial x}\right) = 0, \tag{3.159}$$

then

$$L(-i\omega, ik) = 0 \equiv D(\omega, k). \tag{3.160}$$

From equation (3.156) the dispersion relation for sound waves in one dimension (c being the speed of sound) is equivalent to the simple form

$$\omega = |c| k. \tag{3.161}$$

Here we just consider the positive branch ($c > 0$) $\omega = ck$ since $k > 0$. In this case, both the wave speed (or phase speed) $c = \omega/k$ and the group speed $c_g = \partial \omega/\partial k$ are equal to c, and the waves are nondispersive, since c is wavelength independent. This is not the case in general, and more about these concepts is developed in connection with surface gravity waves in Chapter 11. At the risk of being overly repetitive, to obtain a dispersion relation for a wavelike solution to a constant-coefficient partial differential equation, use a solution in the form of a complex exponential function of space and time. The resulting dispersion relation will be different, of course, for different types of wave. Remaining with acoustic waves, we further illustrate the concept with a simple example concerning the omnidirectional propagation of sound. The Cartesian form of the wave equation governing the space and time evolution of small-amplitude sound waves of speed c in a uniform three-dimensional medium is given by

$$\frac{\partial^2 \phi}{\partial t^2} = c^2 \nabla^2 \phi \equiv c^2 \left(\frac{\partial^2 \phi}{\partial x_1^2} + \frac{\partial^2 \phi}{\partial x_2^2} + \frac{\partial^2 \phi}{\partial x_3^2}\right). \tag{3.162}$$

The dependent variable $\phi(x_1, x_2, x_3, t)$ can represent small pressure deviations from the ambient pressure as a sound wave traverses the medium; it may also represent a velocity component or a displacement in this case. It is worth reminding ourselves that this emphasis on small-amplitude (theoretically, infinitesimal) waves is necessary because the form of the solutions we seek is based on *linear* theory. The speed of the sound waves (assumed constant here) and of waves other than water waves is generally denoted by c in this book, and a particular point in space is defined by the radius vector $\mathbf{r} = \langle x_1, x_2, x_3 \rangle$, and the wavenumber k is replaced by a *wave vector* $\mathbf{k} = \langle k_1, k_2, k_3 \rangle$. Now a plane-wave solution of this equation is one of the form

$$\phi(x_1, x_2, x_3, t) = \text{Re}(a \exp[i(\mathbf{k} \cdot \mathbf{r} - \omega t)])$$
$$= \text{Re}(a \exp[i(k_1 x_1 + k_2 x_2 + k_3 x_3 - \omega t)]), \tag{3.163}$$

where the k_i, $i = 1, 2, 3$, and ω are real quantities; a is a constant (possibly complex), and ω is the angular frequency. In cases where linear stability of the system is of interest we allow ω to be a complex quantity. Now the *wavenumber* $\kappa = |\mathbf{k}| = \sqrt{k_1^2 + k_2^2 + k_3^2}$, and the wavelength $\lambda = 2\pi/\kappa$. By substituting the complex solution $a \exp(i\mathbf{k}\cdot\mathbf{r}-\omega t)$ into equation (3.162), an algebraic equation results, namely, the dispersion relation $\omega = \omega(\kappa)$, where

$$\omega^2 - c^2\kappa^2 = 0, \tag{3.164}$$

or

$$\omega = |c|\,\kappa. \tag{3.165}$$

Not surprisingly, it is a straightforward generalization of its one-dimensional counterpart, equation (3.161). It represents the wave or phase speed of waves propagating parallel and antiparallel to the wave vector $\mathbf{k}$. A plane wave is one component of an integral superposition of such exponential forms if the wave equation is defined in an infinite domain.

D'Alembert Again

In light of the above discussion we now proceed to simplify the D'Alembert solution (3.141) further by considering a single Fourier component (i.e., one harmonic in both time and space):

$$f(x - ct) = a\cos(kx - \omega t) = a\cos k(x - ct), \quad \text{and} \quad g \equiv 0, \tag{3.166}$$

where $\omega = kc$, so that

$$\phi(x, t) = a\cos k(x - ct). \tag{3.167}$$

This represents a periodic progressive wave of amplitude a and wave (or phase) speed $c = \omega/k$. Here we can define a phase function $\theta(x, t)$ by $\theta(x, t) = kx - \omega t = k(x - ct)$. It will be constant for an observer moving with speed such that

$$\frac{d\theta}{dt} = 0, \quad \text{that is,} \quad \frac{dx}{dt} = \frac{\omega}{k} = c. \tag{3.168}$$

The form (3.167) is a solution of equation (3.136) satisfying the initial conditions

$$\phi(x, 0) = a\cos kx, \quad \frac{\partial\phi}{\partial t}(x, 0) = \omega a\sin kx. \tag{3.169}$$

If P is the wave period, the angular frequency is $\omega = 2\pi/P$. If λ is the wavelength, the wavenumber is $k = 2\pi/\lambda$.

Dispersion

With the above discussion in mind, consider now a general linear partial differential equation in two independent variables x and t, written in operator form as

$L(\phi) = 0$, neglecting for now any initial and/or boundary conditions. If L is a homogeneous operator, and x, t do not appear explicitly, the complex substitution

$$\phi(x, t) = a \exp i(kx - \omega t) \tag{3.170}$$

yields, as noted above, the "operator mappings"

$$\frac{\partial}{\partial t} \to -i\omega, \quad \frac{\partial}{\partial x} \to ik, \tag{3.171}$$

so that

$$L(\phi) = 0 \to D(\omega, k) = 0, \tag{3.172}$$

where $D(\omega, k) = 0$ is the dispersion relation. In general $\omega = \omega(k)$ and we can identify each root ω of (3.172) a separate mode. Then

$$\phi(x, t) = a \exp\left[i\left(kx - \omega(k)t\right)\right]. \tag{3.173}$$

When $\omega(k)$ is real, this represents a harmonic wave. If $\operatorname{Im}\omega(k) > 0$, then ϕ becomes exponentially unbounded in t; if $\operatorname{Im}\omega(k) < 0$, then ϕ decays exponentially in t. If we assume that ω is real for all $k \in [0, \infty)$, then the wave speed (or phase speed) is simply

$$c = \frac{\omega(k)}{k}.$$

If c is independent of k, the wave propagation is said to be nondispersive, and $\omega(k) = ck$. Alternatively, if $\omega''(k) = 0$ it is nondispersive. The *group speed* (or velocity if a vector form is needed) c_g is defined as

$$c_g = \omega'(k). \tag{3.174}$$

In general, $c \neq c_g$ since

$$\frac{\omega(k)}{k} \neq \omega'(k). \tag{3.175}$$

Clearly for nondispersive waves, $c_g = c$, i.e. the group and wave speeds are the same. For nondispersive waves, a given superposition of Fourier components will be valid for all time, but is translated in time, that is,

$$\phi(x, t) = \phi(x - ct, 0).$$

This is not the case for dispersive waves, where different frequencies propagate with different phase velocities. If $\omega(k)$ is a complex function, the wave is said to be diffusive.

Group Speed: A Physical Interpretation

Let us consider the combination of two waveforms of the same amplitude and slightly different values of k and ω:

$$\phi_1 = a \cos(kx - \omega t), \quad \phi_2 = a \cos\{(k + \delta k)x - (\omega + \delta\omega)t\}. \tag{3.176}$$

Therefore

$$\phi = \phi_1 + \phi_2 = \left\{ 2a \cos \left(\frac{1}{2}(x\delta k - t\delta \omega) \right) \cos \left[\left(k + \frac{\delta k}{2} \right) x - \left(\omega + \frac{\delta \omega}{2} \right) t \right] \right\}.$$

(3.177)

This is a highly simplified mathematical description of *beats*! Note that the effective amplitude of the carrier wave is

$$A(x, t) = 2a \cos \left\{ \frac{1}{2}(x\delta k - t\delta \omega) \right\},$$

(3.178)

and it varies slowly with period $4\pi/\delta\omega$ and wavelength $4\pi/\delta k$. Compared with the corresponding periods and wavelengths of the initial waveforms, these quantities are large. The locus of constant A is given by

$$x\delta k - t\delta \omega = \text{constant}.$$

(3.179)

Roughly speaking, the "groups" of waves move with speed

$$\frac{dx}{dt} \approx \frac{\delta \omega}{\delta k}.$$

(3.180)

This is an approximation to the group speed; in fact,

$$c_g = \lim_{\delta k \to 0} \frac{\delta \omega}{\delta k} = \frac{d\omega}{dk} = \omega'(k).$$

(3.181)

Group Velocity

Returning to wave propagation in three spatial dimensions, now with $\mathbf{r} = \langle x_1, x_2, x_3 \rangle$ and $\mathbf{k} = \langle k_1, k_2, k_3 \rangle$, we can define the phase by

$$\theta(\mathbf{r}, t) = \mathbf{k} \cdot \mathbf{r} - \omega t.$$

(3.182)

Now wave and group speeds become wave and group velocities, that is, for the wave velocity

$$\mathbf{c}(\mathbf{k}) = \frac{\omega}{k^2} \mathbf{k}, \qquad k = |\mathbf{k}|,$$

(3.183)

and $\omega = \omega(\mathbf{k})$. A dispersion relation is *isotropic* if it is of the form $\omega = \omega(k) = \omega(|\mathbf{k}|)$, that is, independent of the direction $\mathbf{k}$. The group velocity is defined as

$$\mathbf{c}_g(\mathbf{k}) = \nabla_k \omega \equiv \left\langle \frac{\partial \omega}{\partial k_1}, \frac{\partial \omega}{\partial k_2}, \frac{\partial \omega}{\partial k_3} \right\rangle.$$

(3.184)

The trajectory of a point that moves with velocity

$$\frac{d\mathbf{x}}{dt} = \mathbf{c}_g$$

(3.185)

is termed a *ray*.

3.5.3 Applications to Atmospheric Waves

Many different types of waves can be supported by the earth's land, sea, and atmosphere, some of which are named after the scientists who made major contributions to the topic. Some of the waves that can exist are called acoustic, atmospheric, baroclinic, Eady, gravity, internal, Kelvin, lee, Love, planetary, Rayleigh, and Rossby. In advance of a more general exposition of such waves in Chapter 14, the governing differential equations will not be delineated here (with the exception of (3.186)), but for some of these waves we will briefly discuss their dispersion relations and some of the consequences that follow from them. There are also combinations of various types, such as acoustic-gravity waves, which, as the name implies, are supported by the restoring forces of both buoyancy and compressibility in the atmosphere. Stated without proof, the governing wave equation for such waves is [181], [182]

$$c^2 \nabla^2 \frac{\partial^2 \phi}{\partial t^2} - \frac{\partial^4 \phi}{\partial t^4} - \omega_a^2 \frac{\partial^2 \phi}{\partial t^2} + N^2 c^2 \left(\frac{\partial^2 \phi}{\partial x_1^2} + \frac{\partial^2 \phi}{\partial x_2^2} \right) = 0. \tag{3.186}$$

In terms of the quantities present in equation (3.163), the acoustic-gravity wave dispersion relation is

$$k_3^2 = \kappa^2 \left(\frac{N^2}{\omega^2} - 1 \right) + \left(\frac{\omega^2 - \omega_a^2}{c^2} \right), \quad \kappa^2 = k_1^2 + k_2^2, \tag{3.187}$$

which is valid for an isothermal atmosphere (i.e., one in which the sound speed c is constant). There are two naturally occurring frequencies in this model; the first is the so-called *Brunt-Väisälä frequency* N. It is the frequency at which a small parcel of air will oscillate if displaced vertically from its equilibrium position in a stably stratified atmosphere. It is the highest frequency at which gravity waves may propagate. In a horizontally stratified atmosphere

$$N = \left(\frac{g}{\theta} \frac{d\theta}{dz} \right)^{1/2}; \quad \theta = T \left(\frac{p_0}{p} \right)^{R/c_p}, \tag{3.188}$$

where θ is potential temperature, g is the local acceleration of gravity, z is geometric height, T is the absolute temperature (in K) of the fluid particle, R is the gas constant for air (say), and c_p is the specific heat capacity at a constant pressure ($R/c_p \approx 0.286$ for air). In the ocean (where salinity is important)

$$N = \left(-\frac{g}{\rho} \frac{d\rho}{dz} \right)^{1/2} \in \mathbb{R}, \tag{3.189}$$

provided $d\rho/dz < 0$, where ρ is the potential density (which depends on both temperature and salinity).

Returning to a simplified model of the atmosphere described by equation (3.187), the quantity ω_a ($> N$) is called the *acoustic cut-off frequency* and is the other naturally occurring frequency; this is the lowest frequency at which gravity-modified acoustic waves may propagate. ω_a and N are, respectively, defined for an isothermal atmosphere by

$$N = \left(\frac{(\gamma - 1) g}{\gamma H} \right)^{1/2}, \quad \text{and} \quad \omega_a = \frac{1}{2} \left(\frac{\gamma g}{H} \right)^{1/2}, \tag{3.190}$$

where the H is *density scale-height* defined by

$$H = -\left(\frac{1}{\rho}\frac{d\rho}{dz}\right)^{-1},$$

(now $\rho(z)$ is the atmospheric density profile). For this idealized atmosphere, H is a positive constant. Note that $k_3^2 < 0$ when $N^2 < \omega^2 < \omega_a^2$; this corresponds to horizontally propagating waves that are attenuated exponentially in the vertical direction. They are sometimes referred to as *evanescent* or *surface* waves.

Exercise: Derive the dispersion relation (3.187) from equation (3.186).

Much information may be gleaned from such dispersion relations, though in the literature the notation is somewhat different. For simplicity we restrict ourselves to waves for which the first term in the relation (3.187) is much larger than the second. For the earth's atmosphere, this corresponds to compressibility-modified gravity waves with wavelengths of the order of a few kilometers (at most). Furthermore, we impose a uniform wind (sometimes called a zonal wind) U flowing in the x_1 direction. If we further simplify the system to a two-dimensional one and use the following notation in advance of Chapter 14 on atmospheric waves, $\mathbf{r} = \langle x, z \rangle$ and $\mathbf{k} = \langle k, m \rangle$, the determining dispersion relation then takes the form

$$m^2 = k^2 \left(\frac{N^2}{\Omega^2} - 1\right), \tag{3.191}$$

where $\Omega = \omega - kU$ is the *Doppler-shifted frequency*. Equation (3.191) may be rearranged to give

$$\frac{\Omega^2}{N^2} = \frac{k^2}{k^2 + m^2}. \tag{3.192}$$

The most easily observed example of this type of wave is in the lee of mountains: these are *lee waves* originating as a result of air being forced to flow over a mountain under stable atmospheric conditions. The air is set into motion in the form of gravity waves as it moves downstream from the mountain. If their amplitude is large enough, and the temperature and humidity are in the right ranges, clouds will form in the crests of these waves [122]. They will appear stationary with respect to the mountain if the horizontal component of wave velocity c relative to the surface ($c = \omega/k$) is zero. If $k \ll m$ then equation (3.192) implies that $m \approx N/U$, and since the wind supplying energy to these waves is relatively near the ground, the majority of the energy associated with them (traveling at the *group speed* c_g) is propagated upward. This brings up an interesting and seemingly paradoxical property of gravity waves: for $k \ll m$ *the vertical component of their group velocity is antiparallel to the vertical component of the wave velocity*. We demonstrate this for the case of $U = 0$, noting now that the group velocity $\mathbf{c}_g$ is the vector

$$\mathbf{c}_g = \left\langle \frac{\partial \omega}{\partial k}, \frac{\partial \omega}{\partial m} \right\rangle. \tag{3.193}$$

It is readily shown from equation (3.192) that the vertical components of group and wave velocity are respectively given by

$$\frac{\partial \omega}{\partial m} = -\frac{km N}{\left(m^2 + k^2\right)^{3/2}} \approx -\frac{k N}{m^2} \quad \text{and} \quad \frac{\omega}{m} = \frac{k N}{m \left(m^2 + k^2\right)^{1/2}} \approx \frac{k N}{m^2}. \tag{3.194}$$

Gravity waves are not dominant in the overall energy budget in the lower atmosphere, but above an altitude of about 75 km, they are very significant. Note that on a planetary scale, tidal motions are a special case of gravity waves in water with long time and spatial scales. On these large scales in the atmosphere, there are wave motions, generally known as *planetary waves*. In their simplest form they arise as a result of the variation of the Coriolis parameter with latitude, and are known as *Rossby waves* (named after the meteorologist Carl Rossby, who discovered their existence in the 1930s). They are manifested as mid-latitude westerly winds flowing around the earth in the troposphere (where the weather occurs—the lowest 6–10 km of the atmosphere, depending on latitude, being thickest at the equator and thinnest at the poles), and to a lesser extent, the stratosphere immediately above the troposphere. Their pattern is variable; their number, amplitude, and wavelength are changeable; and they influence the movement of air masses from the polar and mid-latitude regions, the balance of heat exchanged between these air masses, and also, to some extent, the development and track of storms.

In terms of the mean flow U the governing dispersion relation is

$$\omega = k \left(\frac{\beta}{k^2 + l^2 + m^2 f_0^2 / N^2} - U \right), \tag{3.195}$$

where in the so-called β-plane approximation, the *Coriolis parameter* $f = 2\Omega \sin \phi$ is written is $f = f_0 + \beta y$ (where f_0 is the value of f at the mid-latitude of the region, ϕ is the latitude, and the quantity Ω *now* represents the angular velocity of the rigidly rotating earth). It turns out that $f_0^2 \ll N^2$, and hence the vertical wavelengths are the order of 1 percent of the horizontal wavelengths. Sometimes these waves are forced by surface features, such as mountains or large land masses, in which case they are stationary with respect to the surface (not unlike the behavior of water waves in the vicinity of rocks in flowing streams).

Thus, when $c = \omega / k = 0$, equation (3.195) implies that

$$m^2 = \frac{N^2}{f_0^2} \left[\frac{\beta}{U} - \left(k^2 + l^2\right) \right], \tag{3.196}$$

and so $m^2 > 0$ when

$$0 < U < \frac{\beta}{\left(k^2 + l^2\right)}. \tag{3.197}$$

Now U is generally a westerly flow, so that it is clear from this that if the flow is either easterly ($U < 0$) or large and westerly ($U > \beta/(k^2 + l^2)$), no vertical propagation occurs ($m^2 < 0$). This result is in fact confirmed by the observed lack of planetary wave activity during the summer months when the mean stratospheric flow is easterly. From the inequality (3.197) it is apparent that for a given value of U, waves of large horizontal wavelength are most readily propagated in these atmospheric regions.

3.6 GENERAL SOLUTION OF THE LINEAR WAVE EQUATION: SOME ASYMPTOTICS

3.6.1 Stationary Phase

In contrast to the earlier discussion of a single Fourier modal solution, the general solution for $L(\phi) = 0$ is an integral superposition of the individual Fourier components,

$$\phi(x,t) = \int_{-\infty}^{\infty} A(k)e^{i(kx-\omega(k)t)}\, dk, \tag{3.198}$$

with initial condition

$$\phi(x,0) = \int_{-\infty}^{\infty} A(k)e^{ikx}\, dk. \tag{3.199}$$

Thus, given $\phi(x,0)$,

$$A(k) = \frac{1}{2\pi} \int_{-\infty}^{\infty} \phi(x,0)e^{-ikx}\, dx. \tag{3.200}$$

Let us use the method of stationary phase to examine the behavior of $\phi(x,t)$ for large times. To do so, define

$$X(k) \equiv \frac{x}{t}k - \omega(k), \tag{3.201}$$

so that integral (3.198) becomes

$$\phi(x,t) = \int_{-\infty}^{\infty} A(k)e^{itX(k)}\, dk. \tag{3.202}$$

Assuming that $X(k)$ is analytic in the complex k-plane for a given value of x/t, points of stationary phase are given by the condition $\partial X(k)/\partial k = 0$, that is,

$$\omega'(k)\left(= c_g\right) = \frac{x}{t}, \tag{3.203}$$

provided $\omega''(k) \neq 0$. Solving for k, we denote points of stationary phase by $k_s = k_s(x/t)$ and so (defining $\alpha = (i\pi/4)\,\text{sgn}\,X''(k_s)$)

$$\phi(x,t) \approx \frac{(2\pi)^{1/2} A(k_s)\exp i\left[tX(k_s) + \alpha\right]}{[t|X''(k_s)|]^{1/2}} \tag{3.204}$$

$$= \frac{(2\pi)^{1/2} A(k_s)\exp i\left[(k_s x - \omega(k_s)t) + \alpha\right]}{[t|\omega''(k_s)|]^{1/2}} \quad \text{as } t \to \infty. \tag{3.205}$$

For m such points [110],

$$\phi(x,t) \approx \sum_{i=1}^{m} \frac{(2\pi)^{1/2} A(k_s)\exp i\left[(k_s x - \omega(k_s)t) - (i\pi/4)\,\text{sgn}\,\omega''(k_s)\right]}{[t|\omega''(k_s)|]^{1/2}}. \tag{3.206}$$

Note that for $\phi(x, t)$ in expression (3.206), the amplitude term is

$$\bar{A}(t) \equiv \frac{(2\pi)^{1/2} A(k_s)}{[t|\omega''(k_s)|]^{1/2}} \propto t^{-1/2} \tag{3.207}$$

as x and $t \to \infty$ such that x/t is fixed. Physically, the energy associated with wavenumber k_s and with "spread" Δk is initially $\propto A^2(k_s)\Delta k$. After time t, the spread between these wavenumbers becomes (for points of stationary phase)

$$|x(k_s) - x(k_s + \Delta k)| = |t\omega'(k_s) - t\omega'(k_s + \Delta k)| \approx t|\omega''(k_s)|\Delta k,$$

so the energy density is proportional to

$$\frac{A^2(k_s)\Delta k}{t|\omega''(k_s)|\Delta k} = \frac{A^2(k_s)}{t|\omega''(k_s)|}, \tag{3.208}$$

consistent with the amplitude of the wave at time t being proportional to

$$\frac{A(k_s)}{[t|\omega''(k_s)|]^{1/2}}.$$

Further analysis [110] shows that when $\omega''(k_s) = 0$, the corresponding contribution of the saddle point k_s to $\phi(x, t)$ is asymptotically given by

$$\phi(x, t) \propto \frac{A(k_s) \exp i[k_s x - \omega(k_s)t] (1/3)! (3)^{1/2}}{[t|\omega'''(k_s)|/6]^{1/3}}, \tag{3.209}$$

provided $\omega'''(k_s) \neq 0$, so $\phi \sim t^{-1/3}$.

3.6.2 Asymptotics for Oscillatory Sources: Wavenumber Surfaces

Now we focus on a specific class of partial differential operator, namely,

$$L(\phi) \equiv L\left(\frac{\partial^2}{\partial t^2}, \frac{\partial^2}{\partial x_1^2}, \frac{\partial^2}{\partial x_2^2}, \frac{\partial^2}{\partial x_3^2}\right)\phi = e^{-i\omega t} f(x_1, x_2, x_3), \tag{3.210}$$

where L is now a polynomial in those second derivatives with constant coefficients. The right-hand side of this equation corresponds to a monochromatic source with $f(\mathbf{x})$ vanishing asymptotically sufficiently fast that its Fourier transform $\hat{f}$ exists, where

$$f(\mathbf{x}) = (2\pi)^{-3/2} \int_{R^3} \hat{f}(\mathbf{k}) e^{i\mathbf{k}\cdot\mathbf{x}} \, d\mathbf{k}. \tag{3.211}$$

In three-dimensional Cartesian coordinates, for example, the differential wavevector element is $d\mathbf{k} = dk_x dk_y dk_z$. Note also that

$$\phi(\mathbf{x}, t) = (2\pi)^{-3/2} e^{-i\omega t} \int_{R^3} \hat{\phi}(\mathbf{k}) e^{i\mathbf{k}\cdot\mathbf{x}} \, d\mathbf{k}. \tag{3.212}$$

The transformed equation is

$$\hat{L}\hat{\phi} = \hat{f}, \tag{3.213}$$

where

$$\hat{L} = L(-\omega^2, -k_1^2, -k_2^2, -k_3^2). \tag{3.214}$$

A particular integral is

$$\hat{\phi} = \hat{L}^{-1}\hat{f}.$$ (3.215)

The complementary function, which must be added to establish the general solution, is a solution of

$$\hat{L}\hat{\phi} = 0,$$ (3.216)

which according to generalized function theory [211] can be written in terms of Dirac's delta distribution as

$$\hat{\phi}_{cf} = H(k_1, k_2, k_3)\delta(\hat{L}) \equiv H(\mathbf{k})S_\omega.$$ (3.217)

Since $H(\mathbf{k})$ is arbitrary, there exists an infinity of solutions. To ensure uniqueness, we need to apply a *radiation condition* requiring that all waves propagate outward from the source. We shall therefore examine the solution

$$\phi(\mathbf{x}, t) = (2\pi)^{-3/2} e^{-i\omega t} \int_{R^3} \frac{\hat{f}(\mathbf{k})}{\hat{L}(\omega, \mathbf{k})} e^{i\mathbf{k}\cdot\mathbf{x}} \, d\mathbf{k},$$ (3.218)

as $|\mathbf{x}| \to \infty$ [33]. The asymptotic behavior of any Fourier integral can be expressed in terms of the singularities of the integrand, which will dominate the solution. Since f vanishes asymptotically to high order, its transform $\hat{f}$ has no singularities, so the singularities of the integrand lie entirely on the so-called *wavenumber surface*

$$S_\omega = \{\mathbf{k} : \hat{L}(\omega, \mathbf{k}) = 0\}.$$ (3.219)

This is the locus of points in $\mathbb{R}^3$ such that plane-wave solutions $\phi(\mathbf{x}, t) = \exp[i(\mathbf{k}\cdot\mathbf{x} - \omega t)]$ (so-called free waves) of the homogeneous equation exist. One way of incorporating the radiation condition is to replace the right-hand side of equation (3.210) by the expression

$$e^{-i(\omega+i\varepsilon)t} f(\mathbf{x}), \quad 0 < \varepsilon \ll |\omega|.$$ (3.220)

This represents a source that has been built up from zero to its present strength gradually during all the time from $t = -\infty$. If the solution were contaminated by the presence of any waves coming in from infinity that had started their journey inward at time t_0, their amplitude would be of the order $\exp(\varepsilon t_0)$, comparable with that of waves generated at the source. Being free waves, their amplitudes would not increase exponentially with time, so that on reaching the neighborhood of the source, they would still have amplitude $\exp(\varepsilon t_0)$. This would by then be negligible compared with $\exp(\varepsilon t)$, the amplitude of waves originating from the source, since $t \gg t_0$, the waves having come in from infinity. Hence if terms of order $\exp(\varepsilon t)$ only are sought in the solution, incoming waves must be absent. In the integral (3.218) we replace $\hat{L}(\omega, \mathbf{k}) = S_\omega$ by $\hat{L}(\omega, \mathbf{k}) = S_{\omega+i\varepsilon}$ and rotate coordinates so that observation point defined by $\mathbf{x}$ is in the direction of one of the axes, k_1, say. Then expanding the integral in (3.218), it becomes

$$\int_{-\infty}^{\infty} dk_2 \int_{-\infty}^{\infty} dk_3 \int_{-\infty}^{\infty} \frac{\hat{f} e^{ik_1 x_1}}{S_{\omega+i\varepsilon}} \, dk_1.$$ (3.221)

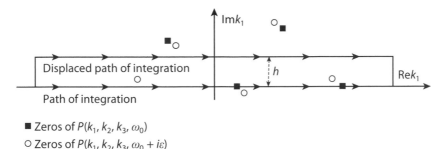

Figure 3.4 Displaced path of integration for (3.221).

For the inner integral we move the path of integration (Figure 3.4) so that $\operatorname{Im} k_1 = h > 0$; the integral over the new path is now $O(e^{-hx})$ for large x, and we can regard quantities of this order as negligible. The only singularities of the integrand between the two paths are poles at $S_{\omega+i\varepsilon} = 0$. The zeros for real k_1 (when $\varepsilon = 0$) represent plane waves. These will be displaced in general for $\varepsilon > 0$ and will only contribute their residues if they are displaced into the region $0 < \operatorname{Im} k_1 < h$. Now $\operatorname{Im} \omega > 0 \Rightarrow \operatorname{Im} k_1 > 0$ if $\hat{L}(\omega, \mathbf{k}) = 0$ specifies $\omega = \omega(\mathbf{k})$ such that

$$\frac{\partial \omega}{\partial k_1} > 0. \tag{3.222}$$

The physical significance of the radiation condition is there will only be a contribution to the wave amplitude in integral (3.218) if the component of $\mathbf{c}_g$ along $\mathbf{x} = (l, 0, 0)$ say, is positive. The contribution to the inner integral of (3.221) from each pole will be

$$2\pi i \, \frac{\hat{f}(\mathbf{k}) e^{ik_1 x_1}}{\partial S_\omega(\mathbf{k}, \omega)/\partial k_1}$$

in the limit $\varepsilon \to 0^+$, and zero otherwise. Hence asymptotically we have

$$\phi \sim 2\pi i \iint_S \frac{\hat{f}(\mathbf{k}) e^{ik_1 x_1}}{\partial S_\omega(\mathbf{k}, \omega)/\partial k_1} \, dk_2 \, dk_3, \tag{3.223}$$

where $S \subset S_\omega$ is that part of S_ω for which the condition (3.222) holds (Figure 3.5).

Using the method of stationary phase, we note that the phase $k_1 x_1$ is stationary on S at all points $k_\alpha^{(i)}$ where the normal to S is parallel to the k_1-axis, that is, where

$$\frac{\partial k_1}{\partial k_2} = \frac{\partial k_1}{\partial k_3} = 0.$$

By rotation of axes we temporarily choose the k_2- and k_3-axes along the principal directions of curvature of the surface (these are the directions in which the radius of curvature attains its absolute maximum and absolute minimum. They are orthogonal). The associated curvatures,

$$\kappa_2 = \frac{\partial^2 k_1}{\partial k_2^2}, \quad \text{and} \quad \kappa_3 = \frac{\partial^2 k_1}{\partial k_3^2},$$

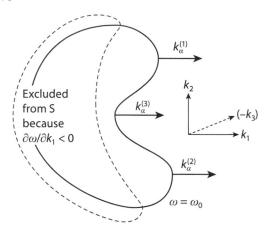

Figure 3.5 Consequence of the radiation condition.

may be used to express the local equation of the surface of S in the neighborhood of a point of stationary phase:

$$k_1 \approx k_1^{(i)} + \frac{1}{2}\kappa_2(k_2 - k_2^{(i)})^2 + \frac{1}{2}\kappa_3(k_3 - k_3^{(i)})^2. \tag{3.224}$$

Now for any nonzero real p,

$$\int_{-\infty}^{\infty} e^{ip(k-k_\alpha)^2}\, dk = \exp\left(i\pi\,[\operatorname{sgn} p]\,/4\right)\sqrt{\frac{\pi}{|p|}}, \tag{3.225}$$

so the contribution $\phi^{(i)}$ to the asymptotic form of ϕ from the point $k_\alpha^{(i)}$ (where the normal to S is in the k_1-direction) is

$$\phi^{(i)} \propto \frac{\hat{f}(k_\alpha^{(i)})\exp i[k_1^{(i)}x_1 + \pi(\operatorname{sgn}\kappa_2 + \operatorname{sgn}\kappa_3)/4]}{[\partial S_\omega(\omega, \mathbf{k})/\partial k_1]_{k_\alpha^{(i)}}\, x_1\,(\kappa_2\kappa_3)^{1/2}}. \tag{3.226}$$

Expressing this in a form invariant under rotation of axes yields a term that behaves as

$$(x|\nabla S_\omega|\sqrt{|\kappa|})^{-1},$$

where $\kappa = \kappa_2\kappa_3$ is the Gaussian curvature at the point $k_\alpha^{(i)}$ of S. We state the result in its full form as a theorem:

Theorem: The solution of

$$L\left(\frac{\partial^2}{\partial t^2}, \nabla^2\right)\phi = e^{-i\omega t} f(\mathbf{x}) \tag{3.227}$$

satisfying the radiation condition, is asymptotically

$$\phi \sim \frac{4\pi^2}{r} e^{-i\omega t} \sum_i \frac{c\hat{f}e^{i\mathbf{k}\cdot\mathbf{r}}}{|\nabla S_\omega|\sqrt{|\kappa|}} + O(r^{-2}) \tag{3.228}$$

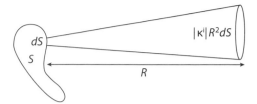

Figure 3.6 Local distribution of energy at distance R.

as $r \to \infty$ along any radius vector l, the summation being taken over all the points of the surface $S_\omega(L = 0)$, where the normal to the surface is parallel to l, and

$$(\mathbf{x} \cdot \nabla S_\omega) \left(\frac{\partial S_\omega}{\partial \omega} \right)^{-1} > 0, \tag{3.229}$$

provided that S_ω has a nonzero Gaussian curvature at each point, where c is a constant such that $|c| = 1$.
Since

$$\mathbf{c}_g = \{c_{g_i}\} = \left\{ \frac{\partial \omega}{\partial k_i} \right\} = \left\langle \frac{\partial \omega}{\partial k_1}, \frac{\partial \omega}{\partial k_2}, \frac{\partial \omega}{\partial k_3} \right\rangle = \left(\frac{\partial \omega}{\partial |\mathbf{k}|} \right) \left(\frac{\mathbf{k}}{|\mathbf{k}|} \right), \tag{3.230}$$

it follows that

$$\mathbf{c}_g = \nabla S_\omega \left(\frac{\partial S_\omega}{\partial \omega} \right)^{-1}. \tag{3.231}$$

The radiation condition (3.229) can be rewritten $\mathbf{x} \cdot \mathbf{c}_g > 0$, that is, all the energy is created at the source and none comes in from infinity. Since $\mathbf{c}_g$ is parallel to ∇S_ω, and $\mathbf{c}_g = d\mathbf{x}/dt$, the energy spreads outward from a source at the group speed directly along radii. A more mathematically rigorous account of the arguments leading to this result can be found in [111].

To understand result (3.228) physically, consider the following (Figure 3.6). The normals from a small area dS around the point $k_\alpha^{(i)}$ fill a cone whose cross-sectional area increases with distance $r = R$ like $|\kappa^{(i)}|R^2 dS$. The energy created in the source region with wavenumbers lying in this elementary area dS is therefore diminished by a factor $|\kappa^{(i)}|^{-1}R^{-2}$ at distance R, which makes the amplitude factor $|\kappa^{(i)}|^{-1/2}R^{-1}$. Note that in cases of vanishing Gaussian curvature, for portions of S_ω that are cylindrical, the r-dependence is $r^{-1/2}$, whereas the amplitude decays as $r^{-5/6}$ in the direction of a cuspidal edge, a rate of decay intermediate between those for cylindrical and spherical waves.

3.7 RAYS AND WAVES IN A SLOWLY VARYING ENVIRONMENT

What is meant by the term "slowly-varying" (or "weakly inhomogeneous")? In the study of wave propagation in an environment that varies spatially, there are two implicit length scales: (i) the wavelength λ and (ii) a typical scale of the inhomogeneity, L. For the system to qualify a weakly inhomogeneous, it is necessary that $\lambda \ll L$. If this is the case, then wave motion is still expected to occur, though it will be nonsinusoidal in character. Nevertheless it will still have associated wave crests and troughs, the amplitude and wavenumber of which will be slowly varying over a wavelength. It is therefore necessary to generalize the previous plane-wave complex solution for an infinite wavetrain

$$\phi(x, t) = ae^{i(kx-\omega t)}. \tag{3.232}$$

Given position and wavenumber vectors $\mathbf{r} = \langle x_1, x_2, x_3 \rangle$ and $\mathbf{k} = \langle k_1, k_2, k_3 \rangle$, respectively there is for such a medium, stratified in the x_3-direction (say), a field variable ϕ expressible in terms of the amplitude and phase such that

$$\phi(\mathbf{r}, t) = a(\mathbf{r}, t) e^{i\theta(\mathbf{r}, t)}.$$

(A particular example of a spatially modulated wave envelope was noted above in connection with beats—see equation (3.178),—though it arose for a uniform environment.)

We can write the Hamilton-Jacobi equation for this system as

$$\omega = \omega(\mathbf{k}, \mathbf{r}) \equiv \mathcal{H}(\mathbf{k}, \mathbf{r}). \tag{3.233}$$

By comparison with the phase term $\theta = \mathbf{k} \cdot \mathbf{r} - \omega t$ for plane waves in homogeneous media, we define the quantities

$$\frac{\partial \theta}{\partial x_1} = k_1; \qquad \frac{\partial \theta}{\partial x_2} = k_2; \qquad \frac{\partial \theta}{\partial t} = -\omega, \tag{3.234}$$

and a *local* k_3-component of the wavevector by

$$\frac{\partial \theta}{\partial x_3} = k_3.$$

The generalization of the dispersion relation (3.172) for such a medium is

$$D(\omega, \mathbf{k}; \mathbf{r}) = 0.$$

We write the third equation in (3.234) as

$$-\frac{\partial \theta}{\partial t} = \mathcal{H}(k_i, x_i),$$

so that after differentiating this with respect to the x_i, we obtain

$$\frac{\partial k_i}{\partial t} + \frac{\partial \mathcal{H}}{\partial k_i}\frac{\partial k_i}{\partial x_i} = -\frac{\partial \mathcal{H}}{\partial x_i}. \tag{3.235}$$

But

$$\frac{\partial \mathcal{H}}{\partial k_i} = \frac{\partial \omega}{\partial k_i} = c_{g_i} = \frac{dx_i}{dt}, \tag{3.236}$$

so (3.235) can be written as

$$\frac{dk_i}{dt} = -\frac{\partial \mathcal{H}}{\partial x_i}. \tag{3.237}$$

This equation describes how the wavenumber associated with a localized wavepacket varies as it moves through a weakly inhomogeneous medium. Note that if $\mathcal{H}$ is independent of the x_i then it expresses the conservation of wavenumber. Taken with the result (3.236), that is,

$$\frac{dx_i}{dt} = \frac{\partial \mathcal{H}}{\partial k_i}, \tag{3.238}$$

the pair (3.237) and (3.238) constitute a form of Hamilton's equations for this system, being the characteristic equations for the corresponding Hamilton-Jacobi equation (3.233).

3.7.1 Some Consequences

The Hamiltonian and Lagrangian

From equation (3.233)

$$\frac{d\omega}{dt} = \frac{d\mathcal{H}}{dt} = \frac{\partial \mathcal{H}}{\partial k_i}\frac{dk_i}{dt} + \frac{\partial \mathcal{H}}{\partial x_i}\frac{dx_i}{dt} = 0. \tag{3.239}$$

Thus the Hamiltonian $\mathcal{H}$ (the total energy of the system) is a constant of the motion; it is independent of time. By analogy with quantum mechanics we can identify ω with the energy of the wavepacket (since $\omega = \mathcal{H}$!) and the k_i with components of the momentum vector.

Also we have that

$$\frac{d\theta}{dt} = \frac{\partial \theta}{\partial t} + \frac{\partial \theta}{\partial x_i}\frac{dx_i}{dt} \equiv \mathcal{L}, \tag{3.240}$$

where $\mathcal{L}$ is the Lagrangian defined by

$$\mathcal{L} = k_i \frac{\partial \mathcal{H}}{\partial k_i} - \mathcal{H} = k_i c_{g_i} - \omega. \tag{3.241}$$

In One Dimension

When $\mathcal{H}$ is independent of x, equation (3.237) implies the conservation of wavenumber (or equivalently, the conservation of wave crests):

$$\frac{\partial k}{\partial t} + c_g \frac{\partial k}{\partial x} = 0. \tag{3.242}$$

Equation (3.242) is just a statement that the wavenumber propagates with speed c_g. Note that $k = 2\pi/\lambda$ is the number of equal-phased lines per unit distance, so it is the density of equal phase lines. ω is the number of such phase lines passing a fixed point, so it is the flux of equal phase lines. Between two points x and $x + dx$ the net rate of outflux is $(\partial \omega/\partial x)$, while the rate of decrease of equal phase lines is $-\partial k/\partial t$, so (3.242) means (for one spatial dimension) that $k(x, t)$ is constant for an observer moving with the group speed $c_g(k)$ [34]. The characteristic equations are

$$\frac{dt}{1} = \frac{dx}{c_g} \left(= \frac{dk}{0} \right), \tag{3.243}$$

and along the characteristic

$$\frac{dx}{dt} = c_g \tag{3.244}$$

k is constant, so an observer moving with the group speed $c_g(k)$ will always keep up with a wave with wavenumber k.

A Comment About Conservation Laws

Generally, conservation laws take the form

$$\frac{\partial A}{\partial t} + \nabla \cdot \mathbf{B} = 0,$$

where A is a density term and $\mathbf{B}$ is a flux term. In particular, for conservation of wave energy density E [30] this relationship is

$$\frac{\partial E}{\partial t} + \nabla \cdot (E\mathbf{c}_g) = 0. \tag{3.245}$$

If the frequency ω is constant, the wave pattern is stationary with respect to time so that

$$\nabla \cdot (E\mathbf{c}_g) = 0. \tag{3.246}$$

As will be seen below, this has significant meaning for ray tubes (or channels in two dimensions).

A Two-Dimensional Example

Suppose we wish to examine rays propagating in the plane (e.g., for a model of rays and waves on the surface of a liquid with ω constant). Returning to more general notation for the phase (which in one spatial dimension is denoted by θ), recall that rays are perpendicular to the phase lines $S(\mathbf{r}) = S(x, y) = $ constant, and it is clear that two such adjacent rays form a ray channel (to which the group velocity is everywhere tangential). Equation (3.246) then implies that the energy fluxes through the ends of any finite length of such a channel of varying cross-sectional width (in two dimensions) are the same, so if those widths are respectively, $d\sigma_0$ and $d\sigma$, then following the channel boundary, we find the conservation of wave energy along the channel expressed as

$$\left[E c_g d\sigma \right] = \left[E c_g d\sigma \right]_0 \tag{3.247}$$

But for wave amplitude $a(x, y)$

$$E \propto |a|^2, \tag{3.248}$$

so that

$$|a|^2 c_g d\sigma = |a_0|^2 [c_g]_0 d\sigma_0; \tag{3.249}$$

hence the amplitude variation along a ray channel is given by

$$a = a_0 \left| \frac{(c_g)_0}{c_g} \frac{d\sigma_0}{d\sigma} \right|^{1/2}. \tag{3.250}$$

This equation describes the amplitude variation along narrow channels, so given a_0, a can be determined anywhere along the channel. The expression $d\sigma_0/d\sigma$ is sometimes known as the ray separation factor, for obvious reasons. Because $\mathbf{k} = \nabla S$, we have

$$|\nabla S|^2 = |\mathbf{k}|^2 = k^2, \tag{3.251}$$

or

$$\left(\frac{\partial S}{\partial x} \right)^2 + \left(\frac{\partial S}{\partial y} \right)^2 = k^2(x, y), \tag{3.252}$$

which is just another form of the familiar eikonal equation. Recall that the phase function is $S = k_0 \psi$, where k_0 is a reference wavenumber, and ψ is the eikonal. In optics, k_0 would be the wavenumber in vacuo, and k/k_0 would correspond to the refractive index n:

$$|\nabla S|^2 = k_0^2 |\nabla \psi|^2 = k^2 \Rightarrow |\nabla \psi|^2 = \frac{k^2}{k_0^2} \equiv n^2. \tag{3.253}$$

Note that on a phase line $S = $ constant, the total derivative condition

$$dS = 0 = \frac{\partial S}{\partial x} dx + \frac{\partial S}{\partial y} dy, \tag{3.254}$$

implies that

$$[y'(x)]_S = -\frac{\partial S/\partial x}{\partial S/\partial y}, \tag{3.255}$$

so on an (orthogonal) ray

$$[y'(x)]_{\text{ray}} = \frac{\partial S/\partial y}{\partial S/\partial x}. \tag{3.256}$$

Then equations (3.252) and (3.256) yield for the ray

$$(1 + y'^2) = \left(\frac{k(x, y)}{\partial S/\partial x} \right)^2, \tag{3.257}$$

and

$$\frac{ky'}{(1 + y'^2)^{1/2}} = \frac{\partial S}{\partial y},$$

(on taking the positive root). Therefore, we see that

$$\frac{d}{dx}\left\{\frac{ky'}{(1+y'^2)^{1/2}}\right\} = \frac{\partial^2 S}{\partial y \partial x} + \left(\frac{\partial^2 S}{\partial y^2}\right)y' = \left(\frac{\partial^2 S}{\partial y \partial x}\frac{\partial S}{\partial x} + \frac{\partial^2 S}{\partial y^2}\frac{\partial S}{\partial y}\right)\left(\frac{\partial S}{\partial x}\right)^{-1}$$

$$= \frac{1}{2}\left[\frac{\partial}{\partial y}(\nabla S)^2\right]\left(\frac{\partial S}{\partial x}\right)^{-1} = \left(\frac{\partial k}{\partial y}\right)(1+y'^2)^{1/2}, \qquad (3.258)$$

where $k = k(x, y(x))$. As noted in (3.253), after a suitable normalization k is akin to the refractive index in optics, and given initial conditions y_0 and y_0', can be solved for the ray trajectory.

Fermat's Principle Again

Let us define the functional for paths between points P_0 and P_1 by modifying the form of equation (3.31) as follows:

$$I = \int_{P_0}^{P_1} f\left(x, y(x), y'(x)\right)dx, \qquad (3.259)$$

and note again that I is an extremum if and only if f satisfies the Euler-Lagrange equation

$$\frac{d}{dx}\left(\frac{\partial f}{\partial y'}\right) - \frac{\partial f}{\partial y} = 0. \qquad (3.260)$$

For the choice

$$f = k(1+y'^2)^{1/2}, \qquad (3.261)$$

equation (3.260) becomes exactly (3.258), that is,

$$\frac{d}{dx}\left[\frac{ky'}{(1+y'^2)^{1/2}}\right] = \left(\frac{\partial k}{\partial y}\right)(1+y'^2)^{1/2}. \qquad (3.262)$$

Thus the eikonal equation and Fermat's principle are equivalent.

Example: Suppose we wish find the equation of a curve over which the Euclidean distance from $(0, 0)$ to $(1, 1)$ is a minimum. Thus we wish to minimize

$$I = \int_0^1 (1+y'^2)^{1/2}\,dx, \quad y(0) = 0, \ y(1) = 1. \qquad (3.263)$$

Here $F = (1+y'^2)^{1/2} = F(y')$ only, so $\partial F/\partial y = 0$. Since

$$\frac{\partial f}{\partial y'} = \frac{y'}{(1+y'^2)^{1/2}} \qquad (3.264)$$

and by the chain rule, the operator

$$\frac{d}{dx} = \frac{\partial}{\partial x} + y'\frac{\partial}{\partial y} + y''\frac{\partial}{\partial y'}, \qquad (3.265)$$

(3.260) becomes

$$\frac{d}{dx}\left(\frac{\partial f}{\partial y'}\right) = y''\frac{\partial^2 f}{\partial y'^2} = 0 \Rightarrow y'' = 0. \qquad (3.266)$$

It follows that $y(x) = ax + b$, and after using our initial conditions, we arrive at the line segment joining the two points, namely, $y = x$.

3.7.2 Wavepackets and the Group Speed Revisited

It has been claimed previously that the energy of a wave packet travels at the group velocity. It is not easy to prove in a rigorous way, but to understand the concept, we will return to the simpler case of a one-dimensional dispersive wave system. Suppose two waves with wavenumbers k_1 and k_2 start from the point $x = 0$ at $t = 0$ with group velocities $c_{g_1}(k_1)$ and $c_{g_2}(k_2)$. After some time some time, these waves will be at x_1 and x_2, where

$$x_1 = c_{g_1}(k_1)t, \text{ and } x_2 = c_{g_2}(k_2)t. \qquad (3.267)$$

This represents approximately a harmonic wave if we neglect the *slow* variations in k_i and $\omega(k_i)$ with x and t. The energy E of the wave between x_1 and x_2 may be taken as approximately proportional to the quantity

$$\int_{x_1=c_{g_1}t}^{x_2=c_{g_2}t} |\phi|^2 \, dx.$$

If we represent $\phi(x, t)$ as an integral superposition in wavenumber space,

$$\phi(x, t) = \int_{-\infty}^{\infty} A(k)e^{i(kx-\omega(k)t)} \, dk, \qquad (3.268)$$

then using the principle of stationary phase (for wavenumber k_i with spread Δk), the associated wave energy is proportional to the quantity $A^2(k_i)\Delta k$, and the energy E has the dependence

$$E \propto \int_{c_{g_1}t}^{c_{g_2}t} \frac{\pi A^2(k_i)}{t|\omega''(k_i)|} \, dx. \qquad (3.269)$$

Since $x = c_g t$, we have, for large times

$$E \propto \int_{c_{g_1}}^{c_{g_2}} \frac{\pi A^2(k_i)}{|\omega''(k_i)|} \, dc_g. \qquad (3.270)$$

In a similar fashion $A(k_i)$ and $\omega''(k_i)$ may be treated as functions of c_g, so the result depends only on $c_{g_{1,2}}$ (i.e., independent of x and t). Consequently the energy between two points of the wave starting at the origin and moving with constant speeds is independent of t, provided these speeds are the local group velocities. This means that the energy "between" k_1 and k_2 travels "between" c_{g_1} and c_{g_2}. The usual limiting process can be carried out as $k_1 \to k_2$.

Chapter Four

Ray Optics: The Classical Rainbow

Rainbows have long been a source of inspiration both for those who would prefer to treat them impressionistically or mathematically. The attraction to this phenomenon of Descartes, Newton, and Young, among others, has resulted in the formulation and testing of some of the most fundamental principles of mathematical physics.

[14]

The rainbow is a bridge between two cultures: poets and scientists alike have long been challenged to describe it. . . .Some of the most powerful tools of mathematical physics were devised explicitly to deal with the problem of the rainbow and with closely related problems. Indeed, the rainbow has served as a touchstone for testing theories of optics. With the more successful of those theories it is now possible to describe the rainbow mathematically, that is, to predict the distribution of light in the sky. The same methods can also be applied to related phenomena, such as the bright ring of color called the glory, and even to other kinds of rainbows, such as atomic and nuclear ones.

[16]

4.1 PHYSICAL FEATURES AND HISTORICAL DETAILS: A SUMMARY

The primary rainbow, which is the lowest and brightest of two that may easily be seen, is formed from two refractions and one reflection in myriads of raindrops (the path for the secondary rainbow of course involves a second reflection). For present purposes we may consider the path of a ray of light through a single drop of rain, because the geometry is the same for all such drops and a given observer. We can appreciate most of the common features of the rainbow by using the ray theory of light; the wave theory of light is needed to discuss the finer features, such as supernumerary bows (discussed below).

The first satisfactory geometrical explanation for the existence and shape of the rainbow was given by René Descartes in 1637. He was unable to account for the colors, however; it was not until 30 years later that Newton remedied this situation (this is beautifully summarized by the statement that "Descartes was able to hang the rainbow in the sky but only Newton was able to paint it." [169]) Descartes used a combination of experiment and theory to deduce that both the primary and secondary bows are caused by refraction and reflection in spherical raindrops.

He correctly surmised that he could reproduce these features by passing light through a large water-filled flask. Since the laws of refraction and reflection had been formulated some 16 years before the publication of Descartes's treatise by the Dutch scientist Snell, Descartes could calculate and trace the fate of parallel rays from the sun impinging on a spherical raindrop. Such rays exit the drop having been deviated from their original direction by varying but large amounts. The ray along the central axis will be deviated by exactly 180°, whereas above this point of entry the angle of deviation decreases until a minimum value of about 138° occurs (for yellow light; other colors have slightly different minimum deviation angles). For rays impinging still higher above the axis the deviation angle increases again. The ray of minimum deviation is called the rainbow (or Descartes) ray. The significant feature of this geometrical system is that the rays leaving the drop are not uniformly spaced: those near the minimum deviation angle are concentrated around it, whereas those deviated by larger angles are spaced more widely. Expressed differently, in a small (say, half a degree) angle on either side of the rainbow angle, there are more rays emerging than in any other one degree interval. It is this concentration of rays that gives rise to the rainbow, at least as far as its light intensity is concerned. In this sense it is similar to (but mathematically simpler than) a caustic formed on the surface of the tea in a cup when appropriately illuminated (More details about caustics will be discussed in Chapter 6).

Thus far we have been describing a generic, colorless type of rainbow. Blue and violet light get refracted more than red light; the actual amount depends on the index of refraction of the raindrop, and the calculations thereof vary a little in the literature, because the wavelengths chosen for red and violet may differ slightly. Thus for a wavelength of 656 nm (1 nm $= 10^{-9}$ m) the cone semi-angle is about 42.3°, whereas for violet light of 405 nm wavelength, the cone semi-angle is about 40.6°, an angular spread of about a 1.7° for the primary bow. Similar (though slightly wider) dispersion occurs for the secondary bow, but the additional reflection reverses the sequence of colors, so the red in this bow is on the inside of the arc. In principle more than two internal reflections may take place inside each raindrop, so higher-order rainbows (tertiary, quaternary, etc.) are possible. It is possible to derive the angular size of such a rainbow after any given number of reflections (Newton was the first to do this). Newton's contemporary, Edmund Halley, noted that the third rainbow arc should appear as a circle of angular radius about 40° around the sun itself The fact that the sky background is so bright in this vicinity, coupled with the intrinsic faintness of the bow itself, would make such a bow almost impossible, if not impossible to see or find without sophisticated optical equipment, but recently several orders beyond the secondary have been identified and photographed [17], [18], [19].

The sky below the primary rainbow is often noticeably brighter than the sky outside it; indeed, the region between the primary and secondary bow is called Alexander's dark band (named for Alexander of Aphrodisias, who studied it in connection with Aristotle's (incorrect) theory of the rainbow). Raindrops scatter incident sunlight in essentially all directions, but a rainbow arises because of a concentration of such scattered light in a particular angular region of the sky. The reason the inside of the primary bow (i.e., inside the cone) is bright is that all the raindrops in the interior of the cone scatter light to the eye also (some occurring from direct reflections at their surfaces), but it is not as intense as the rainbow light and is composed of many colors intermixed. Much of the scattered light, then, comes from raindrops through which sunlight is refracted and reflected: these rays do not emerge between the 42° and 51° angles. Despite the prediction of ray optics this dark angular band is not completely dark, of course, because the surfaces of raindrops reflect light into it; the reduction of intensity, however, is certainly noticeable.

Another commonly observed feature of the rainbow is that when the sun is near the horizon, the nearly vertical arcs of the rainbow near the ground are often brighter than the upper part of the arc. The reason for this appears to be the presence of drops with varying sizes. Drops smaller than, say, 0.2–0.3 mm (about 0.01 inch) are spherical: surface tension is quite sufficient to keep the distorting effects of aerodynamic forces at bay. Larger drops become more oblate in shape, maintaining a circular cross-section horizontally but not vertically. They can contribute significantly to the intensity of the rainbow because of their size (≈ 1 mm or larger) but can only do so when they are "low" on the cone, for the light is scattered in a horizontal plane in the exact way it should to produce a rainbow. These drops do not contribute significantly near the top of the arc because of their noncircular cross-section for scattering. Small drops, in contrast, contribute to all portions of the rainbow. Drop size, as implied above, can make a considerable difference in the intensity and color of the rainbow. The best bows are formed when the drop diameter is $\gtrsim 1$ mm; as the size decreases, the coloration and general definition of the rainbow becomes poorer. Ultimately, when the drops are about 0.05 mm or smaller in diameter, a broad, faint, white arc called a fogbow occurs. When sunlight passes through these very tiny droplets the phenomenon of diffraction becomes important. Essentially, due to the wave nature of light, interactions of light with objects comparable to (or not too much larger than) a typical wavelength causes a light beam to spread out. Thus the rainbow colors are broadened and overlap, giving rise in extreme cases to a broad white fogbow (or cloudbow, since droplets in those typically produce such bows). These white rainbows may sometimes be noted while flying above a smooth featureless cloud bank. The rainbow cone intersects the horizontal cloud layer in a hyperbola if the sun's elevation is less than 42° (or an ellipse if it is greater than this), as familiarity with the conic sections assures us. This phenomenon can also be seen as a "dewbow" on a lawn when the sun is low in the eastern sky. Other related phenomena of interest, such as reflected-light rainbows (produced by reflection from a surface of water behind the observer) and reflected rainbows (produced by reflection from a surface of water in front of the observer) have also been observed and discussed [20], [212].

4.2 RAY THEORY OF THE RAINBOW: ELEMENTARY MATHEMATICAL CONSIDERATIONS

We start by introducing different classes of rays in a more formal way than when the topic was introduced earlier. A ray that is reflected from the outer surface of the

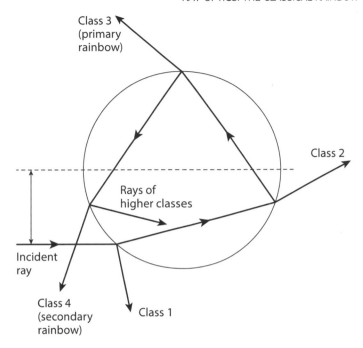

Figure 4.1 Several ray classes.

drop is of Class 1 [16]; of course, part of that "ray" (crudely speaking) is refracted into the drop, and part of that is transmitted as well (Figure 4.1). Such rays—for direct transmission through the drop—are of Class 2; Class 3 rays undergo a single internal reflection at the back of the drop before being refracted and transmitted out, and so on. The primary rainbow is seen in the direction corresponding to the most dense clustering of Class 3 rays leaving the drop after a single internal reflection. Correspondingly higher-order rainbows exist (in principle) at similar "ray clusters" for rays of Class 4 and greater, that is, those undergoing $p - 1 = k$ internal reflections, $p = 3, 4, 5, \ldots$. The class number is always $p + 1$, since equivalently, p is the number of times a ray traverses the droplet.

Referring to Figures 4.2 and 4.3 for guidance, we consider the more general angular deviation D_{p-1} between the incident and emergent rays after $p - 1$ internal reflections ($p > 1$). It was noted in Chapter 1 that, despite the notation θ in the scientific literature for the angle of incidence of a ray on a spherical droplet, the classical notation i will be used here (with the exception of scattering-related descriptions using the impact parameter in Chapter 5). Since each such reflection causes a deviation of $(\pi - 2r)$ radians, where $r = r(i)$ is the angle of refraction (a function of the angle of incidence), and two refractions always occur (one on entering the drop and the other on exiting), the total deviation is

$$D_{p-1}(i) = 2(i - r) + (p - 1)(\pi - 2r)$$

$$\equiv (p - 1)\pi + 2i - 2p \arcsin\left(\frac{\sin i}{n}\right), \tag{4.1}$$

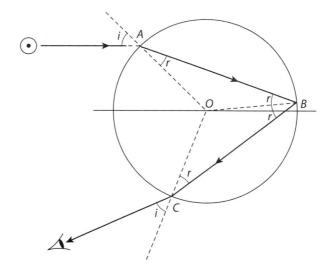

Figure 4.2 Ray trajectory for the primary bow.

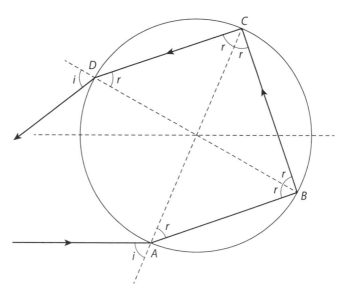

Figure 4.3 Ray trajectory for secondary bow.

on using Snell's law of refraction for refractive index n. Thus for the primary rainbow ($p = 2$),

$$D_1(i) = \pi + 2i - 4r, \tag{4.2}$$

and for the secondary rainbow ($p = 3$),

$$D_2 = 2i - 6r \text{ (modulo } 2\pi), \tag{4.3}$$

and so on. Seeking an extremum of $D_{p-1}(i)$ and using Snell's law of refraction twice (once in differentiated form), it is readily shown that for all p, $D_{p-1}(i)$ is an extremum when

$$\frac{\cos i}{\cos r} = \frac{n}{p}, \tag{4.4}$$

or after some rearrangement,

$$\cos i = \left(\frac{n^2 - 1}{p^2 - 1}\right)^{1/2}, \quad p > 1. \tag{4.5}$$

where n is the index of refraction for the incident ray (assumed monochromatic for the present). For this value of i, i_c say, $D_{p-1} = D_{p-1}^c$, so that for $p = 2$ in particular

$$i_c = \arccos\left(\frac{n^2 - 1}{3}\right)^{1/2}, \tag{4.6}$$

and so

$$D_1^c = \pi + 2i_c - 4\arcsin\left(\frac{\sin i_c}{n}\right) \tag{4.7}$$

$$= \pi + 2\arcsin\left(\frac{4 - n^2}{3}\right)^{1/2} - 4\arcsin\left(\frac{4 - n^2}{3n^2}\right)^{1/2}. \tag{4.8}$$

The ray for which $i = i_c$ is the *rainbow ray* (or Descartes ray) (see Figure 4.4, where it is identified as ray 7).

   For water, $1 < n < 2$ so that expression (4.8) is well defined, but nevertheless in general, care must be taken when considering the arguments of these inverse trigonometric functions. That the angle D_1^c is indeed a minimum for $p = 2$ follows from the fact that (after some rearrangement)

$$D_{p-1}''(i) = \frac{2p(n^2 - 1)\tan r}{n^3 \cos^2 r} \tag{4.9}$$

[21], which is positive since $n > 1$ for $r \in (0, \pi/2)$, so the clustering of deviated rays corresponds to a minimum deflection. Note also that the generalization of equation (4.8) to $p - 1$ internal reflections is

$$D_{p-1}^c(i_c) = (p-1)\pi + 2\arcsin\left(\frac{p^2 - n^2}{p^2 - 1}\right)^{1/2} - 2p\arcsin\left[\frac{1}{n}\left(\frac{p^2 - n^2}{p^2 - 1}\right)^{1/2}\right], \tag{4.10}$$

with appropriate restrictions on n where necessary.

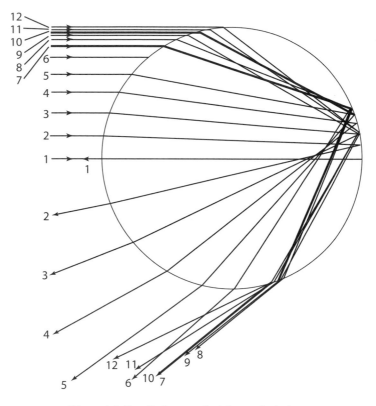

Figure 4.4 Ray 7: the ray of minimum deviation.

The Extrema Are Minima

The deviation of a ray with angle of incidence i after $p - 1 = k$ reflections in a spherical drop of radius a is given by the expression

$$D(i) = 2(i - r) + k(\pi - 2r),\tag{4.11}$$

and clearly $D'(i) = 0$ when

$$\frac{dr}{di} = \frac{1}{k + 1}.\tag{4.12}$$

From Snell's law of refraction for constant refractive index n,

$$\sin i = n \sin r,\tag{4.13}$$

it follows that

$$\frac{dr}{di} = \frac{\cos i}{n \cos r},\tag{4.14}$$

so

$$(k + 1)\cos i = n \cos r.\tag{4.15}$$

On squaring and adding equations (4.13) and (4.15), we obtain the condition for an extremum of $D(i)$ to occur:

$$\cos i = \left(\frac{n^2 - 1}{k^2 + 2k} \right)^{1/2}. \tag{4.16}$$

We also note that

$$D''(i) = -2 (k + 1) \frac{d^2 r}{di^2}. \tag{4.17}$$

From (4.14), after some manipulation

$$\frac{d^2 r}{di^2} = \frac{(1 - n^2) \sin i}{n^3 \cos^3 r} < 0, \tag{4.18}$$

and hence

$$D''(i) = 2 \frac{(k + 1) (n^2 - 1) \sin i}{n^3 \cos^3 r} > 0, \quad k = 1, 2, 3, \dots \tag{4.19}$$

so the extrema are all minima.

Generalization of a Result of Huygens

In 1652 Christiaan Huygens derived a result for $p = 2$ as a function of the (constant) refractive index n only, that is, without any reference to i or i_m (stated in [26], though no reference is provided). Thus he found that

$$D_1^m = 2 \arccos \left[\frac{1}{n^2} \left(\frac{4 - n^2}{3} \right)^{3/2} \right]. \tag{4.20}$$

For $n = 4/3$, it is readily verified that $D_1^m \approx 138°$, as expected. It is also of interest to express the deviation for higher p values in similar terms. This has been carried out [118], and although it is algebraically and trigonometrically intensive, it can be expressed in this form for any positive integer $p = 2, 3, 4, \dots$.

Exercise: Prove the result (4.20)

Exercise: Prove that

$$D_2^m = 2 \arcsin \left\{ (n^2 - 1)^{1/2} \left[\frac{(9 - n^2)^{1/2}}{2n} \right]^3 \right\}. \tag{4.21}$$

Exercise: Prove that

$$D_3^m = 2 \arccos \left\{ \frac{1}{5 (15)^{3/2}} \left[\frac{(16 - n^2)^{3/2}}{n^4} \right] (27n^2 - 32) \right\}. \tag{4.22}$$

4.2.1 Some Numerical Values

Returning to the primary rainbow for which $p = 2$, since the minimum angle of deviation depends on n, we now incorporate dispersion into the model. For red light of wavelength 656 nm in water [22], $n \approx 1.3318$, and for violet light of wavelength 405 nm, $n \approx 1.3435$. These correspond to $D_1^m \approx 137.7°$ (i.e., a rainbow cone semi-angle, the supplement of D_1^m, of $\approx 42.3°$) and $D_1^m \approx 139.4°$ (cone semi-angle of $\approx 40.6°$), respectively. This gives a theoretical angular width of $\approx 1.7°$ for the primary rainbow; in reality it is about 0.5° wider than this, because this is the angular diameter of the sun for an earth-based observer, and hence the incident rays can deviate from parallellism by this amount (and more to the point, so can the emergent rays, as a simple differential argument shows). The abovementioned cone semi-angle for a given D_{p-1}^m is the supplementary angle $180° - D_{p-1}^m$; this is the angle of elevation relative the sun-observer line and for the primary rainbow is about 42°. Light deviated at angles larger than this minimum will illuminate the sky inside the rainbow more intensely than outside it, for this very reason. Thus the outside edge of the primary bow will generally be more sharply defined than the inner edge. For the secondary rainbow ($p = 3$) $D_2^m \approx 231°$ and the cone semi-angle in this case is $D_2^m - 180°$, or approximately 51° for light of an orange color, for which we choose $n = 4/3$ (the angle of minimum deviation for $p = 3$ is $\approx 129°$). Each additional reflection of course is accompanied by a loss of light intensity because of transmission out of the drop at that point, so on these grounds alone, it would be expected that the tertiary rainbow ($p = 4$) would be difficult to observe as noted above. In that case $D_{min} \approx 319°$, and the light concentration therefore is at about an angle of 41° from the incident light direction. As noted already, in principle this will appear behind the observer as a ring around the sun. Due to (i) the increased intensity of sunlight in this region; (ii) the fact that the angles of incidence i_m increase with p and result in a reduction of incident intensity per unit area of the surface, and hence also a reduction for the emergent beam; (iii) higher-order rainbows are wider than orders one and two; (iv) the presence of light reflected from the outer surface of the raindrops (direct glare); (v) light emerging with no internal reflections (transmitted glare); and (vi) the reduced intensity from three reflections, it is not surprising that such rainbows (i.e., $p \geq 4$) have not been reliably reported in the literature until very recently [18], [19].

Thus far we have examined only the variation in the deviation of the incident ray as a function of the angle of incidence (or as a function of the normalized impact parameter). The intensity of the scattered light will also vary with this angle, but even within the limitations of geometrical optics, there are several other factors to consider. One of these is the reduction in intensity of the rays with each successive refraction and reflection; another is the behavior of the coefficients of reflection and refraction as a function of angle of incidence; and a third is to determine how much light energy, as a function of angle of incidence, is deviated into a given solid angle after interacting with the raindrop (i.e., the spreading of light on emerging from the drop). This might be thought of as a classical "scattering cross section" type of problem. Other effects, outside the realm of geometrical optics, are the effects of diffraction and interference. The first two factors are related and will be dealt with first by examining the polarization of the rainbow.

4.2.2 Polarization of the Rainbow

Electromagnetic radiation—specifically, light—is propagated as a transverse wave, and the orientation of this oscillation (or ray) can be expressed as a linear combination of two basis vectors for the space, namely, mutually perpendicular components of two independent linear polarization states [16]. Sunlight is unpolarized (or perhaps more accurately, randomly polarized), being an incoherent mixture of both states, but reflection can and does alter the state of polarization of an incident ray of light. For convenience we can consider the two polarization states of the light incident on the back surface of the drop (i.e., from within) as being, respectively, parallel to and perpendicular to the plane containing the incident and reflected rays. Above a critical angle of incidence (determined by the refractive index) both components are totally reflected, although some of the light does travel around the surface as an "evanescent" wave; this principle will be discussed in more detail and in connection with atmospheric waves in Chapter 14. This critical angle of incidence i_{cr} (not to be confused with the critical angle for the Descartes ray) from within the drop is defined by an expression that is easily obtained from Snell's law of refraction, namely,

$$i_{cr} = \arcsin(n^{-1}), \tag{4.23}$$

where n is the relative refractive index of the water droplet (i.e., n is the ratio of the refractive indices of water and air, both relative to a vacuum). Taking a generic value for n of 4/3 (orange light) we find $i_{cr} \approx 48.6°$. At angles i less than i_{cr} the parallel component of polarization is reflected less efficiently than its perpendicular counterpart, and at the so-called *Brewster angle* it is entirely transmitted, leaving only perpendicularly polarized light to be reflected (partially) from the inside surface of the drop. Since the Brewster angle, as shown below, is close to the angle of total internal reflection i_{cr}, the light that goes on to produce the rainbow is strongly perpendicularly polarized. At the Brewster angle the reflected and refracted rays are orthogonal (and hence complementary), and therefore it follows from Snell's law that

$$i_B = \arctan(n^{-1}), \tag{4.24}$$

(remember this is from within the drop), from which $i_B \approx 36.9°$, differing from i_{cr} by about 12°. If we were to consider this for an air-to-water interface, n^{-1} in equation (4.24) should be replaced by n; the corresponding value i_B above is the supplement of the above angle (i.e., $\approx 53.1°$). The coefficient of reflection of light depends on its degree of polarization: consider first the case of light polarized perpendicular to the plane of incidence, with the amplitude of the incident light taken as unity. It follows from the Fresnel equations that the fraction of the incident intensity reflected is given by [108]

$$R_\perp = \frac{\sin^2(i - r)}{\sin^2(i + r)}, \tag{4.25}$$

and

$$R_{\parallel} = \frac{\tan^2(i - r)}{\tan^2(i + r)} \tag{4.26}$$

if the light is polarized parallel to the plane of incidence. It follows that for a ray entering a drop and undergoing $p - 1 = k$ internal reflections plus two refractions at the spherical boundary (corresponding to the first term below), the fraction of the original intensity remaining in the emergent ray is, for perpendicular polarization,

$$I_{\perp k} = \left[1 - \left(\frac{\sin^2(i - r)}{\sin^2(i + r)}\right)\right]^2 \left(\frac{\sin^2(i - r)}{\sin^2(i + r)}\right)^{2k}, \tag{4.27}$$

and for parallel polarization

$$I_{\parallel k} = \left[1 - \left(\frac{\tan^2(i - r)}{\tan^2(i + r)}\right)\right]^2 \left(\frac{\tan^2(i - r)}{\tan^2(i + r)}\right)^{2k}. \tag{4.28}$$

The total fraction is the sum of these two intensities. For angles of incidence close to zero, for which $i \approx nr$, it follows that the single-reflection intensity coefficients for reflection and refraction become, from expression (4.25)

$$\left(\frac{n - 1}{n + 1}\right)^2 \tag{4.29}$$

and

$$\frac{4n}{(n + 1)^2}, \tag{4.30}$$

respectively. For the choice of $n = 4/3$ under these circumstances it follows that the reflection and refraction coefficients are approximately 0.02 and 0.98, respectively. Notice from equation (4.26) that the fraction of the intensity of reflected light is zero (for parallel polarization) when the denominator in the coefficient vanishes (i.e., when $i + r = \pi/2$); at this point the energy is entirely transmitted, so the internally reflected ray is completely perpendicularly polarized. From this it follows that

$$\sin i = n \sin r = n \cos i, \tag{4.31}$$

whence

$$\tan i = \tan i_B = n \tag{4.32}$$

for external reflection, and equation (4.24) holds for internal reflection. In the latter case, for $n = 4/3$, this yields $i_B \approx 36.9°$, the Brewster polarizing angle, as pointed out above. Note that as $i + r$ passes through $\pi/2$, the tangent changes sign (obviously being undefined at $\pi/2$); this corresponds to a phase change of 180°. From equations (4.27) and (4.28) it can be shown that the primary rainbow is about 96 percent polarized and the secondary bow somewhat less so: approximately 90 percent.

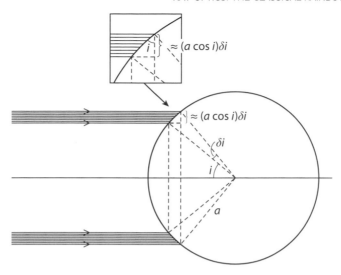

Figure 4.5 A ray bundle with enlarged regions shown in Figures 4.6 and 4.7.

4.2.3 The Divergence Problem

Now we examine essentially the differential scattering cross section for this problem. For the present purposes this is the amount of incident radiation that is scattered by the front surface of the drop into a small solid angle. Consider a thin bundle of rays impinging a spherical raindrop (of radius a), as shown in Figures 4.5, 4.6, and 4.7. The angles of incidence of the rays compasing this bundle lie in the interval $(i, i + \delta i)$. The diagram in Figure 4.5 is cylindrically symmetric about an axis parallel to the direction of the incoming rays, so the area seen by the cylindrical ray bundle is

$$\delta A \approx (2\pi a \sin i)(a \cos i) |\delta i| = \pi a^2 |\delta i| \sin 2i. \tag{4.33}$$

These rays are scattered into an angular interval $(\theta, \theta + \delta\theta)$, which then occupy a solid angle to be determined below. Associated with this small interval is an "onion ring" surface area element δS (See Figures 4.5 and 4.7)

$$\delta S \approx (2\pi a \sin \theta)(a |\delta\theta|) = 2\pi a^2 \sin \theta |\delta\theta|, \tag{4.34}$$

which is equivalent to a solid angle element

$$\delta\Omega = 2\pi \sin \theta |\delta\theta|. \tag{4.35}$$

Now if I_0 is the rate at which the incident light energy falls on a unit area perpendicular to its incoming direction, then the corresponding rate at which it enters a unit solid angle on emerging from the drop is

$$I \approx I_0 \frac{(\pi a^2 \sin 2i) |\delta i|}{(2\pi \sin \theta) |\delta\theta|}, \tag{4.36}$$

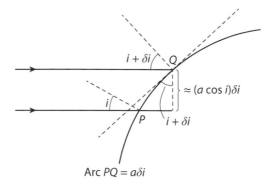

Figure 4.6 Detail In Figure 4.5.

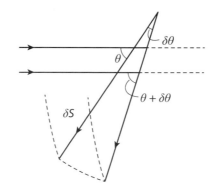

Figure 4.7 The scattering interval $(\theta, \theta + \delta\theta)$.

and passing to the limit, we find that the relative energy rate is

$$\frac{I}{I_0} = \frac{a^2 \sin 2i}{2 \sin \theta} \left| \frac{d\theta}{di} \right|^{-1}. \tag{4.37}$$

Of course, this refers to the emerging intensity at the surface of the drop, so at an external distance r, a factor r^{-2} must be inserted in this expression. We have already noted that the deviation of an incoming ray is denoted by $D_{p-1}(i)$, and that is just $\pi - \theta (i)$ here; so the condition for an extremum in $D_{P-1}(i)$ (defining the rainbow angle) is just $D'_{p-1}(i) = 0$, which immediately implies that $\theta' (i) = 0$, and explains the singularity arising at the rainbow caustic when geometrical optics is used. There is still a little more to be said as far as geometrical optics is concerned, based on equation (4.37). It would appear that it also predicts an infinite intensity when, in degrees, $\theta = 0°$ and $180°$ (i.e. $D(i) = 180°$ and $0°$, respectively). But θ (and hence D) only takes on these values when the angle of incidence $i = 0$, and, specifically, for one internal reflection (for which $\theta = 0°$)

$$\frac{\sin 2i}{\sin \theta} \approx \frac{2i}{\theta}. \tag{4.38}$$

But when $i \approx 0°$

$$\theta = 180° - D(i) \approx 180° - (2i - 4r + 180°) = 4r - 2i. \qquad (4.39)$$

Therefore, since now Snell's law reduces to $i \approx nr$,

$$\frac{\sin 2i}{\sin \theta} \approx \frac{i}{2r - i} \approx \frac{nr}{2r - nr} = \frac{n}{2 - n}. \qquad (4.40)$$

For water, $n \approx 4/3$ and this ratio is then equal to two. From equation (4.39),

$$\frac{d\theta}{di} = 4\frac{dr}{di} - 2 = \frac{4}{n} - 2 = 1 \qquad (4.41)$$

in this limit of small angles of incidence, so there is no significant increase in I in this situation. Finally, it is appropriate to note that a better approximation to the actual intensity distribution in the vicinity of the rainbow angle is provided by so-called *Airy theory*. This, though not a perfect observational fit, incorporates the wave nature of light, and was developed, and in 1838 published by the then Astronomer Royal in Britain, Sir George Biddell Airy [23]. We shall pursue his ideas in Chapter 5.

4.3 RELATED TOPICS IN METEOROLOGICAL OPTICS

4.3.1 The Glory

Mountaineers and hill climbers have noticed on occasion that when they stand with their backs to the low-lying sun and look into a thick mist below them, they may see a set of colored circular rings (or arcs thereof) surrounding the shadow of their heads. Although an individual may see the shadow of a companion, the observer will see the rings only around his or her head. This is the phenomenon of the glory, initially referred to in the scientific literature as the *anticorona* for reasons that will become apparent below. A brief but very useful historical account along with a summary of the theories, can be found in [26]. Early one morning in 1735, a small group of people were gathered on top of a mountain in the Peruvian Andes. They were members of a French scientific expedition sent out to measure a degree of longitude and were led by two gentlemen named Bouguer (a French mathematician and scientist whom we encountered in Chapter 3 in connection with Bouguer's law (3.94) and (3.96)), and La Condamine; a Spanish captain named Antonio de Ulloa also accompanied them. They saw an amazing sight that morning. According to Bouguer this was

> a phenomenon which must be as old as the world, but which no one seems to have observed so far. . . . A cloud that covered us dissolved itself and let through the rays of the rising sun. . . . Then each of us saw his shadow projected upon the cloud. . . . The closeness of the shadow allowed all its parts to be distinguished: arms, legs, the head.

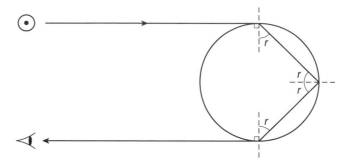

Figure 4.8 Incorrect ray geometry for the glory.

What seemed most remarkable to us was the appearance of a halo or glory around the head, consisting of three or four small concentric circles, very brightly colored, each of them with the same colors as the primary rainbow, with red outermost [270]

Ulloa gave a similar description and also drew a picture. In his account he said

The most surprising thing was that, of the six or seven people that were present, each one saw the phenomenon only around the shadow of his own head, and saw nothing around other people's heads. [270]; (see also [271] and [162] for more details)

During the nineteenth century, many such observations of the glory were made from the top of the Brocken mountain in central Germany, and it became known as the Brocken bow or the "Specter of the Brocken" (being frequently observed on this peak, the highest in the Harz mountains of central Germany). It also became a favorite image among the Romantic writers; it was celebrated by Coleridge in his poem "Constancy to an Ideal Object." Other sightings were made from balloons, the glory appearing around the balloon's shadow on the clouds. Nowadays, while not noted as frequently as the rainbow, it may be seen most commonly from the air, with the glory surrounding the shadow of the airplane. Once an observer has seen the glory, if looked for, it is readily found on many subsequent flights (provided one is on the shadow side of the aircraft!). Some beautiful color photographs have appeared in the scientific literature, many of which can be found at the website http://www.atoptics.co.uk/. The phenomenon can be understood in the simplest terms as essentially the result of light backscattered by cloud droplets, the light undergoing some unusual tranformations en route to the observer (with a correspondingly complicated mathematical description), transformations that are *not* predictable by standard geometrical optics, unlike the basic description of the rainbow. That this must be the case is easily demonstrated by noting a fallacy present in at least one popular meteorological text. The glory, it is claimed, is formed as a result of a ray of light tangentially incident on a spherical raindrop being refracted into the drop, reflected from the back surface and reemerging from the drop in an exactly antiparallel direction into the eye of the observer (see Figure 4.8).

If such a picture were correct, then since the angle of incidence of the ray is 90°, it follows by Snell's law of refraction that the angle of refraction is

$$r = \arcsin\left(n^{-1}\right), \tag{4.42}$$

where n is the refractive index of the raindrop. For an air/water boundary, $n \approx 4/3$ (ignoring the effects of dispersion here; note also that for a water/air boundary the reciprocal of n must be used), and so $r \approx 48.6°$. This means (by the law of reflection) that at the back of the drop, the ray is deviated by less than a right angle, since $2r \approx 97.2°$, and by symmetry the angle of incidence *within* the drop for the exiting ray is also 48.6°, so the total deviation angle (as we saw for the primary rainbow) is

$$D_1(i) = \pi + 2i - 4\arcsin\left(\frac{\sin i}{n}\right) = -4\arcsin\left(\frac{3}{4}\right) \approx -194.4° \text{ or } +165.6°,$$
$$\tag{4.43}$$

(modulo 2π) since $i = 90°$. This means that the exiting ray is about 14° short of being antiparallel. It just won't work as a mechanism for the glory! There are basically two potential ways out of this. We could ask what value of refractive index n would be necessary for the diagram to be correct; thus $i = 90°$; this means that

$$n = \frac{\sin i}{\sin r} = \sqrt{2} \approx 1.4. \tag{4.44}$$

Something between water and glass might do it (perhaps plastic). But absent any evidence that clouds are composed of transparent plastic spheres we discard this suggestion! What remains? The other possibility is that somehow the ray travels around the surface (as a surface wave) for part (or parts) of its trip, the surface portion comprising the missing 14.4° for $r \approx 48.6°$:

$$\theta = 180° - 2(180° - 2r) = 4r - 180° \approx 14.4.° \tag{4.45}$$

The detailed theory for the glory, using these ideas, has been verified experimentally, though it must be pointed out that a complete theoretical explanation of the glory has been one of the most challenging problems of meteorological optics in the past century (see [26] and [54]; for further references, see [25]). However, a lovely description of the phenomenon has been published [137], which points out the more obvious features seen by an observer. Unlike the coronas sometimes seen around the moon or sun when they are shrouded in cloud, which are caused by diffraction, the glory is a primarily backscattering phenomenon (though subtle interference effects do occur) as indicated earlier. Photographs of the glory [137] as an aircraft descends toward its destination illustrate the fact that the angular diameter of the glory is independent of the observer/cloud distance (unlike the shadow of the aircraft inside the glory rings). It is the diameter of the cloud droplets that determine the angular size of the glory, and these are typically 15–25 microns (15–25×10^{-6} m). Unlike the corona discussed below, the light from a glory is strongly polarized [17].

4.3.2 Coronas (Simplified)

It was mentioned at the beginning of this section that the glory used to be referred to as an anticorona. It is sensible to ask: what is a *corona*? By this term is meant the atmospheric phenomenon, not the tenuous outer atmosphere of the sun (the solar corona), which is visible at the time of a total solar eclipse (neither is it a Mexican beer). A corona is a set of diffuse colored rings (with red outermost) similar in appearance to the glory, created by the *diffraction* of light by cloud droplets (usually in altostratus or altocumulus clouds). The size of the rings depends on the drop size; the largest rings are produced by the smallest drops and vice versa. A beautiful phenomenon sometimes visible on a clear night with a full moon (or close to full) is a lunar corona. It consists of several softly colored rings often apparent when only a layer of thin cloud covers the moon without obscuring it. Another related diffraction phenomenon is that of *iridescence* (see below), seen as irregular pale-colored patches of light on clouds that are displaced somewhat from the direction of the sun (or moon). They can be an exquisitely beautiful sight. A corona can also appear around the sun (or indeed, in principle around any sufficiently bright extended source of light), but this should never be looked for directly, because of the danger of permanent damage to eyesight.

What causes a corona? The culprit, so to speak, is diffraction: a consequence of light being obstructed by an obstacle comparable in size to the wavelength of the light. In this case, the obstacle(s) are myriads of cloud droplets, ranging in size from diameters in the range 1–100 microns, with mean diameters of between 10 and 15 microns. While this is an order of magnitude larger than the wavelengths (λ) of visible light (0.45–0.70 microns) it is still sufficient for the effects of diffraction to be apparent, much like the bending of sound around pillars in an auditorium. The much larger raindrops do not give rise to observable diffraction phenomena, though of course, they do "scatter" light to produce the rainbow. The intensity $I(\theta)$ of diffracted light as a function of scattering angle θ for an obscuring disk of radius a is, for small scattering angles θ,

$$I(\theta) = I(0) \left(\frac{2J_1(x)}{x} \right)^2, \tag{4.46}$$

where $I(0)$ is the intensity of light in the forward direction ($\theta = 0$), which is proportional to the area of the disk, and

$$x = \frac{2\pi a \sin \theta}{\lambda}, \tag{4.47}$$

or

$$\theta = \arcsin \left(\frac{x\lambda}{2\pi a} \right). \tag{4.48}$$

More details for a broader angular range can be found in Part III of the book, specifically in Chapter 19 on electromagnetic scattering and Chapter 20 on diffraction. Clearly, for a given value of θ, x is proportional to the ratio of the circumference of the obscuring disk to the wavelength of light impinging on it. The quantity (4.46) is oscillatory, but with a large central amplitude; successive maxima are very much smaller, corresponding to a rapid decrease of ring intensity as we look radially outward from the aureole. In fact, for $x \neq 0$ the first and second maxima are respectively only 1.75 percent and 0.42 percent of the central maximum at $x = 0$. These occur respectively at $x = 5.14$ and $x = 8.42$, from which we can determine that the angles θ_1 and θ_2 for the first and second maxima are given by

$$\theta_1 = \arcsin\left(\frac{x\lambda}{2\pi a}\right) = \arcsin\left(\frac{5.14\lambda}{2\pi a}\right) = \arcsin\left(0.82\frac{\lambda}{a}\right); \qquad (4.49)$$

$$\theta_2 = \arcsin\left(\frac{8.42\lambda}{2\pi a}\right) = \arcsin\left(1.34\frac{\lambda}{a}\right). \qquad (4.50)$$

Of course, it must be remembered that just as a rainbow is the result of scattering of sunlight by myriads of raindrops, so the lunar corona is a result of diffraction by a great many cloud droplets. When these droplets are all more or less the same size, these diffraction effects reinforce each other and the resulting coronae are usually bright and well delineated. This degree of droplet uniformity is present when the cloud is rather new, (i.e., the droplets have the same history), so small environmental differences affecting individual droplets have not had time to accumulate. Such conditions are found most often in altocumulus, cirrocumulus and lenticular clouds. In contrast, if there is a wide range of droplet size in the cloud, the coronae tend to be washed out, because the diffraction patterns overlap, and only the aureole is easily visible. Another variation occurs if the size distribution is narrow, but differs from region to region in the cloud; this may result in coronal arcs of different radii, and hence noncircular, indeed, irregular arcs may be seen. If the patches are sufficiently random, the corona may result in what is referred to as cloud iridescence; this is frequently seen during the day in clouds in the vicinity of the sun.

4.3.3 Rayleigh Scattering—a Dimensional Analysis Argument

Here we use standard methods of dimensional analysis for a plausibility argument (see [122]). Let the intensity I of the radiation scattered from a particle of volume V be described by the expression

$$I = f(V, r, \lambda, n_1, n_2, I_0), \qquad (4.51)$$

where $r, \lambda, n_1, n_2, I_0$ are respectively the distance from the scatterer to the observation point, the wavelength of the scattered radiation, the refractive indices of the exterior and interior media, and the intensity of the incident radiation. The refractive indices, being ratios of speeds of light in different media, are dimensionless, and to be specific, we seek the following algebraic functional decomposition:

$$I \propto V^\alpha r^\beta \lambda^\gamma I_0. \qquad (4.52)$$

A particularly simple result follows for the dimensions of both sides of this equation in terms of the basis vector $[L]$ for length, namely,

$$[L]^{3\alpha}[L]^{\beta}[L]^{\gamma} = [L]^0, \qquad (4.53)$$

because the dimensions of intensity cancel from both sides. Since the amplitude of scattered radiation is proportional to the number of scatterers, which in turn is proportional to the volume of the (composite) particle, and intensity is equal to the square of the amplitude, we have that $\alpha = 2$. Furthermore, $\beta = -2$ since a dipole (e.g., a molecule in which the positive and negative charges are not coincident) radiates energy in all directions (think of the surface of an expanding sphere of light; the energy per unit area decreases as $(\text{radius})^{-2}$). This means that $6 - 2 + \gamma = 0$, or $\gamma = -4$. Thus, in particular,

$$I_{\text{scatt}} \propto \lambda^{-4}, \qquad (4.54)$$

which is effectively *Rayleigh's inverse fourth-power law of scattering*.

Rayleigh scattering theory applies if the particles are small (i.e., $\lesssim 0.1\lambda$); under these circumstances they can be considered as point dipoles. Because the particle is much larger than this size, the light scattered from one part of the particle may be out of phase with that from another part; the resulting interference reduces the intensity of the scattered radiation. There is now a greater intensity in the forward direction, because the cumulative effect of the phase differences is smallest for small scattering angles. Mie scattering theory (or more accurately, the Mie solution) as will be seen in Chapter 18, takes into account all these features, and using Maxwell's equations of electromagnetic theory, the scattering intensity can be expressed as an infinite series of so-called *partial waves*; Rayleigh scattering is represented by the first term in this series, so it is certainly a special (or limiting) case of the Mie solution. For larger particles, more terms have to be included; indeed for rainbows (produced by light scattering from raindrops—see Chapter 5) the number of terms required is of the order of *five thousand*—the ratio of the drop circumference to the wavelength.

Chapter Five

<hr>

An Improvement over Ray Optics: Airy's Rainbow

[Airy's theory] did go beyond the models of the day in that it quantified the dependence upon the raindrop size of (i) the rainbow's angular width, (ii) its angular radius, and (iii) the spacing of the supernumeraries. Also, unlike the models of Descartes and Newton, Airy's predicted a non-zero distribution of light intensity in Alexander's dark band, and a finite intensity at the angle of minimum deviation (as noted above, the earlier theories predicted an infinite intensity there). However, spurred on by Maxwell's recognition that light is part of the electromagnetic spectrum, and the subsequent publication of his mathematical treatise on electromagnetic waves, several mathematical physicists sought a more complete theory of scattering, because it had been demonstrated by then that the Airy theory failed to predict precisely the angular position of many laboratory-generated rainbows.

[149]

5.1 THE AIRY APPROXIMATION

5.1.1 Some Ray Prerequisites

One of the subtle yet common features of a rainbow that ray theory cannot account for is the existence of *supernumerary bows*. These bows are essentially the result of self-interference in waves emerging from the raindrop close to the rainbow angle (i.e., the angle of minimum deviation). In general they will have entered at different angles of incidence and traversed different paths in the denser medium; there is of course a reduction in wavelength inside the drop, but the overall effect of different path lengths along the wavefront is the usual diffraction pattern arising as a result of the destructive/constructive interference between portions of the waves. The spacing between the maxima (or minima) depends on the wavelength of the light and the diameter of the drop; the smaller the drop, the greater the spacing will be. Indeed, if the drops are less than about 0.2 mm in diameter, the first maximum will be distinctly below the primary bow, and several other such maxima may be distinguished if conditions are conducive.

Although this phenomenon is decidedly a wavelike one, we can gain some heuristic insight into the underlying mechanism by temporarily continuing to use

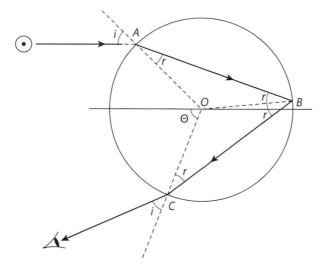

Figure 5.1 Ray trajectory for the primary bow.

a geometrical optics approach to the relevant rays and their associated wavefronts. This can be done by considering no longer the angle of minimum deviation, but the *point of emergence* of a ray from the raindrop as the angle of incidence is increased. A careful examination of Figure 5.1 reveals that as the angle of incidence is increased away from zero, and the point of entry moves clockwise around the drop, the point of emergence C moves first in a counterclockwise direction, reaches an extreme position, and then moves back in a clockwise direction. This extreme point has significant implications for the shape of the wavefront as rays exit the drop in the neighborhood of this point. Referring to Figure 5.1, we wish to find when the acute angle $\Theta(i)$ [$= 4r(i) - i$] is a maximum. This occurs when $\Theta'(i) = 0$, that is, when

$$r'(i) = \frac{1}{4},$$ (5.1)

and

$$r''(i) < 0.$$ (5.2)

The first condition leads to

$$\frac{\cos^2 i}{n^2 - \sin^2 i} = 16,$$ (5.3)

or

$$i \equiv i_C = \arccos\left(\frac{n^2 - 1}{15}\right)^{1/2},$$ (5.4)

(note the uppercase subscript C), so for $n = 4/3$, $i_C \approx 76.8°$, well past the angle of incidence corresponding to the rainbow angle, which is $i_c \approx 59.4°$ for the same value of n. This is an important observation, because it means that there can be pairs of parallel rays emerging from the drop close to the exit point for the rainbow ray, but each ray of the pair emerges on a different side of that ray. Such rays can interfere, and (on the basis of ray theory) this gives rise to the supernumerary bow pattern. It must be recalled however, that it really is interference between two different portions of a light wave that gives rise to supernumerary bows [15], so the ray model above is extremely limited as an explanation of supernumeraries.

Use of the Impact Parameter

It is useful to note a slightly different approach to the geometrical theory of the rainbow [24], [25]. Instead of using the angles of incidence i and refraction r, we can use the impact parameter b (normalized by the drop radius a) so the fundamental variable is $x = \sin i = b/a$. For the primary rainbow in particular ($p = 2$), equation (4.1) becomes

$$\theta(x) = \pi + 2 \arcsin x - 4 \arcsin\left(\frac{x}{n}\right), \tag{5.5}$$

where the deviation angle $D_1(i)$ has now been replaced by $\theta(x(i))$ or just $\theta(x)$, the standard variable used in the study of scattering cross sections. Extrema of $D_1(i)$ can be expressed easily in terms of extrema of $\theta(x)$; thus

$$\theta'(x) = \frac{2}{\left(1 - x^2\right)^{1/2}} - \frac{4}{\left(n^2 - x^2\right)^{1/2}}, \tag{5.6}$$

and

$$\theta''(x) = \frac{2x}{(1 - x^2)^{3/2}} - \frac{4x}{(n^2 - x^2)^{3/2}}. \tag{5.7}$$

By requiring $\theta'(x_c) = 0$, where $x_c = \sin i_c$, it follows that $x_c = \left[(4 - n^2)/3\right]^{1/2}$ from which the result (4.5) is recovered for $p = 2$. Obviously this result can be generalized to higher positive integer values of p. Note also that

$$\theta''(x_c) = \frac{9(4 - n^2)^{1/2}}{2(n^2 - 1)^{3/2}} > 0 \quad \text{for} \quad 1 < n < 2, \tag{5.8}$$

where

$$\theta_R = \theta(x_c) \equiv D_{\min} \equiv D_1(i_c), \tag{5.9}$$

so the angle of deviation is indeed a minimum, as expected. We note two other aspects of this approach. For $x \approx x_c$ we can truncate the Taylor expansion for $\theta(x)$ after the quadratic term:

$$\theta \approx \theta_R + \theta''(x_c)(x - x_c)^2/2, \tag{5.10}$$

an approximation to be employed below.

Another aspect of this formulation addresses dispersion; since $\theta = \theta(x, n)$, it follows from (5.5) that

$$\frac{\partial \theta}{\partial n} = -4 \frac{\partial}{\partial n} \left[\arcsin \left(\frac{x}{n} \right) \right] = \frac{4x}{n \left(n^2 - x^2 \right)^{1/2}}, \tag{5.11}$$

and so at $\theta_R \approx 138°$ (corresponding to $x_c = \left[(4 - n^2)/3 \right]^{1/2} \approx 0.86$ for $n = 4/3$),

$$\frac{\partial \theta}{\partial n} = \frac{2}{n} \left[\frac{4 - n^2}{n^2 - 1} \right]^{1/2}. \tag{5.12}$$

This can be used to estimate the angular spread of the rainbow ($\Delta\theta$) given the variation in n over the visible part of the spectrum (Δn); as noted above, $\Delta\theta$ is slightly less than 2° for the primary bow. Note also that since

$$\frac{d\theta}{d\lambda} \approx \frac{d\theta}{dn} \cdot \frac{dn}{d\lambda} \tag{5.13}$$

for given x_c (i.e., neglecting the small variation of x_c with wavelength λ) and $dn/d\lambda < 0$ in the visible range (normal dispersion), it follows that $d\theta/d\lambda < 0$, and so the red part of the (primary) rainbow emerges at a smaller angle than the violet part, so the latter appears on the underside of the arc with the red outermost.

Phase Variation

The phase accumulated along the critical ray between the surfaces $A'A''$ and $B'B''$ (Figure 5.2) is by symmetry, in terms of x,

$$\eta(x) = \frac{2\pi}{\lambda} \left(|A''A| + |AB| + |BC| + |CB''| \right) \tag{5.14}$$

$$= ka \left[2(1 - \cos i) + 4n \cos r \right] \tag{5.15}$$

$$= 2ka \left[1 - \left(1 - x^2 \right)^{1/2} + 2 \left(n^2 - x^2 \right)^{1/2} \right]. \tag{5.16}$$

In equation (5.15) the first term inside the first set of brackets is the sum of the distances from both A'' and B'' to the surface of the drop; the second term is n times the path length interior to the drop. Noting that

$$\eta'(x) = 2ka \left[\frac{x}{\left(1 - x^2 \right)^{1/2}} - \frac{2x}{\left(n^2 - x^2 \right)^{1/2}} \right], \tag{5.17}$$

and comparing this with (5.6) it is clear that

$$\eta'(x) = kax\theta'(x). \tag{5.18}$$

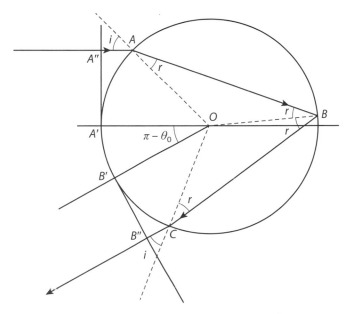

Figure 5.2 Incident and exiting wavefronts.

Near the critical or minimum deviation angle $\theta = \theta_R$, (and hence near x_c) this result may be written in terms of the variable $\xi = x - x_c$:

$$\eta'(\xi) \approx ka\theta'(\xi)(x_c + \xi),\tag{5.19}$$

so that on integration

$$\eta(\xi) = \eta_c + ka\left[x_c(\theta - \theta_0) + \int_0^\xi \mu\theta'(\mu)d\mu\right]\tag{5.20}$$

$$= \eta_c + ka\left[x_c(\theta - \theta_0) + \xi\theta(\xi) - \int_0^\xi \theta(\mu)d\mu\right].\tag{5.21}$$

Using the result (5.10) for $\theta \approx \theta_R$,

$$\theta(\xi) \approx \theta_R + \theta''(x_c)\frac{\xi^2}{2} + O(\xi^3),\tag{5.22}$$

it is readily shown that

$$\eta(\xi) = \eta_0 + ka\left[x_c(\theta - \theta_R) + \theta''(x_c)\frac{\xi^3}{3} + O(\xi^4)\right],\tag{5.23}$$

so that for two rays, each one close to, but on opposite sides of, the critical ray, with equal and opposite ξ values, it follows that their phase difference δ is

$$\delta(\xi) = \eta(\xi) - \eta(-\xi) \approx 2ka\left[\theta''(\theta_R)\frac{\xi^3}{3}\right].\tag{5.24}$$

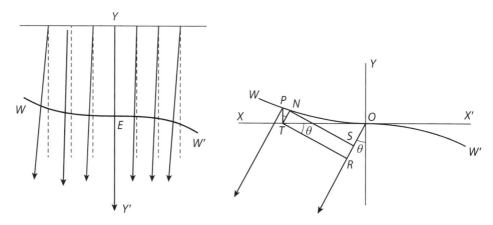

Figure 5.3 The locally cubic wavefront. The dashed vertical lines are normal to the line
containing Y.

When this phase difference is equal to an integer multiple of 2π, say, $2K\pi$, then the
rays interfere constructively in general; however, in this instance it can be shown
that an additional $\pi/2$ must be added because a focal line is passed in the process
([85], chapters 12 and 13). Hence for constructive interference

$$\theta_K - \theta_R \approx \left(\frac{3\pi(K + 1/4)}{ka}\right)^{2/3} \left[\frac{\theta''(x_c)}{8}\right]^{1/3}, \quad K = 1, 2, 3 \ldots . \qquad (5.25)$$

Clearly, in this approximation, $\theta_K - \theta_R \propto (ka)^{-2/3}$, which is quite sensitive to
droplet size. If the droplets are large enough, the supernumerary bows lie inside
the primary bow and thus are not visible; it has been shown that the maximum uniform droplet size rendering them visible is $a \simeq 0.28$ mm [24]. If the droplets are
not uniform in size, the maxima may be washed out by virtue of the spread in sizes;
if the droplets are all very small ($a < 50$ microns) the various colors are not very
dispersed and whitebows or cloudbows result.

5.1.2 The Airy Wavefront

Let the (normalized) wavefront be represented by the function

$$\Psi(x) = e^{ik\eta(x)}, \qquad (5.26)$$

where k is the wavenumber of the light waves. As can be seen from Figure 5.3,
if YY' represents the ray that emerges at minimum deviation, rays to either side
of this are deviated through larger angles as shown. By considering the corresponding wavefront WW' (distorted from the wavefront XX' (as in Figure 5.3),
corresponding to a parallel beam of rays), we are dealing with, locally at least, a
cubic approximation to the wavefront in the neighborhood of the point E (or O),
which is itself a point of inflection. With point E as the origin of coordinates and
YY', XX' as the coordinate axes, we have seen that the equation of the wavefront

WEW' is given by

$$y = \frac{h}{3a^2}x^3.$$

It will be shown below that for $q = p - 1$ *internal reflections,*

$$h = -\frac{(q^2 + 2q)^2 \left[(q + 1)^2 - n^2\right]^{1/2}}{(q + 1)^2 (n^2 - 1)^{3/2}}, \qquad q = 1, 2, 3, \ldots. \qquad (5.27)$$

(In Section 4.2 $k = p - 1$ was used to label the number of internal reflections; in this chapter k will be used to denote the wavenumber of the disturbance, so the former will be labeled by $q = p - 1$.)

At this stage we are in a position to deduce the form of the famous "rainbow integral" introduced by Sir George Airy in 1838 and referred to earlier [23]. Using Figure 5.3b we can derive an expression for the amplitude of the wave in a direction making an angle θ with that of the minimum deviation [24]. By considering the path difference η between the points $P(x, y)$ and O, we have

$$\eta = OS = OR - RS = OR - TN = x \sin\theta - y \cos\theta,$$

or

$$\eta = x \sin\theta - \frac{hx^3}{3a^2}\cos\theta. \qquad (5.28)$$

In relative terms, if the amplitude of a small element δx of the wavefront at O is represented by $\sin \varpi t$, then that from a similar element at P is $\sin(\varpi t + \delta)$, where

$$\delta = \frac{2\pi\eta}{\lambda} = k\eta$$

λ being the wavelength, and k now being the wavenumber of the disturbance, not the number of internal reflections in the drop. To avoid confusion the letter q will be used below to signify the number of internal reflections.

For the whole wavefront the cumulative disturbance amplitude is therefore given by the integral

$$A = A_0 \int_{\mathcal{L}} \sin(\varpi t + k\eta)dx, \qquad (5.29)$$

where A_0 is the (unknown) amplitude per unit length of the wavefront, and $\mathcal{L}$ is the portion of the complete wavefront that is being considered. Because $\sin k\eta$ is an odd function, we may write this equation as

$$A = A_0 \sin \varpi t \int_{\mathcal{L}} \cos k\eta dx. \qquad (5.30)$$

But what should be the limits on this integral? We use the following argument to show that there is little error in taking the integration over $(-\infty, \infty)$. Let x and x_0 be x-locations on the wavefront such that their distances from the far point P is half a wavelength, $\lambda/2$. This ensures that their combined amplitude at P is zero, that is,

$$\eta(x) - \eta(x_0) = (x - x_0)\sin\theta - \frac{h}{3a^2}(x^3 - x_0^3)\cos\theta \qquad (5.31)$$

$$= (x - x_0)\cos\theta\left[\tan\theta - \frac{h}{3a^2}(x^2 + xx_0 + x_0^2)\right] \equiv \frac{\lambda}{2}. \qquad (5.32)$$

In other words,

$$\delta x \equiv x_0 - x = \frac{\lambda}{2\cos\theta}\left[\frac{3a^2}{h(x^2 + xx_0 + x_0^2) - 3a^2\tan\theta}\right]. \qquad (5.33)$$

For the primary bow ($q = 1$) a simple calculation using a generic value of $n = 4/3$ yields

$$h = \frac{81}{14}\left(\frac{5}{7}\right)^{1/2} \approx 4.89, \qquad (5.34)$$

so for a typical (large) raindrop radius $a = 1$ mm, and a wavelength $\lambda \approx 580$ nm or 5.8×10^{-4} mm, using the fact that θ is small, we find that

$$\delta x \approx \frac{\lambda}{2}\left(\frac{a^2}{hx^2}\right) \Rightarrow x^2\delta x \approx 6 \times 10^{-5}\text{ mm}^3. \qquad (5.35)$$

Clearly, δx decreases fairly rapidly as x increases, and so beyond a very small distance from the inflection point O, the distance δx between points that are $\lambda/2$ out of phase decreases, and the combined interference of waves neutralizes the contributions from the rest of the wavefront. In other words, very little error is introduced by taking the limits of the integral to be $(-\infty, \infty)$. Hence we can write (since cosine is an even function)

$$A = A_0\sin\varpi t\int_{-\infty}^{\infty}\cos k\eta dx = A_0\sin\varpi t\int_{-\infty}^{\infty}\cos k\left(x\sin\theta - \frac{hx^3}{3a^2}\cos\theta\right)dx \qquad (5.36)$$

$$= 2A_0\sin\varpi t\int_{0}^{\infty}\cos k\left(\frac{hx^3}{3a^2}\cos\theta - x\sin\theta\right)dx. \qquad (5.37)$$

In terms of the following changes of variable

$$\xi^3 = \frac{4hx^3\cos\theta}{3a^2\lambda}, \qquad (5.38)$$

and

$$m\xi = \frac{4x\sin\theta}{\lambda}, \qquad (5.39)$$

the above integral may be written in the form

$$\left(\frac{3a^2\lambda}{4h\cos\theta}\right)^{1/3}\int_{-\infty}^{\infty}\cos\left[\frac{\pi}{2}(\xi^3 - m\xi)\right]d\xi \equiv MF(-m) \propto \text{Ai}(-m), \qquad (5.40)$$

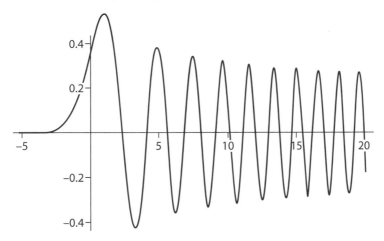

Figure 5.4 The Airy integral Ai $(-m)$. The integral Ai (m) is the mirror image of this graph about the vertical axis.

where M is an obvious constant and as indicated, $F(-m)$ is proportional to Airy's celebrated *rainbow integral* Ai$(-m)$, first published in his paper entitled "On the Intensity of Light in the Neighbourhood of a Caustic." It is now referred to simply as the Airy integral, and as we shall see, it can be written in several different but equivalent forms. Ignoring the harmonic time dependence, we can write the spatial part of $\mathcal{A}$ as

$$\mathbb{A} = 2\mathcal{A}_0 M F(-m), \tag{5.41}$$

the intensity being the square of this quantity. The significance of the parameter m can be noted by eliminating ξ from equations (5.38) and (5.39) to obtain

$$m = \left(\frac{2ka}{\pi}\right)^{2/3} \left(\frac{\sin^3 \theta}{h \cos \theta}\right)^{1/3}, \tag{5.42}$$

which for sufficiently small values of θ (the angle of deviation from the rainbow ray) is proportional to θ.

A graph of the rainbow integral qualitatively resembles the diffraction pattern near the edge of the shadow of a straight edge, which has the following features: (i) low-intensity, rapidly decreasing illumination in regions that geometrical optics predicts should be totally in shadow, and (ii) in the illuminated region (as with diffraction), a series of fringes, which correspond to the supernumerary bows below the primary rainbow. The graph of Ai$(-m)$ is shown in Figure 5.4. The first maximum is the largest in amplitude and corresponds to the primary bow; the remaining maxima decrease rather slowly in amplitude, the period of oscillation decreasing also. The underlying assumption in this approach is that diffraction arises from points on the wavefront in the neighborhood of the Descartes ray (of minimum deviation); provided that the drop size is large compared to the wavelength of light, this is reasonable and is in fact valid for most rainbows. It would not be as useful an assumption for cloud or fog droplets, which are considerably smaller than raindrops, but even then the drop diameter may be five or ten times the wavelength, so the Airy approximation is still useful. Clearly, however, it has a limited domain of validity.

5.1.3 How Are Colors Distributed in the Airy Rainbow?

The previous analysis applies to monochromatic radiation, so an obvious question is how a continuous spectrum of light is distributed when the corresponding Airy bows are combined: how do they mix? The intensity of any particular wavelength is a function of the phase difference

$$k\eta = \frac{2\pi}{\lambda}\left(\frac{hx^3}{3a^2}\cos\theta - x\sin\theta\right), \tag{5.43}$$

so as λ increases the phase difference decreases, resulting in the maxima getting farther apart (in an angular sense) because of the $\cos k\eta$ term in the integral (5.30). This means that the maximum intensities of the various colors will be angularly distributed, and in particular the intensities of the respective primary bows for each portion of the spectrum. This can be quantified as follows.

Since

$$m^3 = \left[\frac{4\sin\theta}{\lambda}\left(\frac{x}{\xi}\right)\right]^3 = \left(\frac{4\sin\theta}{\lambda}\right)^3\left(\frac{3a^2\lambda}{4h\cos\theta}\right), \tag{5.44}$$

it follows that

$$m^3 = \frac{48a^2}{h\lambda^2}\sin^2\theta\tan\theta. \tag{5.45}$$

It can be seen by graphing it that the quantity $\theta^{-3}\sin^2\theta\tan\theta$ varies by less than 1 percent for $\theta \in (0, \pi/6)$, so to a very good approximation

$$m^3 = \frac{48a^2}{h\lambda^2}\theta^3, \tag{5.46}$$

or

$$\theta = \frac{m}{2}\left(\frac{\lambda}{a}\right)^{2/3}\left(\frac{h}{6}\right)^{1/3}. \tag{5.47}$$

Therefore the angular distance between any two successive intensity maxima varies as the two-thirds power of the wavelength-to-radius ratio. Naturally this quantity is larger for red light than for blue, and it increases as the drop size decreases.

Another feature of the rainbow problem that has been neglected thus far is the three-dimensional nature of the diffraction: the wavefront is a surface in three dimensions, not merely a curve in two. The factor multiplying the integrand in equation (5.40) above requires modification. In fact, it is necessary to multiply this factor by $(a/\lambda)^{1/2}$ (see appendix I in [22] for a heuristic discussion of this factor based on Fresnel zones). Since the angle θ will be small, this results in an amplitude proportional to $(a^7/\lambda)^{1/6}$ or equivalently, an intensity proportional to $(a^7/\lambda)^{1/3}$. Thus a relatively strong dependence of intensity on drop size is established in the Airy regime: other things being equal, larger drops give rise to more

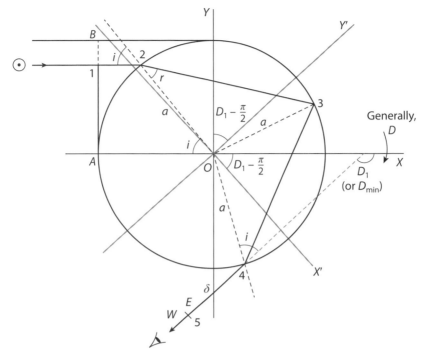

Figure 5.5 Details for the derivation of the cubic wavefront approximation.

intense rainbows. Note also that the λ-dependence is important to determine the intensity distribution with wavelength:

$$I \propto \left[(\lambda a^2) \left(\frac{a}{\lambda} \right)^{1/2} \right]^2 = \left(\lambda^{-1/6} a^{7/6} \right)^2 = \left(\frac{a^7}{\lambda} \right)^{1/3}. \tag{5.48}$$

The diffraction pattern may be thought of as arising from interference of two arms of a cusped wavefront. A set of alternative maxima and minima occur, lying between the direction of the incident light and the rainbow (or Descartes) ray. A change of phase occurs when light passes a focus, so it is to be expected that the rainbow ray would not correspond exactly to the direction of maximum intensity (which is displaced inward), and this is indeed the case, as more complete theory shows. Furthermore, it is not the case that the intensities should be the same along both arms of the cusped wavefront, which implies that the minima will not in general be of zero intensity; in Airy's theory there is assumed to be no variation in intensity along the wave, and as a result the intensity minima are zero.

5.1.4 The Airy Wavefront: A Derivation for Arbitrary p

In the spirit of the phase calculation in equation (5.15), using Figure 5.5 we consider the ray path $1 \to 2 \to 3 \to 4 \to 5$ for the ray of minimum deviation corresponding to a primary bow ($p = 2$ or $q = 1$). The speed of light in air is approximately that in vacuo, and so will be denoted by c; that in the drop is therefore c/n. The entering plane wavefront AB becomes distorted as it passes through and exits the drop, but by definition the travel time for each point on the wavefront AB to the corresponding point on EA is a constant. A small section of the wavefront near the ray of minimum deviation (for which $D = D_1$ or D_{min}) is marked as w. Therefore we shall refer to deviation angles above D_1 as D, where $D - D_1$ is "small." The total travel time for the separate portions of the ray path $1 \rightarrow 2; 2 \rightarrow 3 \rightarrow 4; 4 \rightarrow 5$ is

$$T = \frac{a(1 - \cos i)}{c} + \frac{4na \cos r}{c} + \frac{s}{c}, \tag{5.49}$$

where s is the length of the segment $4 \rightarrow 5$. But note that for the ray at zero angle of incidence, also on the wavefront (at $y = 0$), for which

$$T = \frac{4na}{c}. \tag{5.50}$$

Therefore from (5.49) and (5.50) we find

$$s = a[4n(1 - \cos r) - (1 - \cos i)]. \tag{5.51}$$

For a given deviation angle D, a point on w has coordinates (x, y), where

$$x = a \cos(D - i) + s \cos D, \tag{5.52}$$

$$y = -[a \sin(D - i) + s \sin D]. \tag{5.53}$$

Next, the XY coordinate system is rotated clockwise about the origin by an amount $D_1 - \pi/2$, with the new system labeled as $X'Y'$. The purpose of this is to align the Y' axis parallel to the emerging ray of minimum deviation. Then either by using the rotation matrix $R(\theta)$ with $\theta = -(D_1 - \pi/2)$ or simply making the change

$$D \rightarrow D - D_1 + \frac{\pi}{2} \equiv d + \frac{\pi}{2}, \tag{5.54}$$

the new coordinates (x', y') are given by

$$x' = -[a \sin(d - i) + s \cos d], \tag{5.55}$$

$$y' = -[a \cos(d - i) + s \cos d]. \tag{5.56}$$

Recall that we are considering only rays that exit close to the so-called Descartes or rainbow ray. To the first order of approximation (i.e., to $O(d)$), we may write $\cos d \approx 1$ and $\sin d \approx d$ from which we conclude that

$$x' = -ad \cos i + a \sin i - sd, \tag{5.57}$$

$$y' = -[a \cos i + ad \sin i + s]. \tag{5.58}$$

The segment w of the wavefront is almost perpendicular to the Y' axis, and therefore small changes in y' can induce significant ones in x'; in contrast, the reverse is not true. Hence further approximations are justified in equation (5.57): since d is small (and i is not),

$$x' \approx a \sin i. \tag{5.59}$$

Using equation (5.51),

$$y' = -\left[a \cos i + ad \sin i + a \left\{4n \left(1 - \cos r\right) - \left(1 - \cos i\right)\right\}\right]. \tag{5.60}$$

Suppose now that the ray of minimum deviation D_1 corresponds to angles of incidence and refraction I and R, respectively, and for rays that are close to this ray we may write

$$i = I + \alpha; \quad r = R + \beta; \quad |\alpha| \ll i, \quad |\beta| \ll r. \tag{5.61}$$

(We can do this because the ray of minimum deviation for the primary bow corresponds to $I \approx 1$ radian and $R \approx 0.7$ radians, neither of which are small.) By including the possibility of higher-order bows (corresponding to $q = p - 1$ internal reflections, $q = 1, 2, \ldots$) in equation (5.60)), we have

$$x' = a \sin \left(I + \alpha\right) \approx a \left(\sin I + \alpha \cos I\right), \tag{5.62}$$

and

$$y' = -2a \left\{\cos i + \frac{d}{2} \sin i + n \left(q + 1\right) - n(q + 1) \cos r - \frac{1}{2}\right\}. \tag{5.63}$$

In what follows we express d and β as functions of α in terms of Taylor polynomials of degree three (i.e., up to $O(\alpha^3)$) and will then insert them into equation (5.63):

$$\cos i = \cos(I + \alpha) = \cos I - \alpha \sin I - \frac{\alpha^2}{2} \cos I + \frac{\alpha^3}{6} \sin I + O(\alpha^4). \tag{5.64}$$

Now to this order

$$\frac{d}{2} = \frac{1}{2} \left(D - D_1\right) = \alpha - (q + 1) \beta \tag{5.65}$$

$$= \alpha - (q + 1) \left[\beta' \left(0\right) \alpha + \frac{1}{2} \beta'' \left(0\right) \alpha^2 + \frac{1}{6} \beta''' \left(0\right) \alpha^3\right], \tag{5.66}$$

because $\beta \left(0\right) = 0$. Note that $d/2$ is stationary when $d\alpha = (q + 1)d\beta$, from which it follows that $\beta' \left(0\right) = (q + 1)^{-1}$. Hence

$$\frac{d}{2} \sin i \approx \frac{d}{2} \left(\sin I + \alpha \cos I\right) \tag{5.67}$$

$$\approx - (q + 1) \left[\frac{1}{2} \beta'' \left(0\right) \alpha^2 \sin I + \frac{1}{6} \beta''' \left(0\right) \alpha^3 \sin I + \frac{1}{2} \beta'' \left(0\right) \alpha^3 \cos I\right]. \tag{5.68}$$

Also, noting that $\sin i = n \sin r$ (from (4.13)),

$$\cos r = \cos(R + \beta(\alpha)) \tag{5.69}$$

$$= \cos R + \alpha \left(\frac{d \cos r}{d\alpha}\right)_0 + \frac{\alpha^2}{2}\left(\frac{d^2 \cos r}{d\alpha^2}\right)_0 + \frac{\alpha^3}{6}\left(\frac{d^3 \cos r}{d\alpha^3}\right)_0 + \cdots \tag{5.70}$$

$$= \cos R - \alpha \left(\frac{\sin i}{n}\right)_0 \beta'(0) - \frac{\alpha^2}{2}\left\{\left(\frac{\cos i}{n}\right)_0 \beta'(0) + \left(\frac{\sin i}{n}\right)_0 \beta''(0)\right\} \tag{5.71}$$

$$+ \frac{\alpha^3}{6}\left\{\left(\frac{\sin i}{n}\right)_0 \beta'(0) - 2\left(\frac{\cos i}{n}\right)_0 \beta''(0) - \left(\frac{\sin i}{n}\right)_0 \beta'''(0)\right\}. \tag{5.72}$$

This enable us to write the term $-n(q+1)\cos r$ as

$$-n(q+1)\cos r = -n(q+1)\cos R - (q+1)\beta'(0)$$

$$\times \left[-\alpha \sin I - \frac{\alpha^2}{2}\cos I + \frac{\alpha^3}{6}\sin I\right] \tag{5.73}$$

$$- (q+1)\beta''(0)\left[-\frac{\alpha^2}{2}\sin I - \frac{\alpha^3}{3}\cos I\right]$$

$$+ (q+1)\beta'''(0)\left[\frac{\alpha^3}{6}\sin I\right]. \tag{5.74}$$

Recalling (4.15), we know that

$$n \cos R = (q+1)\cos I, \tag{5.75}$$

and

$$\beta'(0) = \frac{1}{q+1}, \tag{5.76}$$

so adding all the terms together in equation (5.63), it is seen (after some careful book-keeping) that

$$-y' = 2a\left\{\left[1 - (q+1)^2 - (q+1)\beta''(0)\frac{\alpha^3}{6}\right]\cos I + n(q+1) - \frac{1}{2}\right\}. \tag{5.77}$$

On differentiating Snell's law (4.13) twice with respect to i, we find that

$$-\sin i = -n \sin r \left[\beta'(\alpha)\right]^2 + n \cos r \left[\beta''(\alpha)\right], \tag{5.78}$$

and hence

$$-\sin I = -n \sin R \left[\beta'(0)\right]^2 + n \cos R \left[\beta''(0)\right] \tag{5.79}$$

$$= -\frac{n \sin R}{(q+1)^2} + n \cos R \left[\beta''(0)\right]. \tag{5.80}$$

Noting that

$$\sin I = n \sin R, \quad \text{and} \quad n \cos R = (q + 1) \cos I, \tag{5.81}$$

we have

$$\beta''(0) = - \tan I \left[\frac{q^2 + 2q}{(q + 1)^3} \right]. \tag{5.82}$$

Therefore

$$-y' = 2a \left[-\left(q^2 + 2q\right) \cos I + n(q + 1) - \frac{1}{2} \right] + a\alpha^3 \frac{\left(q^2 + 2q\right)}{3(q + 1)^2} \sin I, \tag{5.83}$$

and as before

$$x' = a \sin I + a\alpha \cos I. \tag{5.84}$$

The first terms in these equations for x' and y' are the coordinates of the point of inflection on the curve w. To see this, we can write

$$x' = X_1 + X_2 \alpha, \quad y' = Y_1 + Y_2 \alpha^3, \tag{5.85}$$

eliminate α and easily establish that the point of inflection is indeed at (X_1, Y_1). Then taking this point as the origin of an (x_1, y_1) coordinate system, we now have

$$x_1 = a\alpha \cos I, \tag{5.86}$$

$$y_1 = -a\alpha^3 \frac{\left(q^2 + 2q\right)}{3(q + 1)^2} \sin I = -\frac{q(q + 2) \sin I}{3a^2(q + 1)^2 \cos^3 I} x_1^3. \tag{5.87}$$

Dependence on the angle I may be eliminated by recalling that

$$\cos I = \left(\frac{n^2 - 1}{q^2 + 2q} \right)^{1/2}, \tag{5.88}$$

and so

$$\sin I = \left[\frac{(q + 1)^2 - n^2}{q^2 + 2q} \right]^{1/2}, \tag{5.89}$$

and therefore

$$y_1 = -\frac{\left(q^2 + 2q\right)^2}{3a^2(q + 1)^2 \left(n^2 - 1\right)} \left[\frac{(q + 1)^2 - n^2}{n^2 - 1} \right]^{1/2} x_1^3 \equiv \frac{h}{3a^2} x_1^3, \tag{5.90}$$

as it is traditionally written, where recalling that q is the number of internal reflections,

$$h = -\frac{\left(q^2 + 2q\right)^2 \left((q + 1)^2 - n^2\right)^{1/2}}{(q + 1)^2 \left(n^2 - 1\right)^{3/2}}, \quad q = 1, 2, 3, \ldots. \tag{5.91}$$

Equivalently if this is expressed in terms of the number of ray segments p in the sphere, where $p = q + 1$,

$$h = -\frac{\left(p^2 - 1\right)^2 \left(p^2 - n^2\right)^{1/2}}{p^2 \left(n^2 - 1\right)^{3/2}}, \quad p = 2, 3, 4, \ldots. \tag{5.92}$$

Another Derivation for $q = 1$

We have seen that the deviation angle produced by one internal reflection inside a spherical drop for an angle of incidence θ_i is given by the expression

$$D(\theta_i) = \pi + 2\theta_i - 4\arcsin\left(\frac{\sin\theta_i}{n}\right). \tag{5.93}$$

Hence

$$\frac{dD(\theta_i)}{d\theta_i} = 2 - 4\frac{\cos\theta_i}{\left(n^2 - \sin^2\theta_i\right)^{1/2}}, \tag{5.94}$$

and after some reduction

$$\frac{d^2D(\theta_i)}{d\theta_i^2} = 4\left(n^2 - 1\right)\frac{\sin\theta_i}{\left(n^2 - \sin^2\theta_i\right)^{3/2}}. \tag{5.95}$$

For the rainbow or Descartes ray $D'(\theta_{i_c}) = 0$,

$$\sin^2\theta_{i_c} = \frac{4 - n^2}{3}, \tag{5.96}$$

whence

$$D''(\theta_{i_c}) = \frac{3}{2}\tan\theta_{i_c}, \tag{5.97}$$

so a Taylor series expansion about θ_{i_c} gives

$$D(\theta_i) - D(\theta_{i_c}) = D(\theta_i) - D_m \tag{5.98}$$

$$= \frac{1}{2}D''(\theta_{i_c})\left(\theta_i - \theta_{i_c}\right)^2 + O(\theta_i - \theta_{i_c})^3 \tag{5.99}$$

$$\approx \frac{3}{4}\left(\tan\theta_{i_c}\right)\left(\theta_i - \theta_{i_c}\right)^2 = \frac{du}{dv}, \tag{5.100}$$

(see Figure 5.6).

Since the geometry along the v-axis is the same as that for the incident rays parallel to the axis of symmetry, we may write $v = a\left(\sin\theta_i - \sin\theta_{i_c}\right)$. Very near the rainbow angle we can set $\theta_i - \theta_{i_c} = \delta$ with $|\delta| \ll 1$, so that

$$v \approx a\delta\cos\theta_{i_c} = a\left(\theta_i - \theta_{i_c}\right)\cos\theta_{i_c}, \tag{5.101}$$

whence

$$\frac{du}{dv} \approx \frac{3}{4a^2}\left(\frac{\tan\theta_{i_c}}{\cos^2\theta_{i_c}}\right)v^2. \tag{5.102}$$

Since $u(0) = 0$ this can be integrated to give

$$u = \frac{h}{3a^2}v^3, \tag{5.103}$$

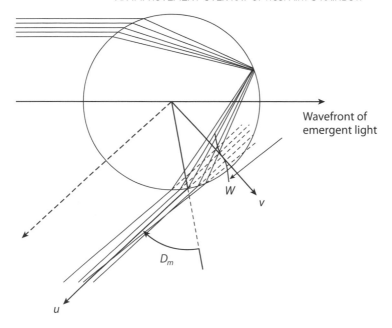

Figure 5.6 The coordinate system for equation (5.103).

where

$$h = \frac{3 \tan \theta_{i_c}}{4 \cos^2 \theta_{i_c}} \approx 4.89, \tag{5.104}$$

for the primary bow [213].

Phase Shift Subtleties: Young's Theory (Ancient and Modern)

In curious statements made in paragraphs twenty and twenty-one of his 1838 paper [23], Airy refers to the "imperfect theory" of light undulations as opposed to the "complete theory." It appears that by the former term, Airy was diplomatically (but somewhat negatively) referring to the theory put forth earlier by Thomas Young (1773–1829) and by the latter, his own theory (further details may be found in [269]).

In his famous Bakerian lecture in 1803, published in 1804 [266], Thomas Young made a statement concerning what we now call *destructive interference*:

From the experiments and calculations which have been premised, we may be allowed to infer, that homogeneous light, at certain equal distances in the direction of its motion, is possessed of opposite qualities, capable of neutralising or destroying each other, and of extinguishing the light, where they happen to be united.

The destructive interference he describes corresponds of course to a phase difference (or phase shift) of π between the two disturbances as a consequence of differences in the length of the optical path taken by the light. In [85] three separate types of phase shift are identified. They are: (i) change of phase at reflection, (ii) phase shift due to the length of the optical path, and (iii) phase shifts due to passing focal lines. By counting the number of focal lines encountered along the entire ray path, a formula can be derived for the resulting total phase shift. For further details, [85] should be consulted, but in essence, for phase shifts of type (iii) applied to the primary bow, the phase shift for impact parameters $b < b_0$ (where b_0 is the impact parameter corresponding to the Descartes rainbow ray) is π, whereas for $b > b_0$ it is $3\pi/2$, and it was this additional phase shift of $\pi/2$ that Young appears to have ignored. In fact much more can be said (emphasis added):

> Young's theory took account of the different path lengths followed by the two rays, thus generating an interference pattern. All solutions based on ray tracing show infinite intensity at the geometrical rainbow angle—with an immediate transition to zero intensity. Airy's theory avoids these problems and also shows that Young's theory of interference is "imperfect" because it incorrectly predicts the locations of supernumerary arcs. Young was unaware of the fact that rays crossing a focal line suffer a phase change of 90°—as explained in the 1950s by van de Hulst. Modifying Young's theory gives much more accurate results for the supernumerary arcs; [in fact so doing] to take account of the additional phase shift of 90° gives results for the supernumerary arcs similar to Airy theory. [Furthermore,] *the modified version of Young's theory gives results for the supernumerary arcs that are identical to the rigorous Debye series*. Predicting supernumerary arcs must take account of the phase shifts due to caustics. Rays with impact parameter $b > b_0$ are subject to an additional phase shift of 90° compared with rays with $b < b_0$. .... The modified version of Young's theory accurately predicts the supernumerary arcs for primary and secondary rainbows; the results for both polarizations are identical to the rigorous Debye series solution (and more accurate than Airy theory). [267]

The Debye series referred to is the Debye expansion mentioned in Chapter 1. For a fascinating account of the back-and-forth between Young and Newton's supporters over the wave versus corpuscular theories light, and their explanations of supernumerary bows, see [169] and [269].

Chapter Six

Diffraction Catastrophes

Optical caustics are surfaces (in space) and curves (in the plane) where light rays are focused. They are as familiar as rainbows and the dancing bright lines of sunlight focused by water waves onto the bottoms of swimming pools. Caustics are the singularities of geometrical optics and can be classified mathematically as the elementary catastrophes of singularity theory. In the plane, the classification gives two singularities: smooth caustic curves, which are "fold" catastrophes, and points where two fold caustics meet on opposite sides of a common tangent, which are "cusp" catastrophes.

[2]

In this section we shall draw on an approach [29] to examine the wavefront associated with a monochromatic wave with $Q(x, y, f(x, y))$ being a point on the wavefront surface and $P(X, Y, Z)$ being a movable observation point. Consider the following distance function l (see Figure 6.1):

$$l(x, y, f; X, Y, Z) = \left[(X - x)^2 + (Y - y)^2 + (Z - f)^2\right]^{1/2}. \qquad (6.1)$$

Assume $Z \gg f$ (meaning that the observation point is "far away") and that (consistent with the paraxial approximation) $|X - x| \ll |Z|, |Y - y| \ll |Z|$. Then

$$l = (Z - f)\left[1 + \frac{(X - x)^2 + (Y - y)^2}{(Z - f)^2}\right]^{1/2}$$

$$\approx (Z - f)\left[1 + \frac{(X - x)^2 + (Y - y)^2}{2(Z - f)^2}\right]$$

$$\approx Z - f + \frac{(X - x)^2 + (Y - y)^2}{2(Z - f)^2}. \qquad (6.2)$$

To calculate the rays through P, the appropriate conditions are $\partial l / \partial x = \partial l / \partial y = 0$, so we can construct a potential

$$\phi = -l = f(x, y) - \frac{x^2 + y^2}{2Z} + \frac{xX + yY}{Z}. \qquad (6.3)$$

We can check this by calculating the rays from the condition that ϕ (and hence l) is stationary (this is just Fermat's principle, of course). Thus

$$\frac{\partial \phi}{\partial x} = \frac{\partial \phi}{\partial y} = 0 \implies \frac{\partial f}{\partial x} = \frac{x - X}{Z}; \quad \frac{\partial f}{\partial y} = \frac{y - Y}{Z}. \qquad (6.4)$$

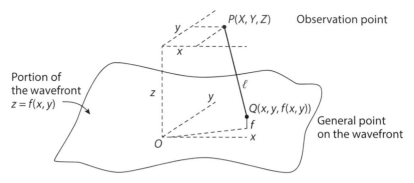

Figure 6.1 Wavefront geometry.

Equation 6.4 is satisfied when $P(X, Y, Z)$ lies on the normal to the wavefront at (x, y), that is, on the ray. Consider now a screen at infinity, such that $Z \to \infty$, but $X/Z, Y/Z \nrightarrow \infty$; (6.3) becomes, neglecting the quadratic terms in x and y,

$$\phi(x, y; u, v) = f(x, y) + \frac{X}{Z}x + \frac{Y}{Z}y \equiv f(x, y) + ux + vy, \qquad (6.5)$$

where u and v are the "angle coordinates", or tangents of the angles OP makes with the OZ axis. The ray conditions $\partial\phi/\partial x = 0$, $\partial\phi/\partial y = 0$ become simply

$$u = -\frac{\partial f}{\partial x}, \qquad v = \frac{\partial f}{\partial y}, \qquad (6.6)$$

meaning that the normal to the wavefront (the ray) is in the direction of the vector $\langle u, v \rangle$. Equation (6.6) is called a *gradient mapping* between the wavefront space (x, y) (or state space) and the observation space (u, v) (or control space). This yields the equation of the curve on the wavefront in (x, y) space that gives rise to the caustic in (u, v) space; it is called the *critical curve* . Equation (6.6) implies that

$$J(u, v) = \begin{vmatrix} \partial u/\partial x & \partial u/\partial y \\ \partial v/\partial x & \partial v/\partial y \end{vmatrix} = 0, \qquad (6.7)$$

that is the Jacobian is zero on a caustic; the gradient map is singular. In general, a small area dS on the wavefront maps into a correspondingly small area $d\sigma$ of angle space according to $d\sigma = |J|dS$; on a caustic $d\sigma$ vanishes to first order, so there is focusing (at infinity). $|J|^{-1}$ is proportional to the intensity of light on the screen $(dS/d\sigma = |J|^{-1})$. As the point (u, v) moves across a caustic, the number of rays passing through it changes.

6.1 BASIC GEOMETRY OF THE FOLD AND CUSP CATASTROPHES

6.1.1 The Fold

Consider a wavefront with a point of inflection, suppressing (without loss of generality) the y coordinate but including u so that now the potential function depends not only on x but also a parameter u:

$$\phi \rightarrow \phi(x; u) = f(x) + ux. \tag{6.8}$$

Furthermore, to study the fold caustic, let $f(x) = x^3$, so that

$$\phi(x; u) = x^3 + ux. \tag{6.9}$$

The phase space (x, u) is of course two dimensional, and the equilibrium "surface" defined by the curve $\phi_x = 0$ is

$$3x^2 + u = 0 \tag{6.10}$$

so that when this is satisfied for $u < 0$,

$$x = \pm \left(\frac{|u|}{3} \right)^{1/2}. \tag{6.11}$$

It is satisfied for $x = 0$ if $u = 0$ and $\phi_x > 0$ when $u > 0$. The singularity set corresponds to $\phi''(x) = 0$, that is, $x = 0$ (this is the bifurcation set) and $\phi'(x) = 0$ (i.e., $u = 0$). The bifurcation set divides the control space into two regions, $u > 0$ and $u < 0$. If $u > 0$, $\phi_x = 0$ has no real solutions, and ϕ has no critical points. If $u < 0$, ϕ has two critical points, a maximum and a minimum, corresponding to two equilibria, one stable, one unstable. They merge into a point of inflection on the bifurcation set.

6.1.2 The Cusp

At the next level of complexity, we have the potential function

$$\phi(x; u, \upsilon) = x^4 + ux^2 + \upsilon x. \tag{6.12}$$

The phase space is $\{(x, u, \upsilon)\}$ and the equilibrium surface is defined by $\phi_x = 0$, or

$$4x^3 + 2ux + \upsilon = 0. \tag{6.13}$$

As for the fold catastrophe, the singularity set is the subset of equilibrium surface for which $\phi''(x) = 0$:

$$6x^2 + u = 0. \tag{6.14}$$

Finding the Bifurcation Set

We utilize equations 6.13 and 6.14 to eliminate x. From the latter,

$$x = \pm \left(-\frac{u}{6} \right)^{1/2}, \tag{6.15}$$

and substituting into the former, we obtain the cusp-shaped curve in control space, given by

$$\upsilon^2 + \frac{8}{27} u^3 = 0. \tag{6.16}$$

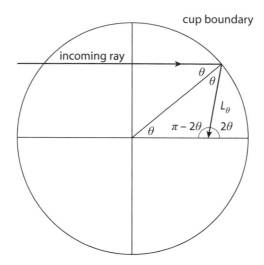

Figure 6.2 Basic geometry for an incoming ray parallel to the horizontal axis L_θ is the reflected ray.

The Teacup Caustic

Referring to Figure 6.2 we parametrize the inside boundary of the reflecting circular ring by $(\cos\theta, \sin\theta)$, $\theta \in (-\pi, \pi]$, and let L_θ be the line along which a ray travels when reflected from an arbitrary point on the boundary. We consider the angular subset $\theta \in (-\pi/2, \pi/2)$. There is an "enveloping curve" to the set of such lines, but where does it touch L_θ? Following [126], we examine the intersection of L_θ and a nearby reflected ray L_ϕ.

As can be seen from Figure 6.2, the point-slope formula for L_θ is

$$y - \sin\theta = \tan\theta \, (x - \cos\theta), \quad \theta \neq \pi/4, \tag{6.17}$$

or in a form valid for all θ,

$$x \sin 2\theta - y \cos 2\theta = \sin\theta. \tag{6.18}$$

Similarly, for L_ϕ,

$$x \sin 2\phi - y \cos 2\phi = \sin\phi, \tag{6.19}$$

so the point of intersection of L_θ and L_ϕ has x-coordinate

$$X(\theta, \phi) = \frac{\sin\theta \cos 2\phi - \sin\phi \cos 2\theta}{\cos 2\phi \sin 2\theta - \sin 2\phi \cos 2\theta},$$

which is clearly singular when $\theta = \phi$. Therefore by L'Hôpital's rule

$$x(\theta) = \lim_{\phi \to \theta} X(\theta, \phi) = \lim_{\phi \to \theta} \frac{\cos\theta \cos 2\phi + 2\sin\phi \sin 2\theta}{2\cos 2\phi \cos 2\theta + 2\sin 2\phi \sin 2\theta}, \tag{6.20}$$

$$= \frac{1}{2}(\cos\theta \cos 2\theta + 2\sin\theta \sin 2\theta) \tag{6.21}$$

$$= \frac{3}{4}\cos\theta - \frac{1}{4}\cos 3\theta, \tag{6.22}$$

from which it follows that

$$y(\theta) = \frac{3}{4}\sin\theta - \frac{1}{4}\sin 3\theta. \tag{6.23}$$

Exercise: Prove the assertions (6.22) and (6.23).

As pointed out in [126], these equations are the parametrization of an epicycloid generated by rolling a circle of radius 1/4 rolling around a circle of radius 1/2 centered at the origin; on the interval $(-\pi/2, \pi/2)$ this is called a *nephroid*. In the optics literature this is associated with the phenomenon of *spherical aberration*.

Osculating Parabolas

From the definition of a parabola, given a reflecting parabolic (instead of circular) boundary with its axis of symmetry parallel to the incoming light rays, then all the reflected rays would intersect at the focus of the parabola. In fact, the focus of the parabola $y_p = f(x)$ that is the best fit to the circle (in the sense that $f'(x)$ and $f''(x)$ at any point on the circle $y_c = f(x)$ are the same as for that parabola) will lie on the enveloping curve—the caustic—of the rays reflected from the circular boundary. *But can this idea be generalized to any smooth reflecting curve?* The answer is yes, and in [126] the following theorem is stated and proved:

The caustic formed by the reflection of incident light rays (parallel to a fixed vector **v**) from a smooth curve is the locus of the foci of the osculating parabolas.

In proving this it transpired that the caustic curve is the singular locus of a planar mapping. Thus we define $\mathbf{x}(\theta) : \mathbb{R} \to \mathbb{R}^2$ as the vector-valued smooth function that parametrizes a curve. Since the tangent vector at $\mathbf{x}(\theta)$ is simply $\mathbf{x}'(\theta)$, the vector **v** can be resolved into its tangential and normal components in terms of the standard inner product $\langle \cdot, \cdot \rangle$, that is,

$$\mathbf{v}_T = \frac{\langle \mathbf{v}, \mathbf{x}'(\theta)\rangle}{\langle \mathbf{x}'(\theta), \mathbf{x}'(\theta)\rangle}\mathbf{x}'(\theta); \quad \mathbf{v}_N = \mathbf{v} - \mathbf{v}_T. \tag{6.24}$$

If $\mathbf{v}_R$ is the direction of the reflected ray, only its normal direction is reversed so we can write

$$\mathbf{v}_R = 2\mathbf{v}_T - \mathbf{v} \tag{6.25}$$

$$= \frac{2\langle \mathbf{v}, \mathbf{x}'(\theta)\rangle}{\langle \mathbf{x}'(\theta), \mathbf{x}'(\theta)\rangle}\mathbf{x}'(\theta) - \mathbf{v} \tag{6.26}$$

(see Figure 6.3).

Since the reflected ray passes through the point $\mathbf{x}(\theta)$, the line of reflection may be parametrized by

$$\mathbf{x}(\theta) + t\mathbf{v}_R = \mathbf{x}(\theta) + t\left(\frac{2\langle \mathbf{v}, \mathbf{x}'(\theta)\rangle}{\langle \mathbf{x}'(\theta), \mathbf{x}'(\theta)\rangle}\mathbf{x}'(\theta) - \mathbf{v}\right). \tag{6.27}$$

This represents a mapping $g(\theta, t) : \mathbb{R}^2 \to \mathbb{R}^2$. The envelope of the rays $\mathbf{v}_R$ is the image of points at which the mapping is singular. To simplify the geometry without

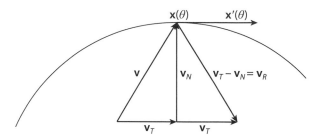

Figure 6.3 Various vectors associated with the point $\mathbf{x}(\theta)$.

loss of generality let the vector $\mathbf{v} = \langle 0, -1 \rangle$, that is, straight down in the usual sense, with the reflecting curve being the graph of $\mathbf{x}(\theta) = \langle \theta, f(\theta) \rangle$. Then

$$g(\theta, t) = \mathbf{x}(\theta) + t\mathbf{v}_R \tag{6.28}$$

$$= \langle \theta, f(\theta) \rangle + t \left\{ \frac{-2 f'(\theta)}{1 + [f'(\theta)]^2} \langle 1, f'(\theta) \rangle - \langle 0, -1 \rangle \right\} \tag{6.29}$$

$$= \left\langle \theta - \frac{2 t f'(\theta)}{1 + [f'(\theta)]^2}, f(\theta) + \frac{t \left(1 - [f'(\theta)]^2 \right)}{1 + [f'(\theta)]^2} \right\rangle \tag{6.30}$$

$$= \langle g_1, g_2 \rangle . \tag{6.31}$$

The Jacobian determinant for g is

$$J_g(\theta, t) = \begin{vmatrix} \partial g_1/\partial \theta & \partial g_1/\partial t \\ \partial g_2/\partial \theta & \partial g_2/\partial t \end{vmatrix} = A^2 \left[A - 2 t f''(\theta) \right], \tag{6.32}$$

where $A = 1 + [f'(\theta)]^2$. This is singular when

$$t = t_s = \frac{1 + [f'(\theta)]^2}{2 f''(\theta)}, \tag{6.33}$$

and for this value of t,

$$g(\theta, t_s) = \left\langle \theta - \frac{f'(\theta)}{f''(\theta)}, f(\theta) + \frac{t \left(1 - [f'(\theta)]^2 \right)}{2 f''(\theta)} \right\rangle . \tag{6.34}$$

If we write down the Taylor polynomial of degree two for $f(\theta)$ and express it in the standard form

$$(\theta - h)^2 = 4 p (f - k), \tag{6.35}$$

the focus is at the point $(h, k + p)$, which is exactly the point specified in equation (6.34), thus proving the theorem.

Exercise: Suppose that the boundary of a semicircular 'cup' is given by

$$f(x) = - \left(1 - x^2 \right)^{1/2}, \quad -1 \leq x \leq 1, \tag{6.36}$$

and that the rays (incident from infinity) are now parallel to the y-axis. Express $f(x)$ in a second degree Taylor polynomial about $x = x_0$, and show that the focus of the resulting parabolic curve is at the point

$$\left(x_0^3, -\frac{1}{2}\left(1 + 2x_0^2\right)\left(1 - x_0^2\right)^{1/2}\right). \tag{6.37}$$

Make the substitution $x_0 = \cos\theta$ to show that the focus is at

$$\left(\frac{3}{4}\cos\theta + \frac{1}{4}\cos 3\theta, -\frac{3}{4}\sin\theta - \frac{1}{4}\sin 3\theta\right). \tag{6.38}$$

Finally, show that the change $\theta \to \theta - \pi/2$ in equation (6.38) yields equations (6.23) and (6.22) respectively, that is,

$$(x(\theta - \pi/2), y(\theta - \pi/2)) \to (y(\theta), x(\theta)). \tag{6.39}$$

The Teacup Caustic: An Elegant Approach

According to Gibson [130], caustics can be viewed as evolutes of orthotomics. Ah! *So* glad he explained that... All seriousness aside, his very elegant analysis is summarized below. When the light source is at infinity, the rays reflected from a circular boundary form what is called a catacaustic, and we can investigate this, following Gibson, by considering the unit circle defined by

$$z(\lambda) = e^{i\lambda}, \tag{6.40}$$

and a pencil of lines parallel to the x-axis. For each parameter λ the unique incident ray through $z(\lambda)$ is reflected in the tangent at that point. The direction of the tangent line is in the direction of the tangent vector $z'(\lambda)$, and since the reflected ray makes a rotation twice this from its incident direction, the reflected ray is in the direction

$$\left[z'(\lambda)\right]^2 = -e^{2i\lambda}. \tag{6.41}$$

Suppose we parametrize in the following way; let

$$Z(\lambda, t) = z(\lambda) + t\left[z'(\lambda)\right]^2 = e^{i\lambda} - te^{2i\lambda}. \tag{6.42}$$

$Z(\lambda, t)$ has components

$$X(\lambda, t) = \cos\lambda - t\cos 2\lambda, \ Y(\lambda, t) = \sin\lambda - t\sin 2\lambda.$$

Now an envelope of the family is called a *caustic* by a reflection of z (the "mirror") with respect to the source (here, at infinity, since the incident rays are parallel). The caustic curve is actually a singular set of the family of parametrized curves, defined by the vanishing of the determinant of the Jacobian matrix $J(Z)$ of the mapping

$$Z(\lambda, t) = (X(\lambda, t), Y(\lambda, t)), \tag{6.43}$$

where

$$J(Z) = \begin{pmatrix} \partial X/\partial\lambda & \partial Y/\partial\lambda \\ \partial X/\partial t & \partial Y/\partial t \end{pmatrix}. \tag{6.44}$$

Here

$$|J(Z)| = 0 \Rightarrow \det \begin{pmatrix} -\sin\lambda + 2t\sin 2\lambda & \cos\lambda - 2t\cos 2\lambda \\ -\cos 2\lambda & -\sin 2\lambda \end{pmatrix} = 0, \quad (6.45)$$

$$\Rightarrow 2t = \cos\lambda. \qquad (6.46)$$

Thus, changing to the dummy variable θ, we find that the caustic is defined by

$$C(\theta) = e^{i\theta} - \frac{1}{2}(\cos\theta)e^{2i\theta} = \frac{1}{4}\left[3e^{i\theta} - e^{3i\theta}\right], \qquad (6.47)$$

or

$$x = \frac{1}{4}[3\cos\theta - \cos 3\theta], \; y = \frac{1}{4}[3\sin\theta - \sin 3\theta], \qquad (6.48)$$

the same as in equations (6.22) and (6.23). These equations define a nephroid, which has two cusps. So what happened to the cardioid we thought we saw? It's okay: it's actually half a nephroid (and looks very close to a cardioid)—this is the *real* caustic (corresponding to reflection of rays from the inside of the cup). There is also a *virtual* caustic corresponding to rays reflected from the outside of the cup (recall the virtual caustic for the Airy rainbow). Note for completeness that the corresponding parametric equations for the cardioid

$$r = 2(1 - \cos\theta), \quad 0 \le \theta \le \pi, \qquad (6.49)$$

(with cusp at $(0, 0)$ and pointing in the negative x-direction) are

$$x = 2\cos\theta(1 - \cos\theta), \quad y = 2\sin\theta(1 - \cos\theta). \qquad (6.50)$$

In Figure 6.4 the upper dashed curve represents the function $y = 5 + x^3$ drawn for $x > 0$ representing a cubic wavefront. The lower dashed graph is the curvature of this function,

$$\kappa = \frac{|y''|}{\left(1 + y'^2\right)^{3/2}} = \frac{6x}{\left(1 + 9x^4\right)^{3/2}},$$

and the solid curve is the radius of curvature of this function, $R = \kappa^{-1}$; clearly it has the shape of a fold. Conversely, a fold will in general correspond to a locally cubic wavefront.

Ray Density Near a Caustic

Consider rays tangent to any small length of the caustic, small enough that it is approximately a circular arc of radius R. The rays are uniformly distributed along the arc [29]. Two such rays, tangent at points A and B, respectively, intersect at the point P, which is at a distance x ($\ll R$) from the caustic (as in

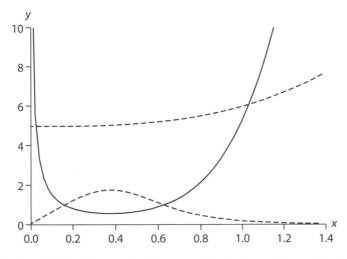

Figure 6.4 Simple example of curvature details for a cubic wavefront.

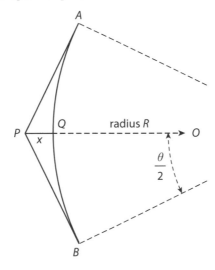

Figure 6.5 Sketch for establishing the ray density as a caustic is approached.

Figure 6.5), and as will be shown below, the length of the caustic arc AB is proportional to $x^{1/2}$. Hence the number of rays crossing the segment PQ is also proportional to $x^{1/2}$, so the density of rays is the derivative of this quantity with respect to x, that is, it is proportional to $x^{-1/2}$.

To establish this result, note from Figure 6.5 that the angle θ subtended by arc AB at the center O is small (though it is exaggerated in the figure), so

$$\cos\frac{\theta}{2} = \frac{R}{R+x} \approx 1 - \frac{\theta^2}{8} \Rightarrow \theta^2 \approx \frac{8x}{R+x}. \qquad (6.51)$$

From this it follows that the arc length AB is

$$R\theta \approx \left[8Rx\,(R+x)^{-1}\right]^{1/2} \approx \left[8Rx\left(1-\frac{x}{R}\right)\right]^{1/2} \approx (8Rx)^{1/2} \propto x^{1/2},$$

(6.52)

which implies the result above (i.e., that the ray density $\propto x^{-1/2}$).

6.2 A BETTER APPROXIMATION

For simplicity we consider a one-dimensional wavefront so the point Q in Figure 6.1 now has coordinates $(x, f(x))$. We further assume that the initial wavefront has uniform unit amplitude and then express the wave disturbance at some point P as an integral superposition of contributions from elements dx of the wavefront. Thus, in view of the form (1.25) (Chapter 1),

$$\psi(\xi, \eta) = \widetilde{C} \int_{-\infty}^{\infty} l^{-1/2} e^{ikl(x;\xi,\eta)} dx,$$

(6.53)

where $\widetilde{C} \in \mathbb{C}$ is to be determined. As $k \to \infty$, the significant contributions to ψ come from near the center of the range, since those from large $|x|$ cancel out by interference (recall from (6.2) that large $|x|$ implies large l); while the wavefront is finite in length, extending the limit to infinity is entirely acceptable here for that reason, as already noted several times. Note that if $\delta l = \lambda/2$, $\delta(kl) = k\delta l = \pi$, and although this changes the sign of the integrand, it barely affects $l^{-1/2} \sim R^{-1/2}$, R being the distance OP. Therefore we can rewrite (6.53) as

$$\psi(\xi, \eta) \approx \widetilde{C} R^{-1/2} \int_{-\infty}^{\infty} e^{ikl(x;\xi,\eta)} dx,$$

(6.54)

where $R = OP = R(\xi, \eta)$. So what value can be ascribed to $\widetilde{C}$? Consider an initially flat wavefront, so from (6.2), for $\psi(0, 0)$ (where $f = 0$, and $Z = Z_0$ when $\xi = 0, \eta = 0$, and $R = Z_0$)

$$l(x; 0, 0) = Z_0 + \frac{x^2}{2Z_0}.$$

(6.55)

But we know that initially $\psi = e^{ikZ_0}$, so on canceling this common factor, we may write, for the initial value of (6.54)

$$1 = \widetilde{C} R^{-\frac{1}{2}} \int_{-\infty}^{\infty} \exp\left(\frac{ikx^2}{2Z_0}\right) dx.$$

(6.56)

Next, we introduce the *Fresnel cosine* and *sine integrals*. They arise, in particular, in connection with diffraction of light by a straightedge and are defined by the expression

$$C(u) + S(u) = \int_0^u \exp\left(\frac{i\pi X^2}{2}\right) dX \tag{6.57}$$

$$= \int_0^u \left[\cos\left(\frac{\pi X^2}{2}\right) + i \int_0^u \sin\left(\frac{\pi X^2}{2}\right)\right] dX. \tag{6.58}$$

If we define

$$x = \left(\frac{\pi Z_0}{k}\right)^{1/2} X, \tag{6.59}$$

and note that here $Z_0 = R$, then (6.56) can be rearranged to give

$$1 = \tilde{C}\left(\frac{\pi}{k}\right)^{1/2} \left[\int_0^\infty \exp\left(\frac{i\pi X^2}{2}\right) - \int_0^{-\infty} \exp\left(\frac{i\pi X^2}{2}\right)\right] dX \tag{6.60}$$

$$= \tilde{C}\left(\frac{\pi}{k}\right)^{1/2} \left[C(\infty) + i S(\infty) - [C(-\infty) + i S(-\infty)]\right]. \tag{6.61}$$

As established below,

$$C(\infty) = S(\infty) = \frac{1}{2}; \quad C(-\infty) = S(-\infty) = -\frac{1}{2}, \tag{6.62}$$

so that

$$1 = \tilde{C}\left(\frac{\pi}{k}\right)^{1/2} (1 + i) = \tilde{C}\left(\frac{2\pi}{k}\right)^{1/2} e^{i\pi/4}, \tag{6.63}$$

whence

$$\tilde{C} = \left(\frac{k}{2\pi i}\right)^{1/2}. \tag{6.64}$$

A further modification of equation (6.3) is in order; in the *paraxial approxima-tion* for which $|X|/Z \ll 1$, so that

$$R = R(X, Z) \equiv OP \approx Z, \tag{6.65}$$

we can write (suppressing the y terms)

$$\phi(x; \xi, \eta) + l \approx R, \tag{6.66}$$

where

$$\xi = \frac{X}{Z}; \quad \eta = \frac{1}{2}\left(\frac{1}{Z_0} - \frac{1}{Z}\right). \tag{6.67}$$

Returning to the expression (6.54), it follows that

$$\psi(\xi, \eta) \approx R^{-1/2} \left(\frac{k}{2\pi i}\right)^{1/2} e^{ikR} D(\xi, \eta), \tag{6.68}$$

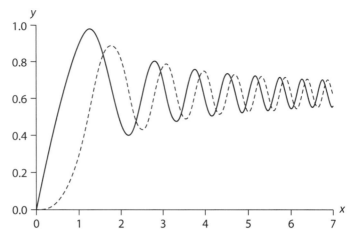

Figure 6.6 $C(x)$ (solid curve) and $S(x)$.

where on using (6.66),

$$D(\xi, \eta) = \int_{-\infty}^{\infty} \exp\left[-ik\phi(x; \xi, \eta)\right] dx \qquad (6.69)$$

is called the *diffraction integral*. Each kind of caustic has its own diffraction integral and corresponding pattern. Two important diffraction integrals are discussed next.

6.2.1 The Fresnel Integrals

These integrals are transcendental functions that arise, for example, in the diffraction of radiation by a plane edge; in view of (6.58), they can be redefined more simply as

$$C(x) = \int_0^x \cos\left(t^2\right) dt, \qquad (6.70)$$

and

$$S(x) = \int_0^x \sin\left(t^2\right) dt; \qquad (6.71)$$

their behavior for $x > 0$ is shown in Figure 6.6.

In the limit as $x \to \infty$ (and generalizing the integrand slightly)

$$C(a) = \int_0^{\infty} \cos\left(at^2\right) dt = \frac{1}{2}\left(\frac{\pi}{2a}\right)^{1/2}, \qquad (6.72)$$

and

$$S(a) = \int_0^{\infty} \sin\left(at^2\right) dt = \frac{1}{2}\left(\frac{\pi}{2a}\right)^{1/2}. \qquad (6.73)$$

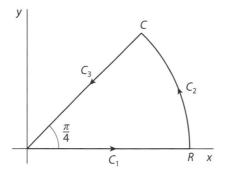

Figure 6.7 Contour for evaluating the Fresnel integrals.

By making the change of variable $\sqrt{a}t \to t$, these integrals become $C(1)$ and $S(1)$, respectively. To prove these results we evaluate the integral

$$E(z) = \int_C e^{iz^2} dz \qquad (6.74)$$

over the wedge-shaped contour C shown in Figure 6.7.

The integrand in (6.74) is analytic inside C so by Cauchy's theorem

$$\int_C e^{iz^2} dz = 0 = \int_{C_1} e^{iz^2} dz + \int_{C_2} e^{iz^2} dz + \int_{C_3} e^{iz^2} dz \qquad (6.75)$$

$$= \int_0^R e^{ix^2} dx + \int_0^{\pi/4} \exp\left[i\left(Re^{i\theta}\right)^2\right] i Re^{i\theta} d\theta$$

$$+ \int_R^0 \exp\left[i\left(xe^{i\pi/4}\right)^2\right] e^{i\pi/4} dx. \qquad (6.76)$$

Therefore

$$\int_0^R e^{ix^2} dx + iR \int_0^{\pi/4} \exp\left[R^2\left(-\sin 2\theta + i\cos 2\theta\right)\right] e^{i\theta} d\theta - e^{i\pi/4} \int_0^R e^{-x^2} dx = 0. \qquad (6.77)$$

Using the well-known result that

$$\lim_{R\to\infty} \int_0^R e^{-x^2} dx = \frac{\sqrt{\pi}}{2}, \qquad (6.78)$$

the above result can be expressed as

$$\lim_{R\to\infty} \int_0^R e^{ix^2} dx = \frac{\sqrt{\pi}}{2} e^{i\pi/4} - \lim_{R\to\infty} i R \mathcal{I}, \qquad (6.79)$$

where

$$\mathcal{I} = \int_0^{\pi/4} \exp\left[R^2\left(-\sin 2\theta + i\cos 2\theta\right)\right] e^{i\theta} d\theta. \qquad (6.80)$$

Since

$$\frac{4\theta}{\pi} \leq \sin 2\theta, \tag{6.81}$$

we can see that

$$|i\,R\mathcal{I}| = |R\mathcal{I}| \leq R \int_0^{\pi/4} \exp\left(-R^2 \sin 2\theta\right) d\theta$$

$$\leq R \int_0^{\pi/4} \exp\left(-4R^2\theta/\pi\right) d\theta \tag{6.82}$$

$$= \pi R \left(\frac{1 - e^{-R^2}}{4R^2}\right) \leq \frac{\pi}{4R}. \tag{6.83}$$

Therefore,

$$\lim_{R \to \infty} |R\mathcal{I}| = 0 \tag{6.84}$$

and hence from the expression (6.77) we have

$$\lim_{R \to \infty} \int_0^R e^{ix^2} dx = e^{i\pi/4} \left(\frac{\sqrt{\pi}}{2}\right) = \frac{1}{2}\left(\frac{\pi}{2}\right)^{1/2} (1 + i), \tag{6.85}$$

whence

$$\lim_{R \to \infty} \int_0^R \sin x^2 dx = \lim_{R \to \infty} \int_0^R \cos x^2 dx = \frac{1}{2}\left(\frac{\pi}{2}\right)^{1/2}. \tag{6.86}$$

Thus

$$C(a) = S(a) = \frac{1}{2}\left(\frac{\pi}{2a}\right)^{1/2}. \tag{6.87}$$

6.3 THE FOLD DIFFRACTION CATASTROPHE

In this section we generalize the fold potential function (6.9) to include a coefficient of the cubic term and use the notation of the previous section, so that

$$\phi(x; \xi, p) \equiv \frac{1}{3}px^3 + \xi x. \tag{6.88}$$

Hence

$$D(\xi) = \int_{-\infty}^{\infty} \exp\left[-ik\left(\frac{1}{3}px^3 + \xi x\right)\right] dx. \tag{6.89}$$

The k dependence can be removed by introducing new dimensionless state and control variables

$$t = (kp)^{1/3}x, \quad s = k^{2/3} p^{-1/3} \xi. \tag{6.90}$$

Since the integrand is odd in x,

$$D(\xi) = 2 \int_0^{\infty} \cos k \left(\frac{px^3}{3} + \xi x\right) dx = 2(kp)^{-1/3} \int_0^{\infty} \cos\left(\frac{1}{3}t^3 + st\right) dt, \tag{6.91}$$

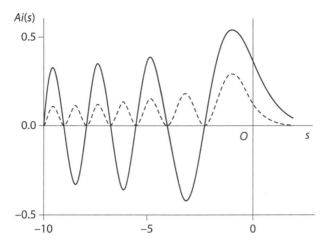

Figure 6.8 Graph of $Ai(s)$ (solid curve) and its square (dashed curve).

that is,

$$D(s) = 2\pi (kp)^{-1/3} Ai(s) \tag{6.92}$$

where $Ai(s)$ is our old friend, the Airy function of the first kind,

$$Ai(s) = \frac{1}{\pi} \int_0^\infty \cos\left(\frac{1}{3}t^3 + st\right) dt. \tag{6.93}$$

Note that s is proportional to the angular distance ξ from the fold (where two rays merge). Also note from the graphs of $Ai(s)$ and its square (Figure 6.8) that (i) there are interference maxima and minima; (ii) there is nonzero intensity on the "dark side" ($s > 0$); (iii) there is finite intensity at $\xi = 0$ (geometrical optics predicts none; clearly, ray theory breaks down on a caustic—the wavefront comes to an end on the caustic with a zero radius of curvature, and folds back on itself); and (iv) the main maximum in the Airy pattern is displaced toward the bright side, from $\xi = 0$.

A Simple Measure of the "Strength" of the Fold Caustic

Equations (6.68) and (6.92) imply that

$$\psi(0) \propto k^{1/2} D(0) \propto k^{1/2} k^{-1/3} Ai(0) \propto k^{1/6}, \tag{6.94}$$

and the exponent of k is called the *Arnold index*. Clearly the "strength" increases with no bound as $k \to \infty$; ($\lambda \to 0$), that is, in the limit of geometrical optics.

6.3.1 The Rainbow as a Fold Catastrophe

In Chapter 4 the deviation angle for a primary bow was found to be

$$D_1(i) = 2i - 4r(i) + \pi. \qquad (6.95)$$

Recall that the critical angle of incidence for the Descartes or rainbow ray is i_c and that for the primary bow,

$$D_1(i_c) \equiv D_c = 2\arccos\left[\frac{1}{n^2}\left(\frac{4-n^2}{3}\right)^{3/2}\right], \qquad (6.96)$$

where

$$i_c = \arccos\left(\frac{n^2-1}{3}\right)^{1/2}, \qquad (6.97)$$

so that

$$\sin i_c = \left(\frac{4-n^2}{3}\right)^{1/2}. \qquad (6.98)$$

In addition to using the angle of incidence as an independent variable, it is frequently valuable to express quantities as functions of the (now dimensional) impact parameter $b(i) = a\sin i$, a being the drop radius:

$$b_c = a\sin i_c = a\left(\frac{4-n^2}{3}\right)^{1/2}. \qquad (6.99)$$

In the language of catastrophe theory, the control space D_1, the direction of the emerging Descartes ray, is folded back on itself at D_c; a fold catastrophe results [29]; note that D_c is independent of a; this means that each droplet, regardless of size, acts in the same way. We shall drop the subscript 1 here, since in this section we consider only the primary bow. Since $D = D(b)$, we have, expanding about b_c,

$$D(b) = D(b_c) + D'(b_c)(b - b_c) + \frac{1}{2}D''(b_c)(b - b_c)^2 + O(b - b_c)^3. \qquad (6.100)$$

Let

$$\delta = b - b_c, \quad \Delta = D(b) - D(b_c) = D - D_c, \qquad (6.101)$$

so

$$\Delta = \frac{1}{2}D''(b_c)\delta^2 + O(\delta^3) \equiv \alpha\delta^2 + O(\delta^3), \qquad (6.102)$$

where now Δ is the *control variable* (in the observation space) and δ is the state variable. As we have seen already, the existence of a fold catastrophe implies the presence of a (virtual) cubic wavefront W in the vicinity of Descartes ray (Figure 6.9). In this neighborhood there is a smooth mapping between the impact parameter δ and distance along this wavefront. We take the distance along W (measured from the virtual Descartes ray) to be $\beta\delta$, β being a constant. From the first equation of (6.6), with $f(\delta)$ describing the height of the wavefront, and $x = \beta\delta$, identifying $u = \Delta$, we have

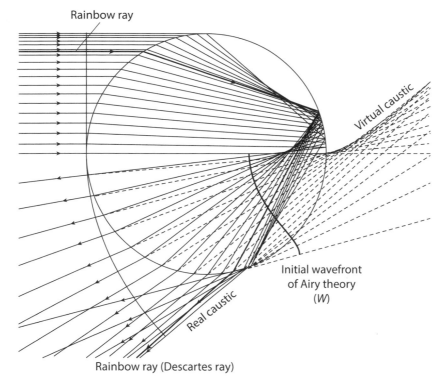

Figure 6.9 Primary rainbow caustics (upper half).

$$\Delta = -\frac{\partial f(\delta)}{\partial x} = -\frac{df}{d\delta}\frac{d\delta}{dx} = -\frac{1}{\beta}\frac{df}{d\delta} \approx \alpha\delta^2, \qquad (6.103)$$

$$f(\delta) = -\frac{1}{3}\alpha\beta\delta^3, \qquad (6.104)$$

where $f(0) = 0$, so (6.5) implies

$$\phi(\delta, \Delta) = -\frac{1}{3}\alpha\beta\delta^3 + \beta\delta\Delta. \qquad (6.105)$$

Hence (6.69) is

$$D(\Delta) = \int_{-\infty}^{\infty} \exp\left(-ik\left[-\alpha\beta\delta^3/3 + \beta\delta\Delta\right]\right)\beta d\delta, \qquad (6.106)$$

which is (6.89) with $x = \beta\delta$, $p = -\alpha\beta^{-2}$, and $\xi = \Delta$. Therefore (6.92) implies that

$$D(s) = 2\pi\left(-k\alpha\beta^{-2}\right)^{-1/3}\text{Ai}(s) = -2\pi\beta^{2/3}(k\alpha)^{-1/3}\text{Ai}(s), \qquad (6.107)$$

where

$$s = k^{2/3}\left(-\alpha\beta^{-2}\right)^{-1/3}\xi = -(k\beta)^{2/3}\alpha^{-1/3}\Delta, \qquad (6.108)$$

the form of which has been encountered several times already.

6.4 CAUSTICS: THE AIRY INTEGRAL IN THE COMPLEX PLANE

> A caustic is a sort of "wound" for straightforward ray theory, which predicts an amplitude rising to infinity at the caustic and then falling discontinuously to zero. The Airy-integral solution ... "heals" the wound, allowing a perfectly finite and continuous transition between the two different regimes. [30]

In other words, a caustic is a boundary between a region with a complicated wave pattern, due to interference between two groups of waves, and a neighboring region containing *no* waves (though more generally, we could extend this definition to n groups of waves in one region and $n - 2$ in the other). If you have not already realized it by now, the key to understanding caustics is a single mathematical concept: the Airy integral. Straightforward ray theory breaks down near a caustic, but the Airy integral "heals the wound." In homogeneous systems analyzed by the method of stationary phase, a local difficulty arises wherever a second derivative of the phase (or in more than one dimension, a principal second derivative) vanishes along with the gradient itself. Near such a point, the Airy integral plays the same role as does the Gaussian integral at a general point. Rays run together near such points, because the group speed (group velocity, in more than one dimension) is stationary. The locus of these points is the caustic, which separates the region without rays from the region twice-covered by rays. Note, however, that it is the assumptions of ray theory that break down in the neighborhood of a caustic, because local gradients of wavenumber become large there. In fact, and rather more generally, the boundary of *any* trapped-wave region is a caustic.

We can do no better here than follow Lighthill's elegant approach to this topic [30] and return to the method of stationary phase for a one-dimensional wave system by considering in more mathematical detail how a limited disturbance must be modified around a wavenumber $k = k_c$, where $c_g = d\omega/dk$ is stationary, that is, where

$$c_g'(k_c) = \omega''(k_c) = 0. \tag{6.109}$$

The locus $x/t = c_g(k_c)$ is a caustic, because values of c_g on one side of the stationary value $c_g(k_c)$ are found for two wavenumbers k, and values on the other side for none. As before, we study the asymptotic behavior of an integral

$$I(t) = \int_0^\infty F(k)e^{it\psi(k)}dk, \tag{6.110}$$

with

$$\psi(k) = \omega(k) - kx/t, \tag{6.111}$$

by again concentrating on where the phase $t\psi(k)$ has vanishing first derivatives (i.e., where $x/t = \omega'(k) = c_g(k)$), but now with the difference that at $k = k_c$, $\psi''(k) = \omega''(k) = 0$. To remedy this, away from the stationary point, the path can be deformed to one on which the imaginary part of ψ is at least δ, $\delta > 0$. This has been mentioned before (Section 3.6.2), but the approach is a little more rigorous in this section. Near the stationary point, however (where this is impossible), a short

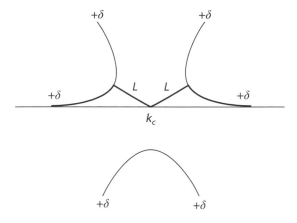

Figure 6.10 Behavior of ψ near $k = k_c$.

link L can be made on which a Taylor expansion is used to estimate the asymptotic value of the integral. Now $\psi'(k_c) = \psi''(k_c) = 0$, but we assume that $\omega'''(k_c) \neq 0$. Contours on which the imaginary part of ψ is δ then take the forms shown in Figure 6.10 when

$$\psi'''(k_c) = \omega'''(k_c) > 0; \qquad (6.112)$$

when $\omega'''(k_c) < 0$, it is only necessary to examine the mirror images of this about the real k-axis [30].

Since now

$$\psi(k) = \psi(k_c) + \frac{1}{6}(k - k_c)^3 \psi'''(k_c) + O(k - k_c)^4, \qquad (6.113)$$

it follows that in the neighborhood of the caustic,

$$\psi(k) - \psi(k_c) = \frac{1}{6}(k - k_c)^3 \psi'''(k_c). \qquad (6.114)$$

Furthermore, since $\psi(k_c) \in \mathbb{R}$

$$\operatorname{Im}\psi(k) = \frac{1}{6}\psi'''(k_c)\operatorname{Im}(k - k_c)^3. \qquad (6.115)$$

Now let $k = k_r + ik_i$ in vicinity of k_c; thus

$$\operatorname{Im}(k - k_c)^3 = 3(k_r - k_c)^2 k_i - k_i^3, \qquad (6.116)$$

and therefore

$$\operatorname{Im}\psi(k) = \frac{\psi'''(k_c)}{6}k_i\left[3(k_r - k_c)^2 - k_i^2\right]. \qquad (6.117)$$

At this stage we let $k_r - k_c = x$ and $k_i = y$ in order to arrive at the result

$$\frac{\operatorname{Im}\psi(k)}{\psi'''(k_c)} = \frac{1}{6}y(3x^2 - y^2) > 0, \qquad (6.118)$$

so that

$$(k - k_c)^3 = |k - k_c|^3 e^{3i\theta}, \tag{6.119}$$

is pure imaginary when $\cos 3\theta = 0$, that is,

$$\theta = \frac{\pi}{6}, \frac{\pi}{2}, \frac{5\pi}{6}, \frac{7\pi}{6}, \ldots, \tag{6.120}$$

and $\sin 3\theta > 0$ only when $\theta = \pi/6$ and $5\pi/6$ in $[0, \pi]$. Hence the straight line segments are each at angle of $\pi/6$ radians to the real k-axis. In the neighborhood of the caustic

$$\frac{x}{t} \approx c_g(k_c).$$

Note that $\psi(k)$ includes a linear term

$$(k - k_c)\psi'(k_c) = (k - k_c)\left[\omega'(k_c) - \frac{x}{t}\right], \tag{6.121}$$

which is a significant addition (compared with the cubic term) when k is very close to k_c. For small values of the second factor in equation (6.121), however, the cubic term in $\psi'''(k_c)$ is the dominant one at the extremities of L. Therefore the link remains effective in raising the imaginary part of ψ to $+\delta$. With error $O[\exp(-t\delta)]$, then, the integral (6.110) can be taken as an integral over the link L, so (6.110) can be written as

$$I(t) = \int_L [F(k_c) + O(|k - k_c|)] \exp\{it [g(k; k_c)]\} dk, \tag{6.122}$$

where

$$g(k; k_c) = \psi(k_c) + (k - k_c)\psi'(k_c) + \frac{1}{6}(k - k_c)^3 \psi'''(k_c) + O(|k - k_c|^4). \tag{6.123}$$

We now need, not the Gaussian integral of the exponential of a quadratic, but the Airy integral of the exponential of a sum of a linear term and a cubic term [30]. Equation (6.122) can be reduced to standard form by the substitution

$$k - k_c = \left[\frac{1}{2}t\psi'''(k_c)\right]^{-1/3} s, \tag{6.124}$$

which simplifies the cubic term to $is^3/3$. Introducing the new variable

$$X = \left[\frac{1}{2}t\psi'''(k_c)\right]^{-1/3} t\psi'(k_c), \tag{6.125}$$

simplifies the linear term to isX, giving

$$I(t) = F(k_c) \exp[it\psi(k_c)] \left[\frac{1}{2}t\psi'''(k_c)\right]^{-1/3} \int \exp\left[i\left(sX + \frac{1}{3}s^3\right)\right] ds, \tag{6.126}$$

with an error factor $1 + O(t^{-1/3})$, since (6.124) implies

$$O(|k - k_c|) = O\left(t|k - k_c|^4\right) = O(t^{-1/3}).$$

The integral in (6.126) is taken along a bent path in the s-plane corresponding to L in the k-plane. At the extremities of this, the modulus of the integrand falls to $O[\exp(-t\delta)]$. The integral therefore differs by an error of only this order from the same integral taken all the way to infinity. Next we define, following the notation of [30],

$$\mathrm{Ai}(X) = \frac{1}{2\pi} \int_{\infty e^{5\pi i/6}}^{\infty e^{\pi i/6}} \exp\left[i(sX + \frac{1}{3}s^3)\right] ds, \tag{6.127}$$

to infer that the asymptotic result for ξ is

$$I \sim 2\pi F(k_c)\,(\exp\left[it\psi(k_c)\right])\left[\frac{1}{2}t\psi'''(k_c)\right]^{-1/3} \mathrm{Ai}(X). \tag{6.128}$$

(This conclusion is also correct when $\psi'''(k_c) < 0$), changing L to its mirror image, but the cube root in expression (6.124) should be taken as negative [30].)

6.4.1 The Nature of Ai(X)

When X has a substantial positive value, the phase $(sX + s^3/3)$ in (6.127) has no stationary point near the real axis. To see this, recall that $X + s^2 = 0$ means that $s = \pm i|X|^{1/2}$, but from (6.124), $\mathrm{Im}\, k > 0 \Rightarrow \mathrm{Im}\, s > 0$, so the integral becomes exponentially small, corresponding to the side of the caustic without waves. However, when X has a substantial negative value, $s^2 - |X| = 0$ and there are two well separated stationary points for real s, given by $s = \pm|X|^{1/2}$. At these points, $\psi''(k)$ is positive and negative, respectively. Then it is natural to seek to estimate the Airy integral as a sum of two Gaussian integrals by deforming the path into one passing through these points at $\pm\pi/4$ to the real axis. So, let

$$f(s) = i\left(sX + \frac{1}{3}s^3\right). \tag{6.129}$$

Then in the neighborhood of $s = |X|^{1/2}$, $X < 0$,

$$f(s) = f(|X|^{1/2}) + (s - |X|^{1/2})f'(|X|^{1/2})$$
$$+ \frac{1}{2}(s - (|X|^{1/2}))^2 f''(|X|^{1/2}) + O(s - (|X|^{1/2}))^3, \tag{6.130}$$

where $f'(|X|^{1/2}) = 0$ and $f''(|X|^{1/2}) = 2is$, that is,

$$f(s) \approx (i|X|^{1/2}X + \frac{1}{3}i|X|^{3/2}) + i(s - |X|^{1/2})^2|X|^{1/2} \tag{6.131}$$

$$= -\frac{2}{3}i|X|^{3/2} + i(s - |X|^{1/2})^2|X|^{1/2}. \tag{6.132}$$

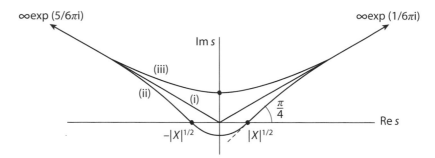

$\infty\exp(5/6\pi i)$ $\infty\exp(1/6\pi i)$

Im s

(iii)

(ii)

(i)

$\dfrac{\pi}{4}$

Re s

$-|X|^{1/2}$ $|X|^{1/2}$

Figure 6.11 The complex s-plane defined by (6.124). Path (i) corresponds to L in Figure 6.10; the deformed path (ii) corresponds to the approximation for $X < 0$ but $|X|$ large (equation (6.143)), and deformed path (iii) corresponds to the approximation for $X > 0$ and large (equation (6.142)).

Now

$$\mathrm{Ai}(X) = \frac{1}{2\pi} \int_{\infty e^{5\pi i/6}}^{\infty e^{\pi i/6}} e^{f(s)}\,ds. \tag{6.133}$$

(see Figure 6.11).

Following [30] we replace this integral by the approximants

$$\int_{\infty e^{3\pi i/4}}^{-\infty e^{3\pi i/4}} e^{f(s)}\,ds, \quad \text{and} \quad \int_{-\infty e^{\pi i/4}}^{\infty e^{\pi i/4}} e^{f(s)}\,ds, \tag{6.134}$$

for $s = -|X|^{-1/2}$ and $s = |X|^{1/2}$, respectively. Thus the contribution from the stationary phase at $s = +|X|^{1/2}$ is

$$I_1 = \exp\left(-\frac{2}{3}i|X|^{3/2}\right) \int_{-\infty e^{\pi i/4}}^{\infty e^{\pi i/4}} \exp\left[i|X|^{1/2}(s - |X|^{1/2})^2\right] ds. \tag{6.135}$$

Now we make a change of variable

$$S = (s - |X|^{1/2})e^{-i\pi/4}, \tag{6.136}$$

so that when $s \to \pm[\infty \exp(i\pi/4)]$, $S \to \pm\infty$. Then

$$I_1 = \exp\left(-\frac{2}{3}i|X|^{3/2} + \frac{i\pi}{4}\right) \int_{-\infty}^{\infty} \exp\left[(i|X|^{1/2} + i\frac{\pi}{2})S^2\right] dS \tag{6.137}$$

$$= \exp\left(-\frac{2}{3}i|X|^{3/2} + \frac{i\pi}{4}\right) \int_{-\infty}^{\infty} \exp(-|X|^{1/2}S^2)\,dS \tag{6.138}$$

$$= \left(\frac{\pi}{|X|^{1/2}}\right)^{1/2} \exp\left(-\frac{2}{3}i|X|^{3/2} + \frac{i\pi}{4}\right). \tag{6.139}$$

In a similar fashion, the contribution from $s = -|X|^{1/2}$ is found to be

$$I_2 = \left(\frac{\pi}{|X|^{1/2}}\right)^{1/2} \exp\left(\frac{2}{3}i|X|^{3/2} - \frac{i\pi}{4}\right); \tag{6.140}$$

note the change of sign in the exponential. Therefore

$$\text{Ai}(X) \sim \frac{I_1}{2\pi} + \frac{I_2}{2\pi} = \frac{1}{2\pi} \left(\frac{\pi}{|X|^{1/2}} \right)^{1/2} \left[\exp\left(-\frac{2}{3}i|X|^{3/2} + \frac{i\pi}{4} \right) \right.$$

$$\left. + \exp\left(\frac{2}{3}i|X|^{3/2} - \frac{i\pi}{4} \right) \right]; \tag{6.141}$$

that is, it consists of two wavelike contributions. These modulate the $\exp[it\psi(k_c)]$ term in (6.128), corresponding to two groups of waves on the side of the caustic with $X < 0$, For large and positive values of X, the path can be similarly deformed so as to pass through an imaginary position of stationary phase $s = +iX^{1/2}$, where $f(iX^{1/2}) = -2X^{3/2}/3$, and the phase $f(s)/i$ has second derivative $2iX^{1/2}$. The path of steepest descent through this point is parallel to the real axis and allows an estimate of $\text{Ai}(X)$ as the Gaussian integral (see, e.g., (6.135))

$$\text{Ai}(X) \sim \frac{1}{2\pi} \exp\left(-\frac{2}{3}X^{3/2} \right) \int_{-\infty}^{\infty} \exp\left[-X^{1/2}(s - iX^{1/2})^2 \right] ds, \ X > 0$$

$$= \frac{1}{2\pi^{1/2}X^{1/4}} \exp\left(-\frac{2}{3}X^{3/2} \right). \tag{6.142}$$

Combining the forms given in (6.127) for large $|X|$ (but $X < 0$), we have

$$\text{Ai}(X) \sim \frac{1}{\pi^{1/2}|X|^{1/4}} \cos\left(\frac{2}{3}X^{3/2} - \frac{\pi}{4} \right), \quad X < 0. \tag{6.143}$$

As pointed out in [30], comparing these asymptotic forms (taken outside their domains of validity to small $|X|$), the correct form of $\text{Ai}(X)$ shows how the Airy integral makes a quick and smooth transition from its oscillatory asymptotic form (6.143) ($X < 0$) to its exponentially vanishing asymptotic form (6.142) ($X > 0$). Recall from (6.111) that

$$\psi(k) = \omega(k) - k\frac{x}{t}. \tag{6.144}$$

Then (6.125) and (6.128) give

$$I \sim \frac{2\pi F(k_c)}{[t\omega'''(k_c)/2]^{1/3}} \text{Ai} \left\{ \frac{t\omega'(k_c) - x}{[t\omega'''(k_c)/2]^{1/3}} \right\} \exp\left[i\left(\omega(k_c)t - k_c x \right) \right]. \tag{6.145}$$

Waves are significant on the side of the caustic where

$$\text{sgn}\left[x - t\omega'(k_c) \right] = \text{sgn}\left[\omega'''(k_c) \right], \tag{6.146}$$

because as noted from (6.127), two groups of waves are present when $X < 0$ (in fact, these two groups of waves of nearby wavenumber beat together). On the other side of the caustic, for $X > 0$, the wave amplitude falls away faster than an exponential in X, the decay distance growing gradually with time. The general theory of stationary phase gives, for one dimension, an amplitude decaying in time like $t^{-1/2}$ as the energy spreads out; but the breakdown of the theory near the caustic is apparent from the fact that $|\omega''(k)|^{-1/2} \to \infty$ there. But of course, the amplitude does not become infinite there; in fact,

$$\text{Ai}(0) = \left[3^{2/3} \Gamma \left(\frac{2}{3} \right) \right]^{-1} \approx 0.35503, \qquad (6.147)$$

which is clearly not infinite! This amplitude does however decay more slowly than $t^{-1/2}$; from (6.128) it appears to decay as $t^{-1/3}$. Nevertheless, there is a subtle feature here: for fixed x/t satisfying the condition (6.146), I asymptotically reverts to the $t^{-1/2}$ form, because the $|X|^{-1/4}$ in (6.143) decays with time in proportion to $(t^{2/3})^{-1/4} = t^{-1/6}$. At the caustic boundary of the wave region, then, a slight enhancement of amplitude (from a value of $O(t^{-1/2})$ to one of $O(t^{-1/3})$) occurs just before its exponential decays to zero.

Chapter Seven

Introduction to the WKB(J) Approximation:

All Things Airy

> In summarizing the various contributions to approximating a quantum
> mechanical wave function by an oscillatory wave depending on a phase
> integral, the physicist J. Calvert noted (i) the alternative notations for the
> approximation, such as WKB, BWK, BWKJ, adiabatic, semiclassical or
> phase integral; (ii) cited work by Wentzel, Kramers, Brillouin, Jeffreys,
> Rayleigh, Liouville, Denham, Langer and Furry, and (iii) wrote "This
> would make it the WKBJRLDLF approximation, I suppose."
>
> [31]

7.1 OVERVIEW

The WKB(J) approximation is widely used in applied mathematics and mathematical physics to find approximate solutions of linear ordinary differential equations (of any order in principle) with spatially varying coefficients (see Chapter 1 for my explanation of the acronym WKB(J)). It is closely associated with the *semiclassical approach* [262] in quantum mechanics in which the wavefunction is considered to have a slowly varying amplitude and/or phase. Consider the differential equation

$$w'' + p(x)w = 0, \tag{7.1}$$

where $p(x)$ is a slowly varying function of x. If p were a constant, we would seek solutions of the form

$$w(x) = a \exp\left[i\theta(x)\right],$$

where a is a constant and in general θ can take on either or both of the forms $\theta(x) = \pm p^{1/2}x$. If p is a slowly varying function of x (in the sense discussed in Chapter 3) then we expect solutions to be of the slightly more general form

$$w(x) = a(x) \exp\left[i\theta(x)\right], \tag{7.2}$$

and also anticipate that the phase term will vary more rapidly than the amplitude term. This condition will be quantified below. With solution (7.2), equation (7.1) takes the form (canceling the exponential factor)

$$a'' + 2ia'\theta' + ia\theta'' - a(\theta')^2 + ap = 0. \tag{7.3}$$

By analogy with the constant-p case, the choice $p = (\theta')^2$ is made, from which

$$\theta(x) = \pm \int^x p^{1/2}(\xi)d\xi.$$

Now we introduce the *approximation* aspect of the WKB(J) method, namely, to neglect the term a'' in favor of $a\theta''$ in equation (7.3), resulting in the simplification

$$2a'\theta' + a\theta'' = 0,$$

or

$$a(x) \propto (\theta')^{-1/2} = \pm p^{-1/4}(x).$$

In terms of arbitrary constants A and B, therefore,

$$w(x) \approx p^{-1/4} \left\{ A \exp \left[i \int^x p^{1/2}(\xi)d\xi \right] + B \exp \left[-i \int^x p^{1/2}(\xi)d\xi \right] \right\}. \quad (7.4)$$

But what should be the asymptotic form of $p(x)$ for the approximation $|a''| \ll |a\theta''|$ to be valid? Suppose that for large $|x|$, $p(x) = O(|x|^n)$ for some n. Then $\theta' = O(|x|^{n/2})$, and $a = O(|x|^{-n/4})$ so

$$\frac{a''}{a\theta''} = O\left(|x|^{-(n+2)/2}\right) \to 0, \quad (7.5)$$

provided $n > -2$.

As an example, let us use the WKB(J) method to construct the leading term of the asymptotic solutions to one form of Airy's differential equation (discussed in more detail later: the other form, by the way, just involves the change $x \to -x$). Note that the sign of p is temporarily changed here, so

$$w'' - xw = 0, \quad (7.6)$$

where $p(x) = x$ clearly satisfies the necessary condition (7.5). Then $a(x) = x^{-1/4}$, and $\theta(x) = (2i/3)x^{3/2}$ so for $x \in \mathbb{R}^+$ the solutions take the form

$$w \sim Ax^{-1/4} \exp \left[\pm \frac{2}{3} x^{3/2} \right],$$

and for $x \in \mathbb{R}^-$,

$$w \sim Ax^{-1/4} \exp \left[\pm \frac{2}{3} i |x|^{3/2} \right],$$

thus exhibiting the expected oscillatory behavior for negative values of x.

The condition $n > -2$ can be relaxed somewhat. For example, if $n = -2$ the method works for Bessel's equation

$$w'' + \frac{1}{x} w' + \left(1 - \frac{v^2}{x^2}\right) w = 0.$$

This can be recast into the form (7.1) via the change of variable

$$y(x) = x^{1/2} w(x)$$

to get

$$y'' + \left(1 + \frac{1 - 4v^2}{4x^2}\right) y = 0. \tag{7.7}$$

Then

$$y(x) \sim A \left(1 + \frac{1 - 4v^2}{4x^2}\right)^{-1/4} \exp\left[\pm i\theta(x)\right],$$

where

$$\theta(x) = \int^x \left(1 + \frac{1 - 4v^2}{4\xi^2}\right)^{1/2} d\xi.$$

Hence

$$w(x) \sim A x^{-1/2} \left(1 + \frac{1 - 4v^2}{4x^2}\right)^{-1/4} \exp\left[\pm i\theta(x)\right].$$

7.1.1 Elimination of the First Derivative Term

Suppose we have the following inhomogeneous differential equation:

$$y''(x) + p(x)y'(x) + q(x)y = f(x). \tag{7.8}$$

The coefficients p and q are assumed to be nonsingular on the interval of interest (which may be infinite). It is useful to seek a change of dependent variable of the form

$$u(x) = y(x)\eta(x) \tag{7.9}$$

such that there is no $u'(x)$ term present, that is, to obtain

$$u''(x) + Q(x)u(x) = F(x). \tag{7.10}$$

Why is is useful to do this? It is frequently the case that, for an infinite or semi-infinite interval, as $|x| \to \infty$, Q approaches a constant value, so the solutions of the homogeneous equation have a well-known form depending on the sign of Q.

Exercise: Show that

$$\eta(x) = \exp\left[\frac{1}{2}\int^x p(\xi)d\xi\right], \tag{7.11}$$

$$Q(x) = q(x) - \frac{1}{2}p'(x) - \frac{1}{4}p^2(x), \tag{7.12}$$

$$F(x) = f(x)\exp\left[\frac{1}{2}\int^x p(\xi)d\xi\right]. \tag{7.13}$$

Note that $u(x)$ and $y(x)$ possess the same zeros, because the transformation (7.9) is nonzero and nonsingular.

Exercise: Use the transformation (7.9), (7.11) for Bessel's equation of order n,

$$x^2 y'' + xy' + \left(x^2 - n^2\right)y = 0, \quad x \neq 0, \tag{7.14}$$

to obtain $u = x^{1/2}y$ and therefore

$$u'' + \left(1 + \frac{1 - 4n^2}{4x^2}\right)u = 0, \tag{7.15}$$

(see also the equivalent form (7.7)). If $n = \pm 1/2$, the exact solution is readily found to be

$$y_\pm(x) \equiv J_{\pm 1/2}(x) = x^{-1/2}(A\cos x + B\sin x), \tag{7.16}$$

and this is also the form for any n as $x \to \infty$:

$$y(x) \sim x^{-1/2}\left(A\cos x + B\sin x\right). \tag{7.17}$$

A more interesting example concerns Legendre's differential equation of order n:

$$\left(1 - x^2\right)y'' - 2xy' + n(n+1)y = 0, \quad |x| \leq 1. \tag{7.18}$$

The solutions are denoted by $y = P_n(x)$. If $x = \cos\theta$, then

$$y'' + \cot\theta\, y' + n(n+1)y = 0, \quad 0 \leq \theta \leq \pi. \tag{7.19}$$

Exercise: For $\theta \in (0, \pi)$ show that

$$u''(\theta) + \left[\left(n + \frac{1}{2}\right)^2 + \frac{1}{4\sin^2\theta}\right]u(\theta) = 0, \tag{7.20}$$

where $u = (\sin\theta)^{1/2}y$. Hence show that for $n \gg (\csc\theta)/2$,

$$P_n(x) \sim \frac{A}{(\sin\theta)^{1/2}}\cos\left[\left(n + \frac{1}{2}\right)\theta + \varepsilon\right], \tag{7.21}$$

ε being a phase term. In fact it transpires that both A and ε are dependent on n; indeed, as $n \to \infty$,

$$A(n) \sim \left(\frac{2}{n\pi}\right)^{1/2}, \quad \varepsilon(n) \sim -\frac{\pi}{4} \tag{7.22}$$

(see [184], p. 231).

7.1.2 The Liouville Transformation

Now we shall examine equations of the form

$$u''(x) + \lambda^2 Q(x)u(x) = 0, \tag{7.23}$$

where $Q(x) > 0$, and $\lambda \to \infty$ with x remaining finite. The Liouville transformation introduces a change of scale for the independent variable via

$$\sigma = \lambda \int^x [Q(\xi)]^{1/2}\, d\xi,$$

so that (**Exercise**) equation (7.23) can be written as

$$\frac{d^2u}{d\sigma^2} + \frac{1}{2}\frac{Q'(x)}{Q^{3/2}}\frac{du}{d\sigma} + u = 0$$

[189].[1] So now we have reintroduced a first-derivative term! But there is a good reason for so doing, as will be seen. Now let us remove the new term in $u'(\sigma)$ by means of (7.9) and (7.11) to obtain

$$Y = u(\sigma)\exp\left\{\frac{1}{4}\int^\sigma \left(\frac{Q'(x)}{Q^{3/2}}\right) d\sigma'\right\},$$

and

$$\frac{d^2Y}{d\sigma^2} + \left[1 - \frac{1}{2\lambda}m'(\sigma) - \frac{1}{4\lambda^2}m^2(\sigma)\right]Y = 0, \tag{7.24}$$

where

$$m(\sigma) = \frac{1}{2}\frac{Q'(\sigma)}{Q^{3/2}(\sigma)}.$$

Note that since $d\sigma' = [Q(\xi)]^{1/2}d\xi$, we may also write

$$Y = u(\sigma)\exp\left\{\frac{1}{4}\int^x \left(\frac{Q'(\xi)}{Q(\xi)}\right) d\xi\right\} = u(\sigma)\,[Q(x)]^{1/4}. \tag{7.25}$$

---

[1] This type of transformation appears in various guises throughout the literature, where it is often referred to as a Liouville-Green transformation. For simplicity it will be referred to as a Liouville transformation in this book. It enables us to write Sturm-Liouville equations in "Liouville normal form" (see Chapter 23).

Question: Under what conditions are solutions to initial-value problems of the following form (related to (7.24)) bounded?

$$\frac{d^2 y}{dx^2} + [1 + f(x)] y = 0; \quad y(0) = c_1; \quad y'(0) = c_2.$$

(The initial point can always be translated to $x = 0$ by a simple change of variable.) To answer this question we use the fact that this equation is readily converted to an integral equation by writing

$$y'' + y = -f(x)y,$$

and so

$$y(x) = c_1 \cos x + c_2 \sin x - \int_0^x f(\xi) y(\xi) \sin (x - \xi) \, d\xi. \qquad (7.26)$$

Exercise: Prove this either by direct construction or substitution.

Repeated use of the triangle inequality yields

$$|y(x)| \leq |c_1 \cos x + c_2 \sin x| + \left| \int_0^x f(\xi) y(\xi) \sin (x - \xi) \, d\xi \right|$$

$$\leq |c_1| + |c_2| + \int_0^x |f(\xi)| \, |y(\xi)| \, d\xi = M + \int_0^x |f(\xi)| \, |y(\xi)| \, d\xi.$$

Next we invoke a variant of Gronwall's inequality, proved below, which states that if $y(x)$, M, and $k \geq 0$, and also

$$y(x) \leq M + k \int_{x_0}^x |f(\xi)| \, |y(\xi)| \, d\xi, \qquad (7.27)$$

then

$$y(x) \leq M \exp \left\{ k \int_{x_0}^x |f(\xi)| \, d\xi \right\}. \qquad (7.28)$$

Proof of (7.28): Let

$$b(x) = \int_{x_0}^x |f(\xi)| \, y(\xi) d\xi.$$

Clearly $b(x_0) = 0$. Then

$$b'(x) = |f(x)| \, y(x) \leq |f(x)| \, [M + kb(x)],$$

that is,

$$\frac{kb'(x)}{M + kb(x)} \leq k |f(x)|,$$

whence

$$M + kb(x) \leq M \exp \left\{ k \int_{x_0}^x |f(\xi)| \, d\xi \right\}.$$

Then if the inequality (7.27) is satisfied, result (7.28) is established, and furthermore, $y(x)$ is bounded, provided also

$$\int_{x_0}^\infty |f(x)| \, dx < \infty.$$

Now we return to the integral in equation (7.26). Clearly it can be rewritten as

$$\sin x \int_0^x f(\xi)\,y(\xi)\cos\xi\,d\xi - \cos x \int_0^x f(\xi)\,y(\xi)\sin\xi\,d\xi$$

$$= \sin x \left[\int_0^\infty - \int_x^\infty\right] f(\xi)\,y(\xi)\cos\xi\,d\xi$$

$$- \cos x \left[\int_0^\infty - \int_x^\infty\right] f(\xi)\,y(\xi)\sin\xi\,d\xi$$

$$\equiv [I_1(0) - I_1(x)]\sin x - [I_2(0) - I_2(x)]\cos x.$$

The integrals $I_1(x)$ and $I_2(x)$ both tend to zero as $x \to \infty$. Therefore,

$$y(x) = c_1\cos x + c_2\sin x - \{[I_1(0) - I_1(x)]\sin x - [I_2(0) - I_2(x)]\cos x\} \quad (7.29)$$

$$= [c_1 + I_2(0)]\cos x + [c_2 - I_1(0)]\sin x + I_1(x)\sin x - I_2(x)\cos x. \quad (7.30)$$

But $y(x)$ is bounded by virtue of the inequality (7.28), so that

$$y \sim [c_1 + I_2(0)]\cos x + [c_2 - I_1(0)]\sin x + o(1), \quad x \to \infty, \quad (7.31)$$

recalling that $f(x) = o\,[g(x)]$ as $x \to X_0$ if

$$\frac{|f(x)|}{|g(x)|} \to 0 = \text{as } x \to X_0$$

(see Appendix A for the order notation $o(x)$).

7.1.3 The One-Dimensional Schrödinger Equation

In its standard form found in many physics textbooks, the time-independent Schrödinger equation for a particle of mass m with energy E moving in a potential $V(x)$ is, in one dimension

$$-\frac{\hbar^2}{2m}\frac{d^2\psi}{dx^2} + V(x)\psi = E\psi. \quad (7.32)$$

The constant $\hbar \approx 1.055 \times 10^{-34}$ Joule-seconds ($\approx 6.582 \times 10^{-16}$ electron-volt-seconds) is known as the *reduced* Planck's constant (i.e., Planck's constant h divided by 2π). Equation (7.32) can be written in the form (7.23) as

$$\psi'' + \lambda^2\,[E - V(x)]\,\psi = 0, \quad (7.33)$$

where $\lambda^2 = 2m/\hbar^2$, and we consider the case of

$$Q(x) = E - V(x) > 0 \quad (7.34)$$

on some domain D of the real line. Then equation (7.24) takes the form

$$\frac{d^2Y}{d\sigma^2} + \left[1 - \frac{1}{2\lambda}\frac{d}{d\sigma}\left(\frac{1}{2}\frac{Q'(\sigma)}{Q^{3/2}(\sigma)}\right) - \frac{1}{4\lambda^2}\left(\frac{1}{2}\frac{Q'(\sigma)}{Q^{3/2}(\sigma)}\right)^2\right]Y = 0. \quad (7.35)$$

Note that this equation can be written as

$$Y''(\sigma) + Y(\sigma)\left[1 + O\left(\frac{1}{\lambda}\right)\right] = 0, \qquad (7.36)$$

so in view of the result (7.31) we know that the leading term of the solution will have the structure

$$Y(\sigma) \sim A\cos\sigma + B\sin\sigma + O\left(\frac{1}{\lambda}\right), \qquad \lambda \to \infty, \qquad (7.37)$$

provided the terms are bounded, that is,

$$\int_{x_0}^{\infty} \left|\frac{d}{d\sigma}\left(\frac{Q'(\sigma)}{Q^{3/2}(\sigma)}\right)\right| d\sigma < \infty, \quad \text{and} \qquad (7.38)$$

$$\int_{x_0}^{\infty} \left|\left(\frac{Q'(\sigma)}{Q^{3/2}(\sigma)}\right)\right|^2 d\sigma < \infty. \qquad (7.39)$$

It follows from (7.25) that

$$\psi(x) \sim [Q(x)]^{-1/4}\left\{A\cos\left[\lambda\int^{x}[Q(\xi)]^{1/2}d\xi\right]\right.$$

$$\left. + B\sin\left[\lambda\int^{x}[Q(\xi)]^{1/2}d\xi\right]\right\} + O\left(\frac{1}{\lambda}\right), \qquad \lambda \to \infty. \qquad (7.40)$$

Obviously, more terms are available as an asymptotic expansion in powers of λ^{-1}; this is discussed shortly.

7.1.4 Physical Interpretation of the WKB(J) Approximation

Suppose that

$$y'' + Q(x)y = 0,$$

where $Q(x)$ is slowly-varying. If we seek a solution of the form $y = \exp[i\theta(x)]$, then

$$-\theta'^2 + i\theta'' + Q = 0. \qquad (7.41)$$

If θ'' is small compared with the other terms, then as already noted, we have the approximations

$$\theta'(x) \approx \pm Q^{1/2} \Rightarrow \theta(x) \approx \pm\int^{x}[Q(\xi)]^{1/2}\,d\xi. \qquad (7.42)$$

Therefore if θ'' is to be small in the above sense,

$$|\theta''| \approx \frac{1}{2}\left|\frac{Q'(x)}{Q^{1/2}(x)}\right| \ll |Q|. \qquad (7.43)$$

In particular for $Q > 0$, note that

$$\cos Q^{1/2}x = \cos(Q^{1/2}x + 2\pi) = \cos\left[Q^{1/2}\left(x + \frac{2\pi}{Q^{1/2}}\right)\right], \qquad (7.44)$$

so the term $2\pi/Q^{1/2}$ may be interpreted as one wavelength of the slowly varying term $Q^{1/2}$. The condition (7.43) is therefore a statement that the change in Q over a wavelength should be small compared with $|Q|$.

Exercise: Show that the iteration

$$\theta'' \approx \frac{1}{2}\frac{Q'(x)}{Q^{1/2}(x)} \tag{7.45}$$

employed in equation (7.41) yields the result

$$\theta(x) \approx \pm \int^x [Q(\xi)]^{1/2}\, d\xi + \frac{1}{4}i \ln Q, \tag{7.46}$$

whence

$$y(x) \approx [Q(x)]^{-1/4}\left\{ A_2 \exp\left[i\int_{x_0}^x [Q(\xi)]^{1/2}\, d\xi \right] + B_2 \exp\left[-i\int_{x_0}^x [Q(\xi)]^{1/2}\, d\xi \right] \right\}, \tag{7.47}$$

as was found previously.

7.1.5 The WKB(J) Connection Formulas

Equation (7.40) represents an approximate general solution to equation (7.33) if the above conditions are satisfied. The method is valuable provided Q does not oscillate too rapidly or is nonzero everywhere in the domain. If it is zero at such point, we are left with the problem of connecting an oscillatory solution for $Q > 0$ to an exponential one for $Q < 0$. It is then necessary to find the relationship between the constants A and B to connect these solutions; naturally enough these are called *connection formulas*.

We consider the case where $Q(x)$ is monotonically increasing from negative values for $x < x_0$ to positive values for $x > x_0$, there being a simple zero at $x = x_0$. For simplicity λ will be set to one here, because we are not now concerned about the asymptotic series. For $x \ll x_0$ we rewrite (7.40) as

$$y(x) \approx |Q(x)|^{-1/4}\left\{ A_1 \exp\left[\int_x^{x_0} |Q(\xi)|^{1/2}\, d\xi \right] + B_1 \exp\left[-\int_x^{x_0} |Q(\xi)|^{1/2}\, d\xi \right] \right\}, \tag{7.48}$$

and for $x \gg x_0$,

$$y(x) \approx [Q(x)]^{-1/4}\left\{ A_2 \exp\left[i\int_{x_0}^x [Q(\xi)]^{1/2}d\xi \right] + B_2 \exp\left[-i\int_{x_0}^x [Q(\xi)]^{1/2}\, d\xi \right] \right\}. \tag{7.49}$$

The requirement that the solutions match appropriately means that given the coefficients A_1 and B_1, we can find A_2 and B_2 and vice versa. One approach is to use an approximate solution that is valid along a path connecting the regions on either side of x_0 for which the WKB(J) approximations are satisfied (but for an alternative approach see [185], [186]). Not too surprisingly, Airy's differential equation is a perfect choice in this situation. Consider (without loss of generality) the alternative form of Airy's differential equation, namely,

$$y''(x) + xy(x) = 0. \tag{7.50}$$

Using a result established later in this Chapter ((7.161) in section 7.1.3) we can write the solution of Airy's equation in the form

$$y(x) = A \int_{-\infty}^{\infty} \exp\left[i\left(\omega x - \frac{\omega^3}{3}\right)\right] d\omega, \tag{7.51}$$

for some constant A [186]. To connect across a turning point it is only necessary here to use the *saddle-point method* to find the asymptotic forms of the Airy integral for large $|x|$. From Chapter 3 we know that for an integral of the form

$$I = \int_C \exp\left[\alpha h(z)\right] dz,$$

where $h'(z_0) = 0$, the saddle-point method yields the approximation

$$I \approx \left(\frac{2\pi}{\alpha |h''(z_0)|}\right)^{1/2} \exp\left[\alpha h(z_0) + i\left(\pm\frac{\pi}{2} - \frac{\phi}{2}\right)\right],$$

where ϕ is the phase for $h''(z_0)$. Returning to (7.51), first for $x \to \infty$, we have $\alpha = x$ and

$$h(\omega) = i\left(\omega x - \frac{\omega^3}{3}\right),$$

hence $h'(\omega_0) = 0$ when $\omega_0 = \pm x^{1/2}$, so that $h''(\omega_0) = \pm 2ix^{-1/2}$, $\phi = \pm\pi/2$. Additionally, if θ is the phase of (here) $\omega - \omega_0$, then for the expression $h''(\omega_0)(\omega - \omega_0)^2$ to be real and negative (for the path of steepest descent), it is necessary that the combined phase is an odd multiple of π:

$$2\theta + \phi = \pm(2n+1)\pi, \quad n = 0, 1, 2, \ldots,$$

or (choosing $n = 0$)

$$\theta = \pm\pi/2 - \phi/2 = \pm\pi/4. \tag{7.52}$$

Noting that $h(\omega_0) = \pm 2ix^{1/2}/3$ and including both values for ω_0, we find that

$$y(x) \sim \frac{2\pi^{1/2}}{x^{1/4}} \cos\left(\frac{2}{3}x^{3/2} - \frac{\pi}{4}\right).$$

For $x \to -\infty$ the large positive parameter is $\alpha = -x$; consequently the sign of h changes, with the result that

$$h(\omega) = -i\left(\omega - \frac{\omega^3}{3x}\right).$$

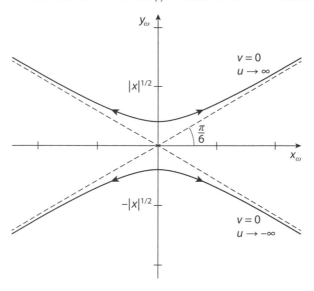

Figure 7.1 Plan view of $u = \operatorname{Re} h(\omega)$; $\omega = x_\omega + i y_\omega$. The saddle points occur at $\omega = \pm |x|^{1/2}$.

Now $h'(\omega_0) = 0$ when $\omega_0 = \pm i\,|x|^{1/2}$ so that $h''(\omega_0) = \pm 2\,|x|^{-1/2}$, $\phi = 0$ or π. If we let $\omega = x_\omega + i y_\omega$ (to avoid confusion with the parameter x in the integral) and $c = 1/(3\,|x|)$, then

$$u = \operatorname{Re} h(\omega) = y_\omega \left(1 + 3cx_\omega^2 - cy_\omega^2\right), \tag{7.53}$$

and

$$v = \operatorname{Im} h(\omega) = x_\omega \left(3cy_\omega^2 - cx_\omega^2 - 1\right). \tag{7.54}$$

We require v to be zero so that $h(\omega)$ is real, and choose the saddle point $\omega_0 = -i\,|x|^{1/2}$ with $\theta = 0$, because on the path through this point u descends into the "valley" (i.e., $u \to -\infty$). (On the path through the other saddle point, $u \to +\infty$.) In fact $v = 0$ on $x = 0$ and the hyperbola $3cy_\omega^2 - cx_\omega^2 - 1$, as sketched in Figure 7.1. The final result is

$$y(x) \sim \frac{\pi^{1/2}}{|x|^{1/4}} \exp\left(-\frac{2}{3}|x|^{3/2}\right). \tag{7.55}$$

In summary then, we have the connection relation for the Airy equation

$$\left(\frac{\pi}{|x|^{1/2}}\right)^{1/2} \exp\left(-\frac{2}{3}|x|^{3/2}\right) \longrightarrow \frac{2\pi^{1/2}}{x^{1/4}} \cos\left(\frac{2}{3}x^{3/2} - \frac{\pi}{4}\right). \tag{7.56}$$

As might be expected, this can be generalized to the case where any $Q(x)$ is approximately linear in the interval under consideration, the result being of the form

$$|Q(x)|^{-1/4} \left\{\exp\left[-\int_x^{x_0} |Q(\xi)|^{1/2}\, d\xi\right]\right\}$$

$$\longrightarrow 2\,[Q(x)]^{-1/4} \left\{\cos\left[\lambda \int_{x_0}^x [Q(\xi)]^{1/2}\, d\xi - \frac{\pi}{4}\right]\right\}. \tag{7.57}$$

Of course, if an exponentially increasing term is present, the Fourier transform would not exist (this case has been discussed in [185]; see also [186]).

7.1.6 Application to a Potential Well

Let us return to the Schrödinger equation (7.33), where now we define

$$Q(x) = \lambda^2 [E - V(x)], \tag{7.58}$$

and consider the case of bound states in a simple generic potential well such that $Q(x) > 0$ for $x \in (a, b)$, and $Q(x) < 0$ for $x < a$ and $x > b$. For bounded solutions in the outer regions the approximate solutions for $\psi(x)$ in the interval (a, b) are (i)

$$\psi(x) \approx \frac{C_1}{[E - V(x)]^{1/4}} \cos\left\{\lambda \int_a^x [E - V(\xi)]^{1/2} d\xi - \frac{\pi}{4}\right\}; \quad x > a, \tag{7.59}$$

and (ii)

$$\psi(x) \approx \frac{C_2}{[E - V(x)]^{1/4}} \cos\left\{\lambda \int_x^b [E - V(\xi)]^{1/2} d\xi - \frac{\pi}{4}\right\}; \quad x < b. \tag{7.60}$$

These two expressions should be the same throughout the region (i.e., as $x \to b^-$ in case (i) and as $x \to a^+$ in case (ii)), so that the sum of the phases in those limits must be the same, and therefore they must be an integral multiple of π. Thus if

$$\Phi_1 = \lambda \int_a^x [E - V(\xi)]^{1/2} d\xi - \frac{\pi}{4}, \tag{7.61}$$

$$\Phi_2 = \lambda \int_x^b [E - V(\xi)]^{1/2} d\xi - \frac{\pi}{4}. \tag{7.62}$$

then

$$\Phi_1 + \Phi_2 = \lambda \int_a^b [E - V(x)]^{1/2} dx - \frac{\pi}{2} = n\pi \tag{7.63}$$

implies that

$$\lambda \int_a^b [E - V(x)]^{1/2} dx = \left(n + \frac{1}{2}\right)\pi, \quad n = 0, 1, 2\ldots, \tag{7.64}$$

or

$$\int_a^b [2m \{E - V(x)\}]^{1/2} dx = \left(n + \frac{1}{2}\right)\pi\hbar. \tag{7.65}$$

Note further that since $\Phi_1 = n\pi - \Phi_2$, so that

$$\cos\Phi_1 = (-1)^n \cos\Phi_2, \tag{7.66}$$

and therefore

$$C_1 \cos\Phi_1 = C_2 \cos\Phi_2 \Rightarrow C_1 = (-1)^n C_2. \tag{7.67}$$

This is similar to the Bohr-Sommerfeld quantization condition in the old (i.e., pre-1925) quantum theory. Note also that the quasi-classical quantum number n is the number of zeros of the wave function. This follows from the fact that the phase of ψ increases from $-\pi/4$ at $x = a$ to $(n + 1/4)\pi$ at $x = b$, so the cosine term vanishes exactly n times in this range. (The wave function ψ decreases monotonically to zero outside this interval, so there are no additional zeros for any finite distance.) For the corresponding potential barrier problems, the WKB(J) method also gives approximate reflection and transmission coefficients, as discussed in Part IV (Chapter 22).

7.2 TECHNICAL DETAILS

In this subsection we return to real variables and reexamine equation (7.33), written as

$$\psi''(x) - \lambda^2 p(x)\psi(x) = 0, \tag{7.68}$$

with the requirement that

$$p(x) \equiv V(x) - E \neq 0 \tag{7.69}$$

on some interval I and

$$\lambda^2 \equiv \frac{2m}{\hbar^2} \gg 1. \tag{7.70}$$

Equation (7.68) can be reduced to a first-order Riccati equation via the transformation

$$\psi(x) = \exp\left(\int^x \phi(s)ds\right), \tag{7.71}$$

from which

$$\psi' = \phi\psi, \text{ and} \tag{7.72}$$

$$\psi'' = \left(\phi^2 + \phi'\right)\psi. \tag{7.73}$$

Hence

$$\phi^2 + \phi' = \lambda^2 p(x). \tag{7.74}$$

Let us first consider a crude approximation: suppose for $x \in I$, both $\phi(x)$ and $p(x)$ are slowly varying and nonzero. This suggests (as noted earlier) that $\phi(x) \approx \lambda p^{1/2}$, and so we choose the form

$$\psi(x) \approx Ae^{\lambda p^{1/2}x} + Be^{-\lambda p^{1/2}x}, \tag{7.75}$$

which has real exponential solutions or oscillatory ones, depending on whether $p(x) > 0$ or < 0 on I. More generally, $p(x) = O(1)$ by hypothesis, and we expect in general that $\phi' = O(\phi)$, which suggests that $\phi = O(\lambda)$, so we seek an expansion of the form

$$\phi(x) = \lambda \sum_{n=0}^{\infty} \phi_n(x)\lambda^{-n} \tag{7.76}$$

(i.e., in decreasing powers of λ). We substitute this infinite series into (7.74) and equate (in particular) the corresponding coefficients powers of λ^2 and λ. Hence

$$\phi_0^2 = p(x), \quad \text{and} \quad 2\phi_0\phi_1 + \phi_0' = 0. \tag{7.77}$$

Clearly

$$\phi_0 = \pm p^{1/2}(x), \quad \text{and} \quad \phi_1 = -\frac{1}{2}\frac{\phi_0'}{\phi_0}, \tag{7.78}$$

whence, ignoring constants of integration,

$$\int^x \phi_1(s)ds = -\frac{1}{2}\ln|\phi_0(x)|, \tag{7.79}$$

and so

$$\int^x \phi(s)ds = \int^x \{\lambda\phi_0(s) + \phi_1(s) + \lambda^{-1}\phi_2(s) + \cdots\}ds \tag{7.80}$$

$$= \pm\lambda\int^x p^{1/2}(s)ds + \ln|\phi_0(x)|^{-1/2} + O(\lambda^{-1}) \tag{7.81}$$

$$= \pm\lambda\int^x p^{1/2}(s)ds + \ln|p(x)|^{-1/4} + O(\lambda^{-1}). \tag{7.82}$$

From (7.71) we obtain a better approximation:

$$\psi(x) = A \exp\left\{\lambda\int^x p^{1/2}(s)ds + \ln|p(x)|^{-1/4} + O(\lambda^{-1})\right\} \tag{7.83}$$

$$+ B \exp\left\{-\lambda\int^x p^{1/2}(s)ds + \ln|p(x)|^{-1/4} + O(\lambda^{-1})\right\}, \tag{7.84}$$

or

$$\psi(x) \approx A|p(x)|^{-1/4} \exp\left\{\lambda\int^x p^{1/2}(s)ds\right\}$$

$$+ B|p(x)|^{-1/4} \exp\left\{-\lambda\int^x p^{1/2}(s)ds\right\} \tag{7.85}$$

$$\equiv A\psi_E(x; \lambda) + B\psi_E(x; -\lambda). \tag{7.86}$$

Obviously this last equation cannot be a valid asymptotic representation for $\psi(x)$ in any interval on which $p(x)$ has a zero (i.e., where $p(x_c) = 0$, $x_c \in I$). Such points $x = x_c$ are referred to as *turning points*. As a generalization of comments made above for the crude approximation, $p(x) > 0$ implies that $\psi(x)$ is a combination of pure exponentials, whereas if $p(x) < 0$, $\psi(x)$ is a combination of pure oscillatory terms. As noted earlier in this chapter, the question naturally arises: if $p(x) < 0$ for $x < x_c$, $p(x_c) = 0$ and $p(x) > 0$ for $x > x_c$ (or vice versa), *how can the oscillatory and exponential solutions be smoothly matched across the turning point?* We seek an equivalent answer to this question from the perspective of asymptotic expansions [32]. It is of interest to find the exact equation that the approximation $\psi_E(x)$ satisfied by the expression (7.86). We use $\psi_E(x)$ in equation (7.68) to find that each of the linearly independent solutions $\psi_E(x; \lambda)$ and $\psi_E(x; -\lambda)$ satisfies

$$\psi_E''(x; \pm\lambda) - \lambda^2 p(x)\psi_E(x; \pm\lambda) \tag{7.87}$$

$$= \left[\frac{5}{16} p^{-2} (p')^2 - \frac{1}{4} p^{-1} p'' \right] \left[p^{-1/4} \exp\{\pm\lambda \int^x p^{1/2}(s)ds\} \right] \tag{7.88}$$

$$= p \left[p^{-3/4} \frac{d^2}{dx^2} \left(p^{-1/4} \right) \right] \psi_E(x; \pm\lambda) \tag{7.89}$$

$$\equiv \lambda^2 p(x) f(x)\psi_E(x; \pm\lambda), \tag{7.90}$$

where

$$f(x) = \lambda^{-2} \left[p^{-3/4} \frac{d^2}{dx^2} \left(p^{-1/4} \right) \right]. \tag{7.91}$$

Hence $\psi_E(x)$ satisfies the equation

$$\psi_E'' - \lambda^2 p(x) \left[1 + f(x) \right] \psi_E = 0. \tag{7.92}$$

This in particular provides a general idea as to the validity of the approximation

$$\psi(x) \approx A\psi_E(x; \lambda) + B\psi_E(x; -\lambda), \tag{7.93}$$

because on any interval for which both $p(x) \neq 0$ and $[p(x)]^{-1/4}$ is (in some sense) slowly varying, $|f(x)| \to 0$ as $\lambda \to \infty$. We will make this a little more precise later (but see the discussion following (7.44)). Of course, there is no reason to stop the approximation at $O(\lambda)$; we can proceed to $O(1)$, $O(\lambda^{-1})$, and so on if necessary. Generalizing (7.86), we may write

$$\psi(x) = A\psi_+(x) + B\psi_-(x), \tag{7.94}$$

where

$$\psi_\pm(x) = |p(x)|^{-1/4} \exp \left\{ \pm\lambda \int^x p^{1/2}(s)ds \right\} \cdot \left\{ \sum_{n=0}^N \lambda^{-n} a_n \right\}, \quad a_0 = 1. \tag{7.95}$$

The last factor in this expression is an asymptotic expansion in the sense of Poincaré [32].

Airy's Differential Equation (Again)

This corresponds to the simple choice $p(x) = x$ in (7.68) so that

$$\psi''(x) - \lambda^2 x\psi(x) = 0. \tag{7.96}$$

To write the solution in a standard form, we let $y = \lambda^{2/3}x$ with $\psi(y) \equiv \psi(x)$ for simplicity of notation. The resulting equation is once more Airy's differential equation in a form appropriate for (7.68):

$$\psi''(y) - y\psi(y) = 0, \tag{7.97}$$

which possesses the linearly independent solutions

$$\psi_1 = \text{Ai}(y) = \text{Ai}(\lambda^{2/3}x),\tag{7.98}$$

and

$$\psi_2 = \text{Bi}(y) = \text{Bi}(\lambda^{2/3}x).\tag{7.99}$$

The WKB(J) approximations to leading order in λ for $p(x) = x$ are

$$\psi_\pm(x) = x^{-1/4}\exp\left(\pm\frac{2}{3}\lambda x^{3/2}\right)\{1 + O(\lambda^{-1})\},\quad x > 0,\text{ and}\tag{7.100}$$

$$\psi_\pm(x) = (-x)^{-1/4}\begin{Bmatrix}\cos\\\sin\end{Bmatrix}\left(\frac{2}{3}\lambda(-x)^{3/2}\right)\{1 + O(\lambda^{-1})\},\quad x < 0.\tag{7.101}$$

7.3 MATCHING ACROSS A TURNING POINT

We now generalize these ideas slightly to the equation

$$\psi''(x) - \lambda^2 p(x)\psi(x) = 0,\tag{7.102}$$

where $p(x)$ is analytic, and without loss of generality assume that the following conditions for a simple turning point apply:

$$p(0) = 0;\quad p'(0) > 0;\tag{7.103}$$

$$p(x) > 0,\quad x > 0;\quad p(x) < 0,\quad x < 0.\tag{7.104}$$

Hence for $x > 0$ the WKB(J) approximations are

$$\psi_\pm(x) = |p(x)|^{-1/4}\exp\left\{\pm\lambda\int^x p^{1/2}(s)ds\right\}.\tag{7.105}$$

For small values of $x > 0$, retaining only the linear term in the Maclaurin series for $p(x)$, we see that

$$\psi_\pm(x) \approx \{p'(0)\}^{-1/4}x^{-1/4}\exp\left(\pm\frac{2}{3}\lambda\left[p'(0)\right]^{1/2}x^{3/2}\right).\tag{7.106}$$

Correspondingly, for small negative values of x we have

$$\psi_\pm(x) \approx \{p'(0)\}^{-1/4}|x|^{-1/4}\begin{Bmatrix}\cos\\\sin\end{Bmatrix}\left(\frac{2}{3}\lambda\left[p'(0)\right]^{1/2}|(x)|^{3/2}\right).\tag{7.107}$$

If we use the same stretching variable as above, namely, $y = \lambda^{2/3}x$, the solutions of equation (7.102) for $|x| \ll 1$ are

$$\psi_1 = \text{Ai}\left(\lambda^{2/3}\left[p'(0)\right]^{1/3}x\right),\tag{7.108}$$

and

$$\psi_2 = \text{Bi}\left(\lambda^{2/3}\left[p'(0)\right]^{1/3}x\right).\tag{7.109}$$

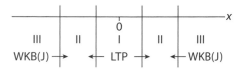

Figure 7.2 Inner and outer WKB(J) regions; LTP is "local turning point".

The Airy functions Ai(y) and Bi(y) are, respectively, exponentially small and large for $x > 0$, so we will seek a solution of (7.102) that is bounded as $x \to \infty$. This implies that

$$\psi(x) = A\psi_1, \quad x > 0; \tag{7.110}$$

$$= \alpha\,\text{Ai}(y) + \beta\,\text{Bi}(y), \quad y = O(1); \tag{7.111}$$

$$= C\psi_1 + D\psi_2; \quad x < 0. \tag{7.112}$$

Therefore the local turning point approximation is valid in the neighborhood of the origin (region I), specifically, for $|x| \ll 1$ and $y = O(1)$. (See Figure 7.2.) Outside this (regions II: $|x| \ll 1$ but $|y| \gg 1$) are two intermediate intervals (regions II: $|x| \ll 1$ and $|y| \gg 1$) Beyond that are two regions for which the WKB(J) approximation is valid (regions III: $x = O(1)$). Regions II can now be identified as intervals in which both the local turning point and the WKB(J) approximations are valid. Since there are two different approximations to $\psi(x)$ in this region, they must match (i.e., they must be asymptotically equivalent). This implies that $\beta \equiv 0$, and by using the asymptotic forms for the Airy functions for $0 < x \ll 1$, but $y \gg 1$, a smoothly varying solution can be found connecting the exponential and oscillatory regions across the turning point. We shall revisit the WKB(J) approximation in Chapter 22.

In summary, the central three regions (II-I-II) can be considered as a "patching region" in which the Airy function solution (the local turning point) is valid. In regions III on either side of this patching region (far from the turning point), the asymptotic forms of the Airy function overlap with the WKB(J) solutions.

7.4 A LITTLE MORE ABOUT AIRY FUNCTIONS

In this subsection, x is real. Recall the form (7.6) of the Airy differential equation

$$y''(x) - xy(x) = 0. \tag{7.113}$$

Asymptotic expressions for the linearly independent solutions Ai(x) and Bi(x) if $x \gg 0$ are, to leading order,

$$\text{Ai}(x) \sim \frac{1}{2}\pi^{-1/2}x^{-1/4}e^{-\xi}; \tag{7.114}$$

$$\text{Bi}(x) \sim \pi^{-1/2}x^{-1/4}e^{\xi}, \tag{7.115}$$

where

$$\xi(x) = \frac{2}{3}x^{3/2}. \tag{7.116}$$

For $x \ll 0$ the corresponding expressions are, to leading order,

$$\text{Ai}(x) \sim \pi^{-1/2}(-x)^{-1/4} \sin\left[\xi(-x) + \frac{\pi}{4}\right]; \tag{7.117}$$

$$\text{Bi}(x) \sim \pi^{-1/2}(-x)^{-1/4} \cos\left[\xi(-x) + \frac{\pi}{4}\right]. \tag{7.118}$$

Corresponding integral representations are

$$\text{Ai}(x) = \frac{1}{\pi}\int_0^\infty \cos\left(\frac{s^3}{3} + sx\right) ds, \tag{7.119}$$

and

$$\text{Bi}(x) = \frac{1}{\pi}\int_0^\infty \left\{\sin\left(\frac{s^3}{3} + sx\right) + \exp\left(sx - \frac{s^3}{3}\right)\right\} ds. \tag{7.120}$$

In fact there are many different representations of Airy integrals [187]; some authors ([26], [27]) use

$$\text{Ai}(x) = \frac{3^{1/3}}{\pi}\int_0^\infty \cos\left(s^3 + 3^{1/3}xs\right) ds, \tag{7.121}$$

whereas others ([24], [28]; see also [188]) use the form (7.119) for Ai$(-x)$. Van de Hulst [85] employs the rainbow integral form

$$f(x) = \int_0^\infty \cos\left[\frac{\pi}{2}\left(sx - \frac{s^3}{3}\right)\right] ds \tag{7.122}$$

(see also (5.40)). Note that (7.119) may also be written as

$$\text{Ai}(x) = \frac{1}{2\pi}\int_{-\infty}^\infty \exp\left[i\left(\frac{s^3}{3} + sx\right)\right] ds, \tag{7.123}$$

from which a useful generalization follows:

$$\text{Ai}(ax) = \frac{1}{2\pi a}\int_{-\infty}^\infty \exp\left[i\left(\frac{s^3}{3a^3} + sx\right)\right] ds. \tag{7.124}$$

For other forms, the monograph [187] should be consulted.

7.4.1 Relation to Bessel Functions

For positive arguments, the Airy functions can be expressed in terms of the modified Bessel functions $I_{1/3}$ and $K_{1/3}$, which are solutions of the modified Bessel equation

$$x^2 y'' + x y' - \left(x^2 + \frac{1}{9}\right) y = 0; \tag{7.125}$$

thus

$$\mathrm{Ai}(x) = \frac{1}{\pi} \left(\frac{x}{3}\right)^{1/2} K_{1/3}\left(\frac{2}{3}x^{3/2}\right), \tag{7.126}$$

and

$$\mathrm{Bi}(x) = \left(\frac{x}{3}\right)^{1/2}\left[I_{1/3}\left(\frac{2}{3}x^{3/2}\right) + I_{-1/3}\left(\frac{2}{3}x^{3/2}\right)\right]. \tag{7.127}$$

For negative arguments,

$$\mathrm{Ai}(-x) = \left(\frac{x}{9}\right)^{1/2}\left[J_{1/3}\left(\frac{2}{3}x^{3/2}\right) + J_{-1/3}\left(\frac{2}{3}x^{3/2}\right)\right], \tag{7.128}$$

and

$$\mathrm{Bi}(-x) = -\left(\frac{x}{3}\right)^{1/2}\left[J_{1/3}\left(\frac{2}{3}x^{3/2}\right) - J_{-1/3}\left(\frac{2}{3}x^{3/2}\right)\right]. \tag{7.129}$$

Exercise: Many ordinary differential equations arising in mathematical physics and applied mathematics are special cases of the equation

$$x^2\frac{d^2y}{dx^2} + (1 - 2a)x\frac{dy}{dx} + \left[b^2c^2x^{2c} + (a^2 - c^2v^2)\right]y = 0, \tag{7.130}$$

where a, b, c, v are constants, and $v \geq 0, b > 0$. Find the general solution of this equation using the following sequence of transformations: (i) let $y(x) = x^a z(x)$. Simplify your result, and then (ii) let $t = x^c$. Hence find the general solution of Airy's differential equation

$$y'' + xy = 0, \tag{7.131}$$

in terms of Bessel functions of the first and second kind (Js and Ys, respectively; see below). (Note that the change $x \to -x$ reduces this to the form (7.113).) You may assume that Bessel's differential equation of order v,

$$\mu^2\frac{d^2f}{d\mu^2} + \mu\frac{df}{dz} + \left(\mu^2 - v^2\right)f = 0,$$

posseses the general solution

$$f(z) = C_1 J_v(\mu) + C_2 Y_v(\mu),$$

where C_1 and C_2 are arbitrary constants.

Solution: The change of dependent variable $y(x) = x^a z(x)$ results in the equation (7.130) being transformed to

$$x^2\frac{d^2z}{dx^2} + x\frac{dz}{dx} + \left(b^2c^2x^{2c} - c^2v^2\right)z = 0. \tag{7.132}$$

This follows because

$$y' = ax^{a-1}z + x^a z', \tag{7.133}$$

and

$$y'' = a(a-1)x^{a-2}z + 2ax^{a-1}z' + x^a z''. \tag{7.134}$$

Similarly, the change of independent variable means that the differential operators

$$x^2 \frac{d^2}{dx^2} + x \frac{d}{dx} = x \frac{d}{dx} \left(x \frac{d}{dx} \right) \longrightarrow c^2 t \left(t \frac{d^2}{dt^2} + \frac{d}{dt} \right); \tag{7.135}$$

if $\mu = bt$, then (7.132) becomes

$$\mu^2 \frac{d^2 z}{d\mu^2} + \mu \frac{dz}{d\mu} + \left(\mu^2 - \nu^2 \right) z = 0. \tag{7.136}$$

This is Bessel's equation of order ν, so in terms of the variable t, equation (7.132) has the general solution

$$z(t) = C_1 J_\nu(bt) + C_2 Y_\nu(bt). \tag{7.137}$$

Since

$$y'' + xy = 0 \tag{7.138}$$

can be written as

$$x^2 y'' + x^3 y = 0, \tag{7.139}$$

we can compare this with the original equation to find that $a = 1/2$, $c = 3/2$, $b = \pm 2/3$ (taking the $+$ sign without loss of generality), and $\nu = 1/3$. Hence from (i) above it follows that

$$y(x) = x^{1/2} \left[C_1 J_{1/3} \left(\frac{2}{3} x^{3/2} \right) + C_2 Y_{1/3} \left(\frac{2}{3} x^{3/2} \right) \right].$$

7.4.2 The Airy Integral and Related Topics

What are the consequences of seeking an integral solution of the form

$$y(x) = \int_a^b e^{\xi x} u(\xi) d\xi, \tag{7.140}$$

for the Airy equation? What functional form might $u(\xi)$ take? Are there any conditions on the limits of integration, or can they be arbitrary?

If we seek an integral solution of the form

$$y(x) = \int_a^b e^{\xi x} u(\xi) d\xi \tag{7.141}$$

to Airy's differential equation

$$y'' + xy = 0, \tag{7.142}$$

where a and b may be infinite and/or complex numbers and ξ is in general a complex variable, direct substitution enables us to write

$$\int_a^b e^{\xi x}\left(\xi^2 u - \frac{du}{d\xi}\right)d\xi + \left[ue^{\xi x}\right]_a^b = 0. \tag{7.143}$$

This will be the case, in particular, provided

$$\left[ue^{\xi x}\right]_a^b = 0, \tag{7.144}$$

and

$$du/d\xi = \xi^2 u. \tag{7.145}$$

This last condition implies of course, that

$$u(\xi) = \exp\left(\xi^3/3\right), \tag{7.146}$$

disregarding a multiplicative constant. This means that the first condition must be such that

$$\left[e^{\xi x + \xi^3/3}\right]_a^b = 0. \tag{7.147}$$

How might this be accomplished? Since the cubic term will dominate the linear one for sufficiently large values of ξ, we might consider a and b to tend to infinity in such a way that a^3 and b^3 are proportional to -1 (i.e., considering a and b now as complex numbers, their arguments should be among the three cube roots of -1). Thus if $z = r\exp(i\theta)$, then

$$z^3 + 1 = 0 \Longrightarrow r^3\exp(3i\theta) = \exp\left[i\pi\left(2n+1\right)\right], \quad n = 0, \pm1, \pm2, \ldots. \tag{7.148}$$

Therefore we may take $r = 1$ and $\theta = (2n+1)\pi/3$, $n = 0, 1, 2, \ldots$ without loss of generality. These roots are

$$\left\{\frac{1}{2} + i\frac{\sqrt{3}}{2}, -1, \frac{1}{2} - i\frac{\sqrt{3}}{2}\right\}. \tag{7.149}$$

This suggests that we choose the paths of integration in the complex ξ-plane to be asymptotic to the arguments $\pi/3$, π, and $5\pi/3$ in order for these conditions to be satisfied. Other representations of the Airy equation ($y'' - xy = 0$, for example) will have correspondingly appropriate solutions. Also note that the integral

$$y(x) = \int_a^b e^{(\xi x + \xi^3/3)}d\xi, \tag{7.150}$$

under the change of variable $\xi = -i\mu$ formally becomes

$$y(x) = -i\int_{ia}^{ib} e^{i(\mu^3/3 - \mu x)}d\mu, \tag{7.151}$$

so that, again formally, the integrals

$$\int_{ia}^{ib}\cos\left(\frac{\mu^3}{3} - \mu x\right)d\mu, \quad \text{and} \quad \int_{ia}^{ib}\sin\left(\frac{\mu^3}{3} - \mu x\right)d\mu, \tag{7.152}$$

are solutions of the Airy equation. Note that the first integral is an even function of μ and the second is an odd function, so if the limits are symmetric about the origin, then the latter is identically zero.

Example: Use the Fourier transform to find one solution of Airy's equation in the form

$$y'' + xy = 0. \tag{7.153}$$

This will involve manipulations similar to those discussed earlier with regard to the WKB(J) connection formulas, except that the integrals will be taken along the real line. We denote the Fourier transform of $y(x)$ by

$$\mathcal{F}(y(x)) = \bar{y}(\omega), \tag{7.154}$$

so that the Fourier transform pair is

$$y(x) = \int_{-\infty}^{\infty} \bar{y}(\omega) e^{-i\omega x} d\omega, \quad \text{and} \quad \bar{y}(\omega) = \frac{1}{2\pi} \int_{-\infty}^{\infty} y(x) e^{i\omega x} dx. \tag{7.155}$$

Fourier transforming the Airy equation, we obtain

$$-\omega^2 \bar{y} + \mathcal{F}(xy) = 0. \tag{7.156}$$

Using the result that

$$\mathcal{F}(xy) = -i d\bar{y}/d\omega, \tag{7.157}$$

this becomes the homogeneous first-order ordinary differential equation

$$\omega^2 \bar{y} + i \frac{d\bar{y}}{d\omega} = 0, \tag{7.158}$$

with solution

$$\bar{y}(\omega) = \bar{y}(0) \exp\left(i\omega^3/3\right), \tag{7.159}$$

$\bar{y}(0)$ being at this stage an undetermined constant, so that

$$y(x) = \bar{y}(0) \int_{-\infty}^{\infty} \exp\left[i\left(\omega^3/3 - \omega x\right)\right] d\omega = 2\bar{y}(0) \int_{0}^{\infty} \cos\left(\frac{\omega^3}{3} - \omega x\right) d\omega, \tag{7.160}$$

using Euler's formula and symmetry arguments. Note that under the transformation $\omega \to -\omega$ the first integral may be written as

$$y(x) = \bar{y}(0) \int_{-\infty}^{\infty} \exp\left[i\left(\omega x - \frac{\omega^3}{3}\right)\right] d\omega. \tag{7.161}$$

Already we see reflections of the arguments developed above (this time for the real integration variable ω), but we can get even more specific. Setting $x = 0$ in the above equation, we obtain

$$y(0) = 2\bar{y}(0) \int_{0}^{\infty} \cos\left(\frac{\omega^3}{3}\right) d\omega, \tag{7.162}$$

where, as we will demonstrate below, in terms of the gamma function, $\Gamma(z)$,

$$\int_0^\infty \cos\left(\frac{\omega^3}{3}\right) d\omega = \frac{3^{-1/6}}{2}\Gamma\left(\frac{1}{3}\right). \tag{7.163}$$

The choice for $y(0)$ is usually made as

$$y(0) = \left[3^{2/3}\Gamma(2/3)\right]^{-1}, \tag{7.164}$$

so that

$$\bar{y}(0) = \left[3^{1/2}\Gamma(2/3)\,\Gamma(1/3)\right]^{-1}. \tag{7.165}$$

Furthermore, since

$$\Gamma(\alpha)\,\Gamma(1-\alpha) = \pi/\sin\pi\alpha, \tag{7.166}$$

with $\alpha = 1/3$ we have that

$$\Gamma(1/3)\,\Gamma(2/3) = 2\pi/3^{1/2}, \tag{7.167}$$

and therefore $\bar{y}(0) = (2\pi)^{-1}$, so in this formulation,

$$y(x) \equiv \text{Ai}(x) = \frac{1}{\pi}\int_0^\infty \cos\left(\frac{\omega^3}{3} - \omega x\right) d\omega.$$

Example: Evaluate the integral

$$I = \int_0^\infty y^p e^{-ky^n} dy, \quad k > 0, \tag{7.168}$$

in terms of the gamma function

$$\Gamma(x) = \int_0^\infty t^{x-1} e^{-t} dt. \tag{7.169}$$

Solution: Let $ky^n = t$, so $kny^{n-1}dy = dt$ implies

$$I = \left(nk^{\frac{p+1}{n}}\right)^{-1}\int_0^\infty t^{\frac{p+1}{n}-1} e^{-t} dt = \left(nk^{\frac{p+1}{n}}\right)^{-1}\Gamma\left(\frac{p+1}{n}\right). \tag{7.170}$$

Thus, for example, if $p = 0$, $k = 1$, and $n = 2$,

$$\int_0^\infty e^{-r^2} dr = \frac{1}{2}\Gamma\left(\frac{1}{2}\right) = \frac{\pi^{1/2}}{2}. \tag{7.171}$$

Similarly, if $p = 0$, $k = 1/3$ and $n = 3$,

$$\int_0^\infty e^{-r^3/3} dr = \left[3\left(\frac{1}{3}\right)^{1/3}\right]^{-1}\Gamma\left(\frac{1}{3}\right) = 3^{-2/3}\Gamma\left(\frac{1}{3}\right). \tag{7.172}$$

7.4.3 Related Integrals

Example: Evaluate the integral

$$\int_0^\infty \cos\left(\frac{\omega^3}{3}\right) d\omega$$

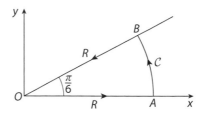

Figure 7.3 The contour for evaluating the Airy integral.

by considering

$$\oint_C e^{i\omega^3/3} d\omega$$

over the closed wedge-shaped contour shown in Figure 7.3.

Solution: By Cauchy's theorem,

$$\oint_C e^{iz^3/3} dz = 0 = \int_{OA} e^{iz^3/3} dz + \int_{AB} e^{iz^3/3} dz + \int_{BO} e^{iz^3/3} dz. \qquad (7.173)$$

On OA, $z = x$, $0 \le x \le R$; on AB, $z = Re^{i\theta}$, $0 \le \theta \le \pi/6$; on BO, $z = re^{i\pi/6}$, $R \ge r \ge 0$. Therefore

$$\int_0^R e^{ix^3/3} dx + \int_0^{\pi/6} e^{iR^3 e^{3i\theta}/3} \left(i Re^{i\theta} \right) d\theta + \int_R^0 e^{ir^3 e^{i\pi/2}/3} \left(e^{i\pi/6} \right) dr = 0, \qquad (7.174)$$

that is,

$$\int_0^R \left(\cos x^3/3 + i \sin x^3/3 \right) dx$$
$$= e^{i\pi/6} \int_0^R e^{-r^3/3} dr - \int_0^{\pi/6} e^{iR^3(\cos 3\theta + i \sin 3\theta)/3} \left(i Re^{i\theta} \right) d\theta. \qquad (7.175)$$

Consider the limit as $R \to \infty$. The first integral on the right, from the previous example becomes

$$3^{-2/3} \Gamma(1/3) \frac{(\sqrt{3} + i)}{2}. \qquad (7.176)$$

For the second integral on the right,

$$\left| \int_0^{\pi/6} e^{iR^3(\cos 3\theta + i \sin 3\theta)/3} \left(i Re^{i\theta} \right) d\theta \right| \le \int_0^{\pi/6} e^{-(R^3 \sin 3\theta)/3} R d\theta, \qquad (7.177)$$

so in terms of the quantity $\phi = 3\theta$ and the inequality $\sin \phi \ge 2\phi/\pi$, we are able to write

$$\int_0^{\pi/6} e^{-(R^3 \sin 3\theta)/3} R d\theta \le \frac{R}{3} \int_0^{\pi/2} e^{-2R^3\phi/(3\pi)} d\phi = \frac{\pi}{2R^2} \left[1 - e^{-(R^3/3)} \right], \qquad (7.178)$$

and this tends to zero as $R \to \infty$. Then we have found that

$$\int_0^\infty \left(\cos x^3/3 + i \sin x^3/3 \right) dx = 3^{-2/3} \Gamma \left(1/3 \right) \left(\frac{\sqrt{3} + i}{2} \right), \qquad (7.179)$$

so on equating the real and imaginary parts of this expression, it follows that

$$\int_0^\infty \cos(x^3/3) dx = \frac{3^{-1/6}}{2} \Gamma \left(\frac{1}{3} \right), \qquad (7.180)$$

and

$$\int_0^\infty \sin(x^3/3) dx = \frac{3^{-2/3}}{2} \Gamma \left(\frac{1}{3} \right). \qquad (7.181)$$

Chapter Eight

Island Rays

Interesting wave patterns can develop around islands. . . . The rays
follow strange trajectories. On the windward side of the island, rays
converge slightly, causing high waves. Waves curving around each side
of the island converge, so waves are lower on the leeward side . . . rays
that curve around the protected side of the island and spin off into deep
water again after turning more than 270 degrees [cross] others on the
way. This is an example of how wave refraction around islands can
contribute to confused seas. The wave-refraction story can get even
more complicated. Depth isn't the only factor that affects wave speed
and can cause refraction. Currents speed waves along, or they can slow
them if the waves are heading in contrary directions. These changes in
wave speed cause wave refraction in the same way as variations in
depth do.

[247]

The previous three chapters were predominantly mathematical in character,
essentially morphing into the wavelike character of ray theory's alter ego in a part
of the book having to do with rays, and therefore seemingly out of place. This is
yet another illustration (to the author's mind at least) of the ray-wave duality that is
a result of these concepts being intimately connected (and therefore not mutually
exclusive). But now we move away from Airy functions and integrals and return
to the ray approximation in the next two chapters. They will lead rather naturally
to the study of waves in earth, sea and sky in Part II of the book. To that end we
first examine a two-dimensional version of ray refraction in a very different context
from optics, namely, rays on the surface of water. Nevertheless, the mathematical
ideas involved are closely related. To see this, recall the eikonal equation (3.252),
namely,

$$\left(\frac{\partial S}{\partial x}\right)^2 + \left(\frac{\partial S}{\partial y}\right)^2 = k^2(x, y), \tag{8.1}$$

and recall from equation (3.258) that

$$\frac{d}{dx}\left[\frac{ky'}{(1+y'^2)^{1/2}}\right] = \left(\frac{\partial k}{\partial y}\right)(1+y'^2)^{1/2}. \tag{8.2}$$

(a)

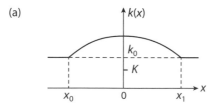

(b)

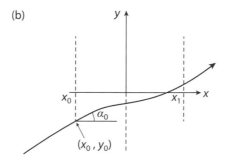

Figure 8.1 (a) $k(x)$ for a ridge; (b) corresponding ray.

8.1 STRAIGHT AND PARALLEL DEPTH CONTOURS

If we assume $k = k(x)$ only, then equation (8.2) reduces to

$$\frac{d}{dx}\left[\frac{ky'}{(1+y'^2)^{1/2}}\right] = 0, \tag{8.3}$$

which implies

$$\frac{ky'}{(1+y'^2)^{1/2}} = K = \text{constant}. \tag{8.4}$$

Figure 8.1 shows the path of a continuously varying ray direction,

$$\sin\alpha = \frac{\Delta y}{\Delta s} \rightarrow \frac{dy}{ds} = \frac{y'}{(1+y'^2)^{1/2}},$$

as $\Delta y, \Delta s \rightarrow 0$, so in this limit

$$k(x)\sin\alpha(x) = K = k_0\sin\alpha_0. \tag{8.5}$$

This of course is just Snell's law of refraction, so $k(x)/k_0$ behaves like a (variable) optical refractive index. Thus, from (8.4) we have

$$\frac{dy}{dx} = \pm\frac{K}{[k^2(x) - K^2]^{1/2}}, \tag{8.6}$$

which allows us to deduce the equation of a ray:

$$y - y_0 = \pm\int_{x_0}^{x}\frac{K\,dx}{[k^2(x) - K^2]^{1/2}}, \tag{8.7}$$

where the sign is determined from context. It is identical in form to equation (3.116). This is the one-dimensional Cartesian analogue of equation (3.99) in Chapter 3, which defines the behavior of the ray angle $\theta(r)$ in polar coordinates, and for this context the polar form will also be discussed below.

Note from equation (8.7) that rays only exist when $k^2(x) > K^2$. As phase lines are orthogonal to rays, their governing differential equation is

$$\frac{dy}{dx} = \pm \frac{(k^2(x) - K^2)^{1/2}}{K},\tag{8.8}$$

and so their solution takes the form

$$K(y - y_0) = \pm \int_{x_0}^{x} [k^2(x) - K^2]^{1/2}\,dx + \text{constant}.\tag{8.9}$$

8.1.1 Plane Wave Incident on a Ridge

In this instance $k_0 \sin(\alpha_0) = K < k(x)$ and $y' > 0$, implying we should take the positive root. It will be shown shortly (Chapter 11) that the governing dispersion relation for surface gravity waves in water of depth $h(x)$ is

$$\omega^2 = gk \tanh kh,\tag{8.10}$$

and for shallow water, $kh \ll 1$. Therefore,

$$\omega^2 \approx ghk^2,\tag{8.11}$$

so

$$k \approx \frac{\omega}{(gh)^{1/2}}.\tag{8.12}$$

Thus, as h decreases, k increases and equation (8.6) implies that y' decreases, so the ray becomes increasingly normal to depth contours. The reverse consequences occur as h increases. Furthermore, in view of (8.11) the result (8.7) may be written

$$|y - y_0| \approx \int_{x_0}^{x} \frac{K\,dx}{\left[(\omega^2/gh(x)) - K^2\right]^{1/2}}.\tag{8.13}$$

In particular the flux equation (3.246) for leading-order amplitude a_0 and equation (8.12) imply that

$$a_0^2 c_g \approx a_0^2 h^{1/2} = \text{a constant},\tag{8.14}$$

or

$$a_0 \propto h^{-1/4}(x),\tag{8.15}$$

known as *Green's law*. This has important implications for the evolution of tsunami waves.

As a tsunami leaves the deep water of the open-ocean and travels into the shallower water near the coast, it transforms . . . a tsunami travels at a speed that is related to the water depth—hence, as the water depth decreases, the tsunami slows. The tsunami's energy flux, which is dependent on both its wave speed and wave height, remains nearly constant. Consequently, as the tsunami's speed diminishes, its height grows. This is called *shoaling*. Because of this shoaling effect, a tsunami that is unnoticeable at sea, may grow to be several metres or more in height near the coast. The increase of the tsunami's waveheight as it enters shallow water is given by

$$\frac{h_s}{h_d} = \left(\frac{H_d}{H_s}\right)^{1/4},\tag{8.16}$$

where h_s and h_d are waveheights in shallow and deep water and H_s and H_d are the depths of the shallow and deep water. So a tsunami with a height of 1 m in the open ocean where the water depth is 4000 m would have a wave height of 4 to 5 m in water of depth 10 m. [210]

Tsunamis

Tsunamis can have wavelengths ranging from 10 to 500 km and wave periods of up to an hour. As a result of their long wavelengths, tsunamis act as shallow-water waves. A wave becomes a shallow-water wave when the wavelength is very large compared to the water depth. Shallow-water waves move at a speed, c, that is dependent upon the water depth and is given by the formula $c = \sqrt{gH}$, where g is the acceleration due to gravity (= 9.8 m/s^2) and H is the depth of water. In the deep ocean, the typical water depth is around 4000 m, so a tsunami will therefore travel at around 200 m/s, or more than 700 km/hr. For tsunamis that are generated by underwater earthquakes, the amplitude of the tsunami is determined by the amount by which the sea-floor is displaced. Similarly, the wavelength and period of the tsunami are determined by the size and shape of the underwater disturbance. As well as traveling at high speeds, tsunamis can also travel large distances with limited energy losses. As the tsunami propagates across the ocean, the wave crests can undergo refraction (bending), which is caused by segments of the wave moving at different speeds as the water depth along the wave crest varies. [210]

Obviously the approximation (8.15) breaks down as the shoreline caustic is approached; indeed, as pointed out in [214], you can surf just before ray tracing breaks down.

Special Case: A Beach

Suppose the beach begins at $x = a$ and ends at shoreline $x = b$. With

$$k(x) = \begin{cases} k_0, & x < a, \\ k_0\left(\frac{1-a/b}{1-x/b}\right), & a < x < b, \end{cases}\tag{8.17}$$

we see that (8.6) yields

$$y' = \frac{K}{\left(k^2 - K^2\right)^{1/2}},$$ (8.18)

where $K = k_0 \sin \alpha_0$. Then on the interval (a, b)

$$k^2(x) - K^2 = k_0^2 \left[\frac{(1 - a/b)^2}{(1 - x/b)^2} - \sin^2 \alpha_0 \right],$$ (8.19)

and (8.18) gives

$$y' = \frac{(1 - x/b) \sin \alpha_0}{\left[(1 - a/b)^2 - [(1 - x/b) \sin \alpha_0]^2 \right]^{1/2}}.$$ (8.20)

Let

$$\beta = \frac{\sin \alpha_0}{1 - a/b}, \quad p = 1 - \frac{x}{b}, \quad \text{and} \quad v = \frac{y}{b}.$$ (8.21)

Then on using the chain rule we find from (8.20) that

$$dv = \frac{-\beta p \, dp}{\left(1 - \beta^2 p^2\right)^{1/2}},$$ (8.22)

so

$$v = \frac{1}{\beta} \left(1 - \beta^2 p^2\right)^{1/2} + v_c.$$ (8.23)

With $v_c = y_c/b$ we see that (8.23) may be rearranged as

$$(x - b)^2 + (y - y_c)^2 = \frac{(b - a)^2}{\sin \alpha_0} = r^2,$$ (8.24)

so the rays are a family of circular arcs centered at $(x, y) = (b, y_c)$ with radius $(b - a) \csc \alpha_0$. Note that if $x = a$ in (8.24), then we may rearrange the result to conclude that

$$y - y_c = \pm (b - a) \cot \alpha_0,$$ (8.25)

where the sign is taken according to the geometry of the problem [43].

Exercise: Sketch the function (8.17).

8.1.2 Wave Trapping on a Ridge

Here we have $k_{max} > K = k_0 \sin (\alpha_0) > k_{min}$. From equation (8.7), rays exist in (a, b) when $k^2(x) > K^2$, in which case (8.7) gives

$$y_b = y_0 + \int_{x_0}^{b} \frac{K \, dx}{[k^2(x) - K^2]^{1/2}}.$$ (8.26)

For a sufficiently smooth $h(x)$ (and hence $k(x)$), near $x = b$,

$$k^2(x) \approx K^2 + (x - b)(k^2)_b',$$

provided $k(b)k'(b) \neq 0$, so the integral is finite but $y'(b) = \infty$. Thus, the line $x = b$ is the envelope of all rays with $0 < \alpha_0 < \pi/2$ and is called a *caustic* (see Chapter 6). Beyond $(b, y(b))$ we have $y' < 0$ until it reaches another caustic at $x = a$. No simple harmonic waves for this value of K are possible in (a, b); this is the phenomenon of *wave trapping* (although external excitations may be possible by meteorological forcing).

8.2 CIRCULAR DEPTH CONTOURS

Earlier, using the definition (3.259), we sought to find extrema of

$$L = \int_{P_0}^{P_1} f\left(x, y(x), y'(x)\right) dx. \tag{8.27}$$

In polar coordinates (r, θ), $h = h(r)$, $k = k(r)$ and now

$$L = \int_{p_0}^{p_1} k(r) \left(1 + r^2 \left(\frac{d\theta}{dr}\right)^2\right)^{1/2} dr, \tag{8.28}$$

so with $\theta' = d\theta/dr$, the Euler-Lagrange equation (3.260) becomes

$$\frac{\partial}{\partial r} \left\{ \frac{\partial}{\partial \theta'} \left[k(1 + r^2(\theta'^2)^{1/2}] \right\} = 0, \tag{8.29}$$

since the integrand is explicitly independent of θ. Therefore

$$\frac{\partial}{\partial \theta'} \left[k(1 + r^2(\theta'^2)^{1/2}\right] = K = \text{constant}, \tag{8.30}$$

or

$$\frac{k(r)r^2\theta'}{(1 + r^2(\theta'^2)^{1/2}} = K \tag{8.31}$$

along any ray, where K characterizes the ray. Hence,

$$\frac{d\theta}{dr} = \pm \frac{|K|}{r(k^2r^2 - K^2)^{1/2}}, \tag{8.32}$$

or

$$\theta - \theta_0 = \pm |K| \int_{r_0}^{r} \frac{dr}{r(k^2r^2 - K^2)^{1/2}}, \tag{8.33}$$

where (r_0, θ_0) is some reference point (Figure 8.2). This is an equation we have encountered before in an optical context (see equation (3.99). To see the significance of the constant K, note that from (8.31) that

$$K = kr \left(\frac{rd\theta}{(dr^2 + r^2d\theta^2)^{1/2}} \right) = kr \left(r \frac{d\theta}{ds} \right) = kr \sin \alpha, \tag{8.34}$$

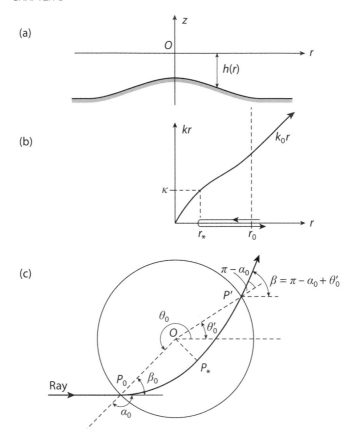

(a)

(b)

(c)

Figure 8.2 Details for submerged shoal.

where α is the local angle between the ray and the (normal) radius vector and s is arc length. In fact,

$$K = k_0 r_0 \sin(\alpha_0). \tag{8.35}$$

This is just another form of Bouguer's law (Chapter 3).

We note that curvature of the depth contours is responsible for the factor r in $K = kr \sin \alpha$. To see this, note that rays only exist if $k^2 r^2 > K^2$, so we can define the critical radius $r = r_\star$ where $k^2 r_\star^2 = K^2$. With $\theta_\star = \theta(r_\star)$, equation (8.33) yields

$$\theta_\star - \theta_0 = \pm |K| \int_{r_0}^{r_\star} \frac{dr}{r(k^2 r^2 - K^2)^{1/2}}. \tag{8.36}$$

Then equation (8.32) implies that $r'(\theta) = 0$ at $(r_\star, \theta_\star)$, that is, when the ray is closest or farthest from the origin.

8.3 CONSTANT PHASE LINES

Consider the ray $F_1(r, \theta) = r - f(\theta) = 0$ and phase line $F_2(r, \theta) = r - g(\theta) = 0$. They are orthogonal, so

$$\nabla[r - f(\theta)] \cdot \nabla[r - g(\theta)] = 0, \tag{8.37}$$

where

$$\nabla \equiv \left(\frac{\partial}{\partial r}, \frac{1}{r}\frac{\partial}{\partial \theta} \right).$$

Then we find

$$\left(1 - \frac{f'}{r} \right) \cdot \left(1 - \frac{g'}{r} \right) = 1 + \frac{f'g'}{r^2} = 0, \tag{8.38}$$

so

$$g'(\theta) = \frac{-r^2}{f'(\theta)}. \tag{8.39}$$

But

$$f'(\theta) = \frac{dr}{d\theta} = \pm\frac{r}{K}\left(k^2 r^2 - K^2\right)^{1/2}, \tag{8.40}$$

so for the phase lines, we see that

$$g'(\theta) = \frac{dr}{d\theta} = \mp\frac{Kr}{\left(k^2 r^2 - K^2\right)^{1/2}}, \tag{8.41}$$

or

$$K(\theta - \theta_0) = \mp\int_{r_0}^{r} \frac{\left(k^2 r^2 - K^2\right)^{1/2}}{r} \, dr. \tag{8.42}$$

8.3.1 Case 1

Here $0 < k(r) < \infty$, and $k(r)$ is monotonic in r. Assuming we are in very shallow water, $\omega \sim k\,(gh)^{1/2}$, so $k \sim h^{-1/2}$ and $kr \sim rh^{-1/2} \to 0$ as $r \to 0$ if $h(0) > 0$. A submerged circular shoal falls into this category. Since k increases as $r \to 0$, the ray passing through (r_0, θ_0) follows a path such that $\theta'(r) < 0$, so

$$\theta - \theta_0 = -K\int_{r_0}^{r} \frac{dr}{r\left(k^2 r^2 - K^2\right)^{1/2}}, \tag{8.43}$$

is valid until a point $P_\star$ is reached, where $r = r_\star$ is a minimum. Thereafter,

$$\theta - \theta_\star = +K\int_{r_\star}^{r} \frac{dr}{r\left(k^2 r^2 - K^2\right)^{1/2}}, \tag{8.44}$$

as r is then increasing. Since the ray is symmetric about the radius vector $\theta = \theta_\star$, we can combine equations (8.43) and (8.44) to give

$$|\theta - \theta_\star| = \int_{r_\star}^{r} \frac{K\,dr}{r\left(k^2 r^2 - K^2\right)^{1/2}}. \tag{8.45}$$

Initially the ray is parallel to the x-axis. Only rays for which $|y| \leq r_0$ are refracted. With θ'_0 the direction of the point at which the ray leaves the shoal, we find that

$$|\theta'_0 - \theta_\star| = \int_{r_\star}^{r_0} \frac{Kdr}{r\left(k^2 r^2 - K^2\right)^{1/2}}. \tag{8.46}$$

A similar result is valid for $0 < y \leq r_0$, so rays from $y > 0$ and $y < 0$ can intersect and interfere, doubling the amplitude for $y = 0$, $x = r_0$, by symmetry [43]. If $y \neq 0$, destructive or constructive interference can occur, depending on wave phases. There exists no deflection ($\beta = 0$) for $\alpha_0 = \pi$ (normal incidence) or $\alpha_0 = \pi/2$ (glancing incidence). Since $\beta > 0$ for $\pi/2 < \alpha_0 < \pi$, there must be a positive maximum for β. (Similarly, there is a maximum for $|\beta|$, $\beta < 0$, for $0 < y < r_0$.) This means, since $\beta = \beta(\alpha_0)$, rays from the same side of the x-axis will intersect also.

8.3.2 Case 2

Here kr decreases to a minimum and then increases. An example of this is a circular island with shoreline at $r = b$, so $h \to 0^+$ as $r \to b^+$, so with $k \sim h^{-1/2}$, $kr \sim bh^{-1/2}$ as $r \to b^+$, so $kr \to \infty$ there. Also, $d\theta/dr \to 0$ as $r \to b^+$ for this reason. At large r, $kr \to kr_0$. Note that there is no change of direction if $K > k_0 a$. If $K < (kr)_{\min}$, rays reach the shore. If $(kr)_{\min} < K < k_0 a$, then rays are refracted and never come closer to shore than $r_\star$.

8.3.3 Case 3

Here waves become trapped on a ring-shaped ridge. The wave undulates between two *circular caustics* defined by $r = r_1$ and $r = r_2$. The ray configuration is repeated after every angular period

$$\Delta\theta = 2K \int_{r_1}^{r_2} \frac{dr}{r\left(k^2 r^2 - K^2\right)^{1/2}}; \tag{8.47}$$

if this is a rational multiple of 2π, the ray paths form a closed curve with eigenvalue condition

$$\frac{\Delta\theta}{2\pi} = \frac{m}{n}, \quad m, n = 1, 2, 3, \ldots. \tag{8.48}$$

8.4 WAVES AND CURRENTS

Briefly we consider a special case of the situation mentioned in the quotation at the beginning of this chapter, namely, the effect of currents on waves [247], by asking the following question about waves and currents moving in *exactly oppo-site* directions: Can waves be stopped by opposing streams? It would certainly appear so, and the answer is "yes," but let's try and make this a little more quantitative [212]. Suppose water is flowing in the positive x-direction over an uneven stream bed, and in particular the depth goes from deep (H_1) to shallow (h) to deep

(H_2; this need not occur in a symmetrical manner). If the extremes are great, then from the conservation of linear fluid momentum we can say that the stream speed will be much less in the deep regions than in the shallow; so much so for really deep regions that we can effectively consider it to be zero for the present purposes. In the shallow region, the stream speed is $V > 0$. Suppose also that waves with speed $c_0 > 0$ in the (still) deep water move into the shallow (flowing) region; how will the speed of the waves change as this happens? If the wavelength in deep water is λ_0, the frequency of the waves is $\nu_0 = c_0/\lambda_0$, and this will also be their frequency in the deep water on the far side of the shallows. If the speed of the waves in the shallows relative to the stream is c, then relative to the still water (or a stationary observer) it is $c + V$; if their wavelength is now λ, then their frequency is now $(c + V)/\lambda$. If waves are neither created nor destroyed between these two places, then the same number of waves must pass each point, which means that the frequencies are identical. Thus

$$\frac{c_0}{\lambda_0} = \frac{c + V}{\lambda}. \tag{8.49}$$

At this point, we can consider two options: the first is the case for which all the depths H_1, H_2, and h are large compared with both λ_0 and λ; in this case the wave speed is $(gk)^{1/2}$, where the wavenumber $k = 2\pi/\lambda$ (see equation (11.27)), so

$$\lambda_0 = \frac{2\pi}{g} c_0^2 \quad \text{and} \quad \lambda = \frac{2\pi}{g} c^2. \tag{8.50}$$

On substituting these expressions into equation (8.49) we find, after a little rearrangement, the following quadratic equation for the desired speed c:

$$c^2 - cc_0 - Vc_0 = 0. \tag{8.51}$$

This has roots

$$c = \frac{1}{2} c_0 \left[1 \pm \left(1 + \frac{4V}{c_0} \right)^{1/2} \right], \tag{8.52}$$

or in terms of the dimensionless quantities $\tilde{c} = c/c_0$ and $\tilde{V} = V/c_0$,

$$\tilde{c} = \frac{1}{2} \left[1 \pm \left(1 + 4\tilde{V} \right)^{1/2} \right]. \tag{8.53}$$

Does this allow for the possibility of waves moving in both directions, one for each root? Figure 8.3 shows both branches of $\tilde{c}$.

Note that the c-intercepts are just 0 and 1, and the vertex of the parabola is located at $(-1/4, 1/2)$. The portion of the graph (including the vertex) for which $\tilde{V} < 0$ and $0 < \tilde{c} < 1$ corresponds to waves moving against the stream; clearly here in dimensional terms $c < c_0$; the wave speed relative to the stream decreases. Note also that there is no solution c if $\tilde{V} < -1/4$ (or $V < -c_0/4$). This can be interpreted to mean that surface gravity waves will not be able to traverse an opposing stream if the speed of the stream exceeds $c_0/4$. At the vertex, $\tilde{c} = 1/2$, so $c = c_0/2$, and individual waves are still able to move upstream, but the group speed is half this, and wave groups at this speed are unable to advance. Individual waves move upstream

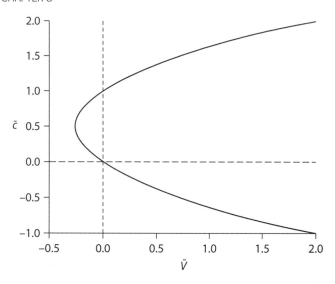

Figure 8.3 The wave speed $\tilde{c}$ as a function of the stream speed $\tilde{V}$.

through this limiting group and like old soldiers, just fade away, for there are no more groups ahead to accommodate them. But the energy associated with these wavelets surely cannot just disappear, can it? As wave groups move upstream, they cannot get past this limit point in the (V, c) parameter space, so groups moving faster than $c_0/4$ must eventually slow down and pile up there, concentrating the energy in such a way that the waves must increase in height (but notice now we are on shaky ground, because the formula we have used for wave speed is for very small amplitude waves). If this is true, then eventually they will break, and their energy will be dissipated rapidly by frictional processes not considered here. And lo, the waves are no longer present beyond this point. Note that for $\tilde{V} > 0$, $c > c_0$, as indicated from the upper branch (the lower branch corresponds to negative values of $\tilde{c}$) and hence $\tilde{c}$, and by hypothesis, $c > 0$ here.

The complementary topic of scattering of surface gravity waves will be discussed in Chapter 17 using the governing differential equations to infer properties that cannot be determined from a purely ray-theoretic approach.

Chapter Nine

Seismic Rays

Seismic tomography is an imaging technique that uses seismic waves generated by earthquakes and explosions to create computer-generated, three dimensional images of Earth's interior. If the Earth were of uniform composition and density seismic rays would travel in straight lines. . . . But our planet is broadly layered, causing seismic rays to be refracted and reflected across boundaries. Layers can be inferred by recording energy from earthquakes (seismic waves) of known locations, not too unlike doctors using CT scans . . . to "see" into hidden parts of the human body. How is this done with earthquakes? The time it takes for a seismic wave to arrive at a seismic station from an earthquake can be used to calculate the speed along the wave's ray path. By using arrival times of different seismic waves scientists are able to define slower or faster regions deep in the Earth. Those that come sooner travel faster. Later waves are slowed down by something along the way. Various material properties control the speed and absorption of seismic waves. Careful study of the travel times and amplitudes can be used to infer the existence of features within the planet.

[248]

9.1 SEISMIC RAY EQUATIONS

When the curvature of the earth may be neglected, the resulting flat-earth approximation and the assumption of lateral homogeneity mean that the speeds of the various seismic waves may be considered functions of depth only. *Surface* seismic waves generally occur within the top 30 km or so of the earth's crust, and there are two basic types of such wave. *Rayleigh waves* have horizontal components of displacement in the direction of propagation, and *Love waves* have horizontal displacements perpendicular to their direction of propagation and arise in stratified or layered media. The theory is of these waves is rather more complicated than is necessary to develop in this chapter, but we shall encounter the basic physical and mathematical principles for elastic surface waves in Part II (Chapter 10). Furthermore, surface waves in fluids will be revisited in Chapter 11, and also when we examine ocean acoustic *waveguides* in Chapter 12, where a simple plane-parallel stratified model is developed, which is applicable to either geophysical or oceanographical situations (such as the thermocline).

Some Numerical Values

The elasticity of a homogeneous isotropic solid is characterized in terms of two constants k and μ, both defined in units of stress (or force per unit area). k is the *bulk modulus* (or modulus of compressibility, or simply the incompressibility). For granite, $k \approx 2.7 \times 10^{11}$ dynes/cm^2; for water, $k \approx 2.0 \times 10^{10}$ dynes/cm^2. The other constant, μ, is the *modulus of rigidity* (or simply, the rigidity). For granite, $\mu \approx 1.6 \times 10^{10}$ dynes/cm^2; for water, $\mu \equiv 0$. In an elastic solid of density ρ, two types of waves can propagate; so-called P (primary) waves with speed

$$c_P = \left(\frac{k + 4\mu/3}{\rho} \right)^{1/2}, \tag{9.1}$$

and S (shear) waves with speed

$$c_S = \left(\frac{\mu}{\rho} \right)^{1/2} < c_P. \tag{9.2}$$

For granite, $c_P \approx 5.5$ km/s and $c_S \approx 3.0$ km/s; for water, $c_P \approx 1.5$ km/s and $c_S = 0$. P waves are waves of compression and dilation, much like sound waves in a gas, being longitudinal in character. The other type, S waves, cannot propagate in gases or liquids, because they involve shearing and twisting motion; only solids can support such forces, in which the particles move transversely to the direction of the waves (in a vertical or a horizontal plane). The fact that such waves can exist only in solids can be used to detect the presence of liquid zones (or lack thereof) in the earth's interior.

Just as sound, electromagnetic, or water waves can undergo reflection, refraction, diffraction, interference, etc., so too can seismic waves, but they outdo the former types in terms of their behavior on encountering a boundary or surface of discontinuity in the earth. If a P wave hits such a surface obliquely, it is decomposed into a reflected P wave and a refracted P wave: no surprises so far. However, it also generates a reflected S wave and a refracted S wave. This can been understood from the nature of the waves themselves: at a point of (oblique) incidence, the rock boundary is compressed *and* sheared, producing four resultant waves via what is called *wave conversion*. Only S waves can be polarized, however. Such waves encounter structural discontinuities that refract, reflect, and polarize them. SH waves cause elements of the rock (or "particles"!) to oscillate in a horizontal plane only; when the particles oscillate in a vertical plane containing the direction of motion, the waves are designated SV waves. The pressures acting on the rock in the interior of the earth cause the rock density to increase with depth, but both the rigidity (μ) and incompressibility (k) increase faster than the density (ρ), provided of course, the rock is not molten (in which case $\mu = 0$) [109]. This means that, in general, both c_P and c_S increase with depth.

How Far Away Is the Epicenter?

Consider just for simplicity the propagation of the two types of *body wave* after an earthquake has occurred. (Surface waves are extremely important of course, since they are the ones that do the most damage, traveling in essentially two dimensions as opposed to three.) Suppose that the primary tremors are recorded at times t_1 and t_2, respectively. Let the distance of the earthquake source be s miles (assuming that the waves travel in straight lines and that the speeds are constant, neither of which is in general true of course!). Then since $s = c_P t_1 = c_S t_2$ (say), since $c_P > c_S$ it follows that $t_2 > t_1$ and hence the time difference

$$\Delta t = t_2 - t_1 = s \left(\frac{1}{c_S} - \frac{1}{c_P} \right) = s \left(\frac{c_P - c_S}{c_P c_S} \right). \tag{9.3}$$

Therefore

$$s = \left(\frac{c_P c_S}{c_P - c_S} \right) \Delta t.$$

In practice this type of calculation would be extended to include the variable speeds and records at other locations to triangulate the location and depth of the epicenter.

9.2 RAY PROPAGATION IN A SPHERICAL EARTH

A crude model of the earth's structure might be characterized as follows. Below an outer *crust* (both oceanic and continental) varying between approximately 10 and 50 km thick, the *mantle* extends to a depth of about 2,900 km, below which the *outer core* extends to a depth of 5,200 km, beyond which lies the *inner core*, all the way to the center, some 6,400 km from the surface. The rocky, brittle crust is what fractures during earthquake activity. The boundary between the crust and mantle is referred to as the *Mohorovičić discontinuity*, or *Moho*. The mantle contains about two-thirds of the earth's mass, and it is in this region that convection drives plate tectonic activity. S and P wave activity in this region enables the convective activity to be mapped. Convection in the outer core, coupled with the earth's rotation, drives and maintains the earth's magnetic field (the so-called *dynamo problem*). This region accounts for about 30 percent of the earth's mass. The inner core is solid iron and suspended in the molten outer core.

For simplicity we shall consider the speeds of the body waves, P and S, to be functions of radial position r only. Across each layer the quantity $(r_k \sin i_k) / V_k$ is constant, $k = 1, 2, 3 \ldots$. To see this, note that in an *optical* context, at the interface of spherical shells 1 and 2, Snell's law of refraction applies to the rays,

$$n_1 \sin i_1 = n_2 \sin \alpha_1 \approx n_2 \sin i_2, \tag{9.4}$$

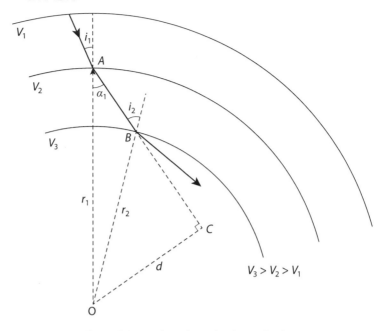

Figure 9.1 Ray in a piecewise-layered sphere.

the approximation being valid if the shells are sufficiently thin, and where n_1 and n_2 are the relative constant refractive indices in the respective media. But is this concept even relevant in the current non-optical context? It is if we replace the speed of light in the medium with the speed(s) V_k of the elastic wave(s) in the various shells, and recall that in optics the relative refractive index is inversely proportional to the speed of light in the medium. Hence we can rewrite equation (9.4) as

$$\frac{\sin i_1}{V_1} = \frac{\sin i_2}{V_2}. \tag{9.5}$$

From Figure (9.1) we see that

$$r_1 \sin \alpha_1 \approx r_1 \sin i_1 = r_2 \sin i_2, \tag{9.6}$$

so equation (9.5) becomes

$$\frac{r_1 \sin i_1}{V_1} = \frac{r_2 \sin i_2}{V_2} \equiv p, \tag{9.7}$$

where p is sometimes known as the *horizontal slowness* (being inversely proportional to speed; recall the secret Pythagorean theorem in Chapter 2). Clearly this is conserved along the ray path. In the limit of infinitely many infinitesimally thin layers merging into a radially symmetric continuum, this result becomes Bouguer's theorem once again, but in the form

$$r \sin i \, (r) = pV(r). \tag{9.8}$$

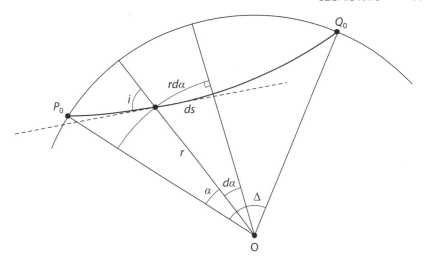

Figure 9.2 Ray geometry for continuous properties. The angle subtended by $P_0 O Q_0$ is Δ.

In planar geometry (i.e., as $r \to \infty$ in equation (9.7)) and $r \to z$

$$\sin i(z) = pV(z),\qquad(9.9)$$

and at the altitude z_c when the tangent to the ray is horizontal in this geometry,

$$pV(z_c) = 1.\qquad(9.10)$$

The result (9.9) also follows from the x-component of equation (3.75):

$$n\frac{dx}{ds} = \text{constant},\qquad(9.11)$$

which leads to (9.9).

Since the speeds of P and S waves are in general different, these waves will have different ray paths in both continuously varying and piecewise-constant media. Let us proceed with the former type of medium (therefore having no interfaces and consequent wave reflections) and establish some useful integral identities. Suppose that the ray path extends, via the earth's interior, from point P_0 to the point Q_0 at the surface. From Figures 9.2 and 9.3,

$$p = \frac{r \sin i}{V} \approx \frac{r}{V}\left(\frac{rd\alpha}{ds}\right),\quad \text{and}\quad ds^2 \approx dr^2 + (rd\alpha)^2,\qquad(9.12)$$

so

$$d\alpha = \frac{pV\,dr}{r\left(r^2 - p^2V^2\right)^{1/2}} = \frac{p\,dr}{r\left(\eta^2 - p^2\right)^{1/2}},\qquad(9.13)$$

where $\eta = r/V$. If r_c is the value of r at the deepest (or halfway) point of the trajectory, then by the symmetry assumed here, the angle subtended at the center O by the complete ray path $P_0 Q_0$ in Figure 9.2 is

$$\Delta = 2p \int_{r_c}^{r_0} \frac{p}{r\left(\eta^2 - p^2\right)^{1/2}}\,dr.\qquad(9.14)$$

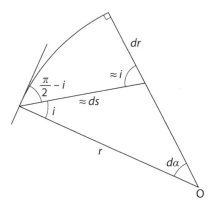

Figure 9.3 Detail from Figure 9.2 (greatly exaggerated).

Since $V = ds/dt$ it follows that the time of travel

$$T = \int_0^s \frac{ds'}{V} = 2 \int_{r_c}^{r_0} \frac{\eta^2 dr}{r \left(\eta^2 - p^2\right)^{1/2}}. \tag{9.15}$$

Exercise: Show also that

$$T = p\Delta + 2 \int_{r_c}^{r_0} \frac{\left(\eta^2 - p^2\right)^{1/2} dr}{r}. \tag{9.16}$$

Obviously these results depend (in particular) on knowing the value of p for a particular ray. But since

$$p = \left[\frac{r \sin i}{V}\right]_{P,Q} = \left[\frac{r_0 \sin i_0}{V}\right]_{P_0,Q_0} \tag{9.17}$$

it can in principle be determined by surface measurements.

9.2.1 A Horizontally Stratified Earth

If the the ray path lengths are small enough (i.e., less than about 1,000 km), then it is a reasonable simplification to assume that the medium is horizontally stratified (rather than radially so). We have already noted that in the limit of the radius of curvature becoming infinite, the horizontal slowness or ray parameter p is redefined as

$$p = \frac{\sin i(z)}{V(z)} = \frac{1}{V(z)} \frac{dx}{ds}, \tag{9.18}$$

where ds is a ray path segment shown in Figure 9.4. In addition, the parameter η is redefined as $\eta = [V(z)]^{-1}$. Note that we also have

$$\frac{dz}{ds} = \cos i(z) = \left(1 - p^2 V^2\right)^{1/2} = \frac{\left(\eta^2 - p^2\right)^{1/2}}{\eta}, \tag{9.19}$$

and

$$\frac{dx}{dz} = \left(\frac{dx}{ds}\right)\left(\frac{ds}{dz}\right) = \frac{p}{\left(\eta^2 - p^2\right)^{1/2}}. \tag{9.20}$$

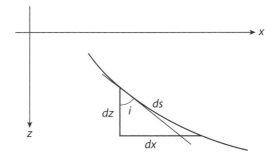

Figure 9.4 Segment of ray path.

If z_m is the depth to which a ray travels from the surface (say) before turning up toward the surface again, then the horizontal distance traveled for $0 \leq z \leq z_m$ is another integral of the type encountered already, namely,

$$x = \int_0^z \frac{p}{\left[\eta \left(z'\right)^2 - p^2\right]^{1/2}} dz', \tag{9.21}$$

and the time to travel this distance is

$$t = \int \frac{ds}{v(z)} = \int_0^z \frac{dz'}{v\left(z'\right) \cos\theta\left(z'\right)} = \int_0^z \frac{\eta^2\left(z'\right)}{\left[\eta\left(z'\right)^2 - p^2\right]^{1/2}} dz'. \tag{9.22}$$

Hereafter the dependence on the dummy variable z' will be understood. Also, note that, similar to the radially stratified case,

$$t = \int_0^z \left[\frac{p^2}{\left(\eta^2 - p^2\right)^{1/2}} + \left(\eta^2 - p^2\right)^{1/2}\right] dz' = px + \int_0^z \left(\eta^2 - p^2\right)^{1/2} dz'. \tag{9.23}$$

By path symmetry, then, the total horizontal surface-to-surface distance traveled and time of travel are therefore

$$X = 2p \int_0^{z_m} \frac{dz}{\left(\eta^2 - p^2\right)^{1/2}}; \tag{9.24}$$

$$T = 2 \int_0^{z_m} \frac{\eta^2}{\left(\eta^2 - p^2\right)^{1/2}} dz. \tag{9.25}$$

These expressions are the planar versions of the results (9.14) and (9.15).

9.2.2 The Wiechert-Herglotz Inversion

We apply the operator

$$\mathcal{B}\left(X\right) = \int_{p_0}^{p_1} \frac{X}{\left(p^2 - \eta_1^2\right)^{1/2}} dp, \tag{9.26}$$

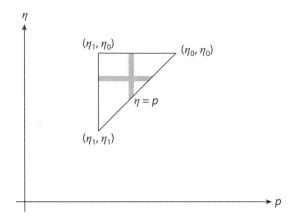

Figure 9.5 Geometry for the integral (9.29).

to equation (9.24), where

$$p_0 = \eta_{p_0} = \eta_0 = \frac{1}{V_0},\tag{9.27}$$

and

$$p_1 = \eta_{p_1} = \eta_1 = \frac{1}{V_1}.\tag{9.28}$$

This corresponds to an integration across ray paths from the ray with zero range to the ray whose vertex is at a depth where $V_1 = \eta_1^{-1}$. Then

$$\mathcal{B}(X) = \int_{\eta_0}^{\eta_1} dp \int_{\eta_0}^{\eta_1} \frac{2p\,(dz/d\eta)}{\left(p^2 - \eta_1^2\right)^{1/2}\left(\eta^2 - p^2\right)^{1/2}}\,d\eta.\tag{9.29}$$

Interchanging the order of integration (see Figure 9.5), we see that

$$\int_{\eta_0}^{\eta_1} \frac{X}{\left(p^2 - \eta_1^2\right)^{1/2}}\,dp = \int_{\eta_0}^{\eta_1} d\eta \int_{p}^{\eta_1} \frac{2p\,(dz/d\eta)}{\left(p^2 - \eta_1^2\right)^{1/2}\left(\eta^2 - p^2\right)^{1/2}}\,dp.\tag{9.30}$$

The left-hand side can be integrated by parts and the right-hand side integrated to yield

$$\left[X \operatorname{arccosh}\left(\frac{p}{\eta_1}\right)\right]_{\eta_0}^{\eta_1} - \int_{\eta_0}^{\eta_1} \frac{dX}{dp} \operatorname{arccosh}\left(\frac{p}{\eta_1}\right)dp$$

$$= \int_{\eta_0}^{\eta_1} \frac{dz}{d\eta}\left\{\arcsin\left[\frac{2p^2 - \left(\eta^2 + \eta_1^2\right)}{\left(\eta^2 - \eta_1^2\right)}\right]\right\}_{p=\eta}^{\eta_1}\,d\eta.\tag{9.31}$$

Since $X = 0$ when $p = \eta_0$, the first term on the left vanishes, so (9.31) becomes

$$\int_{\eta_0}^{\eta_1} \text{arccosh}\left(\frac{p}{\eta_1}\right)\left(\frac{dX}{dp}\right) dp = \pi \int_{\eta_0}^{\eta_1} \left(\frac{dz}{d\eta}\right) d\eta, \tag{9.32}$$

or

$$\int_0^{X_1} \text{arccosh}\left(\frac{p}{\eta_1}\right) dX = \pi \int_0^{Z_1} = \pi Z_1, \tag{9.33}$$

where Z_1 is the vertex of the ray path corresponding to $\eta = \eta_1$. Finally, substituting for p and η_1 we have

$$Z_1 = \frac{1}{\pi} \int_0^{X_1} \text{arccosh}\left(\frac{V_1}{V_p}\right) dX. \tag{9.34}$$

Recall from equation (9.14) that

$$\Delta = 2p \int_{r_c}^{r_0} \frac{p}{r \left(\eta^2 - p^2\right)^{1/2}} dr. \tag{9.35}$$

This is similar in form to the horizonally stratified version (9.24). Since for the spherical case $\eta = r/V$, we can replace $dz/d\eta$ with $r^{-1}\left(\partial r/\partial \eta\right)$ in (9.24) and the integral, (9.33), after reversal of the limits, is replaced by

$$\int_0^{\Delta_1} \text{arccosh}\left(\frac{p}{\eta_1}\right) d\Delta = \pi \int_{r_1}^{r_0} \frac{dr}{r}, \tag{9.36}$$

or

$$\frac{r_0}{r_1} = \exp\left[\frac{1}{\pi}\int_0^{\Delta_1} \text{arccosh}\left(\frac{p}{\eta_1}\right) d\Delta\right]. \tag{9.37}$$

This result is also of use for the inverse problem in optics (see [73] for further references). From experimentally obtained (T, Δ) plots this result can be used sequentially to obtain velocity-depth behavior for P and S waves. These in turn can provide information on the wavespeed-determining quantities ρ, k, and μ in equations (9.1) and (9.2). Obviously it is not possible to obtain a unique solution for the set $\{\rho, k, \mu\}$ without additional information or constraints. Recall that the above derivation makes the assumption that the velocity is a continuously increasing function of depth.

9.2.3 Further Properties of X in the Horizontally Stratified Case

For future reference note that

$$\frac{dt}{dx} = \left(\frac{dt}{dz}\right)\left(\frac{dz}{dx}\right) = \frac{1}{pV^2} = p. \tag{9.38}$$

when $z = z_m$. This can be written more succinctly as [119]

$$p = \frac{dT}{dX}. \tag{9.39}$$

So how does X (in particular) vary with the ray parameter p? From the definition (9.24) we see that

$$\frac{dX}{dp} = 2 \int_0^{z_m} \frac{dz}{\left(\eta^2 - p^2\right)^{1/2}} + 2p \frac{d}{dp} \int_0^{z_m} \frac{dz}{\left(\eta^2 - p^2\right)^{1/2}}. \tag{9.40}$$

The second integral can be integrated by parts; if we define $f(\eta) = dz/d\eta$ and $\eta(0) = \eta_0$, noting that $\eta(z_m) = p$, then

$$\int_0^{z_m} \frac{dz}{\left(\eta^2 - p^2\right)^{1/2}} = \int_0^{z_m} f(\eta) \frac{d\eta/dz}{\left(\eta^2 - p^2\right)^{1/2}} dz = \int_{\eta_0}^{p} \frac{f(\eta)}{\left(\eta^2 - p^2\right)^{1/2}} d\eta \tag{9.41}$$

$$= \left[f(\eta) \operatorname{arccosh}\left(\frac{\eta}{p}\right) \right]_{\eta_0}^{p} - \int_0^{z_m} f'(\eta) \left[\operatorname{arccosh}\left(\frac{\eta}{p}\right) \right] \frac{d\eta}{dz} dz. \tag{9.42}$$

Hence

$$\frac{dX}{dp} = 2 \int_0^{z_m} \frac{dz}{\left(\eta^2 - p^2\right)^{1/2}} + 2p \frac{d}{dp} \left\{ -f(\eta_0) \operatorname{arccosh}\left(\frac{\eta_0}{p}\right) \right.$$

$$\left. - \int_0^{z_m} f'(\eta) \left[\operatorname{arccosh}\left(\frac{\eta}{p}\right) \right] \frac{d\eta}{dz} dz \right\},$$

$$= 2 \int_0^{z_m} \frac{dz}{\left(\eta^2 - p^2\right)^{1/2}} + 2p \left[\frac{\eta_0 f(\eta_0)}{p \left(\eta_0^2 - p^2\right)^{1/2}} + \int_0^{z_m} \frac{\eta f'(\eta)}{p \left(\eta^2 - p^2\right)^{1/2}} \frac{d\eta}{dz} dz \right],$$

$$= \frac{2\eta_0 f(\eta_0)}{p \left(\eta_0^2 - p^2\right)^{1/2}} + 2 \int_0^{z_m} \frac{\left[f(\eta) + \eta f'(\eta)\right] d\eta}{\left(\eta^2 - p^2\right)^{1/2}} \frac{d\eta}{dz} dz. \tag{9.43}$$

The logarithmic derivative of the velocity field is

$$\zeta(z) = \frac{1}{V} \frac{dV}{dz} = -V \frac{d\eta}{dz}, \tag{9.44}$$

from which it follows that

$$\eta f(\eta) = \frac{1}{V} \frac{dz}{d\eta} = -\frac{1}{\zeta(z)}, \tag{9.45}$$

and therefore

$$\left[f(\eta) + \eta f'(\eta) \right] \frac{d\eta}{dz} = \frac{d}{dz} \left[\eta f(\eta) \right] = \frac{1}{\zeta^2} \frac{d\zeta}{dz}. \tag{9.46}$$

This enables us to write (9.43) as

$$\frac{dX}{dp} = -\frac{2}{\zeta_0 \left(\eta_0^2 - p^2\right)^{1/2}} + 2\int_0^{z_m} \frac{\zeta'(z)}{\zeta^2 \left(\eta^2 - p^2\right)^{1/2}} dz \qquad (9.47)$$

$$= -\frac{2}{\zeta_0 \left(\eta_0^2 - p^2\right)^{1/2}} + 2\int_0^{z_m} \frac{1}{\zeta^2 \left(\eta^2 - p^2\right)^{1/2}} \left(\zeta^2 - \frac{1}{v}\frac{d^2v}{dz^2}\right) dz. \quad (9.48)$$

Mathematically, the simplest case for which to find $X(p)$ is when $d\zeta/dz = 0$, that is,

$$V(z) = V(0)\exp(az) \equiv V_0\exp(az), \qquad (9.49)$$

where $a > 0$ has been chosen because V gradually increases with depth (increasing z) in many geophysical situations. For this special case (considered in [119])

$$X = -\frac{2}{a}\int_{p(0)}^{p(z_m)} \frac{dp}{\left(\eta_0^2 - p^2\right)^{1/2}} = \frac{2}{a}\left[\arccos\left(\frac{p}{\eta_0}\right)\right]_{p(0)}^{p(z_m)} = \frac{2}{a}\arccos(pV_0).$$

$$(9.50)$$

Therefore

$$p = \frac{1}{V_0}\cos\left(\frac{aX}{2}\right), \qquad (9.51)$$

and hence from equation (9.39),

$$T = \frac{2}{aV_0}\sin\left(\frac{aX}{2}\right). \qquad (9.52)$$

Graphs proportional to the quantities X and T are shown below for this idealized case are shown in Figures 9.6 and 9.7.

We consider briefly some general properties of expression (9.47), namely

$$\frac{dX}{dp} = -\frac{2}{\zeta_0 \left(\eta_0^2 - p^2\right)^{1/2}} + 2\int_0^{z_m} \frac{\zeta'(z)}{\zeta^2 \left(\eta^2 - p^2\right)^{1/2}} dz.$$

In what follows reference is made to Figure (9.8).

Proceeding from the surface $z = 0$ and noting that the wave speed v generally increases gradually with depth (still a relatively simple case to consider), ζ will be a positive quantity of moderate magnitude, while $\zeta'(z)$ is likely to be rather small. Under such circumstances the integrated term will dominate, and $X'(p)$ will be negative. As p increases, this term will decrease further (Figure 9.8d), and we can then infer that the positive quantity $p(X)$ will decrease slowly with X (Figure 9.8e). Since $p = dT/dX$ the generic graph of $T(X)$ may also be inferred

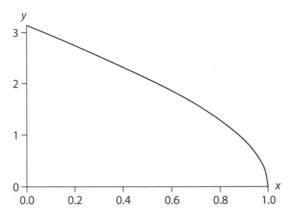

Figure 9.6 Graph of $y = aX$ versus $x = pv_0$.

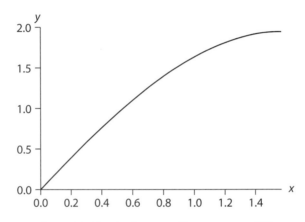

Figure 9.7 Graph of $y = av_0 T$ versus $x = aX/2$.

(Figure 9.8f), yielding, as expected in this case, T increasing with X. The speed of the wave at the deepest point ($z = z_m$) is $v_m = 1/p = (dT/dX)^{-1}$; it increases with X because it penetrates into deeper regions where the wave speeds are higher. This type of argument can serve as the basis for examining the behavior of more pathological velocity profiles, for example, (i) where $X'(P)$ changes sign; (ii) $X'(P)$ is discontinuous (corresponding to discontiuities in the travel time curve $T(X)$); (iii) *shadow zones* exist, where no rays are received in some X-interval, and so forth, but these will not be pursued here. More details of the implications from such diagrams can be found in [119], [251].

As noted earlier, the crust-Moho-mantle-outer core-inner core structure of the earth means that continuously varying models are severely limited in their ability to describe all aspects of ray and wave propagation in our planet. And clearly, with this degree of complexity, the travel times of P and S waves in the crust and mantle are complicated, to say the least, and vary considerably. Some surface

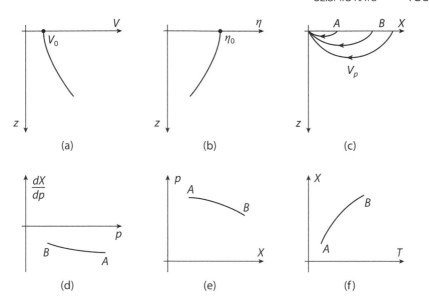

Figure 9.8 Seismological properties associated with the derivative $X'(p)$.

waves will be present, and there will generally be reflections and refractions from velocity discontinuities throughout the interior and surface layers. If either type of wave, P or S, is incident on the outer surface, a reflected P and a reflected S wave will be produced, whereas if either type is incident on the mantle-core boundary, it will produce reflected P and S waves *and* a refracted P wave. Furthermore, since there is a decrease in the velocity of P waves across this boundary, there will be a shadow zone followed by a focusing region [251]. Not so surprisingly, perhaps, a P wave incident on the outer core/inner core boundary will generate both a reflected and a refracted P wave (as in the analogous optical counterpart). The outer core only supports P waves (this fact naturally leads to the conclusion that it is in liquid form).

PART II
Waves

Chapter Ten

Elastic Waves

> The field equations of physics (yielding the dynamic and thermodynamic variables) for a given continuous medium arise from three conservation equations: conservation of mass, momentum and energy. For an elastic medium these equations of motion are called the *Navier equations*. From them, a rich variety of stress waves is obtained (unlike waves in fluids, which are rather limited). ... When we deal with an elastic solid we are concerned with the dynamic variables: stress and strain. The energy equation for an elastic solid yields a relation between the stress and strain. For small-amplitude deformations there is a linear relation between the stress and strain, which is given by Hooke's law: stress is proportional to strain for the one-dimensional case. For two and three dimensions the stress tensor is a linear function of the strain tensor.
>
> [249]

Mathematical models of elastic behavior usually posit elastic media as continuous and deformable (within certain limits) The assumption of continuity means that the molecular and crystalline structure of the medium is ignored in favor of a more abstract (but highly useful) geometric concept, one that takes advantage of the notion of a large differential volume element dV, in which the number of molecules is large enough to average or smooth out the effects of individual molecules. In the study of elastic media the existence of two states is assumed: before deformation and after deformation. The first is the equilibrium state, which is subsequently deformed into the latter. Suppose that the volume elements for the equilibrium and deformed configurations are denoted, respectively, by $dV_X = dX_1 dX_2 dX_3$ and $dV_x = dx_1 dx_2 dx_3$. Then a deformation is given by the transformation $\mathbf{X} \to \mathbf{x}$, where $x_i = x_i(X_j), i = 1, 2, 3$ and $j = 1, 2, 3$. Furthermore, there is a unique inverse transformation, and the volume elements are related via the Jacobian J of the mapping $\mathbf{X} \to \mathbf{x}$; thus

$$dV_x = |J| \, dV_X, \tag{10.1}$$

where

$$J = \frac{\partial (x_1, x_2, x_3)}{\partial (X_1, X_2, X_3)}. \tag{10.2}$$

10.1 BASIC NOTATION

The position vector $\mathbf{x} = \langle x_1, x_2, x_3 \rangle$ of any point in a body after it has been elastically deformed is related to its initial position vector $\mathbf{X} = \langle X_1, X_2, X_3 \rangle$ by the vector sum $\mathbf{x} = \mathbf{X} + \mathbf{u}$, where the displacement vector $\mathbf{u} = \langle u_1, u_2, u_3 \rangle$. In an Eulerian frame of reference, $u_i = u_i(\mathbf{x}, t)$, $i = 1, 2, 3$. For small deformations, the *linear strain tensor* is defined as

$$e_{ij} = \frac{1}{2}\left(\frac{\partial u_i}{\partial x_j} + \frac{\partial u_j}{\partial x_i}\right). \tag{10.3}$$

Clearly this is a symmetric tensor; the rotation tensor r_{ij}, in contrast, is antisymmetric:

$$r_{ij} = \frac{1}{2}\left(\frac{\partial u_i}{\partial x_j} - \frac{\partial u_j}{\partial x_i}\right). \tag{10.4}$$

The *vorticity vector* can be defined in terms of the Levi-Civita alternating tensor ε_{ijk} as $\varpi = \langle \varpi_1, \varpi_2, \varpi_3 \rangle$, where $\varpi_k = \varepsilon_{ijk} r_{ij}$. Another important tensor is the *dilatation* $\triangle = e_{kk}$; this is the change in volume of an elastic element due to the deformation (in vector notation, $\triangle = \nabla \cdot \mathbf{u}$, and $\varpi = \nabla \times \mathbf{u}$). The *symmetric stress tensor* σ_{ij} is defined in terms of e_{ij} and the Kronecker delta δ_{ij} as follows:

$$\sigma_{ij} = \lambda e_{kk}\delta_{ij} + 2\mu e_{ij}. \tag{10.5}$$

λ and μ are the *Lamé constants* for the medium; μ is also known as the *shear modulus* (or modulus of rigidity). These constants can be expressed in terms of important quantities in engineering science, namely *Young's modulus E* and *Poisson's ratio v* as follows:

$$\lambda = \frac{Ev}{(1+v)(1-2v)}, \quad \mu = \frac{E}{2(1+v)}. \tag{10.6}$$

Introducing the *bulk modulus k* (defined below in terms of E and v), it can be shown that

$$E = \frac{\mu(2\mu + 3\lambda)}{\lambda + \mu}, \quad v = \frac{\lambda}{2(\lambda + \mu)}, \quad \text{and} \quad k = \frac{E}{3(1-2v)} = \lambda + \frac{2}{3}\mu. \tag{10.7}$$

Exercise: Derive (10.7).

The governing equation of motion for an elastic medium of constant density ρ can be stated succinctly as

$$\rho\frac{\partial^2 u_i}{\partial t^2} = \frac{\partial \sigma_{ij}}{\partial x_j}, \tag{10.8}$$

or

$$\rho\frac{\partial^2 u_i}{\partial t^2} = \lambda\delta_{ij}\frac{\partial e_{kk}}{\partial x_j} + 2\mu\frac{\partial e_{ij}}{\partial x_j} = (\lambda + \mu)\frac{\partial}{\partial x_i}\left(\frac{\partial u_j}{\partial x_j}\right) + \mu\frac{\partial^2 u_i}{\partial x_j \partial x_j}. \tag{10.9}$$

In vector form these are known as *Navier equations* for the linear displacement field $\mathbf{u}$, and in the absence of body forces they are

$$\rho\frac{\partial^2 \mathbf{u}}{\partial t^2} = (\lambda + \mu)\nabla(\nabla \cdot \mathbf{u}) + \mu\nabla^2\mathbf{u}. \tag{10.10}$$

Using the vector operator identity

$$\nabla \times \nabla \times \mathbf{s} = \nabla (\nabla \cdot \mathbf{s}) - \nabla^2 \mathbf{s}, \tag{10.11}$$

equation (10.10) can be rewritten as

$$\rho \frac{\partial^2 \mathbf{u}}{\partial t^2} = (\lambda + 2\mu) \nabla (\nabla \cdot \mathbf{u}) - \mu \nabla \times (\nabla \times \mathbf{u}). \tag{10.12}$$

Dividing (10.12) by ρ, applying the *divergence operator* $\nabla \cdot$ to both sides, and utilizing the commutativity of the operators $\nabla \cdot$ and ∇^2, we obtain

$$\frac{\partial^2 \Delta}{\partial t^2} = \frac{(\lambda + 2\mu)}{\rho} \nabla^2 \Delta \equiv c_P^2 \nabla^2 \Delta, \tag{10.13}$$

which is the wave equation for the divergence of the velocity field (or dilatation Δ) that propagates at speed

$$c_P = \left(\frac{\lambda + 2\mu}{\rho} \right)^{1/2}, \tag{10.14}$$

the speed of so-called P (for *primary*) body waves. If we return to equation (10.12) and now divide it by ρ and apply the *curl operator* $\nabla \times$ to both sides, we obtain

$$\frac{\partial^2 \varpi}{\partial t^2} = \frac{\mu}{\rho} \nabla^2 \varpi \equiv c_S^2 \nabla^2 \varpi. \tag{10.15}$$

Thus the speed of so-called shear (or *secondary*) waves is

$$c_S = \left(\frac{\mu}{\rho} \right)^{1/2} < c_P, \tag{10.16}$$

so these waves propagate less fast than dilatational waves; in crude terms, expansions and contractions will be felt before shearing occurs. Although the basic characteristics of these waves have been mentioned in Chapter 9, it is worth summarizing them here. There are two types of *body wave* (as opposed to *surface waves*) that can propagate in elastic media (i.e., through the solid earth). P waves are waves of compression and dilation, much like sound waves in a gas, being longitudinal in character (Figure 10.1). In contrast, S waves cannot propagate in gases or liquids, because they involve shearing and twisting motion; only solids can support such forces, in which the particles move transversely to the direction of the waves (in a vertical or a horizontal plane; Figure 10.2). The fact that such waves can exist only in solids can be used to detect the presence of liquid zones (or lack thereof) in the earth's interior. In contrast, seismic *surface* waves generally occur within the top 30 km or so of the earth's crust, and there are also two basic types of such wave. *Love waves* have horizontal displacements perpendicular to their direction of propagation (Figure 10.3), and arise in stratified or layered media. *Rayleigh waves* have horizontal components of displacement in the direction of propagation (Figure 10.4). For pure granite, $c_P \approx 5.5$ km/s and $c_S \approx 3.0$ km/s; the corresponding speeds for steel are similar. Note that in water, $c_P \approx 1.5$ km/s and $c_S = 0$. The pressures acting on rock in the interior of the earth cause the rock density to increase with depth, but according to Bolt [109] both the rigidity μ and

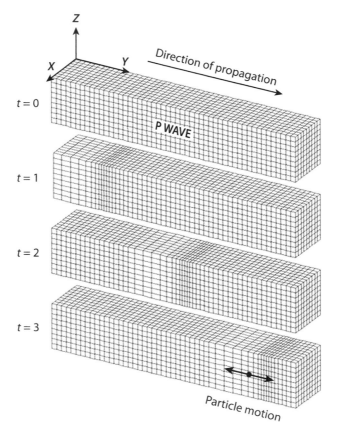

Figure 10.1 Time sequence of propagation of a P wave.

incompressibility k increase faster than the density ρ, provided of course, the rock is not molten (in which case $\mu = 0$). This means that, in general, both c_P and c_S increase with depth.

The Helmholtz Representation

We can write a smooth vector field $\mathbf{u}$ in terms of a scalar potential ϕ and a vector potential $\mathbf{A}$ as

$$\mathbf{u} = \nabla\phi + \nabla \times \mathbf{A}, \tag{10.17}$$

subject to the condition that $\nabla \cdot \mathbf{A} = 0$ (this constraint will be encountered again in Chapter 19, dealing with electromagnetic waves). From the form (10.12) for the Navier equations, using the commutation properties of the operators this can be written

$$\nabla\left[\rho\frac{\partial^2\phi}{\partial t^2} - (\lambda + 2\mu)\nabla^2\phi\right] + \nabla \times \left[\rho\frac{\partial^2\mathbf{A}}{\partial t^2} - \mu\nabla^2\mathbf{A}\right] = \mathbf{0}. \tag{10.18}$$

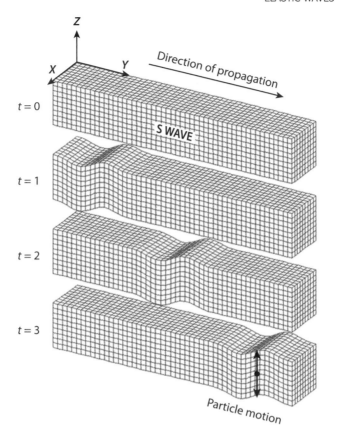

Figure 10.2 Time sequence of propagation of an S wave.

In particular, we can take ϕ and $\mathbf{A}$ to satisfy the uncoupled equations

$$\frac{\partial^2 \phi}{\partial t^2} = c_P^2 \nabla^2 \phi \quad \text{and} \quad \frac{\partial^2 \mathbf{A}}{\partial t^2} = c_S^2 \nabla^2 \mathbf{A}, \tag{10.19}$$

respectively, but these do not provide the most general solution to (10.12), of course. Nevertheless, they will serve our purposes well.

10.2 PLANE WAVE SOLUTIONS

Considerable insight into the nature of these body waves may be gained by seeking solutions of the form $\mathbf{u} = \mathbf{A} \exp\{i(\mathbf{k} \cdot \mathbf{x} - \omega t)\}$. These are waves of constant amplitude $\mathbf{A}$ and wavenumber $\mathbf{k} = \langle k_1, k_2, k_3 \rangle$ that are constant on all planes $\mathbf{k} \cdot \mathbf{x} - \omega t = $ constant. The waves are in the direction $\mathbf{k}$ and propagate individually with phase speed $\omega/|\mathbf{k}|$.

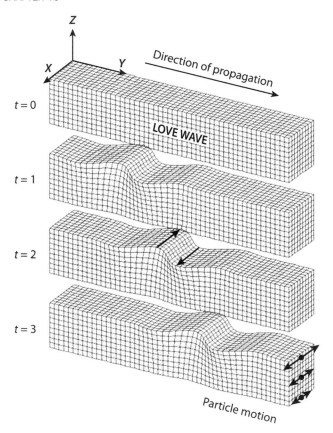

Figure 10.3 Time sequence of propagation of a Love wave.

Exercise: Use equation (10.12) to show that the plane wave solution above satisfies the equation

$$\rho\omega^2 \mathbf{A} = (\lambda + \mu)(\mathbf{A} \cdot \mathbf{k})\mathbf{k} + \mu |\mathbf{k}|^2 \mathbf{A}. \tag{10.20}$$

Taking the scalar product of both sides of (10.20) with $\mathbf{k}$ gives

$$(\mathbf{A} \cdot \mathbf{k})\left[\rho\omega^2 - (\lambda + 2\mu)|\mathbf{k}|^2\right] = 0. \tag{10.21}$$

The least interesting solution (or so it first appears) of this equation is the class of waves for which $(\mathbf{A} \cdot \mathbf{k}) = 0$, that is, for which the amplitude and wave vectors are orthogonal. From the previous equation this implies that $\omega = \pm(\mu/\rho)^{1/2}|\mathbf{k}|$, corresponding to transverse shear (or rotational) waves with speed c_S. The other solution, for which $\omega = \pm[(\lambda + 2\mu)/\rho]^{1/2}|\mathbf{k}|$, has amplitude $\mathbf{A}$ satisfying $\mathbf{A} = (\mathbf{A} \cdot \mathbf{k})\mathbf{k}/|\mathbf{k}|^2$, that is, $\mathbf{A}$ is parallel to the wave vector $\mathbf{k}$. Therefore these waves are dilatational and longitudinal and propagate with speed c_P.

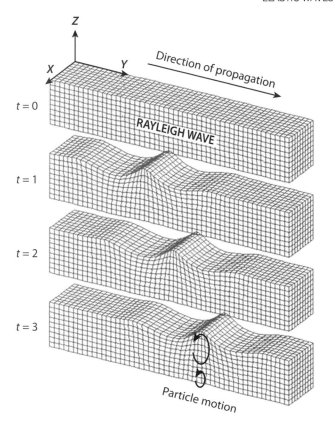

Figure 10.4 Time sequence of propagation of a Rayleigh wave.

10.3 SURFACE WAVES

We now consider the propagation of surface waves on a stress-free surface (i.e., with no normal or shear stress). We define the semi-infinite surface to be $x_2 = 0$ (the medium occupying the region $x_2 \leq 0$), and rotate the Cartesian coordinate system such that the wave vector $\mathbf{k}$ is oriented along the horizontal x_1-direction. Such waves, as their name implies, will be confined to a localized region on and below the surface. This necessitates at least four independent parameters to accommodate the stress-free conditions. These will be incorporated by seeking solutions of the Navier equations in the following form, corresponding to dilatational and shear components [36]:

$$\mathbf{u} = \mathbf{U} \exp\left[i\left(kx_1 - \omega t\right) + \alpha x_2\right] + \mathbf{V} \exp\left[i\left(kx_1 - \omega t\right) + \beta x_2\right] \quad (10.22)$$

$$\equiv \mathbf{U} \exp\theta(\alpha) + \mathbf{V} \exp\theta(\beta), \quad (10.23)$$

where α and β are both positive numbers representing the spatial decay rates of the two vector components $\mathbf{U}$ and $\mathbf{V}$.

Substituting this into the Navier equations yields the following vector equation, similar to (10.20):

$$-\rho\omega^2\mathbf{W} = (\lambda + \mu)(\mathbf{W} \cdot \mathbf{p})\mathbf{p} + \mu|\mathbf{p}|^2\mathbf{W}. \tag{10.24}$$

In this equation (which actually represents two vector equations), $\mathbf{W}$ can be either $\mathbf{U}$ or $\mathbf{V}$ and $\mathbf{p} = \langle ik, \alpha\rangle$ or $\langle ik, \beta\rangle$. Proceeding as above we obtain

$$(\mathbf{W} \cdot \mathbf{p})\left[\rho\omega^2 + (\lambda + 2\mu)|\mathbf{p}|^2\right] = 0. \tag{10.25}$$

If $\mathbf{W} \cdot \mathbf{p} = 0$, then $ikU_1 + \alpha U_2 = 0$, or in terms of an arbitrary constant U, $U_1 = -\alpha U$ and $U_2 = ikU$, from which the dispersion relation is

$$\omega^2 = c_S^2\left(k^2 - \alpha^2\right). \tag{10.26}$$

Also, for the rotational (or shear) component,

$$\omega^2 = c_P^2\left(k^2 - \beta^2\right). \tag{10.27}$$

Hence the displacement is

$$\mathbf{u} = \langle u_1, u_2, 0\rangle = U\langle -\alpha, ik\rangle\exp\theta(\alpha) + \langle V_1, V_2\rangle\exp\theta(\beta), \tag{10.28}$$

where

$$\theta(\alpha) = i(kx_1 - \omega t) + \alpha x_2, \tag{10.29}$$

and so forth. The stress components are readily determined, and applying the stress-free conditions for σ_{12} and σ_{22} on $x_2 = 0$ gives the following pair of equations:

$$-U(k^2 + \alpha^2) + (\beta V_1 + ikV_2) = 0; \tag{10.30}$$

$$ik\lambda V_1 + (\lambda + 2\mu)\beta V_2 + 2i\mu\alpha kU = 0. \tag{10.31}$$

It would be useful to find a relationship between V_1 and V_2; the choice $\beta V_1 = ikV_2$ (so the amplitude is parallel to the wave vector, corresponding to longitudinal waves) reduces this set of equations to

$$-U\left(k^2 + \alpha^2\right) + 2ikV_2 = 0; \tag{10.32}$$

$$\left[(\lambda + 2\mu)\beta - \frac{\lambda k^2}{\beta}\right]V_2 + 2i\mu\alpha kU = 0. \tag{10.33}$$

Nontrivial solutions exist, provided that

$$\left(k^2 + \alpha^2\right)\left[(\lambda + 2\mu)\beta^2 - \lambda k^2\right] + 4k^2\mu\alpha\beta = 0. \tag{10.34}$$

Using the definitions (10.14), (10.16), (10.26), and (10.27), this expression can be squared and rewritten as

$$\left(1 - \frac{\omega^2}{2k^2c_S^2}\right)^4 = \left(1 - \frac{\omega^2}{k^2c_S^2}\right)\left(1 - \frac{\omega^2}{k^2c_P^2}\right) = \left(1 - \frac{\omega^2}{k^2c_S^2}\right)\left[1 - \frac{c_S^2}{c_P^2}\left(\frac{\omega^2}{k^2c_S^2}\right)\right]. \tag{10.35}$$

This rather fearsome-looking dispersion relation contains a lot of information about what are generally referred to as *Rayleigh waves*. These are a form of surface waves (see [113]), and writing $\Omega = \omega^2/k^2 c_s^2$, this takes the form

$$\left(1 - \frac{\Omega}{2}\right)^4 = (1 - \Omega)(1 - m\Omega), \tag{10.36}$$

where

$$m = \frac{c_S^2}{c_P^2} = \frac{\mu}{\lambda + 2\mu} < \frac{1}{2}. \tag{10.37}$$

Canceling a nonzero factor (since for Rayleigh waves, $\omega^2 - k^2 c_S^2 \neq 0$), the dispersion relation reduces to the cubic polynomial

$$f(\Omega) = \Omega^3 - 8\Omega^2 + (24 - 16m)\Omega - 16(1 - m), \tag{10.38}$$

or equivalently,

$$\Omega^3 - 8\Omega^2 = 16(1 - m) - (24 - 16m)\,\Omega. \tag{10.39}$$

Written in this latter form, and noting that $(1 - m) > 0$ and $(3 - 2m) > 0$, it is clear that the right-hand side of this equation is a straight line with negative slope. There are three real roots for $0.5 > m > m^*$, where $m^* \approx 0.3215$, as may be found numerically or using Cardan's formula. For $m < m^*$ the unique real root must lie in the interval $(0, 1)$ as evidenced by the fact that $f(0) < 0$ and $f(1) > 0$. Furthermore, from the values for $f(0)$ and $f(1)$ it is clear that the root is close to $\Omega = 1$. In the particular case of $\lambda = \mu$ (sometimes referred to as Poisson's relation), $m = 1/3 > m^*$ and then

$$f(\Omega) = \Omega^3 - 8\Omega^2 + \frac{56}{3}\Omega - \frac{32}{3}. \tag{10.40}$$

Now $f(\Omega) = 0$ has roots $\Omega = 4, 2 + 2(3)^{-1/2}$, and $2 - 2(3)^{-1/2}$. Only the smallest root (≈ 0.8453) corresponds to surface waves. This is because, from above the decay coefficient

$$\alpha = \left(k^2 - \frac{\omega^2}{c_S^2}\right)^{1/2} = |k|\left(1 - \frac{\omega^2}{k^2 c_S^2}\right)^{1/2} = |k|\,(1 - \Omega)^{1/2} > 0. \tag{10.41}$$

Similarly,

$$\beta = \left(k^2 - \frac{\omega^2}{c_P^2}\right)^{1/2} = |k|\left(1 - \left[\frac{c_S^2}{c_P^2}\right]\frac{\omega^2}{k^2 c_S^2}\right)^{1/2} = |k|\,(1 - m\Omega)^{1/2} > 0, \tag{10.42}$$

since $m < 1/2$. Hence both wave modes decay exponentially as $x_2 \to -\infty$. Consider next the smaller of the two remaining roots. The quantity

$$1 - m\Omega = 1 - 2m\left[1 + (3)^{-1/2}\right] < 0 \quad \text{when} \quad m < \frac{(3)^{1/2}}{2\left[(3)^{1/2} + 1\right]} \approx 0.3170 < m^*.$$
$$\tag{10.43}$$

These correspond to pure imaginary α and β, that is, the scalar potentials (see below) are oscillatory in x_2 and therefore do not correspond to surface waves.

They arise as spurious solutions that come about from squaring expression (10.34); nevertheless they do have physical significance. They correspond, respectively, to (i) a compressional wave incident on the free surface in such a way that only reflected shear waves occur, and (ii) the reverse case of an incident shear wave and a reflected compressional wave (see [42] for further details).

10.4 LOVE WAVES

Since the particle motions for these surface waves lie in a horizontal plane but are perpendicular to the direction of the wave (i.e., the waves are horizontally polarized), we write

$$\mathbf{u}(\mathbf{r}, t) = \langle 0, u(x_2) \rangle \exp\left[i\left(kx_1 - \omega t\right)\right].\tag{10.44}$$

Equation (10.10) then reduces to the simple scalar form

$$\frac{\partial^2 u}{\partial t^2} = c_S^2 \nabla^2 u.\tag{10.45}$$

We can model the earth's crust as a uniform layer of thickness h above a homogeneous half-space $x_3 > h$ (representing the upper mantle). This is a great oversimplification of course, but it will be sufficient to establish the basic characteristics of Love waves. We denote the speeds c_S of secondary waves in the layer and half-space as c_1 and c_2 and the corresponding densities and Lamé constants as ρ_1, ρ_2 and μ_1, μ_2, respectively. Then the resulting equations for u in each region are

$$\frac{d^2 u}{dx_3^2} - b_i^2 u = 0, \quad i = 1, 2,\tag{10.46}$$

where

$$b_i = \left(k^2 - \frac{\omega^2}{c_i^2}\right)^{1/2}.\tag{10.47}$$

The general solution of the above equation for u is

$$u(x_2) = A_i \exp\left(-b_i x_2\right) + B_i \exp\left(b_i x_2\right), \quad i = 1, 2.\tag{10.48}$$

The boundary conditions for the disturbance to be classified as a surface wave require that (i) $u(x_2) \to 0$ as $x_2 \to \infty$, (ii) the outer surface $x_2 = 0$ be stress-free, and (iii) both the displacement and stress are continuous at the interface $x_2 = h$. To implement condition (i) a branch for $\pm b_2$ must be chosen, so we take $\mathrm{Re}(-b_2) < 0$, thus forcing $B_2 = 0$. Condition (ii) implies that $\sigma_{j2}(0) = 0$ (or $\sigma_{2j}(0) = 0$), $j = 1, 2$, and in this case the only nontrivial condition is that $u'(0) = 0$, which implies that $A_1 = B_1$. Hence, in the crustal layer

$$u(x_3) = 2A_1 \cosh b_1 x_2,\tag{10.49}$$

and in the upper mantle

$$u(x_3) = A_2 \exp\left(-b_2 x_2\right).\tag{10.50}$$

The final conditions (iii) correspond to

$$2A_1 \cosh b_1 h = A_2 \exp(-b_2 h) \qquad (10.51)$$

and

$$2\mu_1 b_1 A_1 \sinh b_1 h = -\mu_2 b_2 A_2 \exp(-b_2 h). \qquad (10.52)$$

The requirement that equations (10.51) and (10.52) possess nontrivial solutions yields the dispersion relation

$$\tanh b_1 h = \frac{\mu_2 b_2}{\mu_1 b_1}. \qquad (10.53)$$

Clearly $\mu_2 b_2 < \mu_1 b_1$ for solutions of (10.53) to exist. Solutions correspond to dispersive waves with phase speed determined by equation (10.53). Both Love waves and Rayleigh waves generally propagate in the outer crustal layers and upper mantle.

P waves, a generalization of acoustic (or pressure) waves are the fastest waves in the categories we have considered, and so register first on seismograms. Their speeds are typically 5–7 km/s in the earth's crust, but they exceed 8 km/s in the earth's mantle and core (compare this with approximately 1.5 km/s in water and 0.3 km/s in air). In contrast, S waves do not travel through fluids and so do not exist in the earth's outer core (which is inferred to be primarily liquid iron); their speeds are typically 3–4 km/s in the crust, 4.5 km/s or more in the mantle, and about 2.5–3 km/s in the solid inner core. The surface Love and Rayleigh waves propagate along the surface of the crust, and have maximum amplitude at the surface, which therefore decreases with depth. Both types of wave are dispersive (i.e., their speeds are frequency dependent, with lower frequencies propagating at higher velocity). The penetration depths of these waves is also frequency dependent, with lower frequencies penetrating to greater depth. In terms of appearance and particle motion, Rayleigh waves are analogous to surface water waves. The speed of Love waves can be slightly higher than that of Rayleigh waves: the former is in the range 2.0–4.4 km/s, whereas the latter is about 2.0–4.2 km/s. But these latter waves lead us naturally to their watery counterparts.

Chapter Eleven

Surface Gravity Waves

In fluid dynamics, gravity waves are waves generated in a fluid medium or at the interface between two media when the force of gravity or buoyancy tries to restore equilibrium. An example of such an interface is that between the atmosphere and the ocean, which gives rise to wind waves. When a fluid element is displaced on an interface or internally to a region with a different density, gravity will try to restore it toward equilibrium, resulting in an oscillation about the equilibrium state or wave orbit. Gravity waves on an air-sea interface of the ocean are called surface gravity waves or surface waves, while gravity waves that are within the body of the water (such as between parts of different densities) are called internal waves. Wind-generated waves on the water surface are examples of gravity waves, as are tsunamis and ocean tides. Wind-generated gravity waves on the free surface of the Earth's ponds, lakes, seas and oceans have a period of between 0.3 and 30 seconds (3 Hz to 0.03 Hz). Shorter waves are also affected by surface tension and are called gravity-capillary waves and (if hardly influenced by gravity) capillary waves.

[265]

The study of water waves is essentially the part of hydrodynamics that deals with an inviscid, incompressible, irrotational fluid. As a first approximation the compressibility of the water is small enough to be neglected ... these properties of water waves allow for simplification in the equations of motion. Moreover, water waves are essentially two-dimensional, giving an additional simplification. Wave motion in water and other fluids involves a free surface—a surface on which no external forces act. Therefore we say that water waves are surface waves. The water particles are subject to the action of gravity, which is an external force that cannot be neglected in the equations of motion. Surface wave problems have interested many illustrious mathematicians and physicists, such as Euler, the Bernoulli brothers, Lagrange, Cauchy, Poisson, and Poincaré, and also the British school of Airy, Stokes, Kelvin, Rayleigh, Lamb, and the American school of K. Friedrichs, F. John, J. Keller, J. Stoker, and so on.

[249]

11.1 THE BASIC FLUID EQUATIONS

By comparison with waves in elastic media, the basic wave types in this chapter are simpler (at least physically). The compressibility of the medium is neglected (though it is of course not zero), and for sufficiently long waves the motion is confined to a two-dimensional surface. We consider (with little loss of generality) a two-dimensional Cartesian system (x, z). The undisturbed free surface of the fluid is the plane $z = 0$. The lower boundary of the "channel" is at $z = -h$, where h may tend to infinity. The surface perturbation is denoted by $z = \eta(x, z, t)$, and the velocity field in the fluid is $\mathbf{u}(x, z, t) = \langle u_1(x, z, t), u_2(x, z, t)\rangle$. Consistent with the assumptions of ideal fluid dynamics the fluid is assumed to be inviscid and the flow incompressible and irrotational [34], [35]. Therefore

$$\nabla \cdot \mathbf{u} = 0 \text{ and } \nabla \times \mathbf{u} = \mathbf{0}. \tag{11.1}$$

These equations imply the existence of a velocity potential $\phi(x, z, t)$ such that $\mathbf{u} = \nabla\phi$, with ϕ satisfying Laplace's equation

$$\nabla^2\phi = 0. \tag{11.2}$$

The lower boundary condition is a rigid one (i.e., the vertical component of velocity vanishes there). In terms of the velocity potential this is

$$\left[\frac{\partial\phi}{\partial z}\right]_{-h} = 0. \tag{11.3}$$

The dynamical boundary condition relates a generalized pressure difference across the surface to the velocity field via Euler's equation of motion

$$\frac{\partial\mathbf{u}}{\partial t} = -\nabla\left(\frac{p}{\rho} + \frac{1}{2}\mathbf{u}^2 + gz\right), \tag{11.4}$$

or equivalently

$$\frac{\partial\phi}{\partial t} + \left(\frac{p}{\rho} + \frac{1}{2}\mathbf{u}^2 + gz\right) = f(t). \tag{11.5}$$

This is the famous *Bernoulli equation* for unsteady irrotational flow. In this equation, ρ is the fluid density, g is the gravitational acceleration and $f(t)$ is a spatially independent function that can be absorbed into the velocity potential. In so doing, the linearized version—that is, $|\mathbf{u}|$ is assumed to be small enough that its square and higher powers can be neglected—becomes

$$\frac{\partial\phi}{\partial t} + \frac{p}{\rho} + gz = 0. \tag{11.6}$$

We are primarily interested in the behavior at the free surface, where $z = \eta$, and if the effects of a *surface tension* force T are also included, this equation is generalized to

$$\frac{\partial\phi}{\partial t} - \frac{T}{\rho}\kappa + g\eta = 0 \text{ (at } z = \eta), \tag{11.7}$$

where κ is the surface curvature. The stabilizing effect of this force will be discussed shortly. For small slopes, $\kappa \approx \eta_{xx}$, and so instead we use the equation

$$\frac{\partial \phi}{\partial t} - \frac{T}{\rho} \frac{\partial^2 \eta}{\partial x^2} + g\eta = 0 \text{ (at } z = \eta\text{).} \tag{11.8}$$

Replacing $\partial \phi / \partial t$ by the first term in its Taylor expansion about $z = 0$, the dynamic boundary condition for small amplitude waves becomes

$$\frac{\partial \phi}{\partial t} - \frac{T}{\rho} \frac{\partial^2 \eta}{\partial x^2} + g\eta = 0 \text{ (at } z = 0\text{).} \tag{11.9}$$

The remaining boundary condition is the kinematic one. This simply relates the vertical component of the velocity at the surface to that of the surface (again using the same Taylor approximation):

$$\frac{\partial \eta}{\partial t} = \left[\frac{\partial \phi}{\partial z} \right]_{z=0}. \tag{11.10}$$

As we know, a simple ansatz is often sought in the form of a Fourier mode with wavenumber $k(= 2\pi/\lambda)$ and angular frequency ω so that the shape η of the free surface is given by

$$\eta(x, t) = a \cos(kx - \omega t), \tag{11.11}$$

a being the (constant) amplitude. Using separation of variables with

$$\phi(x, z, t) = G(x, t) Z(z), \tag{11.12}$$

condition (11.10) implies that $G(x, t) = \sin(kx - \omega t)$ and consequently,

$$\frac{d^2 Z}{dz^2} - k^2 Z = 0, \tag{11.13}$$

and so

$$Z(z) = A e^{kz} + B e^{-kz}. \tag{11.14}$$

We apply the lower boundary condition (11.3), which now takes the form $Z'(-h) = 0$, to obtain

$$Z(z) = B \left(e^{-kz} + e^{2kh} e^{kz} \right). \tag{11.15}$$

Applying the kinematic condition (11.10) to this defines B as

$$B = \frac{\omega a}{k \left(e^{2kh} - 1 \right)}, \tag{11.16}$$

which leads to

$$Z(z) = \frac{\omega a \cosh k (z + h)}{k \sinh kh}, \tag{11.17}$$

and

$$\phi(x, z, t) = \frac{\omega a \cosh k (z + h)}{k \sinh kh} \sin(kx - \omega t). \tag{11.18}$$

So far, the dynamic boundary condition (11.9) has not been applied. Once this has been done, a lot more information about the structure of this relatively simple linear system can be determined via the dispersion relation connecting ω and k in a very specific manner, that is,

$$D(\omega, k) = 0. \tag{11.19}$$

11.2 THE DISPERSION RELATION

After a little algebra incurred by substituting (11.18) into (11.9), the dispersion relation (11.19) unfolds as the expression

$$\omega^2 = k\left(g + \frac{T}{\rho}k^2\right)\tanh kh. \tag{11.20}$$

The wave speed, or speed of an individual wave (as defined by a particular crest, for example) is defined by $c = \omega/k$, so from (11.20) it follows that

$$c = \left[\left(\frac{g}{k} + \frac{T}{\rho}k\right)\tanh kh\right]^{1/2}, \tag{11.21}$$

or in terms of the wavelength λ,

$$c(\lambda) = \left[\left(\frac{g\lambda}{2\pi} + \frac{2\pi T}{\lambda\rho}\right)\tanh\left(\frac{2\pi h}{\lambda}\right)\right]^{1/2}. \tag{11.22}$$

It is interesting to consider some special cases of (11.22), albeit in a very nonrigorous fashion:

11.2.1 Deep Water Waves

Although we shall qualify this condition a little later (Section 11.2.4), "deep" here means that the wavelength is small compared with the depth of the water,

$$\frac{h}{\lambda} \gg 1, \tag{11.23}$$

which means that

$$\tanh\frac{2\pi h}{\lambda} \approx 1. \tag{11.24}$$

Under these circumstances,

$$c(\lambda) \approx \left(\frac{g\lambda}{2\pi} + \frac{2\pi T}{\rho\lambda}\right)^{1/2}, \tag{11.25}$$

which represents the wave speed for disturbances (with apologies to readers who dislike anthropomorphisms) that "feel" the effects of gravity and surface tension but do not "feel" the bottom of the channel, reservoir, or the like. Furthermore, if the waves are long enough such that the effect of surface tension is negligible compared with that of gravity, that is,

$$\frac{g\lambda}{2\pi} \gg \frac{2\pi T}{\rho\lambda}, \tag{11.26}$$

then

$$c(\lambda) \approx \left(\frac{g\lambda}{2\pi}\right)^{1/2}. \tag{11.27}$$

This is the relationship between speed and wavelength for ocean waves, for example, which are completely dominated by gravity. At the other extreme, for short waves,

$$\frac{g\lambda}{2\pi} \ll \frac{2\pi T}{\rho\lambda}, \tag{11.28}$$

and

$$c(\lambda) \approx \left(\frac{2\pi T}{\rho\lambda}\right)^{1/2}. \tag{11.29}$$

Such waves (ripples) are completely dominated by surface tension, and the shorter they are, the faster they move. Now we proceed to the other extreme and examine shallow water waves.

11.2.2 Shallow Water Waves

In this case the depth of the water is small compared with the wavelength:

$$\frac{h}{\lambda} \ll 1. \tag{11.30}$$

Under these circumstances,

$$\tanh\frac{2\pi h}{\lambda} \approx \frac{2\pi h}{\lambda}. \tag{11.31}$$

These waves do "feel" the bottom. For most problems of interest in this wave situation, the second term in equation (11.22) is negligible, so that the following result is valid for gravity waves in shallow water:

$$c \approx (gh)^{1/2}. \tag{11.32}$$

This is an important result: it means that the wave speed is independent of wavelength and implies that all the waves travel with the same speed; in principle at least, any complex initial wave configuration may retain an identifiable shape for quite some time. In practice, however, nonlinear effects will soon destroy that initial shape unless those effects are balanced by dispersion or dissipation, in which case some kind of solitary wave evolves [122]. Actually, the strong inequalities we have employed to distinguish between deep water waves and their shallow water counterparts do not need to be enforced so strictly; sometimes it is sufficient to demand that $h < \lambda/2$ for shallow water waves, for example.

Let us return to equation (11.25) for deep-water capillary-gravity waves, for there is quite a bit more information that can be extracted. In the extreme cases given by equations (11.27) and (11.29), we have seen that the square of the speed behaves in a (i) linear and (ii) a rectangular hyperbolic fashion, respectively, as functions of wavelength. In the intermediate region (i.e., where the terms $g\lambda/2\pi$ and $2\pi T/\rho\lambda$ are comparable), both restoring forces are comparable, and the respective graphs of $c(\lambda)$, being continuous, must coincide. There is clearly a minimum speed c since $c''(\lambda) > 0$ at the critical wavelength λ_c given by the solution of

$$c'(\lambda) = 0,$$

which is

$$\lambda_c = 2\pi \left(\frac{T}{g\rho}\right)^{1/2}. \tag{11.33}$$

In MKS units $\lambda_c \approx 0.0172$ m for water. For wavelengths less than or greater than this, the dominant restoring force tends to be, respectively, surface tension or gravity. The corresponding minimum speed is

$$c_{min} = \left(\frac{4gT}{\rho}\right)^{1/4} \approx 0.23 \text{ m/s}. \tag{11.34}$$

This means that any breeze or gust of wind with speed less than 0.23 m/s will not generate any propagating waves other than a transient disturbance. Wind speeds above this minimum value will in principle generate two sets of waves, with wavelengths on each side of λ_c; one set with $\lambda < \lambda_c$ (capillary waves) and one set with $\lambda > \lambda_c$ (gravity waves). Note that these results may be derived without the use of calculus: use of the arithmetic mean-geometric mean inequality gives the required result. We do this by writing equation (11.25) for brevity as

$$c^2 = \alpha\lambda + \frac{\beta}{\lambda}, \quad \alpha > 0, \beta > 0. \tag{11.35}$$

It follows by the arthmetic mean–geometric mean inequality that the sum of these two terms is never less than $2(\alpha\beta)^{1/2} = 2gT/\rho$. Since the minimum of c^2 occurs when the minimum of c does, the corresponding result stated above for c_{min} is established. It is appropriate at this point to add a comment on wave refraction: we now have a simple model explaining why ocean waves tend to line up parallel to the beach, even if far out to sea they are approaching it obliquely. To see this, fix the wavelength for any particular wave of interest. Far out, the wave is in deep water ($\lambda \ll h_{deep}$) and so $c \propto \lambda^{1/2}$. Nearer in, the wave is in shallow water ($\lambda \gg h_{shallow}$) and so $c \propto (h_{shallow})^{1/2}$, which is of course smaller than $\lambda^{1/2}$, so the part of the wavefront nearest the beach slows down compared with that further out, and the whole wavefront tends to slew around.

11.2.3 Instability

As we ought to be aware by now, linear theory deals with waves that are, strictly speaking, of *infinitesimal* amplitude. In reality of course, this is impossible to achieve: all waves are nonlinear according to this definition. It is a fortunate practical reality that many of the predictions of linear theory still work for waves of small amplitude, but ultimately, if conditions are right, the system will become nonlinear. This can occur when the forces driving the wave motion are such that the amplitude of a wave increases, ultimately vitiating the assumption of small amplitude. Sometimes as this happens, mechanisms tending to oppose wave growth, such as dispersion or dissipation, may become more important and actually balance the growth tendency in such a way that the wave maintains its shape for a considerable distance (under these circumstances it is called a *solitary wave*). In general, the increase in wave amplitude will lead to the ubiquitous but poorly understood phenomenon of turbulence. What is of interest here is the transition to nonlinearity as predicted by linear theory. In a sense this represents the other side of the coin from much of our earlier analyses, where plane wave solutions with time harmonic dependence were sought (i.e., waves with time dependence of the form $\exp(-i\omega t)$, ω being a real number). The fundamental idea in (temporal) linear stability theory is that ω is permitted to be a complex quantity, (i.e., $\omega = \omega_r + i\omega_i$, where of course the respective real and imaginary parts ω_r and ω_i are real). This is a significant difference: now the temporal dependence will be oscillatory with exponential growth (if $\omega_i > 0$ and $\omega_r \neq 0$) or oscillatory with exponential decay ($\omega_i < 0$ and $\omega_r \neq 0$), obviously because

$$e^{-i\omega t} = e^{-i\omega_r t} e^{\omega_i t}.$$

If $\omega_r = 0$, then the wave just grows or decays exponentially with no oscillatory behavior; sometimes the case $\omega_i > 0$ and $\omega_r \neq 0$ is referred to as *overstability*. If there is no dissipation in the system (which is unlikely in practice), then both signs for ω_i arise from the equations; in contrast, if $\omega_i < 0$ for each wave, then any disturbance to the equilibrium state decays away in time and so the system supporting it is said to be *linearly stable*. This means that the equilibrium is returned. This being so, without loss of generality we will focus on linear instability, wherein at least one wave exists for which $\omega_i > 0$.

Let us examine first the following physical situation in some mathematical detail. A deep layer of inviscid fluid of density ρ_2 flows with uniform speed U over another deep layer of density $\rho_1 > \rho_2$, which is at rest. The interfacial layer has coefficient of surface tension T. Here the reason for the adjective "deep" is that then we may safely neglect the effects of any boundaries, because they are too far away to influence the physics significantly. (This is a gross oversimplification, of course: "deep" = semi-infinite!) We know that the governing dispersion relation between angular frequency ω and wavenumber k (here $k > 0$ without loss of generality) is obtained by seeking solutions to the governing differential equation for the dependent variable $\eta(x, t)$ in the form

$$\eta(x, t) = Ae^{i(kx - \omega t)}, \tag{11.36}$$

where A is a (possibly complex) constant amplitude. For the interfacial stability problem stated above, η (or more accurately, Re η) represents the (small) perturbation of the interface along the interface (x-direction) and in time. The gravitational acceleration is vertically downward, naturally enough. The dispersion relation is quadratic in ω and after a considerable amount of algebra it can be written in the form [34]

$$(\rho_1 + \rho_2)\,\omega^2 - 2\rho_2 U k\omega + \rho_2 U^2 k^2 - k[k^2 T + g\,(\rho_1 - \rho_2)] = 0. \quad (11.37)$$

Note that, given the standard form (11.36), in which k here is a positive real number and ω may be a complex one, the system will be stable (i.e., not growing in time) *if and only if* ω *is real*. Therefore we are interested in the conditions under which this will (or will not) occur, and those conditions can be determined by examining the discriminant of the quadratic equation (11.37), which we temporarily write in the form

$$A\omega^2 - B\omega + C = 0,$$

(the definitions of A, B, and C being obvious). Clearly a necessary and sufficient condition for stability is that $B^2 \geq 4AC$, or

$$(\rho_1 + \rho_2)\,[k^2 T + g\,(\rho_1 - \rho_2)] \geq \rho_1\rho_2 U^2 k. \quad (11.38)$$

Values of wavenumbers k for which (11.38) is satisfied correspond to wavelengths for which the system is stable. However, generally disturbances consist of a range (usually infinite, in principle) of wavenumbers, so that to be stable for all values of $k > 0$, it is necessary that

$$\rho_1\rho_2 U^2 \leq \min_k\{(\rho_1 + \rho_2)\,[kT + gk^{-1}\,(\rho_1 - \rho_2)]\}. \quad (11.39)$$

This minimum can be determined by the standard techniques of calculus, but it is much less cumbersome (and therefore more elegant) to employ the arithmetic-geometric mean inequality to obtain

$$kT + gk^{-1}\,(\rho_1 - \rho_2) \geq 2\,[(\rho_1 - \rho_2)\,gT]^{1/2},$$

so that from (11.39) we find

$$U^2 \leq 2\,(\rho_1^{-1} + \rho_2^{-1})\,[(\rho_1 - \rho_2)\,gT]^{1/2}, \quad (11.40)$$

for stability to all disturbances for which $k > 0$ (incidentally, these results are similar for $k < 0$, as may be seen by changing the sign of x in (11.36) and taking account of any corresponding sign changes in spatial derivatives; negative values of k just mean that the disturbance propagates in the opposite direction). From equation (11.40) it is apparent that if either T or g is zero, instability must occur, no matter how small the relative speed U; the presence of both surface tension and gravity is needed to prevent the instability. However, note from the inequality (11.38) that either T or g (but not both) being zero will guarantee stability for a given speed U for some range of k (and hence λ), meaning that the expression $B^2 - 4AC$ is seen to become negative for a range of k. For sufficiently short waves (or large enough k values) or sufficiently long waves (small enough k values) the system is stable, but the system is generally unstable for an intermediate range of k.

Some interesting results can also be obtained if we systematically "turn off" various parameters in the problem as originally stated. Thus, if we set each of

ρ_2, U, and T equal to zero in equation (11.37), we obtain the very simple dispersion relation

$$\omega^2 = gk, \tag{11.41}$$

(or more generally $\omega^2 = g\,|k|$), which of course is exactly that for deep water long-wavelength surface gravity waves. Since $k > 0$ the system is stable, and we can justifiably regard such waves as special cases of hydrodynamic stability. If only U and T are set equal to zero in equation (11.37), then a slight generalization of equation (11.41) results:

$$\omega^2 = gk\left(\frac{\rho_1 - \rho_2}{\rho_1 + \rho_2}\right). \tag{11.42}$$

In this case the result represents the dispersion relation for a special case of *internal gravity waves* propagating along the interface between two adjacent layers of fluid (the less dense fluid (ρ_2) overlaying the denser one (ρ_1)).

Now we examine the case of surface capillary-gravity waves by setting $\rho_2 = 0$ (and letting $\rho_1 = \rho$) along with $U = 0$. The resulting dispersion relation is

$$\omega^2 = gk + T\frac{k^3}{\rho},$$

which is a special case of the result (11.20) for deep water.

The final special case that we consider here occurs when $\rho_1 = \rho_2 \equiv \rho$ and $T = 0$ in equation (11.37), resulting in a special case of the *Kelvin-Helmholtz instability* (due here to shear in the form of a vortex sheet, i.e. a discontinuity in the speed U; this is of course present in the problem as originally posed). Under these circumstances there are two complex conjugate roots of the dispersion relation,

$$\omega = \frac{1}{2}Uk(1 \pm i),$$

the positive root guaranteeing instability for all values of k. This instability arises when, simply put, the destabilizing effects of changes of wind speed with altitude are sufficient to overcome the stabilizing effects of buoyancy when the denser fluid is at the bottom. The study has an interesting history; in 1871 Lord Kelvin used this type of analysis to model wind-generated ocean waves. Somewhat later, in 1890, Helmholtz applied the theory to billow clouds, which are atmospheric indicators of the presence of strongly sheared winds. In Chapter 14 this topic will be examined in more detail.

Example: From equation (11.22), setting $T = 0$, it is seen that

$$c(\lambda) = \left[\frac{g\lambda}{2\pi}\tanh\left(\frac{2\pi h}{\lambda}\right)\right]^{1/2}.$$

Furthermore, we have seen that if the water is shallow ($\lambda \gg h$), $c \approx (gh)^{1/2}$, independent of the wavelength. Use the *alternating series estimation theorem*

(stated below) to show that if $\lambda > 10h$, the estimate $c^2 \approx gh$ is accurate to within $0.014g\lambda$. The *alternating series estimation theorem* states:
If $s = \sum_{n=1}^{\infty} (-1)^{n-1} b_n$ is the sum of an alternating series that satisfies (i) $0 \leq b_{n+1} \leq b_n$ and (ii) $\lim_{n \to \infty} b_n = 0$, then the remainder R_n satisfies $|R_n| = |s - s_n| \leq b_{n+1}$, where s_n is the nth partial sum.

Solution: Let $x = 2\pi h/\lambda$ and $f(x) = \tanh x$; we are interested in small values of x. In the Maclaurin series for $\tanh x$, $f(0) = f''(0) = f^{(iv)}(0) = \ldots = 0$, so

$$f(x) = x - \frac{x^3}{3} + \frac{2x^5}{15} + O\left(x^7\right). \tag{11.43}$$

The series is alternating, because it represents an odd function. Typically, $b_n = c_n x^{2n-1}$, the c_n being numerical coefficients, so $|b_{n+1}/b_n| = |c_{n+1}/c_n| x^2$, where $|c_{n+1}/c_n| < 1$, so $b_{n+1} < b_n$ (recall that $x < 1$), and from the nature of the coefficients c_n, it follows that $\lim_{n \to \infty} b_n = 0$. Hence the conditions of the above theorem are satisfied, and so the error in the approximation $f(x) = \tanh x \approx x$ is less than the first neglected term. This means that if s, s_1 and R_1, are respectively, the sum of the series, the first term (which is the first partial sum), and the remainder after the first term, then

$$|R_1| = |s - s_1| < b_2 = x^3/3 = (2\pi h/\lambda)^3 /3. \tag{11.44}$$

Now if $\lambda > 10h$, then

$$(2\pi h/\lambda)^3 /3 < (2\pi/10)^3/3 = \pi^3/375; \tag{11.45}$$

hence the error in the approximation $c \approx (gh)^{1/2}$ is less than

$$(g\lambda/2\pi) \times \left(\pi^3/375\right) = \pi^2 g\lambda/750 \approx 0.0132g\lambda < 0.014g\lambda. \tag{11.46}$$

Seiches

Problem: Consider a rectangular "lake" of length L and breadth b containing undisturbed water of depth h. When wind-induced waves are present, so-called sloshing motions or *seiches* can occur (these are standing waves). The effects of surface tension are negligible and can be ignored. The angular frequencies ω and wavenumbers k satisfy the dispersion relation

$$\omega^2 = gk \tanh kh, \tag{11.47}$$

where in this two-dimensional problem

$$k^2 = \left(\frac{m\pi}{L}\right)^2 + \left(\frac{n\pi}{b}\right)^2, m = 0, 1, 2, \ldots; n = 0, 1, 2 \ldots, \tag{11.48}$$

and $g = 9.8$ m/s^2. Suppose that $L = 30$ km, $b = 2$ km, and $H = 100$ m, and the wind sets up the mode $m = 1$ and $n = 0$. Find the period of oscillation $(2\pi/\omega)$ in minutes.

Suppose that later the mode of oscillation is $m = 2$ and $n = 1$; what is the period of oscillation now?

11.2.4 Group Speed Again

Consider again a group of sinusoidal waves with a continuous but narrow range Δk of wave numbers centered around $k = k_0$:

$$\phi(x, t) = \int_{k_0 - \Delta k}^{k_0 + \Delta k} A(k) e^{i(kx - \omega(k)t)} \, dk, \tag{11.49}$$

with

$$0 < \frac{\Delta k}{k_0} \ll 1,$$

and as before, $A(k)$ is the wavenumber spectrum. Performing the usual Taylor expansion, we have

$$\omega(k) = \omega(k_0) + (k - k_0)\omega'(k_0) + O(k - k_0)^2, \tag{11.50}$$

and also

$$A(k) = A(k_0) + (k - k_0)A'(k_0) + O(k - k_0)^2. \tag{11.51}$$

Let

$$\alpha = \frac{k - k_0}{k_0}, \tag{11.52}$$

so $k = k_0(\alpha + 1)$, and we denote $\omega(k_0)$ by ω_0, recalling that $\omega'(k_0) = c_g$. If $A(k_0)$ is sufficiently smooth, we can approximate $A(k)$ by $A(k_0)$ and write

$$kx - \omega t \approx k_0(\alpha + 1)x - \left(\omega_0 + \alpha k_0 c_g\right)t = k_0 x - \omega_0 t + \alpha k_0 (x - c_g t). \tag{11.53}$$

Now using the abbreviations

$$k = k_0 \pm \Delta k \Rightarrow \alpha = \pm \frac{\Delta k}{k_0} = \pm \delta, \tag{11.54}$$

we can write

$$\varphi(x, t) \approx A(k_0) e^{i(k_0 x - \omega_0 t)} k_0 \int_{-\delta}^{\delta} \exp\left[i\alpha k_0 (x - c_g t)\right] d\alpha. \tag{11.55}$$

Since

$$\int_{-\delta}^{\delta} \exp\left[i\alpha k_0 (x - c_g t)\right] d\alpha = \left[\frac{e^{i\alpha k_0 (x - c_g t)}}{i k_0 (x - c_g t)}\right]_{-\delta}^{\delta} \equiv \left[\frac{e^{i\alpha k_0 p}}{i k_0 p}\right]_{-\delta}^{\delta} = \frac{2}{k_0 p} \sin(k_0 p \delta)$$

$$= \frac{2}{k_0 p} \sin(p\Delta k), \tag{11.56}$$

it follows that

$$\varphi(x, t) \approx 2A(k_0)\frac{\sin\left[\Delta k(x - c_g t)\right]}{x - c_g t}e^{i(k_0 x - \omega_0 t)} = \widetilde{A}e^{i(k_0 x - \omega_0 t)}, \tag{11.57}$$

where

$$\widetilde{A} = 2A(k_0)\frac{\sin(\Delta k(x - c_g t))}{x - c_g t}. \tag{11.58}$$

Thus (11.57) represents a locally sinusoidal wave train with a slowly modulated amplitude envelope, advancing with speed c_g. Again, since the envelope (11.58) pertains to wave groups, c_g is referred to as the group speed (or group velocity, $\mathbf{c}_g = \nabla_k \omega$ in higher dimensions).

What follows is an interlude about puddles. I'm fortunate to be able to walk to my office; it's about a mile each way, and most of the time I don't get caught in the rain. Occasionally, of course, I do, and then the most intriguing part of my walk arises after the rain has all but passed and before the trees have had a chance to dry off. Puddles abound, and drops of water falling from the trees into these puddles create circular ripples that move outward in a well-defined manner. I've noticed that behind the waves is an expanding circular central region of calm water. How fast does this region expand? This requires an interesting calculation concerning the group speed of capillary-gravity waves in deepish water (a nice precise term). First though, what do we *really* mean by "deep" and "shallow" water? We have used (and will continue to do so) the strong inequalities $kh \ll 1$ (or $\lambda \gg h$) for shallow water, and $kh \gg 1$ (or $\lambda \ll h$) for deep water, but how large is "large"? Recall that these inequalities arise from considering the extremes of the function $\tanh kh$, so if we require that $\tanh kh > 0.95$, say, for our waves to be considered deep water waves, then this means that $kh \gtrsim 2$ or $\lambda \lesssim 3h$.

Let's assume for the moment that the puddle is deep in the sense that its depth exceeds half that of the shortest wavelength that I observe on its surface. Then we can write

$$\omega^2 = k\left(g + \frac{T}{\rho}k^2\right)\tanh kh \approx gk + bk^3,$$

where $b = T/\rho$. Differentiating this expression implicitly, we find that the group velocity of deep water capillary-gravity waves is

$$c_g(k) = \frac{d\omega}{dk} = \frac{g + 3bk^2}{2\left(gk + bk^3\right)^{1/2}}. \tag{11.59}$$

Note that if we set $g = 0$ then for pure capillary waves $c_g = 3c/2$.

Referring to equation (11.59) we infer that if this group velocity has a minimum value for some wavenumber k, then the last wave group to move out from the center of the disturbance will travel with this speed, and since the energy also travels with the group speed, the region

$$r < \left(c_g\right)_{\min}t,$$

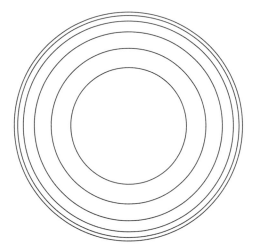

Figure 11.1 Typical pattern arising from a raindrop falling on a puddle.

is calm. A further differentiation shows that

$$\frac{dc_g}{dk} = 0 \quad \text{when} \quad k = \left[\left(\frac{2}{3^{1/2}} - 1 \right) \frac{g\rho^{1/2}}{\gamma} \right] \equiv k^*.$$

Furthermore it is readily shown that this critical wavenumber corresponds to a minimum of c_g. Thus there is a region a calm water, and it expands with speed

$$(c_g)_{\min} = c_g(k^*) \approx 1.09 \left(\frac{g\gamma}{\rho} \right)^{1/4}.$$

Now it's time to put in the numbers. We use $g = 9.81$ m/s^2, $\gamma = 0.074$ N/m, and $\rho = 10^3$ kg/m^3 to find that $(c_g)_{\min} \approx 0.18$ m/s (or mixing our units, about 0.4 mph). The wavelength corresponding to this minimum speed is just $\lambda^* = 2\pi/k^* \approx 4.4$ cm. So for the deep water approximation to be valid, $h \gtrsim 2$ cm, which may well have been the case. See Figure 11.1 for a sketch of this phenomenon.

Question: Is there a corresponding minimum group speed for shallow water waves?

11.2.5 Wavepackets

A general disturbance can be represented in the form of a Fourier integral,

$$\eta(x, t) = \int_{-\infty}^{\infty} A(k)e^{i(kx-\omega t)}dk, \tag{11.60}$$

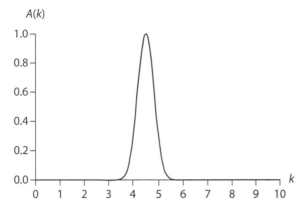

Figure 11.2 A Gaussian amplitude $A(k)$.

where the real part of this expression is to be taken in due course. Consider an initial disturbance, at $t = 0$:

$$\eta(x, 0) = \int_{-\infty}^{\infty} A(k)e^{ikx}\,dk, \tag{11.61}$$

and suppose that $A(k)$ is Gaussian, that is,

$$A(k) = A_0 e^{-\sigma(k-k_0)^2}, \tag{11.62}$$

where A_0, k_0, and σ are real constants. Writing $k_1 = k - k_0$, we have that

$$\eta(x, 0) = \int_{-\infty}^{\infty} A_0 e^{ikx} e^{-\sigma(k-k_0)^2}\,dk = A_0 \int_{-\infty}^{\infty} e^{i(k_1+k_0)x} e^{-\sigma k_1^2}\,dk_1$$

$$= A_0 \int_{-\infty}^{\infty} e^{ik_0x} e^{-\sigma(k_1^2 - ik_1 x/\sigma)}\,dk_1 = A_0 e^{ik_0x} e^{-x^2/4\sigma} \int_{-\infty}^{\infty} e^{-\sigma(k_1 - ix/2\sigma)^2}\,dk_1$$

$$= A_0 e^{ik_0x} e^{-x^2/4\sigma} I, \tag{11.63}$$

where

$$I = \left(\frac{\pi}{\sigma}\right)^{1/2}, \tag{11.64}$$

as may be shown.

Exercise: Establish this result. Note that it is not simply a question of formally modifying the result for the integral of a Gaussian function; complex integration around a suitable rectangular contour is required.

The result (11.64) indicates that

$$\operatorname{Re}\eta(x, 0) = \operatorname{Re} A_0 \left(\frac{\pi}{\sigma}\right)^{1/2} e^{ik_0x} e^{-x^2/4\sigma} = A_0 \left(\frac{\pi}{\sigma}\right)^{1/2} (\cos k_0 x) e^{-x^2/4\sigma}. \tag{11.65}$$

Now if $\eta(x, 0)$ contains a large number of wave crests, $|k|$ is large, from which it follows that $|A(k)| \ll |A_0|$ *unless* $k \approx k_0$. Representative graphs of $A(k)$ and $\operatorname{Re}\eta(x, 0)$ are shown in Figures 11.2 and 11.3.

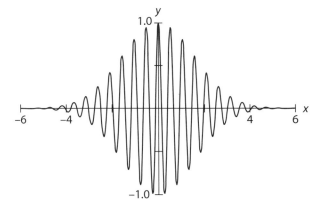

Figure 11.3 The quantity Re $\eta(x, 0)$.

When $k \approx k_0$,

$$\omega (k) = \omega (k_0) + (k - k_0)\, \omega'(k_0) + O(k - k_0)^2 \approx \omega (k_0) + (k - k_0)\, c_g, \quad (11.66)$$

where once again $c_g = \omega'(k_0)$, so that

$$\eta(x, t) \approx \exp\left[i\left\{k_0 x - \omega (k_0)\, t\right\}\right] \int_{-\infty}^{\infty} A(k) \exp\left[i (k - k_0)(x - c_g t)\right] dk. \quad (11.67)$$

Exercise: Establish result (11.67).

The integral is a function of $x - c_g t$ only, which establishes that the envelope of the wave (and the wave packet, at least to this level of rigor) moves with the group speed c_g. As will be seen in the next section, the distinction between wave and group speed is an important one when it comes to formulating a mathematical model of ship (and duck) waves.

11.3 SHIP WAVES

11.3.1 How Does Dispersion Affect the Wave Pattern Produced by a Moving Object?

For those who live near ponds, rivers, or lakes, these fascinating undulations are perhaps the most commonly visible and yet most overlooked form of wave patterns. The name is a little deceptive, because ducks and swans (in particular) do just as good a job in producing them, perhaps even better, inasmuch as the waves and wakes they produce are small enough in scope that they can be appreciated in their totality by the observer. However, we shall continue to refer to the moving source of waves as a ship, even though we know that ducks and swans qualify! We model the ship as a point source for concentric waves of all wavelengths at all points of its motion at speed V_s. This is of course quite a restrictive assumption, but it has

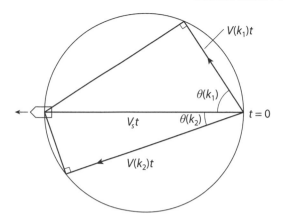

Figure 11.4 The angles $\theta(k_i)$.

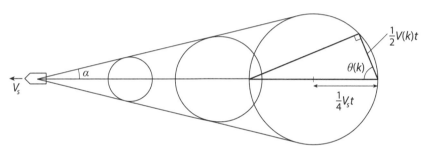

Figure 11.5 Wavecrest envelope.

the advantage of analytical tractability. For deep water gravity waves, as we have already noted,

$$\omega(k) = (gk)^{1/2}; \quad c = 2c_g = \left(\frac{g}{k}\right)^{1/2}. \tag{11.68}$$

This implies that wave crests travel twice as fast as the associated energy. Let us examine the wave pattern that appears steady from the point of view of an observer on the ship (Figure 11.4). The component of V_s in the direction of motion of the crests must equal the wave speed c, so

$$c(k) = V_s \cos\theta(k). \tag{11.69}$$

But as we have noted, the energy of the waves travels at the group speed c_g, so the wave crests that appear to be motionless must lie on the circle of radius $V_s t/4$ (Figure 11.5). We then see that

$$\alpha = \arcsin\left(\frac{V_s t/4}{V_s t/2 + V_s t/4}\right) = \arcsin 1/3 \approx 19.5°, \tag{11.70}$$

which is independent of V_s for deep water. The wavelength of waves propagating at an angle θ to the direction of motion is given by [36] as

$$c = \left(\frac{g}{k}\right)^{1/2} = V_s \cos\theta, \tag{11.71}$$

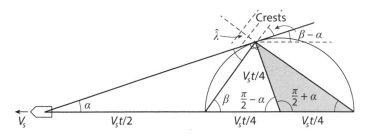

Figure 11.6 Geometry for wavelength relations.

that is,

$$V_s \cos\theta = \left(\frac{g\lambda}{2\pi}\right)^{1/2}, \tag{11.72}$$

and so

$$\lambda = 2\pi V_s^2 g^{-1} \cos^2\theta. \tag{11.73}$$

The maximum wavelength occurs for $\cos^2\theta = 1$ (i.e., for $\theta = 0$). Thus we find that

$$\lambda_{\max} = \frac{2\pi V_s^2}{g}. \tag{11.74}$$

As a boat or duck travels in deep water, the waves in its wake are confined to a wedge-shaped region (with wedge angle $\approx 39°$); this is known as the *Kelvin wedge*, named for William Thomson, 1st Baron Kelvin (1824–1907), an Ulster-born mathematical physicist who contributed to many fields, including fluid dynamics. The wave crests at the edge of the Kelvin wedge make an angle β with the direction of motion . Note from Figure 11.6 that

$$\frac{\pi}{2} + \alpha + 2\left(\frac{\pi}{2} - \beta\right) = \pi, \tag{11.75}$$

so

$$\beta = \frac{1}{2}\left(\frac{\pi}{2} + \alpha\right) \approx \frac{1}{2}(90° + 19.5°) \approx 55°, \tag{11.76}$$

and $\beta - \alpha \approx 35°$. The ratio of the wavelength of these side-arm waves to $\lambda_{\max}$ is given from equation (11.73) as

$$\cos^2\theta = \cos^2\left[\frac{1}{2}\left(\frac{\pi}{2} - \alpha\right)\right]$$

$$= \frac{1}{2}\left[1 + \cos\left(\frac{\pi}{2} - \alpha\right)\right] = \frac{1}{2}(1 + \sin\alpha) = \frac{2}{3} = \frac{\widehat{\lambda}}{\lambda_{\max}}. \tag{11.77}$$

Thus

$$\widehat{\lambda} = \frac{2}{3}\lambda_{\max}. \tag{11.78}$$

If the ship moves at speed $-V_s$ along the x-axis and lies, instantaneously, at the origin O, consider the waves generated when the ship was at point $(X, 0)$, and

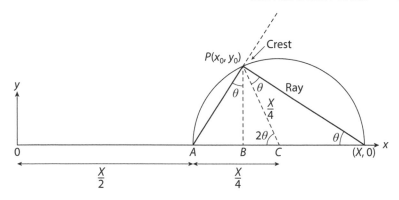

Figure 11.7 Geometry for equations (11.79).

propagating at an angle θ to the x-axis. Let the wave crest that appears stationary from the ship lie at the point (x_0, y_0).

Exercise: Show from Figure 11.7 that

$$x_0 = X(\theta)\left(1 - \frac{1}{2}\cos^2\theta\right), \quad \text{and} \quad y_0 = \frac{1}{2}X(\theta)\sin\theta\cos\theta. \qquad (11.79)$$

Since $X = X(\theta)$, as different parts of the wave crest will be due to waves emitted at different times, they correspond to different values of X. The slope of the wave crest at point A is $\tan(\pi/2 - \theta)$, or

$$\frac{dy}{dx} = \cot(\theta). \qquad (11.80)$$

But on the crest AP,

$$\frac{dy}{dx} = \frac{dy_0}{d\theta} \cdot \frac{d\theta}{dx_0}. \qquad (11.81)$$

Therefore

$$\frac{dy_0}{d\theta} = \frac{d}{d\theta}\left[\frac{1}{4}X(\theta)\sin 2\theta\right] = \frac{1}{4}\left[2X(\theta)\cos 2\theta + X'(\theta)\sin 2\theta\right], \qquad (11.82)$$

and

$$\frac{dx_0}{d\theta} = X(\theta)\cos\theta\sin\theta + X'(\theta)\left(1 - \frac{1}{2}\cos^2\theta\right). \qquad (11.83)$$

Furthermore it is readily shown that

$$\frac{dX}{d\theta} = -X\tan(\theta), \qquad (11.84)$$

which implies that $X = A\cos\theta$ for a constant A. Then finally we obtain the parametric equations for the ship wave pattern:

$$x_0 = A\cos\theta\left(1 - \frac{1}{2}\cos^2\theta\right), \text{ and } y_0 = \frac{A}{2}\cos^2\theta\sin\theta. \qquad (11.85)$$

11.3.2 Whitham's Ship Wave Analysis

In [115] two equivalent problems are stated: (1) that of an object (the "ship") moving with constant speed $\mathbf{U}$ on water at rest, and (2) a uniform stream $\mathbf{U}$ flowing past a fixed object on the surface. Then it is useful to consider waves with wavenumber vector $\mathbf{k} = \langle k_1, k_2 \rangle$ with respect to these different frames of reference. Surfaces of constant phase are defined by $\phi(\mathbf{x}, t) = $ constant ($\mathbf{x} = \langle x_1, x_2 \rangle$), so that (as we have seen in Chapter 3) a local frequency ω and wavenumber vector can be defined by the equations

$$\omega = -\frac{\partial \phi}{\partial t}; \quad \mathbf{k} = \nabla \phi. \tag{11.86}$$

The wave (or phase) velocity $\mathbf{c}$ is parallel to the vector $\mathbf{k}$, and in terms of the unit vector $\hat{\mathbf{k}} = \mathbf{k}/|\mathbf{k}| = \mathbf{k}/k$,

$$\mathbf{c} = -\frac{\partial \phi/\partial t}{|\nabla \phi|} \hat{\mathbf{k}}. \tag{11.87}$$

Furthermore, from (11.86) we recover (in a slightly different form) the law of wave conservation (3.237)

$$\frac{\partial \mathbf{k}}{\partial t} + \nabla \omega = 0. \tag{11.88}$$

Note that this equation implies that $\nabla \times \mathbf{k} = \mathbf{0}$:

$$\frac{\partial k_1}{\partial x_2} = \frac{\partial k_2}{\partial x_1}. \tag{11.89}$$

We denote the wave velocities, group velocities, and local frequencies in frames 1 and 2, respectively, by the appropriate subscripts. Thus

$$\mathbf{c}_2 = \left(\mathbf{U} \cdot \hat{\mathbf{k}}\right) \hat{\mathbf{k}} + \mathbf{c}_1, \tag{11.90}$$

$$\mathbf{c}_{g2} = \mathbf{c}_{g1} + \mathbf{U}, \tag{11.91}$$

$$\omega_2 = \mathbf{c}_2 \cdot \mathbf{k} = (\mathbf{U} + \mathbf{c}_1) \cdot \mathbf{k} = \mathbf{U} \cdot \mathbf{k} + \omega_1. \tag{11.92}$$

As we know, in the absence of surface tension effects,

$$\omega_1 = (gk)^{1/2}, \mathbf{c}_1 = \left(\frac{g}{k}\right)^{1/2} \hat{\mathbf{k}}, \text{ and } \mathbf{c}_{g1} = \frac{1}{2}\left(\frac{g}{k}\right)^{1/2} \hat{\mathbf{k}}. \tag{11.93}$$

For a steady pattern in frame of reference 2, $\omega_2 = 0$, so that

$$\mathbf{U} \cdot \mathbf{k} + \omega_1 = 0, \tag{11.94}$$

that is, if $\mathbf{U} = \langle U, 0 \rangle$, then

$$U k_1 + (gk)^{1/2} = 0. \tag{11.95}$$

As noted from Figure 11.8,

$$\mathbf{k} = \langle -k \sin \chi, k \cos \chi \rangle \tag{11.96}$$

so

$$\sin \chi = \left(\frac{g}{kU^2}\right)^{1/2}. \tag{11.97}$$

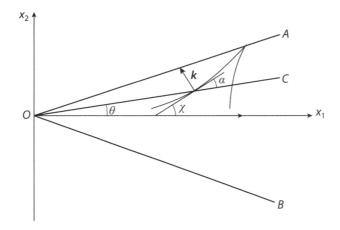

Figure 11.8 Ship wave pattern geometry.

Note also from (11.95) the implicit relation $k_1 = g(k_2)$, so (11.89) implies that

$$\frac{\partial k_2}{\partial x_1} = g'(k_2) \frac{\partial k_2}{\partial x_2}. \tag{11.98}$$

Hence

$$k_2 = F\left[x_1 - g'(k_2)x_2\right] \tag{11.99}$$

for some differentiable function F, indicating that k_2 (and also k_1) are constant on the straight-line characteristics

$$\frac{dx_2}{dx_1} = -g'(k_2) = -\frac{dk_1}{dk_2} = \frac{d(k\sin\chi)}{d(k\cos\chi)} \tag{11.100}$$

$$= \frac{\tan\chi + k(d\chi/dk)}{1 - k(d\chi/dk)\tan\chi} = \tan(\chi - \alpha), \tag{11.101}$$

(provided $\cos\chi \neq 0$), where

$$\tan\alpha = -k\frac{d\chi}{dk} = \frac{1}{2}\tan\chi, \tag{11.102}$$

as may be determined using equation (11.97). The characteristics are in the direction of $\mathbf{c}_{g2}$, making an angle $\theta = \chi - \alpha$ with the direction of the stream, such that

$$\tan\theta = \frac{x_2}{x_1} = -g'(k_2). \tag{11.103}$$

Let us now reexamine the phase angle generically defined by

$$\phi = \mathbf{k} \cdot \mathbf{r} = kr\cos\left(\frac{\pi}{2} + \alpha\right) = kr\cos\left(\frac{\pi}{2} + \chi - \theta\right) = -kr\sin(\chi - \theta) \tag{11.104}$$

in the spirit of stationary phase. This will vary quite rapidly over most of its range, so the most significant contributions will arise from regions where

$$\frac{d\phi}{dk} = -r\sin(\chi - \theta) - kr\frac{d\chi}{dk}\cos(\chi - \theta) = 0. \tag{11.105}$$

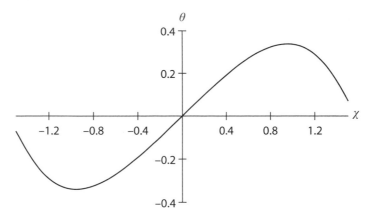

Figure 11.9 Plot of $\theta\,(\chi)$.

We find exactly the criterion (11.102) above, or equivalently,

$$\tan\chi = 2\tan(\chi - \theta). \tag{11.106}$$

Solving this for θ, we obtain

$$\theta = \arctan\left(\frac{\tan\chi}{\tan^2\chi + 2}\right). \tag{11.107}$$

This expression defines the range of angles for which there will be constructive interference behind the ship (i.e., a wake). Note that this result is independent of U and r, so this is quite a general result. As can be seen from Figure 11.9, no real solution exists for $|\theta| \lesssim 0.35 \approx 20°$.

Of course, we can calculate this quite precisely; by differentiating $\tan\theta$ with respect to χ and using (11.107), we find that the derivative is zero when

$$\chi = \chi_m = \arctan\sqrt{2}, \ \text{or} \ \chi_m \approx 54.7°, \tag{11.108}$$

and hence

$$\theta = \theta_m = \arctan\left(\frac{\sqrt{2}}{4}\right) \approx 19.5°. \tag{11.109}$$

This is the semi-angle of the Kelvin wedge. Note from (11.107) that $\theta \to 0$ both as $\chi \to 0$ and as $\chi \to \pi/2$, which also implies the existence of an extremum. The wedge region $|\theta| \in [0, \theta_m]$ bounded by AOB (Figure 11.8) is therefore covered twice, and two values of the wave vector $\mathbf{k}$ correspond to each characteristic OC, so the phase lines are cusped at OA (since $\chi_m \neq \pi/2$, the wave crest cannot turn back smoothly).

The parametric equations of wavecrests (i.e., lines of constant phase) may be found as follows. Since $\phi = -kr\sin\alpha$, from (11.97) we have

$$r = -\frac{\phi U^2 \sin^2\chi}{g\sin\alpha} \equiv -A\sin\chi\cos\chi\left(4 + \tan^2\chi\right)^{1/2}. \tag{11.110}$$

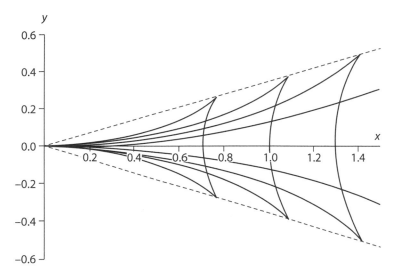

Figure 11.10 Calculated ship wave pattern: lines of constant phase.

Since $x = r \cos \theta$, using (11.107) we find that

$$x = -A \sin \chi (1 + \cos^2 \chi). \qquad (11.111)$$

Similarly,

$$y = r \sin \theta = -A \sin^2 \chi \cos \chi. \qquad (11.112)$$

These lines of constant phase are shown in Figure 11.10.

Again, as noted above, the size and structure of the ship (or whatever is the source of the wake) have not been incorporated in this analysis. The presupposition is that the bow generates such wavetrains, but the stern will do also (and presumably points in between), so there will be some interaction between several sets of Kelvin wedges, though it will not be apparent from the linear theory approach adopted here.

11.3.3 A Geometric Approach to Ship Waves and Wakes

The analysis that follows is adapted from the intuitive treatment by Tricker in his classic book on the subject of waves and wakes [272] (see also [122]). It can be regarded as a more physically and geometrically oriented complement to the analysis in the previous section. (I am a firm believer in the principle that, to understand a concept as fully as possible, examining it from as many complementary levels of description as possible is invaluable.) In Figure 11.11, suppose that a ship (considered here as a point P—a ship with very few passengers!) travels with uniform speed V from O to P. Let this be our basic unit of time (i.e., we measure time in multiples of the time to travel distance OP). Then the distance OP in the figure is the speed of the ship in these units.

Suppose that the ship, while at O, generated a wave of such a wavelength that its (wave) speed is represented by OR. Then by the time the ship has reached P

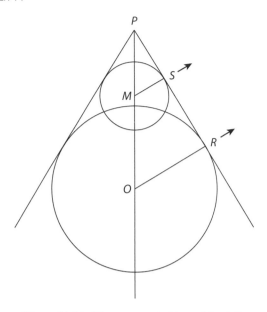

Figure 11.11 Waves generated by a ship at P.

the wave has spread out into a circle of radius OR. Similarly, waves generated elsewhere, such as at M, would have spread out into circles, the radii of which are proportional to the time it takes for the ship to move from M to P. Clearly, they will all define and reinforce the common tangent line PR (see Figure 11.12).

Obviously, a ship may be expected to generate a wide range of wavelengths (and hence speeds) as it moves through the water, just as a stone does when thrown into water. However, the wake of a ship is formed by those waves that are able to keep up with the ship, so those waves will correspond to a particular wavelength out of the spectrum of possible waves generated by the ship's passage through the water. It is the waves of exactly the right speed and in the right position to reinforce one another that will contribute to the visible wake. Since PR is perpendicular to OR, PRO must lie on a semicircle with PO as diameter. PR is the wavefront of the waves generated as the ship passes along OP, and these keep pace with it as they travel in the direction OR. The energy which these contain, however, will travel at the group velocity, which for surface gravity waves is half the wave velocity, and will, therefore, be found along the line PS (where $OS = SR$; see Figure 11.13). Exactly the same arguments apply to waves of appropriate speed in other directions (e.g., OR'), which have their energy at the halfway point S' (see Figure 11.14).

As we consider more such directions, it becomes apparent that the energy of such waves lies on the smaller circle with MO as diameter (this follows from the fact that the equation $r = a \cos \theta$ defines a circle of diameter a in polar coordinates; and $OS' = OM \cos \theta$). The energy of all the waves in the wake, then, will be concentrated along the envelope of these small circles, drawn for all points along the path of the ship. *It is these common tangent lines that are noticed when we see the oblique part of the wake.* From Figure 11.15, note that $MO = \frac{1}{2}OP$. Thus if

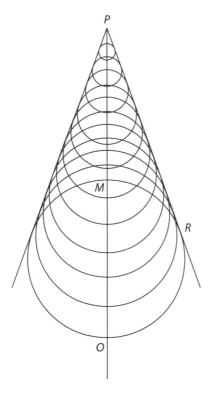

Figure 11.12 The envelope of such waves.

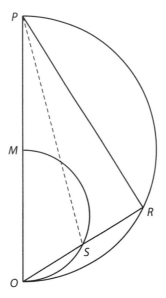

Figure 11.13 A wavefront in the direction OR.

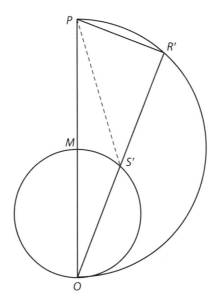

Figure 11.14 A wavefront in direction OR'.

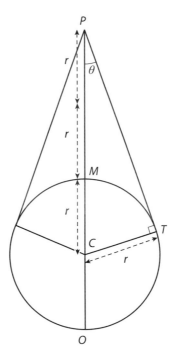

Figure 11.15 The side-arms of the wake.

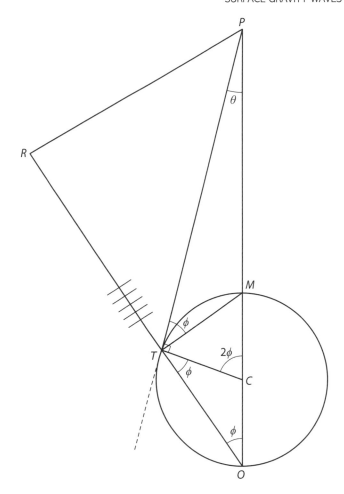

Figure 11.16 Geometric details for the side arms of the ship wake.

$CM = r = CT$, $CP = 3r$, and so the semi-apex angle of the wake is

$$\theta = \arcsin\left(\frac{1}{3}\right), \qquad (11.113)$$

that is, $\theta \approx 19.47°$ or $19°28'$, so the angle between the two arms of the wake is 2θ or approximately $39°$, as found previously.

Provided the water is deep (i.e., the ratio of wavelength:depth is small), this angle is independent of the speed of the ship. The wavefronts of the waves in the side arms of the wake are not parallel to the arms themselves (i.e., they are not parallel to PT in Figure 11.16). The waves at T were generated when the ship was at O, and so the wavefronts will be perpendicular to OT and thus parallel to MT. Had the waves not fallen back because of the effect of the group velocity, they would have reached the position PR. Suppose that they cross the line of the wake at an angle ϕ; the geometry in this figure shows that $\phi \approx 35°14'$. The remainder of

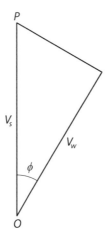

Figure 11.17 Speed relationship (11.114).

the wake is composed of waves that move behind and in the same direction as the ship; their wavelengths are such that they travel at the speed of the ship.

Another geometrical argument can be used to determine this wavelength in terms of the wavelength of the waves in the lateral arms of the wakes. In Figure 11.17 the velocity of the ship (and the waves astern) is denoted by V_s, and the velocity of the side arm waves is denoted by V_w. Clearly,

$$V_w = V_s \cos \phi, \qquad (11.114)$$

and in terms of the corresponding (gravity-dominated) wavelengths, in deep water,

$$V_w = \left(\frac{g \lambda_w}{2\pi} \right)^{1/2}, \qquad (11.115)$$

and

$$V_s = \left(\frac{g \lambda_s}{2\pi} \right)^{1/2}, \qquad (11.116)$$

so that

$$\frac{\lambda_w}{\lambda_s} = \frac{V_w^2}{V_s^2} = \cos^2 \phi = \frac{1 + \cos 2\phi}{2} = \frac{1 + \sin \theta}{2} = \frac{2}{3}. \qquad (11.117)$$

This ratio may be determined directly from photographs of ship wakes (see Figure 11.18).

Note that this theory is approximate in the sense that some things have been neglected. The length, width, and shape of the hull have been ignored (as noted already), along with the waves generated by various other parts of the hull. Certainly a similar pattern of waves will be generated by the stern, and these two wave trains will interfere. A further point to note is that the generation of ship waves requires energy, and this limits the maximum speed that the ship can attain (over and above the loss of energy due to friction with the water). If two geometrically

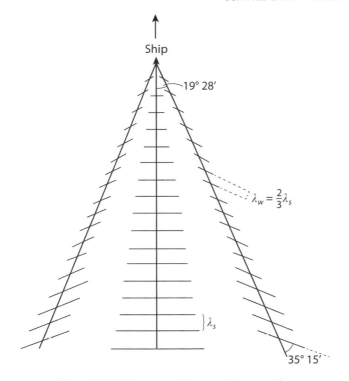

Figure 11.18 Summary of wave-wake properties.

similar ships have dimensions L_1 and L_2, respectively, and if, traveling at speeds V_1 and V_2 they generate waves of wavelengths λ_1 and λ_2, respectively, then on deep water:

$$\frac{V_1^2}{V_2^2} = \frac{g\lambda_1}{g\lambda_2}. \tag{11.118}$$

If, as is reasonable to expect, the wavelengths are proportional to the linear dimensions of the ships, then

$$\frac{V_1^2}{L_1} = \frac{V_2^2}{L_2}, \tag{11.119}$$

which is called *Froude's law* of corresponding speeds.

11.3.4 Ship Waves in Shallow Water

All waves travel with the same speed $(gD)^{1/2}$ in shallow water of depth D. The group velocity is (therefore) the same as the wave velocity. Waves generated as the vessel moves through various points of its path will now spread out with this speed regardless of wavelength, and will therefore all lie on circles having a common

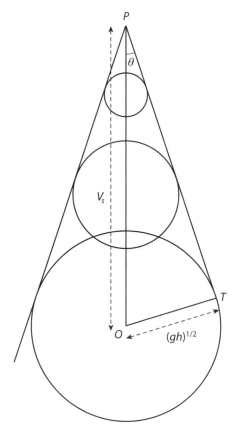

Figure 11.19 Geometry for a wake in shallow water.

tangent through P, the present position of the ship. Clearly from Figure 11.19,

$$\sin \theta = \frac{(gD)^{1/2}}{V_s},$$ (11.120)

so that as the speed increases, the wake becomes narrower. In contrast to the deep water case, the wavefronts of the wake lie along PT so the angle ϕ is now zero. Obviously for this theory to make sense, the ship speed must exceed $(gD)^{1/2}$ ($\sin \theta < 1$ for a wake to occur), so there is no wave capable of keeping up with the ship. But what happens if $V_s < (gD)^{1/2}$? Speed boats moving in shallow harbors or estuaries are definitely affected by this "critical speed." The boat starts off with no problem, but when this critical speed is neared, the waves created by the vessel build up into a large wave just ahead of it, so in a sense it is going uphill all the time and doing so with difficulty! It is like trying to break the equivalent of the sound barrier. (It can be done if the boat is capable of a sudden burst of power that carries it quickly beyond the critical speed before the bow wave has had time to form).

11.4 A DISCRETE APPROACH

In this section we examine how the wave patterns combine both ahead of and behind the duck or ship (let's say a duck-ship) in deep water. Successive wavelets or disturbances will be computed (in a sort of pseudo-Huygens' construction) using properties of the group speed at discrete time steps for both long (gravity) waves behind a duck-ship and short (capillary) waves ahead of a rock or fishing line in a stream.

11.4.1 Long Waves

We have already noted two useful extremes for surface gravity waves in this chapter. For long waves $c \approx (gh)^{1/2}$ (i.e., the wave speed is independent of wavelength, and so there is no dispersion): waves travel with the same speed regardless of wavelength. It is also the case that $c_g = c$. In fact tsunamis fall into this category of shallow water waves, and they will be addressed in Chapter 13, but here we shall focus on the other extreme, namely, short waves. Then $c \approx (g\lambda/2\pi)^{1/2}$, so the longer the wave, the faster it moves. This time the group speed is $c_g = c/2$. At the risk of overemphasizing the point, this result is extremely important in discussing the wave patterns made by ships and ducks. To set the scene, consider a group G of surface gravity waves that after a certain time t has moved out a distance r from the source of the "splash," where $c_g = r/t$. Now suppose that we direct our gaze to follow a particular wave crest as it moves outward from the initially disturbed water. Since it travels at twice the group speed, it will enter and pass though various wave groups in its journey radially outward, so when it encounters the group G, this wave crest has (approximately) the wave speed $c = 2r/t$, but from elementary calculus this is equal to the derivative dr/dt, so the governing differential equation for the position of the crest as a function of time is very simple:

$$\frac{dr}{dt} = 2\frac{r}{t}, \tag{11.121}$$

the solution of which is

$$r = Ct^2, \tag{11.122}$$

(C is a constant, or at least a quantity independent of time; we cannot determine from this approach whether it will be a different constant for different wave crests.) Equation (11.122) tells us that the radial distance covered by an emerging wave crest increases as the square of the time elapsed, and the speed of the wave is given by $r/t = Ct$, so the crest must accelerate as it moves outward; also the area of the circular wave crest increases as the fourth power of time. Our next task is to see how successive wave crests combine to form patterns if the original disturbance is now considered to be a moving source of waves. We suppose that the (very small) ship is moving at a constant speed along an imaginary x-axis from left to right, and that at this moment we set $t_0 = 0$ and place the ship at the origin of the Cartesian coordinate system, so that earlier time intervals correspond to $t_1 = -1, t_2 = -2, \ldots t_n = -n, n = 1, 2, 3, \ldots$, representing the number of time units in the past. The ship therefore travels a unit distance in a unit of our time here, so its speed is 1. When $n = 0$ (i.e., now!), the waves have had no time to spread

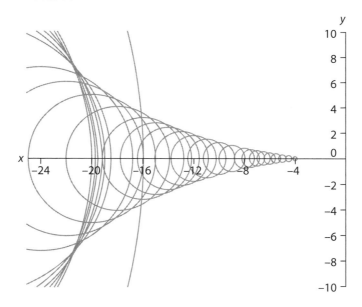

Figure 11.20 Computed arrowhead pattern: surface gravity waves.

out, so they all have radius $r = 0$. However, waves generated at all previous times, $t_n = -n$, have expanded into circles with radii proportional to n^2, as indicated by equation (11.122). Of course we are simplifying things by considering discrete intervals of time, whereas the ship is producing a continuous stream of waves as it moves, but the general pattern produced will be essentially the same and will give rise to the same wave envelope.

How are we to represent the equations of these individual circles (corresponding to a particular wave crest as before)? It is not hard to see that the circle produced at $t = -n$, when the ship's bow was at the point $(-n, 0)$, has the equation

$$(x + n)^2 + y^2 = \left(Cn^2\right)^2 = An^4, \tag{11.123}$$

where A is a scaling constant that will be chosen below to elucidate the spatial development of the pattern. The overall impression is that of an arrowhead (Figure 11.20) Looking at the diagram, one can just about make out the concentration of circular arcs for higher values of n, forming the back of the arrowhead (for the choice of A). For even higher values of n, the wave crests are found *ahead* of the wake formed by lower n values, because they are in fact traveling faster than the ship (or supercharged duck; one such crest is indicated in Figure 11.20).

11.4.2 Short Waves

If we now examine very short waves or ripples such that h/λ is large, we know that the wave speed is given to a very good approximation by

$$c = \left(\frac{2\pi \gamma}{\rho \lambda}\right)^{1/2}, \tag{11.124}$$

where γ is the coefficient of surface tension, and ρ is the water density. Since in this approximation c is inversely proportional to the square root of λ, shorter waves travel faster than longer waves. In this case the group speed exceeds the wave speed, in fact we know that $c_g = 3c/2$, so individual ripples get overtaken by groups of this type of *capillary* wave. Note that the terms "ripple" and "capillary wave" are synonymous here. They refer to surface waves of such short wavelength that the restoring force of gravity (as in "capillary-gravity waves") is negligible compared to that of surface tension. We can adapt the same argument used above for gravity waves to this new class of wave, driven not by gravity but by surface tension. If an individual ripple has radius r at time t, it is part of a group whose speed is (on average) r/t, but the ripple's speed is only 2/3 of the group speed and is represented by the derivative dr/dt, so that now

$$\frac{dr}{dt} = \frac{2r}{3t},$$
(11.125)

which has the general solution

$$r = Bt^{2/3},$$
(11.126)

B being another constant, generally different for each ripple. For such ripples, the circle produced at $t = -n$, when the ship's bow was at the point $(-n, 0)$ has the equation

$$(x + n)^2 + y^2 = \left(Bt^{2/3}\right)^2 = Dn^{4/3}.$$
(11.127)

Compared with duck waves, the envelope of these waves is very different; it's a smooth rounded curve and qualitatively describes the wave pattern ahead of, say, a fishing line, stick, or even a rock in a stream. The usual pattern of ship waves is often present behind the line or rock as well, being formed by the longer wavelength gravity waves described earlier. Without loss of generality, we may gain insight into this pattern by considering again our traveling point source of waves (or, if you wish, the fishing line moving through still water). From equation (11.125) the speed of the crest is $2r/3t$, or, using (11.126), $2B/(3t^{1/3})$, a quantity that obviously decreases with time. From our discussion of surface gravity waves, note that the relative speed of the ship, duck, stick, or fishing line is 1 unit distance per unit time (i.e., $v = 1$). Therefore when $2B/(3t^{1/3}) \leq 1$, or

$$t \geq (2B/3)^3 \equiv t^*,$$
(11.128)

the ripples from succeeding disturbances reinforce just ahead and behind the ship (at $t = t^*$ the ripple speed is equal to that of the ship), as indicated in Figure 11.21.

11.5 FURTHER ANALYSIS FOR SURFACE GRAVITY WAVES

In this section we get to play with some of the mathematical methods alluded to in earlier sections. In cylindrical coordinates (with imposed circular symmetry) the

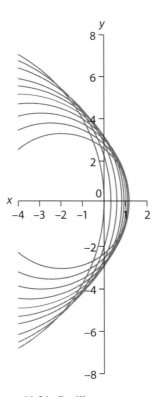

Figure 11.21 Capillary wave pattern.

time-harmonic part of the velocity potential $\Phi(r, z)$ satisfies Laplace's equation, that is,

$$\frac{\partial^2 \Phi}{\partial r^2} + \frac{1}{r} \frac{\partial \Phi}{\partial r} + \frac{\partial^2 \Phi}{\partial z^2} = 0, \qquad (11.129)$$

where the z-axis is positive downward. Separation of variables yields Bessel's equation for the radial part, and the solution regular at the origin is proportional to $J_0(kr)$, where k is radial wavenumber related to the separation constant. Specifically,

$$\Phi(r, z; t) = A J_0(kr) e^{-kz} e^{-i\omega(k)t}, \qquad (11.130)$$

where for linear surface waves on deep water, $\omega(k) = (gk)^{1/2}$. The surface displacement (at $z = 0$) is taken as

$$\eta(r, t) = a J_0(kr) e^{-i\omega t}, \qquad (11.131)$$

so then using the dynamic boundary condition

$$\frac{\partial \Phi}{\partial t} = -g\eta, \qquad (11.132)$$

we may write

$$\Phi(r, z; t) = -i \frac{g}{\omega} \eta(r, t) e^{-kz} = -i \frac{ag}{\omega} e^{-kz} e^{-i\omega t}, \qquad (11.133)$$

thereby indicating a phase difference of $\pi/2$ between the velocity potential and surface displacement. Now we specialize to the case for an initial condition [40], [131]

$$\eta(r, 0) = a, r < R, \tag{11.134}$$

$$= 0, r > R. \tag{11.135}$$

This corresponds to lifting up a plate of radius R from the infinite expanse of fluid at time $t = 0$. We now examine the subsequent temporal and spatial behavior of this initial surface disturbance. From the Fourier-Bessel integral theorem we may write (for suitably well-behaved functions $G(r)$)

$$G(r) = \int_0^\infty k J_0(kr) dk \int_0^\infty \xi J_0(k\xi) G(\xi) d\xi, 0 < r < \infty. \tag{11.136}$$

For the initial condition (11.134) we have

$$G(r) = \eta(r, 0) = a \int_0^\infty k J_0(kr) dk \int_0^R \xi J_0(k\xi) d\xi \tag{11.137}$$

$$= aR \int_0^\infty k J_0(kr) J_1(kR) dk \tag{11.138}$$

(on using a well-known result for the second integral). For $t > 0$ it follows by applying the same approach to (11.133) that

$$\eta(r, t) = aR \int_0^\infty J_0(kr) J_1(kR) e^{-i(gk)^{1/2}t} dk, \text{ and} \tag{11.139}$$

$$\Phi(r, z; t) = -iagR \int_0^\infty J_0(kr) J_1(kR) \frac{e^{-i(gk)^{1/2}t}}{(gk)^{1/2}} dk. \tag{11.140}$$

We first examine the integral for $\eta(r, t)$, which can be rewritten as a double integral using the result

$$J_0(x) = \frac{1}{\pi} \int_0^\pi e^{ix \cos \alpha} d\alpha \tag{11.141}$$

to find (on reversing the order of integration) that

$$\eta(r, t) = \frac{aR}{\pi} \int_0^\pi d\alpha \int_0^\infty J_1(kR) \exp\left(i \left[kr \cos \alpha - (gk)^{1/2} t\right]\right) dk \tag{11.142}$$

$$\equiv \frac{aR}{\pi} \int_0^\pi I(\alpha) d\alpha. \tag{11.143}$$

In terms of the new variable $p = (kR)^{1/2}$ and a scaled time $\tau = (gR)^{1/2} t/2r$, we have

$$I(\alpha) = \frac{2}{R} \int_0^\infty p J_1(p^2) \exp\left(i \left[p^2 \cos \alpha - 2p\tau\right] r/R\right) dp. \tag{11.144}$$

Since we are considering the evolution of the initial disturbance at large distances from the source (i.e., $r \gg R$), it will be appropriate to use the principle of stationary phase for $\chi'(p) = 0$, for the function

$$\chi(p) = p^2 \cos \alpha - 2p\tau. \tag{11.145}$$

This occurs when $p = \tau \sec \alpha \equiv p_0$. Since

$$\chi(p) = \cos \alpha \left[(p - p_0)^2 - p_0^2 \right], \tag{11.146}$$

$$\mathcal{I}(\alpha) \approx \frac{2}{R} \exp\left(-i r R^{-1} p_0^2 \cos \alpha\right) \tag{11.147}$$

$$\times \int_{p_0 - \delta}^{p_0 + \delta} p J_1\left(p^2\right) \exp\left(i \left[(p - p_0)^2 \cos \alpha\right] r / R\right) dp \tag{11.148}$$

$$\approx \frac{2}{R} \exp\left(-i r R^{-1} p_0^2 \cos \alpha\right) \frac{p_0}{q} J_1\left(p_0^2\right) \int_{-q\delta}^{q\delta} \exp\left(i t^2\right) dt, \tag{11.149}$$

$$\approx \frac{2}{R} \exp\left(-i r R^{-1} p_0^2 \cos \alpha\right) \frac{p_0}{q} J_1\left(p_0^2\right) \int_{-\infty}^{\infty} \exp\left(i t^2\right) dt, \tag{11.150}$$

where $q = (r R^{-1} \cos \alpha)^{1/2}, t = q(p - p_0)$ and the interval of integration has been narrowed to that containing the dominant contribution, that is, $[p_0 - \delta, p_0 + \delta]$. This last approximation is justified by taking the limit $q \to \infty$ for constant δ provided $\cos \alpha \neq 0$, and using the result

$$\int_{-\infty}^{\infty} \exp\left(i t^2\right) dt = \pi^{1/2} \exp\left(i \pi / 4\right). \tag{11.151}$$

Hence

$$\mathcal{I}(\alpha) \approx \frac{2 p_0 \pi^{1/2}}{q R} J_1\left(p_0^2\right) \exp i \left(-r R^{-1} p_0^2 \cos \alpha + \frac{\pi}{4}\right). \tag{11.152}$$

Provided that $\tau \ll 1$ (and we exclude those α such that $\cos \alpha \approx 0$), p_0 is small (in some sense), and it is reasonable to replace $J_1(p_0^2)$ by the first term in its Maclaurin expansion (i.e., by $p_0^2/2$). Finally we have

$$\mathcal{I}(\alpha) \approx \left(\frac{\pi \sec^7 \alpha}{r R}\right)^{1/2} \left(\frac{g R t^2}{4 r^2}\right)^{3/2} \exp\left[i \left(-\frac{g t^2}{4 r \cos \alpha} + \frac{\pi}{4}\right)\right]. \tag{11.153}$$

Now we can proceed to evaluate (11.142). Noting that if $\alpha \in (\pi/2, \pi)$ then $p_0 < 0$ and this is outside the range of the integral (11.144), so we restrict the range of α in (11.142) to $(0, \pi/2)$. If V_0 is the volume of water displaced by the cylindrical disk of height a at time $t = 0$, then $V_0 = \pi R^2 a$. The result (11.142) may then be expressed using (11.153) as

$$\eta(r, t) \approx \frac{V_0}{r^2} \left(\frac{g t^2}{4 \pi r}\right)^{3/2} \int_0^{\pi/2} (\sec^7 \alpha)^{1/2} \exp i \left(-\frac{g t^2}{4 r \cos \alpha} + \frac{\pi}{4}\right) d\alpha. \tag{11.154}$$

The objective here is to determine the asymptotic functional form of the surface displacement η as $t \to \infty$ for large but fixed r. Noting the dependence of this integral on the quantity $w = g t^2/4 r$, this requirement is equivalent to determining η as $w \to \infty$. Applying the method of stationary phase to the exponent factor

$\chi(\alpha) = \sec\alpha$, which is clearly stationary for $\alpha = \alpha_0 = 0$ in the range of interest, we have

$$\eta(r, t) \approx \frac{V_0}{r^2}\left(\frac{gt^2}{4\pi r}\right)^{3/2}\int_0^{\pi/2}(\sec^7\alpha)^{1/2}\exp i\left(-\frac{gt^2}{4r\cos\alpha} + \frac{\pi}{4}\right)d\alpha, \quad (11.155)$$

and so

$$\eta(r, t) \approx \frac{V_0}{r^2}\left(\frac{w}{\pi}\right)^{3/2}\exp i\left(-w + \frac{\pi}{4}\right)\int_0^{\delta}\exp\left(-i w\alpha^2/2\right)d\alpha \quad (11.156)$$

$$\approx \frac{V_0}{r^2}\left(\frac{w}{\pi}\right)^{3/2}\left(\frac{\pi}{2w}\right)^{1/2}e^{-iw} = \frac{V_0}{\sqrt{2\pi}r^2}we^{-iw}, \quad (11.157)$$

on using the result

$$\int_0^{\infty}\exp\left(-i w\alpha^2/2\right)d\alpha = \left(\frac{\pi}{2w}\right)^{1/2}e^{-i\pi/4}. \quad (11.158)$$

Reverting to the variables (r, t), we have the physically important result

$$\mathrm{Re}\,[\eta(r, t)] \approx \frac{gV_0t^2}{4\sqrt{2\pi}r^3}\cos\left(\frac{gt^2}{4r}\right), r > 0. \quad (11.159)$$

Thus with circular symmetry, the disturbance amplitude decays spatially as r^{-3}. Focusing our attention only on the oscillatory part, consider consecutive maxima occurring at $w_n = 2n\pi$ and $w_{n+1} = 2(n+1)\pi$ (at constant t), where n is large enough that we may use differentials. Recall that since $w \gg 1$ in this approximation, that condition is satisfied. Thus

$$|w_{n+1} - w_n| = |\delta w| = 2\pi \equiv \frac{gt^2}{4r^2}|\delta r|. \quad (11.160)$$

Thus the local wavelength $\lambda(r)$—the radial distance between crests)—is given by

$$\lambda(r) = |\delta r| \approx \frac{8\pi r^2}{gt^2}. \quad (11.161)$$

Therefore to leading order, at any given time this increases as r^2; correspondingly at any given r it decreases as t^{-2}.

What, it may be asked, is the result of superimposing a continuous sequence of these radially expanding waves in time and space? If we consider the source of the circular waves to be a point moving rectilinearly with speed V_s—a rather poor approximation to a ship, boat, or even a duck, as mentioned repeatedly—then we can use the above analysis to find the stationary patterns of waves moving with the "ship". We shall examine the cumulative effects at any point $P(r, \theta)$ of all the waves produced by the point source from a long time ago ($t = -\infty$) to the present moment ($t = 0$). The source is considered to be at the origin O at $t = 0$, and $|t|$ seconds earlier at the point $Q(V_st, 0)$, where $t < 0$. Denote the resulting pattern $\Sigma(r)$ by the integral of the surface elevations η produced at all earlier instants:

$$\Sigma(r) = \int_{-\infty}^{0}\eta(r, t)dt = C\int_{-\infty}^{0}\frac{w_P\exp(-i w_P)}{r_P^2}dt, r_P = |QP|. \quad (11.162)$$

(The constant C is of no consequence in what follows.) If $\theta \in (0, \pi/2)$, then by the law of cosines,

$$r_P^2 = r^2 + V_s^2 t^2 + 2rV_s t \cos \theta, \quad (t < 0). \qquad (11.163)$$

Here $w_P = gt^2/4r_P = f(t)$, and recalling that the approximation we are using was derived under the assumption $w \gg 1$, we are again able to use the method of stationary phase. Thus when $f'(t) = 0$,

$$\frac{g}{4} \left[\frac{2t}{r_P} - \frac{t^2}{r_P^3} \left(V_s^2 t + rV_s \cos \theta \right) \right] = \frac{gt}{4r_P^3} \left(2r^2 + V_s^2 t^2 + 3rV_s t \cos \theta \right) = 0.$$
$$(11.164)$$

The quadratic in t has roots

$$t_\pm = \frac{3r}{2V_s} \left(-\cos \theta \pm \left[\cos^2 \theta - \frac{8}{9} \right]^{1/2} \right). \qquad (11.165)$$

Recall that $t \in (-\infty, 0)$, so the only further criterion needed is for the roots to be real, that is, $|\theta| < \theta_0 = \arccos(\sqrt{8}/3) \approx 19°28'$. This is once more the semiwedge angle bounding the interference pattern produced by the motion of the tiny boat! The pattern itself is essentially a combination of the two main contributions to the integral (11.162) for all locations, namely,

$$f(t_+) \exp[-if(t_+)] + f(t_-) \exp[-if(t_-)]. \qquad (11.166)$$

Note that on the wedge lines $\theta = \pm\theta_0$, the two real roots coincide.

Chapter Twelve

Ocean Acoustics

The transmission of sound in the ocean is dependent on several environmental factors, the most important of which are the depth and configuration of the bottom, the physical properties of the bottom material, the sound velocity structure within the ocean, the distribution and character of sounder scatterers within the ocean, and the shape of the ocean surface; ... Because of the peculiar distribution of the depths in the ocean, the solution of underwater-sound problems divides into two categories. ... The primary difference between the two is simply one of dimensions; the average depths of water for deep-water transmission are around 10,000 to 20,000 feet whereas those for shallow-water transmission are less than 300 feet.

[250]

12.1 OCEAN ACOUSTIC WAVEGUIDES

A sound speed profile which increases monotonically with depth below the ocean surface is upward-refractive, acting as a duct in which sound may be transmitted to long ranges with little attenuation. A well-known example is the mixed layer, in which the temperature is uniform and the sound speed approximately scales with the hydrostatic pressure, increasing linearly with depth. The depth of the mixed layer depends on surface conditions, but is of the order of 100 m. Deeper channels are found in ice-covered polar waters, where the temperature and sound speed profiles both show a minimum at the surface. A typical surface duct in the Arctic Ocean may extend to depths of 1000 m or more and is capable of supporting very-low-frequency (VLF) (1–50 Hz) acoustic transmissions with no bottom interactions. [138]

12.1.1 The Governing Equation

For scalar waves propagating in a infinite nonhomogeneous two-dimensional region, the governing equation for $U = U(x, z, t)$ is

$$\nabla^2 U \equiv \frac{\partial^2 U}{\partial x^2} + \frac{\partial^2 U}{\partial z^2} = \frac{1}{c^2(z)} \frac{\partial^2 U}{\partial t^2}, \tag{12.1}$$

$$-\infty < x < \infty, \quad -\infty < z < \infty. \tag{12.2}$$

As usual in this type of problem, we seek time-harmonic waves of the form

$$U(x, z, t) = u(x, z)e^{-i\omega t}, \tag{12.3}$$

so that u satisfies the Helmholtz equation

$$\nabla^2 u + \frac{\omega^2}{c^2(z)} u = 0. \tag{12.4}$$

Since $x \in (-\infty, \infty)$, we introduce the Fourier transform of $u(x, z)$, $\mathcal{F}_x(u)$ by

$$\mathcal{F}_x(u) = \tilde{u}(k, z) = \frac{1}{(2\pi)^{1/2}} \int_{-\infty}^{\infty} u(x, z) e^{-ikx} dx. \tag{12.5}$$

Therefore

$$\mathcal{F}_x \left(\frac{\partial^2 u}{\partial x^2} \right) = -k^2 \tilde{u}(k, t),$$

provided that, as assumed here

$$\lim_{|x| \to \infty} u \text{ and } \frac{\partial u}{\partial x} = 0,$$

and

$$\mathcal{F}_x \left(\frac{\partial^2 u}{\partial z^2} \right) = \frac{d^2 \tilde{u}}{d z^2}.$$

Therefore equation (12.1) reduces to the simple form

$$\frac{d^2 \tilde{u}}{dz^2} + \left[\frac{\omega^2}{c^2(z)} - k^2 \right] \tilde{u} = 0. \tag{12.6}$$

Next we consider a piecewise constant symmetrical profile for the sound speed to mimic the structure of some types of ocean waveguide. With z positive and increasing downward, let

$$\begin{aligned} c(z) &= c_0, & z < 0; \\ &= c_1, & 0 < z < H; \\ &= c_0, & z > H. \end{aligned}$$

We also define

$$k_0^2 = \frac{\omega^2}{c_0^2} - k^2, \quad \text{and} \quad k_1^2 = \frac{\omega^2}{c_1^2} - k^2, \tag{12.7}$$

and then apply a so-called *radiation condition* by requiring that waves outside the middle layer move away from that layer, that is, move toward $z = \pm\infty$ (in other words, waves cannot come in from infinity) [38].

Consider first the case of $k_0^2 > 0$ (i.e., $\omega^2 > k^2 c_0^2$), so that the general solution of equation (12.6) outside $[0, H]$ is

$$\tilde{u}(k, z) = A e^{-ik_0 z} + B e^{ik_0 z}, \tag{12.8}$$

or recalling the time factor,

$$\tilde{U}(k, z, t) = A \exp\left[-i\left(k_0 z + \omega t\right)\right] + B \exp\left[i\left(k_0 z - \omega t\right)\right]. \tag{12.9}$$

In the upper layer $z < 0$, the phase factor multiplying B is $\phi_-(z, t) = k_0 z - \omega t$. As t increases, z must increase in order to follow a constant ϕ_-, meaning that the wave propagates downward. Clearly this does not satisfy the radiation condition, so we set $B = 0$. The is no such problem for $\phi_+(z, t) = k_0 z + \omega t$, so

$$\tilde{u}(k, z) = A e^{-ik_0 z}, \quad z < 0. \tag{12.10}$$

For the bottom layer ($z > H$), similar considerations yield

$$\tilde{u}(k, z) = B e^{ik_0 z}, \quad z > H. \tag{12.11}$$

For the middle layer, let $k_1^2 > 0$. Because we expect waves to propagate in both z-directions in this region, we write

$$\tilde{u}(k, z) = C \cos k_1 z + D \sin k_1 z, \quad 0 < z < H. \tag{12.12}$$

Boundary conditions: We require the continuity of $\tilde{u}$ and $\tilde{u}'$ at $z = 0, H$. These four conditions imply, respectively, that

$$A = C; \tag{12.13}$$

$$-ik_0 A = k_1 D; \tag{12.14}$$

$$C \cos k_1 H + D \sin k_1 H = B e^{ik_0 H}; \tag{12.15}$$

$$-k_1 C \sin k_1 H + k_1 D \cos k_1 H = ik_0 B e^{ik_0 H}. \tag{12.16}$$

A necessary and sufficient condition for the existence of nontrivial solutions to this system of equations is that the determinant of the coefficient matrix be zero. After some algebra the condition is found to be

$$\tan k_1 H = -\frac{2 i k_0 k_1}{k_0^2 + k_1^2}, \tag{12.17}$$

which is an implicit dispersion relation for k. Only when (12.17) is satisfied can waves be supported in this system, meaning that this is an eigenvalue relation: only z-directed waves with certain values of k, k_* say, will satisfy the boundary conditions, and these waves will propagate with phase speeds $c = \omega / k_*$.

Exercise: Establish the result (12.17). (Note that k_0 and k_1 are either real or imaginary if $\omega \in \mathbb{R}$ and $k \in \mathbb{R}$.)

Question: Thus far it has been assumed that $k_0 \in \mathbb{R}$ and $k_1 \in \mathbb{R}$. What does this imply about the solutions of (12.17)?

12.1.2 Low Velocity Central Layer

Obviously $c_1 < c_0 \Rightarrow k_1^2 > k_0^2$, hence we seek guided waves with phase speeds in the interval

$$c_1 < c(\omega) = \frac{\omega}{k} < c_0, \tag{12.18}$$

which implies

$$k_1 \in \mathbb{R}, \quad k_0 = i K_0, \tag{12.19}$$

where

$$K_0 = \left(k^2 - \frac{\omega^2}{c_0^2}\right)^{1/2} \in \mathbb{R}. \tag{12.20}$$

Since for $z < 0$,

$$\tilde{u}(k, z) = Ae^{-ik_0 z} = Ae^{K_0 z}, \tag{12.21}$$

the solution decays away exponentially in the half-space $z < 0$. For $z > H$,

$$\tilde{u}(k, z) = Be^{ik_0 z} = Be^{-K_0 z}, \tag{12.22}$$

which means that the guided waves are trapped near the low velocity layer (these are sometimes referred to as evanescent waves, but more commonly just surface waves). Under these circumstances, equation (12.17) becomes

$$\tan k_1 H = \frac{2k_1 K_0}{k_1^2 - K_0^2}. \tag{12.23}$$

or, writing this expression out in full, the dispersion relation takes the form

$$\tan \left[H\left(\frac{\omega^2}{c_1^2} - k^2 \right)^{1/2} \right] = \frac{2(\omega^2 - k^2 c_1^2)^{1/2}(k^2 c_0^2 - \omega^2)^{1/2}}{\omega^2 (c_0^2 + c_1^2) - 2k^2 c_0^2 c_1^2}. \tag{12.24}$$

For real and well-defined solutions to exist, obvious restrictions follow from (12.18) and

$$k^2 \neq \frac{\omega^2}{2} \left(\frac{1}{c_0^2} + \frac{1}{c_1^2} \right). \tag{12.25}$$

12.1.3 Leaky Modes

Now we consider the central region to have a higher sound speed, so $c_1 > c_0 \Rightarrow k_0^2 > k_1^2$. Hence, we seek guided waves in the interval

$$c_0 < c(\omega) < c_1 \tag{12.26}$$

which in turn implies that

$$k_0 \in \mathbb{R}, \quad k_1 = iK_1,$$

where

$$K_1 = \left(k^2 - \frac{\omega^2}{c_1^2} \right)^{1/2} \in \mathbb{R}. \tag{12.27}$$

Since $\tan i\theta = i \tanh \theta$, (12.17) becomes

$$i \tanh K_1 H = \frac{-2k_0 K_1}{K_1^2 - k_0^2}. \tag{12.28}$$

This equation possesses no real solutions unless $K_1 = 0$ (or $k^2 c_0^2 = \omega^2$). Hence, $k \in \mathbb{C}$ for this dispersion relation to be satisfied. Returning to equation (12.9), note that

$$\tilde{U}(k, z, t) = \tilde{u}(k, z)e^{i(kz - \omega t)} = \tilde{u}(k, z) \exp\left[i \{ \mathrm{Re}(k)z - \omega t \} \right] \times \exp\left[-\mathrm{Im}(k)z \right], \tag{12.29}$$

and this decays exponentially with horizontal distance if $\text{Im}(k) > 0$. This is because wave energy refracts out of the central (high sound speed) layer. To demonstrate this, note that

$$\tilde{u}e^{-i\omega t} \sim Ae^{-i(k_0 z + \omega t)}, \quad z < 0$$

and so increasing t implies decreasing z (the wave moves upward out of the layer), and

$$\tilde{u}e^{-i\omega t} \sim Be^{i(k_0 z - \omega t)}, \quad z > H,$$

so increasing t implies increasing z (the wave moves downward out of the layer). Thus waves propagate out of, and consequently wave energy is continuously radiated away from the central layer. They are often referred to in the geophysical (and other) literature as *leaky modes*. To find them numerically the fundamental idea is to look for singularities of the $|F(k)|$ in the complex k-plane, where in view of (12.28)

$$F(k) = \left(i \tanh K_1 H + \frac{2k_0 K_1}{K_1^2 - k_0^2} \right)^{-1}.$$

Such leaky modes have been used to analyze waves propagating along a subduction zone (a plate sliding downward in the earth's mantle) [38].

In summary, waves straying out of a low-c central channel are refracted back into the channel (waveguide effect), and waves are refracted away from a high-c channel (the mirage theorem again). This last case is so identified because in most circumstances in the atmosphere or ocean, lower or higher sound speeds are associated with higher or lower densities, respectively (similar to the optical case).

12.2 ONE-DIMENSIONAL WAVES IN AN INHOMOGENEOUS MEDIUM

Most ocean-surface waveguides can be accurately represented by an inverse square sound speed profile, which may be monotonic increasing (upward refracting) or decreasing (downward refracting) with depth, and whose detailed shape is governed by just three parameters. An analysis of the sound field below the sea surface in the presence of such a profile shows that it consists of a near-field component, given by a branch-line integral, plus a sum of uncoupled normal modes representing the trapped radiation which propagates to longer ranges. The modal contribution is identically zero in the case of the downward refracting profiles. The properties of the modes emerge from a straightforward theoretical development involving first- and second-order asymptotics: each mode shows an oscillatory region immediately below the surface, terminating at the extinction depth, below which the mode decays exponentially to zero; the extinction depth increases rapidly with both mode number and the reciprocal of the acoustic frequency; a reciprocal relationship exists between the extinction depth and the mode strength; and there is no mode cutoff, nor are there any evanescent modes. [138]

Despite the above claim of monotonicity, there are circumstances for which nonmonotonic sound speed profiles are appropriate [139], [140]. We now examine a continuously varying version of the low sound speed central channel (though this is less well-defined for obvious reasons). From (12.4), the Helmholtz equation for the linear time-harmonic acoustic pressure perturbation $u(x, z)$ is

$$\nabla^2 u + \frac{\omega^2}{c^2(z)} u = 0, \tag{12.30}$$

but now the sound speed profile for $z \in (-\infty, \infty)$ is

$$c^2(z) = \frac{c_0^2}{1 + \mu^2(z)}, \tag{12.31}$$

where c_0 is a positive constant, and μ is a bounded function that tends to zero as $|z| \to \infty$. We decompose $u(x, z)$ into the Fourier-component form

$$u(x, z) = e^{ikx} v(z), \tag{12.32}$$

from which (12.30) reduces to

$$\frac{d^2 v}{dz^2} + \left[\frac{\omega^2}{c_0^2} \{ 1 + \mu^2(z) \} - k^2 \right] v = 0, \tag{12.33}$$

that is,

$$\frac{d^2 v}{dz^2} + [\lambda - V(z)] v = 0, \tag{12.34}$$

where

$$\lambda \equiv \kappa^2 = \frac{\omega^2}{c_0^2} - k^2, \tag{12.35}$$

and

$$V(z) = -\frac{\omega^2 \mu^2(z)}{c_0^2}. \tag{12.36}$$

For $\mu^2 > 0$, this is a Schrödinger equation with an attractive potential $V(z)$. As will be discussed in Chapter 23, we can expect the possibility of bound-state solutions under these circumstances that are localized about the depth at which $\mu^2(z)$ has a maximum. This will not occur when $\mu^2 < 0$ (i.e., where c^2 has a local maximum), which corresponds to waves being refracted away from the central region.

12.2.1 Integral Formulation

Now we examine the above situation in rather more mathematical detail for a particular choice of $\mu(z)$ [39] and position a two-dimensional line source of strength Q and frequency ω at the depth of unique minimum sound speed in an infinite medium for that profile. Specifically, choose

$$\mu^2(z) = 2 \text{sech}^2 \left(\frac{\omega z}{c_0} \right), \tag{12.37}$$

and define $k_0 = \omega/c_0 > 0$. Then returning to (12.30) we have

$$\frac{\partial^2 u}{\partial x^2} + \frac{\partial^2 u}{\partial z^2} + k_0^2 (1 + 2 \text{sech}^2 k_0 z) u = -Q \delta(x) \delta(z). \tag{12.38}$$

As expected, we impose a radiation boundary condition: there are to be no incoming waves as $|x|, |z| \to \infty$. Define the Fourier transform in the usual manner as

$$u(x, z) = \frac{1}{2\pi} \int_{-\infty}^{\infty} \tilde{u}(k, z) e^{ikx} dk, \tag{12.39}$$

with inverse

$$\tilde{u}(k, z) = \int_{-\infty}^{\infty} u(k, z) e^{-ikx} dx. \tag{12.40}$$

Using conditions $u \to 0$, and $\partial u / \partial x \to 0$ as $|x| \to \infty$, we obtain

$$\frac{d^2\tilde{u}}{dz^2} + (k_0^2 - k^2 + 2k_0^2 \text{sech}^2 k_0 z) \tilde{u} = -Q\delta(z). \tag{12.41}$$

If we integrate this equation across the singularity at $z = 0$ and define $\tilde{u}_>$ and $\tilde{u}_<$ as the solutions for positive and negative z, respectively, then

$$\frac{d\tilde{u}_>}{dz} \Big|_0 - \frac{d\tilde{u}_<}{dz} \Big|_0 = -Q, \tag{12.42}$$

and $\tilde{u}$ is continuous at $z = 0$. For $z \neq 0$, $\tilde{u}$ satisfies the homogeneous version of the differential equation (12.41). In terms of a new independent variable $y = k_0 z$, for $z \neq 0$, (12.41) becomes

$$\frac{d^2\tilde{u}}{dy^2} + \left[\eta^2 + 2\text{sech}^2 y\right] \tilde{u} = 0, \tag{12.43}$$

where

$$\eta(k) = \left(1 - \frac{k^2}{k_0^2}\right)^{1/2}. \tag{12.44}$$

Since $\tilde{u}$ must contain only waves moving away from the source, the solutions $\tilde{u}_>$ and $\tilde{u}_<$ can be identified with the following solutions using the *method of Darboux* (see box below), so that

$$\tilde{u}_> = A(k) e^{i\eta y} (i\eta - \tanh y); \tag{12.45}$$

$$\tilde{u}_< = B(k) e^{-i\eta y} (i\eta + \tanh y). \tag{12.46}$$

Continuity of the solution at $y = 0 \Rightarrow A(k) = B(k)$, so (12.42) becomes

$$\frac{d\tilde{u}_>}{dy} \Big|_0 - \frac{d\tilde{u}_<}{dy} \Big|_0 = -\frac{Q}{k_0}, \tag{12.47}$$

and therefore from (12.45) and (12.46) we obtain

$$A = \frac{Q}{2k_0(\eta^2 + 1)} = \frac{Qk_0}{2(2k_0^2 - k^2)}. \tag{12.48}$$

We can also rewrite equations (12.45) and (12.46) as

$$\tilde{u} = A(k) e^{i\eta|y|} (i\eta - \tanh|y|) = A e^{i\eta k_0|z|} (i\eta - \tanh k_0|z|),$$

and thus

$$u(x, z) = \frac{Qk_0}{4\pi} \int_{-\infty}^{\infty} e^{ikx + i\eta k_0|z|} \left(\frac{i\eta - \tanh k_0|z|}{2k_0^2 - k^2}\right) dk \tag{12.49}$$

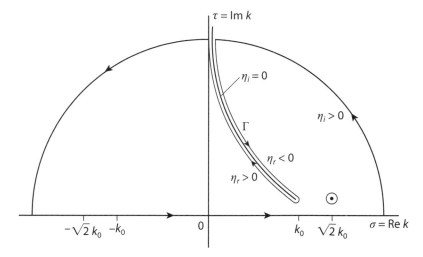

Figure 12.1 Contour for integral (12.49).

for the acoustic pressure field. This is a Fourier integral over the complete set of functions e^{ikx} (i.e., propagating solutions). However, there is a contribution from a *bound state* solution, as mentioned above. To identify this, we replace (12.49) by a contour integral (see Figure 12.1), closed (for $x > 0$ and $z > 0$) in the upper-half complex k-plane, provided $\mathrm{Im}(k_0\eta) > 0$, in which case the $e^{i\eta k_0|z|}$ contribution from the arc will vanish exponentially. Since $k_0\eta = (k_0^2 - k^2)^{1/2}$ we must choose the branch cut to ensure $\mathrm{Im}(k_0\eta) > 0$ everywhere. The appropriate choice of contour is easily determined if we consider k_0 to be slightly complex, that is, $k_0 \to k_0 + i\varepsilon$ with $\varepsilon > 0$ (corresponding to a dissipation or switch-on mechanism from $t = -\infty$). If we write $k = \sigma + i\tau$, so that, defining

$$k_0\eta \equiv \eta_r + i\eta_i = \sqrt{(k_0 + i\varepsilon)^2 - (\sigma + i\tau)^2}, \tag{12.50}$$

it follows that

$$\eta_r^2 - \eta_i^2 = k_0^2 - \varepsilon^2 - \sigma^2 + \tau^2, \tag{12.51}$$

and

$$\eta_r\eta_i = k_0\varepsilon - \sigma\tau. \tag{12.52}$$

It is required that $\eta_i > 0$; now the locus of points for which $\eta_i = 0$ is (from (12.52)) the rectangular hyperbola $\sigma\tau = k_0\varepsilon$. Equation (12.51) implies that on this hyperbola $\eta_r = 0$ when $\sigma = k_0$ and $\tau = \varepsilon$. Because $\eta_r^2 > 0$, we must confine σ and τ to the portion of the hyperbola for which $\sigma < k_0$ and $\tau > \varepsilon$. Thus the branch cut Γ in the first quadrant proceeds from (k_0, ε) to $(0, \infty)$ in the (σ, τ)-plane (see Figure 12.1). $k_0\eta$ will thus be real along Γ but discontinuous across Γ. If we move away from the cut Γ to a point above it, either by increasing σ at constant τ or increasing τ at constant σ, then in (12.52), $k_0\varepsilon - \sigma\tau < 0 \Rightarrow \eta_r < 0$ above Γ, since $\eta_i > 0$. Similarly, $\eta_r > 0$ below Γ.

The Method of Darboux (Summary)

Suppose that $u(x)$ is the general solution of

$$u'' - [\lambda + V(x)]u = 0, \tag{12.53}$$

and u_p is any particular solution for $\lambda = \lambda_p$. It can be shown that

$$z = u' - u \left(\frac{u'_p}{u_p} \right) \tag{12.54}$$

is the general solution of

$$z'' - \left[\lambda - \lambda_p + u_p \left(\frac{1}{u_p} \right)'' \right] z = 0, \tag{12.55}$$

that is, of

$$z'' - \left[\lambda + u - 2 \left(\ln u_p \right)'' \right] z = 0. \tag{12.56}$$

Suppose that $u = 0$ in equation (12.53), so the general solution is

$$u = A \exp \left(\lambda^{1/2} x \right) + B \exp \left(-\lambda^{1/2} x \right). \tag{12.57}$$

A particular solution for $\lambda = 1$ and $A = B = 1/2$ is

$$u_p = \cosh x, \tag{12.58}$$

so that equation (12.56) becomes

$$z'' - \left[\lambda - 2\mathrm{sech}^2 x \right] z = 0, \tag{12.59}$$

for which the general solution is

$$z = A \left[\exp \left(\lambda^{1/2} x \right) \right] \left(\lambda^{1/2} - \tanh x \right) - B \left[\exp \left(-\lambda^{1/2} x \right) \right] \left(\lambda^{1/2} + \tanh x \right). \tag{12.60}$$

Exercise: Establish these results.

12.2.2 Poles

Poles occur in the integrand at $k = \pm\sqrt{2}k_0$, or for slightly complex k, at $k = \pm\sqrt{2}(k_0 + i\varepsilon)$, moving them into the first and third quadrants, respectively. We may subsequently rewrite the Fourier representation of the pressure field, given that (in symbolic shorthand form)

$$\oint_C = \int_{-\infty}^{\infty} + \int_{\mathrm{arc}} + \int_{\Gamma}, \tag{12.61}$$

that is,

$$u(x, z) = \int_{-\infty}^{\infty} = \oint_C - \int_{\text{arc}} - \int_{\Gamma} = 2\pi i \,[\text{Res } k]_{\sqrt{2}k_0} - 0 - \int_{\Gamma}$$

$$= 2\pi i \,[\text{Res } k]_{\sqrt{2}k_0} - \frac{Qk_0}{4\pi} \int_{\Gamma} e^{ikx + i\eta k_0 z} \left(\frac{i\eta - \tanh k_0 z}{2k_0^2 - k^2} \right) dk. \quad (12.62)$$

For the evaluation of the residue note that $k_0\eta = ik_0$ when $k = \sqrt{2}k_0$, and after some rearrangement of $\tanh k_0 z$, the contribution to $u(x, z)$ from the first term in (12.62) is

$$u_1 = \frac{Q}{4\sqrt{2}} \exp i \left(\sqrt{2}k_0 x - \frac{\pi}{2} \right) \text{sech } k_0 z. \quad (12.63)$$

To determine the contribution from the second term note that around the branch cut, the integral corresponds to the following behavior for η_r from (12.51) (recalling that $\eta_i = 0$ on Γ), namely, that $\eta_r^2 \to \infty$ at the upper point $(0, \infty)$ while being zero at (k_0, ε). Since $\eta_r < 0$ above the cut, and $\eta_r > 0$ below the cut, we see that η must run from $-\infty$ to 0 and then 0 to ∞ along Γ. Now $k = k_0 \left(1 - \eta^2\right)^{1/2}$, so the integral along Γ can be written as one in η. Thus contribution becomes

$$u_2 = \frac{Qk_0}{4\pi} \int_{-\infty}^{\infty} e^{i\eta k_0 z} (i\eta - \tanh k_0 z) F(\eta) d\eta, \quad (12.64)$$

where

$$F(\eta) = \frac{\eta \exp \left(i k_0 \sqrt{1 - \eta^2} x \right)}{\left(k_0^2 - \eta^2 \right)^{1/2} \left(k_0^2 + \eta^2 \right)}. \quad (12.65)$$

The functions

$$\chi(\eta, z) = k_0 e^{i\eta k_0 z} (i\eta - \tanh k_0 z) \quad (12.66)$$

for real η do not of themselves constitute a complete set of functions. The bound-state contribution (12.63) must also be included and is found by evaluating the quantity $\chi(ik_0, z)/k_0$. More general details for this type of problem can be found in [139], [140].

Discrete and Continuous Spectra in Wave Theory

Frequently (as here) it is necessary to evaluate integrals associated with wave problems in the complex wavenumber (k) or frequency (ω) plane, and this may involve, as above, branch line integrals and pole residues. Sometimes essential singularities are also present. Commonly, branch line integrals are associated with freely propagating waves, and the poles with surface or "trapped" waves (evanescent in a direction orthogonal to the boundary). Such examples are common (see, e.g., [90]), but in this regard we can do little better than quote the abstract from the article [120]:

Solutions to source-excited field problems are frequently represented as super-positions of source-free field solutions. The latter are in general of two types: eigenmodes and non-eigenmodes which are related to the zeros of the total impedance or alternatively the poles of the scattering coefficient of a system. The eigenmodes are everywhere finite and comprise a complete orthogonal set. The non-eigenmodes become infinite in the infinitely remote spatial limits of a region and are not in general members of a complete orthogonal set; examples are "radio-active states," "damped resonances," and "leaky waves." Despite their physically singular behavior, the nonmodal solutions can be employed to represent field solutions in certain ranges.

12.3 MODEL FOR A STRATIFIED FLUID: CYLINDRICAL GEOMETRY

A mathematical model of acoustic wave propagation in a stratified medium (for which the sound speed c varies with depth z) was proposed in [141], [142] in a study of underwater sound propagation and ducting. For time-harmonic behavior the governing equation is of the standard form

$$\nabla^2 P + \frac{\omega^2}{c^2(z)} P = 0; \quad 0 \le z \le H, \tag{12.67}$$

where $P(x, y, z)$ is the pressure perturbation, z is downward, and the the lower boundary is at $z = H$. In cylindrically symmetric geometry (r, z) the following boundary conditions are imposed:

(1) $P(0) = 0$; (a free or pressure-release surface);
(2) $P(H) \cos \alpha + P'(H) \sin \alpha = 0$;
(3) $P(r) \to A \exp(ikr) + g(r)$, where $g(r) \to 0, r = (x^2 + y^2)^{1/2} \to \infty$.

This last condition is a particular case of the Sommerfeld radiation condition in cylindrical geometry, that is,

$$\lim_{r \to \infty} r^{1/2} \left| \frac{\partial u}{\partial r} - iku \right| = 0. \tag{12.68}$$

We separate variables such that $P(r, z) = R(r)\phi(z)$ to obtain (in particular) the equation satisfied by ϕ:

$$\frac{d^2\phi}{dz^2} + \left[\frac{\omega^2}{c^2(z)} - k^2 \right] \phi = 0, \tag{12.69}$$

k^2 being the separation constant. This is a self-adjoint equation possessing a discrete sequence of real eigenvalues $k_n^2 \to -\infty$ with eigenfunctions ϕ_n for the boundary-value problem as posed. The radiation condition (3) requires the choice of $H_0^{(1)}(kr)$ for the solution of

$$\frac{1}{r} \frac{d}{dr} \left(r \frac{dR}{dr} \right) + k^2 R = 0. \tag{12.70}$$

The limit $H \to \infty$ is of particular interest for spectral analysis of equation (12.69), and therefore we make the additional assumption that $c(z)$ is bounded above as $z \to \infty$ so that

$$\lim_{z \to \infty} \frac{\omega^2}{c^2(z)} = a^2 > 0. \tag{12.71}$$

In view of this, equation (12.69) can be expressed as

$$\frac{d^2\phi}{dz^2} + [q(z) + \lambda]\phi = 0, \tag{12.72}$$

where

$$q(z) = \frac{\omega^2}{c^2(z)} - a^2 = \frac{\omega^2}{c^2(z)} - \frac{\omega^2}{c^2(\infty)}; \tag{12.73}$$

$$\lambda = a^2 - k^2 = \frac{\omega^2}{c^2(\infty)} - k^2. \tag{12.74}$$

If $|q(z)|$ is integrable, that is,

$$\int_0^\infty |q(z)| \, dz < \infty, \tag{12.75}$$

then (12.72) possesses asymptotic expansions with leading terms [141]

$$M(z) \sim \exp\left(i\lambda^{1/2}z\right), \, z \to \infty, \tag{12.76}$$

$$N(z) \sim \exp\left(-i\lambda^{1/2}z\right), \, z \to \infty. \tag{12.77}$$

It can be seen from condition (3) near the start of this section that $M(z)$ satisfies the above radiation condition at $z = \infty$.

An interesting argument has been made regarding the nature of the spectrum of equation (13.8) [141], and it is briefly summarized here. The eigenvalue equation for that equation and boundary conditions (1) and (2) above is

$$\begin{vmatrix} M(0) & N(0) \\ M(H)\cos\alpha + M'(H)\sin\alpha & N(H)\cos\alpha + N'(H)\sin\alpha \end{vmatrix} = 0. \tag{12.78}$$

Then if

$$-\max\{q(z)\} < \lambda < 0, \tag{12.79}$$

$\lambda(H)$ attains discrete values that have limits independent of α as $H \to \infty$. There are three basic reasons for this, the first being that the problem is self-adjoint for $H < \infty$ and therefore has a discrete spectrum with only a finite number of eigenvalues in this range. Furthermore, for finite fixed values of z, both M and N are entire functions of λ [152]; finally, for $\lambda < 0$, $M(z) \to \infty$ and $N(z) \to 0$ as $z \to \infty$. From (12.78) it follows that

$$N(0) = \frac{M(0)\{N(H)\cos\alpha + N'(H)\sin\alpha\}}{M(H)\cos\alpha + M'(H)\sin\alpha}, \tag{12.80}$$

and therefore the right-hand side of this equation can be made arbitrarily small if H is chosen to be sufficiently large. The roots of (12.80) therefore approach those of $N(0) = 0$ as $H \to \infty$, and since $N(0)$ is an integral function of λ those roots cannot have a finite point of accumulation.

By contrast, in the same limit $H \to \infty$, the values of $\lambda \in [0, \infty)$ lie in the continuous spectrum of equation (12.69), that is, the number of eigenvalues between any two fixed values of λ increases without limit as $H \to \infty$. This follows directly from the oscillation properties of eigenfunctions: the nth eigenfunction has exactly n nodes (zeros) in the interior of $(0, H)$. Given the asymptotic behavior of M and N these properties become increasingly similar to those of $\sin \lambda^{1/2} z$ for $\lambda > 0$ and sufficiently large values of z. Then for a given range of λ the change in the number of zeros of $\sin \lambda^{1/2} z$ on $(0, H)$ tends to infinity as $H \to \infty$, and so too must the number of eigenvalues lying in this range.

A useful physical interpretation of these results is possible if with each value of λ we associate an angle θ between a real "ray" (for $k^2 > 0$) and the z-axis such that

$$k = \frac{\omega}{c(z)} \sin \theta(z). \tag{12.81}$$

Then for some value z_0 of z this can be written as

$$k = \left[\frac{\omega}{c(z_0)} \right] \left[\frac{c(z_0)}{c(z)} \right] \sin \theta(z) = \frac{\omega}{c(z_0)} \sin \theta(z_0), \tag{12.82}$$

where

$$\sin \theta(z_0) = \frac{c(z_0)}{c(z)} \sin \theta(z). \tag{12.83}$$

This is just another form of Snell's law of refraction for acoustic rays! Modes with $\lambda \geq 0$ (i.e., $a^2 \geq k^2$) can be associated with constructive interference among rays that are reflected at the bottom surface when $H < \infty$. Correspondingly, modes for which $-\max\{q(z)\} < \lambda < 0$ are associated with constructive interference among rays that are refracted back toward the surface at some ray-dependent critical depth. Below that they decay exponentially with depth. Just like surface gravity waves in a channel of depth H, they are independent of the lower surface, provided the channel is deep enough. The reader's attention is also drawn to a related paper using a resolvent Green's function technique:

A method is given for developing velocity-depth profiles for which the one-dimensional wave equation can be solved exactly. This is done by transforming an equation whose solutions are known into the wave equation, and then stating the solutions to the wave equation in terms of the original equation. The allowed velocity-depth variations are then given as a Schwarzian derivative and an invariant first derived by Malet. This technique, when applied to the geometric differential equation, leads to the Epstein profile. This method has been developed independently by Bose in constructing solvable Schrödinger potentials. and more recently by Heading in developing refractive index profiles for electromagnetic wave propagation. [209]

The reader is encouraged to follow up the references therein (which may also be found in a summary of the article [201]).

12.4 THE SECH-SQUARED POTENTIAL WELL

Consider the one-dimensional time-independent Schrödinger equation for the potential well defined on the real line [41]

$$V(x) = -\lambda(\lambda - 1) H^2 \text{sech}^2 Hx, \tag{12.84}$$

where $\lambda > 1$ (and H is *not* Planck's constant!). (In some of the quantum mechanics literature this is known as the *modified Pöschl-Teller* potential.) This is a different approach to the potential well problem discussed earlier (12.36) (and since the independent variable is not the cylindrical z-axis, we revert to the variable x). Then

$$u''(x) + \left[k^2 + \lambda(\lambda - 1) H^2 \text{sech}^2 Hx\right] u(x) = 0. \tag{12.85}$$

If we introduce the change of independent variable $y = \cosh^2 Hx$, followed by a new dependent variable

$$p(y) = y^{-\lambda/2} u, \tag{12.86}$$

equation (12.85) takes the form of a hypergeometric differential equation,

$$y(1-y) p'' + \left\{\left(\lambda + \frac{1}{2}\right) - (\lambda + 1)y\right\} p' - \frac{1}{4}\left(\lambda^2 + \frac{k^2}{H^2}\right) p = 0. \tag{12.87}$$

12.4.1 Positive Energy States

The solution we seek is to be valid on the interval $|x| \in [0, \infty)$ or $y \in [1, \infty)$. If

$$a = \frac{1}{2}\left(\lambda + i\frac{k}{H}\right); \quad b = \frac{1}{2}\left(\lambda - i\frac{k}{H}\right), \tag{12.88}$$

then the general solution of (12.87) is, in terms of arbitrary constants A and B,

$$p(y) = A\,{}_2F_1\left(a, b, \frac{1}{2}; 1 - y\right) + B(1-y)^{1/2}\,{}_2F_1\left(a + \frac{1}{2}, b + \frac{1}{2}, \frac{3}{2}; 1 - y\right). \tag{12.89}$$

Note that if the energy $k^2 > 0$ (i.e. for real values of k) then a and b are complex conjugates; also, at $x = 0$ ($y = 1$) in terms of the original variable,

$$u(0) = A + B(1-y)^{1/2}. \tag{12.90}$$

Following standard practice in many texts on quantum mechanics (e.g., [41]), we choose two linearly independent real solutions, one even and the other odd in the original variable x. These are found by selecting $A = 1$, $B = 0$ for the former and $A = 0$, $B = 1$ for the latter, respectively. Then the even solution is

$$u_e(x) = \left(\cosh^\lambda Hx\right) {}_2F_1\left(a, b, \frac{1}{2}; -\sinh^2 Hx\right), \tag{12.91}$$

and the odd solution is

$$u_o(x) = \left(\cosh^\lambda Hx \sinh Hx\right) {}_2F_1\left(a + \frac{1}{2}, b + \frac{1}{2}, \frac{3}{2}; -\sinh^2 Hx\right). \tag{12.92}$$

Using the asymptotic behavior of the hyperbolic and hypergeometric functions as $x \to -\infty$, it can be shown that (i)

$$u_e(x) \sim 2^{-\lambda} e^{\lambda H|x|} \Gamma\left(\frac{1}{2}\right)(A + B),$$ (12.93)

where

$$A = \frac{\Gamma(b-a)}{\Gamma(b)\,\Gamma(1/2-a)} 2^{2a} e^{-2aH|x|};$$ (12.94)

$$B = \frac{\Gamma(a-b)}{\Gamma(a)\,\Gamma(1/2-b)} 2^{2b} e^{-2bH|x|},$$ (12.95)

and (ii)

$$u_o(x) \sim \pm 2^{-(\lambda+1)} e^{(\lambda+1)H|x|} \Gamma\left(\frac{3}{2}\right)(C + D),$$ (12.96)

where

$$C = \frac{\Gamma(b-a)}{\Gamma(b+1/2)\,\Gamma(1-a)} 2^{2a+1} e^{-(2a+1)H|x|};$$ (12.97)

$$D = \frac{\Gamma(a-b)}{\Gamma(a+1/2)\,\Gamma(1-b)} 2^{2b+1} e^{-(2b+1)H|x|}.$$ (12.98)

Exercise: Show that equation (12.93) holds, given the definitions (12.99) and (12.95).

The positive sign in (12.96) is used when $x > 0$, and the negative sign when $x < 0$. For real values of k these solutions can be restructured the following more convenient form, after some reduction:

$$u_e(x) \sim C_e \cos(k|x| + \phi_e); \quad u_o(x) \sim \pm C_o \cos(k|x| + \phi_o),$$ (12.99)

where the phases are

$$\phi_e = \arg\left\{\frac{\Gamma(ik/H)\exp(-ik\ln 2/H)}{\Gamma([\lambda + ik/H]/2)\,\Gamma([1 - \lambda + ik/H]/2)}\right\},$$ (12.100)

and

$$\phi_o = \arg\left\{\frac{\Gamma(ik/H)\exp(-ik\ln 2/H)}{\Gamma([\lambda - 1 + ik/H]/2)\,\Gamma([2 - \lambda + ik/H]/2)}\right\}.$$ (12.101)

The amplitude factors can be determined directly from the expressions (12.93) and (12.96) but are left implicit here. Suppose that we combine the solutions (12.99) as

$$u(x) = Au_e(x) + Bu_o(x).$$ (12.102)

For $x > 0$ we arrive at the forms

$$2u = AC_e\{\exp[i(kx + \phi_e)] + \exp[-i(kx + \phi_e)]\}$$

$$+ BC_o\{\exp[i(kx + \phi_o)] + \exp[-i(kx + \phi_o)]\}$$ (12.103)

for $x > 0$, and

$$2u = AC_e \left\{ \exp\left[i\left(kx - \phi_e\right)\right] + \exp\left[-i\left(kx - \phi_e\right)\right] \right\}$$
$$- BC_o \left\{ \exp\left[i\left(kx - \phi_o\right)\right] + \exp\left[-i\left(kx - \phi_o\right)\right] \right\} \quad (12.104)$$

for $x < 0$. In terms of reflection and transmission coefficients R and T, we seek the asymptotic forms

$$u \sim e^{ikx} + R e^{-ikx}, \quad x \to -\infty, \quad (12.105)$$
$$u \sim T e^{ikx}, \quad x \to \infty. \quad (12.106)$$

By comparing the two sets of equations we deduce that

$$R = \frac{1}{2}\left(e^{2i\phi_e} + e^{2i\phi_o}\right); \quad T = \frac{1}{2}\left(e^{2i\phi_e} - e^{2i\phi_o}\right), \quad (12.107)$$

with corresponding reflection and transmission intensity coefficients

$$|R|^2 = \cos^2\left(\phi_e - \phi_o\right); \quad |T|^2 = \sin^2\left(\phi_e - \phi_o\right), \quad (12.108)$$

which naturally satisfy the conservation requirement

$$|R|^2 + |T|^2 = 1. \quad (12.109)$$

The explicit forms of these intensity coefficients may be found directly from equations (12.100) and (12.101). Using the abbreviation $w = k/2H$, the phase difference is

$$\phi_e - \phi_o = \arg \Gamma \left(\frac{\lambda + 1}{2} + iw\right) + \arg \Gamma \left(1 - \frac{\lambda}{2} + iw\right) \quad (12.110)$$

$$- \arg \Gamma \left(\frac{\lambda}{2} + iw\right) - \arg \Gamma \left(\frac{1 - \lambda}{2} + iw\right). \quad (12.111)$$

In view of the relation

$$\Gamma\left(\xi\right)\Gamma\left(1 - \xi\right) = \frac{\pi}{\sin \pi \xi} \quad (12.112)$$

it follows that

$$\arg \Gamma(\xi) - \arg \Gamma\left(1 - \xi\right) = -\arg \sin \pi \xi. \quad (12.113)$$

Therefore, following [41], if we define

$$\xi_1 = \frac{\lambda + 1}{2} + iw; \quad \xi_2 = \frac{\lambda}{2} + iw, \quad (12.114)$$

the expression (12.110) can be rewritten as

$$\phi_e - \phi_o = \arg \sin \pi \left(\frac{\lambda}{2} + iw\right) - \arg \sin \pi \left(\frac{\lambda + 1}{2} + iw\right). \quad (12.115)$$

Using properties of sine functions with complex arguments (i.e., $\sin(x + iy)$, etc.), this is equivalent to

$$\phi_e - \phi_o = \arctan\left(\tan\frac{\pi\lambda}{2}\tanh\pi w\right) + \arctan\left(\cot\frac{\pi\lambda}{2}\tanh\pi w\right), \quad (12.116)$$

or

$$\phi_e - \phi_o = \arctan\left[\frac{\sinh(\pi k/H)}{\sin\pi\lambda}\right] \equiv \arctan\zeta, \quad (12.117)$$

after some interesting trigonometric manipulations! Clearly, if λ is an integer, this reduces to

$$\phi_e - \phi_o = \frac{\pi}{2}, \quad (12.118)$$

in which case perfect transmission occurs because $|T|^2 = 1$, $|R|^2 = 0$. This is a generalization of the obvious result when $\lambda = 1$ (see (12.84)). The other extreme occurs in the limit of vanishing energy (so that $k \to 0^+$), because then $\phi_e - \phi_o = 0$ and $|T|^2 = 0$, $|R|^2 = 1$. In general we may write these coefficients in terms of ζ; thus

$$|R|^2 = \frac{1}{1+\zeta^2}; \quad |T|^2 = \frac{\zeta^2}{1+\zeta^2}. \quad (12.119)$$

12.4.2 Bound States

Now we consider negative energy so that k is imaginary, and we write $E = k^2 = -\kappa^2$, whence a and b are now real, so from (12.88) we choose

$$a = \frac{1}{2}\left(\lambda - \frac{\kappa}{H}\right); \quad b = \frac{1}{2}\left(\lambda + \frac{\kappa}{H}\right). \quad (12.120)$$

It is again appropriate to use the asymptotic expressions (12.93) and (12.96), but the exponents $-2aH|x|$ and $-(2a+1)H|x|$ in the first terms of these asymptotic forms can be positive for large enough values of κ. To ensure the existence of bounded solutions it is necessary that the coefficients of these terms are identically zero or in other words, at the poles of the gamma functions in the demoninators of those first terms. Since these can only exist at zero or the negative integers, we must have, for even solutions, the eigenvalue condition

$$\frac{1}{2} - a = \frac{1-\lambda}{2} + \frac{\kappa}{2H} = -n, \quad (12.121)$$

or

$$\frac{\kappa}{H} = \lambda - 1 - 2n, \quad (12.122)$$

and for the odd solutions, the eigenvalue condition

$$1 - a = 1 - \frac{\lambda}{2} + \frac{\kappa}{2H} = -n, \quad (12.123)$$

or

$$\frac{\kappa}{H} = \lambda - 2 - 2n. \quad (12.124)$$

Then the energy levels become

$$E = E_n = -\kappa^2 = -H^2 (\lambda - 1 - 2n)^2 \qquad (12.125)$$

for even solutions, and

$$E_n = -\kappa^2 = -H^2 (\lambda - 2 - 2n)^2 \qquad (12.126)$$

for odd solutions.

As can be seen, this approach yields rather more information on the bound states than that developed using the method of Darboux, but it must be pointed out that not all these states may be physically reasonable in the waveguide problem.

Chapter Thirteen

Tsunamis

13.1 MATHEMATICAL MODEL OF TSUNAMI PROPAGATION (TRANSIENT WAVES)

A tsunami is a series of ocean waves generated by sudden displacements in the sea floor, landslides, or volcanic activity. In the deep ocean, the tsunami wave may only be a few inches high. The tsunami wave will increase in height to become a fast moving wall of turbulent water several meters high on reaching the coastline. They can cause great destruction and are most frequent in the Pacific Ocean and Indonesia primarily because of the large number of active submarine earthquake zones around the Pacific Rim.

[143]

A tsunami is a series of ocean waves with very long wavelengths (typically hundreds of kilometres) caused by large-scale disturbances of the ocean, such as earthquakes; landslide; volcanic eruptions; explosions; meteorites. These disturbances can either be from below (e.g., underwater earthquakes with large vertical displacements, submarine landslides) or from above (e.g., meteorite impacts).

[143]

Tsunami is a Japanese word with the English translation: "harbour wave". In the past, tsunamis have been referred to as "tidal waves" or "seismic sea waves". The term "tidal wave" is misleading; even though a tsunami's impact upon a coastline is dependent upon the tidal level at the time a tsunami strikes, tsunamis are unrelated to the tides. (Tides result from the gravitational influences of the moon, sun, and planets.) The term "seismic sea wave" is also misleading. "Seismic" implies an earthquake-related generation mechanism, but a tsunami can also be caused by a non-seismic event, such as a landslide or meteorite impact.

Tsunamis are also often confused with storm surges, even though they are quite different phenomena. A storm surge is a rapid rise in coastal sea-level caused by a significant meteorological event—these are often associated with tropical cyclones.

[210]

We assume a two-dimensional ocean of (undisturbed) constant depth h and no rigid boundaries other than the seafloor. In this chapter we develop the basic features discussed earlier and incorporate a fluctuation on the ocean floor that will generate and drive the tsunami wave. We shall examine the space and time development of the resulting disturbance but deviate a little from the notation used in chapter 11, simply to relieve the reader from frequent page-turning back to that chapter. In addition, it affords some notational continuity with the sources cited below. Thus the variable Φ (instead of ϕ) will be used to signify the velocity potential and, instead of η, the free surface displacement will be denoted here by $\xi(x, z, t)$. Consequently the governing equations are [43], [112]

$$\nabla^2 \Phi = 0, \ \Phi = \Phi(x, z, t), \tag{13.1}$$

and on the free upper surface,

$$\frac{\partial \xi}{\partial t} = \frac{\partial \Phi}{\partial z}, \ z = 0, \tag{13.2}$$

and

$$\frac{\partial \Phi}{\partial t} + g\xi = \frac{-P_a(x, t)}{\rho}, \ z = 0, \tag{13.3}$$

where $P_a(x, t)$ is the prescribed atmospheric pressure. On the seafloor we introduce a displacement $H(x, t)$ due to an earthquake or other disturbance such that now

$$z = -h + H(x, t). \tag{13.4}$$

If the ground motion is known, we may write the lower boundary implicitly as $F(x, z, t) = 0$, where

$$F(x, z, t) = z + h - H(x, t). \tag{13.5}$$

Thus we may write the total derivative of F as

$$\frac{dF}{dt} + \mathbf{v} \cdot \nabla F = 0, \tag{13.6}$$

so that on $F = 0$ continuity of the normal component of velocity requires that

$$\frac{\partial \Phi}{\partial z} = \frac{\partial H}{\partial t} + \frac{\partial \Phi}{\partial x}\frac{\partial H}{\partial x}; \ z = -h + H, \tag{13.7}$$

or neglecting quadratically small terms,

$$\frac{\partial \Phi}{\partial z} = \frac{\partial H}{\partial t} \equiv W(x, t) \ \text{on} \ z \approx -h. \tag{13.8}$$

Initial conditions will be prescribed as needed in the following Laplace transform approach. To that end we define the Laplace transform of the function $f(t)$ by

$$\bar{f}(s) = \int_0^\infty e^{-st} f(t)dt, \tag{13.9}$$

and its inverse transform by

$$f(t) = \frac{1}{2\pi i} \int_\Gamma e^{st} \bar{f}(s)ds, \tag{13.10}$$

where Γ is a vertical line to the right of all singularities in the complex s-plane. Then we can transform equations (13.1) and (13.8) to get

$$\nabla^2 \bar{\Phi}(x, z, s) = 0, \ -h < z < 0 \tag{13.11}$$

and

$$\frac{\partial \bar{\Phi}}{\partial z} = \bar{W}(x, s), \quad z = -h. \tag{13.12}$$

Similarly, transforming (13.2) and (13.3), we have

$$-\xi(x, 0) + s\bar{\xi}(x, s) = \frac{\partial \bar{\Phi}}{\partial z}(x, 0, s), \tag{13.13}$$

and

$$-\Phi(x, 0, 0) + s\bar{\Phi}(x, 0, s) + g\bar{\xi}(x, s) = -\frac{\bar{P}_a}{\rho}. \tag{13.14}$$

After a little rearrangement (13.13) and (13.14) become the single condition

$$\frac{\partial \bar{\Phi}}{\partial z} + \frac{s^2}{g}\bar{\Phi} = -\frac{s\bar{P}_a}{\rho g} - \xi(x, 0) + \frac{s}{g}\bar{\Phi}(x, 0, 0). \tag{13.15}$$

The potential $\Phi(x, 0, 0)$, can be interpreted physically as follows. If all is calm for $t < 0$, and at $t = 0$ an impulsive pressure $P_a(x, t) = I(x)\delta(t)$ is applied on the free surface, then integrating (13.3) across $t = 0$ yields

$$\Phi(x, 0, 0^+) - \Phi(x, 0, 0^-) + g\int_{0^-}^{0^+} \xi dt = -\frac{I(x)}{\rho}\int_{0^-}^{0^+} \delta(t)dt = -\frac{I(x)}{\rho}. \tag{13.16}$$

Hence

$$\Phi(x, 0, 0^+) = \frac{I(x)}{\rho}. \tag{13.17}$$

13.2 THE BOUNDARY-VALUE PROBLEM

The boundary-value component of the problem is defined by the equations (13.11), (13.12), and (13.15). We expect no motion to be felt far from the source; so

$$\Phi(x, t) \to 0 \text{ as } |x| \to \infty \Rightarrow \Phi(x, s) \to 0 \text{ as } |x| \to \infty.$$

Since for the problem as defined there are no finite boundaries in the x-variable, we use and define the Fourier transform pair by

$$\tilde{f}(k) = \int_{-\infty}^{\infty} e^{-ikx} f(x)dx \tag{13.18}$$

and

$$f(x) = \frac{1}{2\pi}\int_{-\infty}^{\infty} e^{ikx} \tilde{f}(k)dk. \tag{13.19}$$

Hence the *Fourier-Laplace transform* of $\Phi(x, z, t)$, namely, the quantity $\hat{\Phi}(k, s; z)$, satisfies

$$\frac{d^2\hat{\Phi}}{dz^2} - k^2\hat{\Phi} = 0, \quad -h < z < 0, \tag{13.20}$$

$$\frac{d\hat{\Phi}}{dz} = \hat{W}, \quad z = -h; \tag{13.21}$$

and

$$\frac{d\hat{\Phi}}{dz} + \frac{s^2}{g}\hat{\Phi}(k, 0, 0) = F(k, s; z) = 0, \tag{13.22}$$

where

$$F(k, s; z) = -\frac{s\hat{P}_a}{\rho g} - \bar{\xi}(k, 0) + \frac{s}{g}\tilde{\Phi}(k, 0, 0). \tag{13.23}$$

The general solution of (13.20) is

$$\hat{\Phi} = A\cosh[k(z + h)] + B\sinh[k(z + h)] \tag{13.24}$$

with arbitrary constants A and B. Applying the boundary conditions (13.21) and (13.22), we find that

$$\hat{\Phi} = \frac{\text{sech}\,(kh)}{s^2 + gk\tanh(kh)}\left[gF\cosh[k(z + h)] + \frac{\tilde{W}}{k}(s^2\sinh(kz) - gk\cosh(kz))\right]. \tag{13.25}$$

Exercise: Derive (13.25).

Hence by formal inversion over the contour Γ_1 (to be defined),

$$\Phi(x, z, t) = \frac{1}{2\pi}\int_{-\infty}^{\infty} e^{ikx}dk\left[\frac{1}{2\pi}\int_{\Gamma_1} e^{st}\hat{\Phi}(k, z, s)ds\right], \tag{13.26}$$

and from (13.3),

$$\xi(x, t) = -\frac{P_a}{\rho g} - \frac{1}{g}\frac{d\Phi}{dt}(x, 0, t)$$

$$= -\frac{P_a}{\rho g} + \frac{1}{2\pi g}\int_{-\infty}^{\infty} e^{ikx}dk\left[\frac{1}{2\pi}\int_{\Gamma_1} -se^{st}\hat{\Phi}(k, 0, s)ds\right]. \tag{13.27}$$

13.3 SPECIAL CASE I: TSUNAMI GENERATION BY A DISPLACEMENT OF THE FREE SURFACE

We examine a transient disturbance due to an initial displacement on the free surface. Let

$$P_a(x, t) = W(x, t) = \Phi(x, 0, 0) = 0, \quad \text{and} \quad \xi(x, 0) \equiv \xi_0(x) \neq 0, \tag{13.28}$$

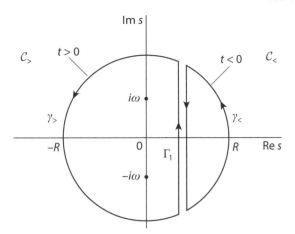

Figure 13.1 Inverse Laplace transform contours.

which implies (from condition (13.23)) that $\hat{W} = 0$ and $F = -\tilde{\xi}_0(k, 0)$. Equation (13.27) provides the free surface height

$$\xi(x, t) = \frac{1}{4\pi^2 i} \int_{-\infty}^{\infty} e^{ikx} \tilde{\xi}_0(k) dk \int_{\Gamma_1} \frac{s e^{st} ds}{s^2 + gk \tanh(kh)}. \qquad (13.29)$$

The integrand has two pure imaginary poles at

$$s = \pm i (gk \tanh(kh))^{1/2} = \pm i\omega. \qquad (13.30)$$

For $t < 0$ the contour $C_<$ is closed in the right half of the s-plane as shown in Figure 13.1, and the integrand is analytic within and on $C_<$ (there being no poles, and $\mathrm{Re}\, s > 0 \Rightarrow e^{st} \to 0$ as $s \to \infty$ on $\gamma_<$). Hence by Jordan's lemma and Cauchy's theorem the integral around this contour is zero, so $\xi = 0$ for $t < 0$. For $t > 0$ the contour $C_>$ is closed as shown, and again by Jordan's lemma, the contour around the semicircular arc $\gamma_>$ vanishes as $R \to \infty$, so that, informally,

$$\int_{C_>} = \int_{\Gamma_1} + \int_{\gamma_>} = 2\pi i \sum \text{Residues},$$

where for $t > 0$

$$\lim_{R \to \infty} \frac{1}{2\pi i} \int_{C_>} = \frac{1}{2\pi i} \int_{\Gamma_1} \frac{s e^{st} ds}{s^2 + \omega^2} = \cos \omega t. \qquad (13.31)$$

Using this result, (13.29) yields

$$\xi(x, t) = \frac{1}{2\pi} \int_{-\infty}^{\infty} \tilde{\xi}_0(k) e^{ikx} (\cos \omega t) \, dk. \qquad (13.32)$$

Since ω is an even function of k, so is $\cos \omega (k) t$. Now we can partition $\xi_0(x)$ into even and odd components,

$$\xi_e = \frac{1}{2} [\xi_0(x) + \xi_0(-x)]; \quad \xi_o = \frac{1}{2} [\xi_0(x) - \xi_0(-x)], \qquad (13.33)$$

so that

$$\tilde{\xi}_0(k) = \int_{-\infty}^{\infty} \xi_0(x)e^{-ikx}dx = \int_{-\infty}^{\infty} (\xi_e + \xi_o)(\cos kx - i \sin kx)\,dx$$

$$= 2\int_0^{\infty} [\xi_e \cos kx - i\xi_o \sin kx]\,dx \equiv \tilde{\xi}_0^e + \tilde{\xi}_0^o, \tag{13.34}$$

where $\tilde{\xi}_0^e$ is real and an even function of k, while $\tilde{\xi}_0^o$ is imaginary and an odd function of k. For purposes of illustration let $\tilde{\xi}_0$ be an even function of x; then (13.32) becomes

$$\xi(x,t) = \frac{1}{\pi}\int_0^{\infty} \tilde{\xi}_0^e(k)\cos kx \cos \omega t\,dk$$

$$= \frac{1}{2\pi}\,\mathrm{Re}\int_0^{\infty} \tilde{\xi}_0^e(k)\left[e^{i(kx-\omega t)} + e^{i(kx+\omega t)}\right]dk. \tag{13.35}$$

The first term is an integral superposition of plane waves moving to the right, and the second term signifies a corresponding waveform moving to the left. To gain a better understanding of this expression in physical terms, we need to apply the method of stationary phase. To that end a brief mathematical hiatus is in order. We shall return to (13.35) in due course.

Stationary Phase Revisited

Consider the integral

$$I(t) = \int_a^b f(k)e^{ig(k)t}dk, \tag{13.36}$$

where f and g are smooth real functions of the real variable k. When t is large, $\cos(gt)$ and $\sin(gt)$ oscillate rapidly as k varies, and there is very little net area under the curve because of cancellation (the *Riemann-Lebesgue lemma* discussed below provides mathematical justification for this comment). If $g'(k_0) = 0$ for some k_0, there is a significant contribution to $I(t)$ from the integrand in the immediate neighborhood of k_0. The corresponding exponential part of the integrand may be written as

$$e^{ig(k_0)t}e^{i[g(k)-g(k_0)]t}.$$

The real part of the second term, $\cos[g(k) - g(k_0)]t$, varies slowly, while the imaginary part crosses the k-axis at $k = k_0$. Now to $O\left[(k - k_0)^2\right]$,

$$g(k) \approx g(k_0) + \frac{1}{2}(k - k_0)^2 g''(k_0), \tag{13.37}$$

and we can replace the interval (a, b) by $(-\infty, \infty)$ with little error. Then

$$I(t) \approx e^{ig(k_0)t}f(k_0)\int_{-\infty}^{\infty} \exp\left[\frac{1}{2}(k - k_0)^2 g''(k_0)\right]dk. \tag{13.38}$$

Now

$$\int_{-\infty}^{\infty} \exp\left(\pm i\beta^2 t\right) d\beta = \left(\frac{\pi}{t}\right)^{1/2} e^{\pm i\pi/4}, \tag{13.39}$$

where the $+$ or $-$ sign is chosen according to $g''(k_0) > 0$ or < 0, respectively. The error can be shown to be $O(t^{-1})$, so to this order of approximation,

$$I(t) \approx e^{ig(k_0)t} f(k_0) \left(\frac{2\pi}{t|g''(k_0)|}\right)^{1/2} e^{\pm i\pi/4}, \tag{13.40}$$

which is the general form we require, decaying with time as $t^{-1/2}$.

Returning to (13.35), consider $x > 0$. For the first integral (making the obvious comparison with (13.36)) it is clear that

$$G(k) = k\left(\frac{x}{t}\right) - \omega(k), \tag{13.41}$$

and we know that

$$\omega(k) = (gk \tanh kh)^{1/2}, \tag{13.42}$$

so under the shallow-water approximation (appropriate for tsunami wave propagation) $(kh \ll 1)$,

$$\omega'(k) = c_g \approx c \approx (gh)^{1/2}, \tag{13.43}$$

and there exists a stationary point such that $G'(k_0) = 0$ when

$$\frac{x}{t} = \omega'(k_0) = c_g(k_0), \tag{13.44}$$

provided $x/t < (gh)^{1/2}$ (see Figure 13.2).
   Question: Why is this last restriction necessary?
   For the second integral,

$$G'(k) = \frac{x}{t} + \omega'(k) > 0, \quad k \in (0, \infty), \tag{13.45}$$

so there is no such stationary point, and this integral accordingly has the asymptotic behavior $\sim t^{-1}$ as $t \to \infty$. Then behind the leading edge of the wave (i.e., $x < (gh)^{1/2}t$),

$$\xi(x, t) \approx \tilde{\xi}_0^e(k_0) \left[\frac{1}{2\pi t|\omega''(k_0)|}\right]^{1/2} \cos\left(k_0 x - \omega(k_0)t + \frac{\pi}{4}\right) + O(t^{-1}), \tag{13.46}$$

since $\omega'' < 0 \Rightarrow g'' > 0$. Also, as noted above, ahead of the leading edge,

$$\xi \sim O(t^{-1}), \quad x > (gh)^{1/2}t. \tag{13.47}$$

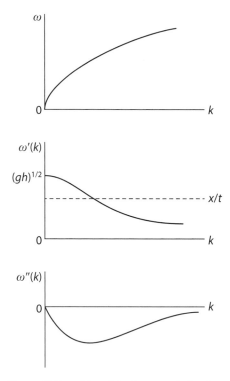

Figure 13.2 $\omega(k)$ and its first two derivatives.

The (Generalized) Riemann-Lebesgue Lemma

Lemma: If

$$\int_{-\infty}^{\infty} |f(t)|\, dt < \infty, \tag{13.48}$$

then

$$\lim_{\lambda \to \infty} \int_{-\infty}^{\infty} e^{i\lambda t} f(t)\, dt = 0. \tag{13.49}$$

Proof: Let $\varepsilon < 0$, and let $X(\varepsilon) > 0$ be such that

$$\int_{-\infty}^{-X} |f(t)|\, dt < \frac{\varepsilon}{4}, \quad \text{and} \quad \int_{X}^{\infty} |f(t)|\, dt < \frac{\varepsilon}{4}.$$

Then we have

$$\left| \int_{-\infty}^{\infty} e^{i\lambda t} f(t)\, dt \right| < \frac{\varepsilon}{2} + \left| \int_{-X}^{X} e^{i\lambda t} f(t)\, dt \right|. \tag{13.50}$$

Because of the Weierstrass approximation theorem [202], we may select a polynomial on $[-X, X]$ such that

$$\int_{-X}^{X} |f(t) - g(t)| \, dt < \frac{\varepsilon}{4}.$$

Then, integration by parts gives

$$\left| \int_{-X}^{X} e^{i\lambda t} g(t) \, dt \right| = \left| \left[\frac{e^{i\lambda x} g(x)}{i\lambda} \right]_{-X}^{X} - \frac{1}{i\lambda} \int_{-X}^{X} e^{i\lambda x} g'(x) \, dx \right|.$$

Because $g(x)$ is a polynomial, both it and its derivative are bounded on the compact interval $[-X, X]$, so this term may be bounded by $\varepsilon/4$ by making λ suitably large. Then the result (13.49) follows.

Returning to the integrals (13.35), we examine simplifications resulting from the assumption of *deep* water. It has already been noted that tsunami wave propagation is a shallow water phenomenon, but for completeness a temporary digression is in order.

13.3.1 A Digression: Surface Waves on Deep Water (Again)

For deep water, $kh \gg 1$ so that

$$\omega \approx (gk)^{1/2}, \quad \text{and} \quad \omega'(k) = \frac{1}{2} \left(\frac{g}{k} \right)^{1/2}, \tag{13.51}$$

and

$$G'(k_0) = 0 \Rightarrow \frac{x}{t} = \frac{1}{2} \left(\frac{g}{k_0} \right)^{1/2} = \frac{g}{2\omega_0} \tag{13.52}$$

(see Chapter 11).

Therefore

$$\omega_0 = \frac{gt}{2x}; \quad k_0 = \frac{\omega_0^2}{g} = \frac{gt^2}{4x^2}, \tag{13.53}$$

so that

$$\left(\frac{d\omega}{dk} \right)_0 = \frac{1}{2} \left(\frac{g}{k_0} \right)^{1/2}, \tag{13.54}$$

and

$$\left(\frac{d^2\omega}{dk^2} \right)_0 = -\frac{g^{1/2}}{4k_0^{3/2}} = -\frac{g^{1/2}}{4} \left(\frac{4x^2}{gt^2} \right)^{3/2} = -\frac{2x^3}{gt^3}. \tag{13.55}$$

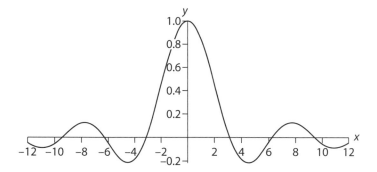

Figure 13.3 The function sinc x.

Then from equation (13.46)

$$\xi(x,t) \approx \frac{\widetilde{\xi}_0^e(k_0)}{2\pi^{1/2}} \left(\frac{g^{1/2}t}{x^{3/2}} \right) \cos \left(\frac{gt^2}{4x} - \frac{\pi}{4} \right),$$ (13.56)

since $\cos\theta$ is an even function. This generalizes a result in [34].

Let us examine a particularly simple (if unrealistic) application of this result. Suppose $\xi_0(x)$ is the "box" function

$$\xi_0(x) = H(x+a) - H(x-a), \qquad a > 0,$$ (13.57)

where $H(\theta)$ is the Heaviside step function

$$H(\theta) = 0, \quad \theta < 0; \quad H(\theta) = 1, \quad \theta \geq 0.$$ (13.58)

This is a rectangle of unit height on the interval $[-a, a]$. Therefore

$$\widetilde{\xi}_0^e(k) = \int_{-\infty}^{\infty} \xi_0(x)e^{-ikx}dx = \frac{2\sin ka}{k} \equiv 2a \operatorname{sinc} ka.$$ (13.59)

Here the function sinc x is the *sine cardinal function*, defined by

$$y = \begin{cases} \operatorname{sinc} x = \dfrac{\sin x}{x}, & x \neq 0, \\ 1, & x = 0, \end{cases}$$ (13.60)(13.61)

(see Figure 13.3).

This function appears in many applications, such as diffraction theory in optics, spectroscopy, information theory, and digital signal processing.

Exercise: Verify the result (13.59).

Therefore from (13.53),

$$\widetilde{\xi}_0^e(k_0) = \frac{2\sin(k_0 a)}{k_0} = \frac{8x^2}{gt^2} \sin \left(\frac{gat^2}{4x^2} \right),$$ (13.62)

so that from (13.56),

$$\xi(x,t) \approx \frac{4}{t} \left(\frac{x}{\pi g} \right)^{1/2} \sin \left(\frac{gat^2}{4x^2} \right) \cos \left(\frac{gt^2}{4x} - \frac{\pi}{4} \right)$$ (13.63)

(for $x < \sqrt{ght}$). To illustrate the shape of this surface elevation we ignore all constants in this expression by considering the function

$$Z(x,t) = t^{-1}x^{1/2} \sin\left(\frac{t^2}{x^2}\right) \cos\left(\frac{t^2}{x} - \frac{\pi}{4}\right). \tag{13.64}$$

Exercise: Examine graphs of (13.64) for fixed t (in x) and fixed x (in t).

Physics Associated with (13.46)

Behind the leading edge an observer moving at speed x/t sees a train of sinusoidal waves of wave numbers k_0, and frequency $\omega(k_0)$, whose group speed is x/t and whose amplitude decays as $O(t^{-1/2})$ or even faster in deep water (see 13.63). For "large" values of x/t (but less than $(gh)^{1/2}$), k_0 is small, so the corresponding wavelength λ_0 is large; in contrast, a faster-moving observer sees even longer waves, which are also of larger amplitude since $|\omega''(k_0)|$ is smaller (see Figure 13.2). The precise shape of $\xi_0(x)$ determines $\tilde{\xi}_0(k)$, of course, and hence the amplitude of the dispersed waves; for example, if

$$\xi_0(x) = \frac{b}{x^2 + b^2}, \tag{13.65}$$

a symmetrical lump of "area" π and characteristic spread b, then

$$\tilde{\xi}_0(k_0) \propto e^{-k_0|b|}, \tag{13.66}$$

which is small if the spread b is large (unless k_0 is small). As b increases, the amplitude for a given k_0 obviously decreases.
Exercise: Verify (13.66).

13.3.2 How Fast Does the Wave Energy Propagate?

The total energy associated with the wave is [43]

$$E = \int_{-\infty}^{\infty} C\xi^2(x)\,dx, \tag{13.67}$$

where the constant $C = \rho g/2$ for surface gravity waves on a channel of unit width. Consider then the initial waveform

$$\xi_0(x) = \frac{1}{2\pi}\int_{-\infty}^{\infty} e^{ikx}\tilde{\xi}_0(k)\,dk. \tag{13.68}$$

We make use of *Plancherel's theorem* [215], which in this context states that the integral of the square of the Fourier transform of a function $\xi(x) \in \mathcal{L}_2(-\infty, \infty)$ is equal to the integral of the square of the function itself (i.e., $\xi(x)$ is square-integrable). Thus we have

$$E = E_0 = C\int_{-\infty}^{\infty} \xi_0^2(x)\,dx = \frac{C}{4\pi^2}\int_{-\infty}^{\infty} |\tilde{\xi}_0(k)|^2\,dk = \frac{C}{2\pi^2}\int_0^{\infty} |\tilde{\xi}_0(k)|^2\,dk, \tag{13.69}$$

where $\tilde{\xi}_0(k)$ is assumed here to be an even function. The energy contained in wavenumbers between k_0 and $k_0 + dk$ is

$$\Delta E_0 \approx \frac{C}{2\pi^2} |\tilde{\xi}_0(k_0)|^2 dk. \tag{13.70}$$

Asymptotically this energy is found in the interval between $x = t\omega'(k_0)$ and $x = t\omega'(k_0 + dk)$, an interval of length

$$|t\omega'(k_0) - t\omega'(k_0 + dk)| \approx t|\omega''(k_0)|dk, \tag{13.71}$$

so the energy per unit length (or energy density) is

$$E_d = \frac{C}{2\pi^2} \frac{|\tilde{\xi}_0(k)|^2}{t|\omega''(k_0)|} dk. \tag{13.72}$$

For a sinusoidal wave of the form $\xi_0 = A[\text{Re}(e^{ikx})]$, the average energy per unit length is (**Exercise**)

$$E_d = \frac{CA^2}{2\pi} \cos^2 kx\, dx = \frac{CA^2}{4\pi} \int_0^{2\pi} (1 + \cos 2kx)\, dx = \frac{CA^2}{2}, \tag{13.73}$$

so equating expressions (13.72) and (13.73), the amplitude A is given by

$$A = \frac{1}{\pi} \left[\frac{|\tilde{\xi}_0(k)|^2}{t|\omega''(k_0)|} \right]^{1/2}, \tag{13.74}$$

which, apart from a factor $(\pi/2)^{1/2}$ is the amplitude derived in (13.46), so we have established a physical basis for this result. From either version of that asymptotic result, at any large time t we consider the waves sandwiched between two observers moving at respective speeds $c_{g1} \equiv c_g(k_1)$ and $c_{g2} \equiv c_g(k_2)$. Geometrically, in a time-space diagram this corresponds to waves between the two rays $x_1/t = c_{g1}$ and $x_2/t = c_{g2}$. The total energy between these rays is

$$\int_{x_1}^{x_2} E_d(x)dx \propto \int_{x_1}^{x_2} \frac{|\tilde{\xi}_0(k)|^2}{t|\omega''(k_0)|} dx. \tag{13.75}$$

For fixed t, $x = \omega'(k_0)t$ and $\omega''(k_0) < 0$, so that

$$dx \approx t\omega''(k_0)\, dk_0 = -t |\omega''(k_0)|\, dk_0,$$

from which we may conclude that

$$\int_{x_1}^{x_2} E_d(x)dx \propto \int_{k_2}^{k_1} |\tilde{\xi}_0(k)|^2 dk_0 = K, \tag{13.76}$$

a constant, where we have used the fact that $x_2 > x_1$ implies $k_2 < k_1$. This means that the total energy of the waves is conserved between two observers moving at their respective local group speeds. We can approach this result from a kinematic viewpoint also.

Question: Why does $x_2 > x_1$, imply $k_2 < k_1$?

13.3.3 Kinematics Again

From (13.44), dropping the subscript zero, $x = \omega'(k)t$, so differentiating partially with respect to x and t, we find that

$$1 = t\frac{d\omega'(k)}{dk}\frac{\partial k}{\partial x} = t\omega''\frac{\partial k}{\partial x}, \tag{13.77}$$

and

$$0 = t\frac{d\omega'(k)}{dk}\frac{\partial k}{\partial t} + \omega'(k) = t\omega''\frac{\partial k}{\partial t} + \omega'. \tag{13.78}$$

Therefore

$$\frac{\partial k}{\partial x} = \frac{1}{t\omega''}; \quad \frac{\partial k}{\partial t} = \frac{-\omega'}{t\omega''}, \tag{13.79}$$

so that

$$\frac{\partial k}{\partial t} + \omega'(k)\frac{\partial k}{\partial x} = 0, \tag{13.80}$$

or equivalently the conservation equation

$$\frac{\partial k}{\partial t} + \frac{\partial \omega}{\partial x} = 0. \tag{13.81}$$

From (13.80) we determine the solution

$$k(x,t) = f[x - c_g(k)t], \tag{13.82}$$

so k remains constant if the argument $x - c_g(k)t$ does, meaning that k is constant along paths in the (x, t)-plane satisfying $dx/dt = c_g$; this follows from

$$dk = \frac{\partial k}{\partial t}dt + \frac{\partial k}{\partial x}dx = 0 = \left(\frac{\partial k}{\partial t} + \frac{dx}{dt}\frac{\partial k}{\partial x}\right)dt, \tag{13.83}$$

and comparing with (13.80). If $k = $ constant, $c_g = $ constant, so these paths are straight lines

$$x - c_g t = \text{constant}, \tag{13.84}$$

as noted above. It is also of interest to see that a conservation law follows from equation (13.72), namely,

$$\frac{\partial}{\partial t}\left(\frac{E_d}{\omega}\right) + \frac{\partial}{\partial x}\left(c_g\frac{E_d}{\omega}\right) = 0. \tag{13.85}$$

This is admittedly a little artificial, because ω is constant along a ray. If it were not, this would express what was encountered in Chapter 3, namely, the *conservation of wave action* (E_d/ω). In the present case of course,

$$\frac{\partial E_d}{\partial t} + \frac{\partial}{\partial x}(c_g E_d) = 0. \tag{13.86}$$

13.4 LEADING WAVES DUE TO A TRANSIENT DISTURBANCE

As noted earlier, tsunamis are shallow water waves, wherein $\omega \approx k(gh)^{1/2}$, so $c \approx (gh)^{1/2}$ for the fastest waves with $k \approx 0$.

This statement needs clarification. The dispersion relation for surface gravity waves in a channel of depth h is given by (13.42), from which we see that

$$c = \left(\frac{g}{k}\tanh kh\right)^{1/2} \propto \left(\frac{\tanh \beta}{\beta}\right)^{1/2}, \tag{13.87}$$

with $\beta = kh$. Expanding $\tanh \beta$ readily indicates that c achieves its maximum value in the limit $\beta \to 0^{+}$.

In the neighborhood of the wavefront (from (13.41)) the derivative of the phase function $G(k)$ is

$$G'(k) \approx \frac{x}{t} - (gh)^{1/2}. \tag{13.88}$$

This is small (and $c_g \approx c$), so the phase is nearly stationary; however,

$$\omega''(k) \approx -(gh)^{1/2}h^2 k \tag{13.89}$$

is also very small (verify this, dear Reader), and an approximation better than (13.46) is needed. To this end, we take the above expansion in β a little further by writing (13.42) as

$$\omega(k) = \left(\frac{g}{h}\right)^{1/2}(\beta \tanh \beta)^{1/2} = \left(\frac{g}{h}\right)^{1/2}\left(\beta - \frac{1}{6}\beta^3 + \frac{19}{360}\beta^5 + O(\beta^7)\right), \tag{13.90}$$

or in terms of k,

$$\omega(k) = \left(\frac{g}{h}\right)^{1/2}\left(kh - \frac{k^3 h^3}{6} + \frac{19}{360}k^5 h^5 + O(kh)^7\right) \tag{13.91}$$

$$\approx (gh)^{1/2}\left(k - \frac{k^3 h^2}{6} + \frac{19}{360}k^5 h^4\right). \tag{13.92}$$

Therefore

$$\omega'(k) \approx (gh)^{1/2}\left(1 - \frac{k^2 h^2}{2} + O(kh)^4\right), \tag{13.93}$$

$$\omega'' \approx -(gh)^{1/2}(kh^2 + O(k^3 h^4)), \tag{13.94}$$

and so forth, so that

$$G(k) = k\frac{x}{t} - \omega(k) \approx k\frac{x}{t} - (gh)^{1/2}\left(k - \frac{k^3 h^2}{6}\right) \tag{13.95}$$

$$= k\left[\frac{x}{t} - (gh)^{1/2}\right] + (gh)^{1/2}\frac{h^2 k^3}{6}. \tag{13.96}$$

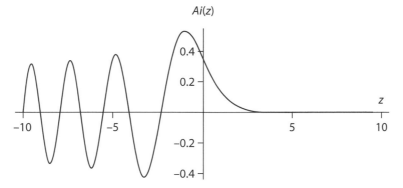

Figure 13.4 The graph of Ai(z) for real argument z.

Note that the term in square brackets is very small near the leading edge of the wave. From (13.35), using only the first integral, since $\tilde{\xi}_0^e(k)$ is real, we can write

$$\xi(x,t) = \frac{1}{2\pi} \int_0^\infty \tilde{\xi}_0^e(k) \cos(kx - \omega t)dk + O(t^{-1}) \tag{13.97}$$

$$\approx \frac{\tilde{\xi}_0^e(0)}{2\pi} \int_0^\infty \cos\left[(x - (gh)^{1/2}t) + \left(\frac{(gh)^{1/2}h^2t}{6}\right) k^3 \right] dk \tag{13.98}$$

to leading order. Now we make a change of variable to z, where

$$z^3 = \frac{2[x - (gh)^{1/2}t]^3}{(gh)^{1/2}h^2t}, \tag{13.99}$$

and

$$z\alpha = k[x - (gh)^{1/2}t]. \tag{13.100}$$

Then to leading order $\xi(x,t)$ becomes, in terms of the Airy function Ai(z)

$$\text{Ai}(z) = \frac{1}{\pi} \int_0^\infty \cos\left(z\alpha + \frac{\alpha^3}{3}\right) d\alpha, \tag{13.101}$$

$$\xi(x,t) \approx \frac{\tilde{\xi}_0^e(0)}{[4(gh)^{1/2}h^2t]^{1/3}} \text{Ai}(z), \tag{13.102}$$

or

$$\xi(x,t) \approx \left(\frac{1}{4(gh)^{1/2}h^2t}\right)^{1/3} \tilde{\xi}_0^e(0) \, \text{Ai}\left\{ \left(\frac{2}{(gh)^{1/2}h^2t}\right)^{1/3} [x - (gh)^{1/2}t] \right\}. \tag{13.103}$$

The Airy integral Ai(z) is oscillatory for $z < 0$ and decays exponentially for $z > 0$ (see Figure 13.4).

Thus for fixed t, the argument of the Airy function in equation (13.103) is proportional to $x - (gh)^{1/2}t$, which is the distance from the wavefront defined by

$x = (gh)^{1/2}t$. At a fixed instant the amplitude is small ahead of the front, and the highest peak some distance behind. From (13.103) it is clear that the amplitude decreases like $t^{-1/3}$, but the same factor also appears in the argument of the Airy function, so the waveform is stretched out in time. As seen earlier, the rest of the wavetrain decays as $t^{-1/2}$, so the head lives longer than the rest of the body [43]! Furthermore, the amplitude of leading wave is proportional to $\tilde{\xi}_0^e(0)$, but recall that

$$\tilde{\xi}_0^e(k) = \int_{-\infty}^{\infty} e^{-ikx} \xi_0^e(x) dx \Rightarrow \tilde{\xi}_0^e(0) = \int_{-\infty}^{\infty} \xi_0^e(x) dx,$$

and this is exactly the area of initial displacement in this two-dimensional model.

13.5 SPECIAL CASE 2: TSUNAMI GENERATION BY A DISPLACEMENT OF THE SEAFLOOR

In this section we model the generation of a tsunami by an idealized earthquake—a tilting of the seafloor. It is commonly reported that the arrival of a tsunami at a coast is often preceded by the withdrawal of water from the beaches (the sea-god breathes in!), and the first crest may not be the largest. To try and replicate these features we assume there is no initial disturbance on the free surface, that is,

$$\xi(x, 0) = \Phi(x, 0, 0) = P_a(x, 0, t) = 0 \Rightarrow F = 0, \tag{13.104}$$

(see (13.28)). On $z = -h$, the seafloor displacement $H(x, t)$ is prescribed, so $W = \partial H/\partial t$ is known. From equation (13.25)

$$\hat{\Phi} = \frac{\hat{W}}{k \cosh kh} \left(\frac{s^2 \sinh kz - gk \cosh kz}{s^2 + gk \tanh kh} \right), \tag{13.105}$$

and the free-surface displacement is (see (13.27))

$$\xi(x, t) = \frac{-1}{4\pi^2 i} \int_{-\infty}^{\infty} \frac{e^{ikx}}{\cosh kh} dk \int_{\Gamma_1} \frac{s\hat{W} e^{st}}{s^2 + \omega^2} ds, \tag{13.106}$$

where $\omega = (gk \tanh kh)^{1/2}$ as before. To incorporate seafloor motion corresponding to a sudden displacement, we set

$$H(x, 0^-) = 0, \quad H(x, 0^+) = H_0(x), \tag{13.107}$$

and represent the vertical seafloor "speed" by

$$\frac{\partial \Phi}{\partial z} = W(x, t) = H_0(x)\delta(t) \Rightarrow \hat{W}(k, s) = \tilde{H}_0(k). \tag{13.108}$$

Question: Do you see why?
As before,

$$\frac{1}{2\pi i} \int_{\Gamma_1} \frac{s e^{st}}{s^2 + \omega^2} ds = \cos(\omega t), \quad t > 0, \tag{13.109}$$

so (13.106) can be written as

$$\xi(x, t) = \frac{1}{4\pi} \int_{-\infty}^{\infty} \frac{\tilde{H}_0(k)}{\cosh kh} \left[e^{i(kx+\omega t)} + e^{i(kx-\omega t)} \right] dk. \tag{13.110}$$

As before, we may write a function as a sum of its even and odd parts. In particular,

$$H_0(x) = H_0^e(x) + H_0^o(x); \tag{13.111}$$

$H_0^e(x)$ has properties similar to Special Case I for $\xi_0^e(x)$ apart from the $(\cosh kh)^{-1}$ factor, which limits the influence of short waves. Accordingly we focus attention on the odd part. Let $H_0^o = dB(x)/dx$ so that

$$\widetilde{H}_0^o(k) = ik\widetilde{B}(k). \tag{13.112}$$

Now the Fourier transform of an odd real function is imaginary and odd (**Question:** Why?), hence $\widetilde{B}(k)$ is real and even in k. Therefore

$$\xi(x,t) = \frac{1}{2\pi} \int_{-\infty}^{\infty} \frac{ik\widetilde{B}(k)e^{ikx}}{\cosh(kh)} \frac{1}{2} \left(e^{i\omega t} + e^{-i\omega t} \right) dk \tag{13.113}$$

$$= \frac{1}{2\pi} \frac{d}{dx} \left[\mathrm{Re} \int_0^{\infty} \frac{\widetilde{B}(k)e^{ikx}}{\cosh(kh)} \left(e^{i\omega t} + e^{-i\omega t} \right) dk \right]. \tag{13.114}$$

For large t and away from the leading edge of the waves, the method of stationary phase can again be used, but there is a difference in the dependence on t as we shall see shortly. In the neighborhood of the leading wave for $x > 0$, the second integral dominates (as before), and the most important contribution comes from waves for which $k \approx 0$ (i.e., very long waves). If we examine the term in square brackets in (13.114), we find

$$\mathrm{Re} \int_0^{\infty} \frac{\widetilde{B}(k)e^{i(kx-\omega t)}}{\cosh(kh)} dk \approx \mathrm{Re}\ \widetilde{B}(0) \int_0^{\infty} \exp i \left(k\frac{x}{t} - \omega(k) \right) t\, dk, \tag{13.115}$$

so as with (13.103),

$$\mathrm{Re} \int_0^{\infty} \frac{\widetilde{B}(k)e^{i(kx-\omega t)}}{\cosh(kh)} dk \approx \pi \widetilde{B}(0) \left(\frac{2}{(gh)^{1/2} h^2 t} \right)^{1/3}$$

$$\times \mathrm{Ai} \left[\left(\frac{2}{(gh)^{1/2} h^2 t} \right)^{1/3} \left(x - (gh)^{1/2} t \right) \right]. \tag{13.116}$$

Then it follows from (13.114) that

$$\xi(x,t) \approx \frac{1}{2} \widetilde{B}(0) \left(\frac{2}{(gh)^{1/2} h^2 t} \right)^{2/3} \frac{d}{dx} \left(\mathrm{Ai} \left[\left(\frac{2}{(gh)^{1/2} h^2 t} \right)^{1/3} \left(x - (gh)^{1/2} t \right) \right] \right), \tag{13.117}$$

clearly dependent on the *derivative* of the Airy function [112]. A very important feature to note is that now the leading wave attenuates with time as $t^{-2/3}$, much faster than in Special Case I ($\sim t^{-1/3}$), because in tilting, the seafloor movement is, in a sense, half positive and half negative (recall that we focused on the odd function $H_0^o(x)$ here), thereby reducing the net effect. The graph of $\mathrm{Ai}'(z)$ is shown in Figure 13.5, where it can be seen that the first crest is smaller than the ones behind it.

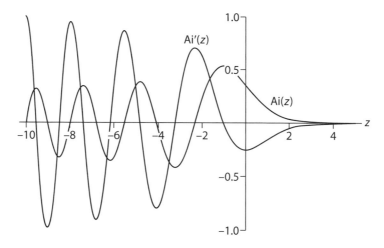

Figure 13.5 The Airy function Ai(z) and its derivative.

Suppose also for simplicity that

$$H_0^o(x) = -\sin x, \quad |x| < \frac{\pi}{2}; \quad \text{and} \tag{13.118}$$

$$H_0^o(x) = 0, \quad |x| \geq \frac{\pi}{2}. \tag{13.119}$$

Then in the interval $(-\pi/2, \pi/2)$, $B(x) = \cos x$ (ignoring the integration constant), and

$$\widetilde{B}(k) = \frac{1}{2}(\pi + \sin k\pi), \tag{13.120}$$

so $\widetilde{B}(0) > 0$. With this choice of $H_0^o(x)$ the seafloor tilts to the right, and wave front propagating to right is led by a trough. Also, subsequent crests increase amplitude (so for $x < 0$, the wavefront is led by a crest). If ground tilts to left (e.g., $H_0^o(x) = \sin x$, $|x| < \frac{\pi}{2}$, etc.), $\widetilde{B}(0) < 0$, and the right-going wave is led by an elevation.

Chapter Fourteen

Atmospheric Waves

Drop a stone into a pool of water. The spreading ripples are gravity waves. The waves occur between any stable layers of fluids of different density. When the fluid boundary is disturbed, buoyancy forces try to restore the equilibrium. The fluid returns to its original shape, overshoots and oscillations then set in which propagate as waves. Gravity or buoyancy is the restoring force hence the term—gravity waves. These waves (internal gravity or buoyancy waves) abound in the stable density layering of the upper atmosphere. Their effects are visibly manifest in the curls of the stratosphere's nacreous clouds, in the moving skein-like and billow patterns of the mesosphere's noctilucent clouds and in the slowly shifting bands of the thermosphere's airglow. What triggers them? The "stones into the pond" are disturbances far below in the troposphere, for example, wind flow over mountain ranges and violent thunderstorms. Jet stream shear and solar radiation are other sources. An initial small amplitude at the tropopause increases with height until the waves break in the mesosphere and lower thermosphere. Their wavelengths can range up to thousands of kilometres. Their periods range from a few minutes to days. They do more than give clouds interesting shapes. They are vital in their role of transferring energy, momentum and chemical species between the different atmospheric layers and in the subsequent influence on upper atmosphere winds, turbulence, temperature and chemistry.

[263]

Waves provide a common mathematical framework which unites a wide variety of geophysical phenomena, and indeed unites many of the broader range of physical disciplines as well. Early studies of elastic, hydrodynamic, acoustic, seismic, optical, and electron waves helped to establish a familiarity with wave behavior that sped the development of a vast literature on ionospheric radio propagation and, more recently, on plasma waves in general. These topics in turn have had their influence on the older studies, by providing a deeper physical understanding and an expanded range of mathematical techniques that were born in response to the challenging new situations encountered.

[144]

Wave motion arises when "parcels" of fluid are displaced from equilibrium by restoring forces, which can include compressibility, gravity (or buoyancy), inertia, rotation, or electromagnetic forces, and combinations of these. In this chapter we shall examine aspects of such forces with the exception of electromagnetic forces acting over large scales (which can arise, for example because of the earth's magnetic field). It is not generally realized how much wave motion–related phenomena occur in the atmosphere, but even a casual observer of the skies can become enthralled by the patterns to be seen above them [216], or in images of other planets in the solar system that possess atmospheres.

14.1 GOVERNING LINEARIZED EQUATIONS

Relative to a frame of reference fixed on the earth, rotating with angular velocity Ω, the momentum equation is, in the absence of any other external forces [34],

$$\frac{d\mathbf{u}}{dt} = -2\Omega \times \mathbf{u} - \frac{1}{\rho}\nabla p + \mathbf{g}, \tag{14.1}$$

where $\mathbf{u} = \langle u, v, w \rangle$ is the velocity vector field (including any zonal flow; see below). The vector $\mathbf{g}$ is the frame-based accelation due to gravity and incorporating the centripetal acceleration $\Omega \times (\Omega \times \mathbf{r})$. The term

$$2\Omega \times \mathbf{u} = f_0 \mathbf{k} \times \mathbf{u}, \tag{14.2}$$

where $f_0 = 2|\Omega| \sin\phi$, is the *Coriolis parameter*, and ϕ is the latitude. The differential operator

$$\frac{d}{dt} = \frac{\partial}{\partial t} + \mathbf{U} \cdot \nabla = \frac{\partial}{\partial t} + U\frac{\partial}{\partial x}, \tag{14.3}$$

is the Lagrangian time derivative in the presence of a uniform zonal flow $\mathbf{U} = \langle \widehat{U}, 0, 0 \rangle$. The continuity equation is

$$\frac{d\rho}{dt} + \rho\nabla \cdot \mathbf{u} = 0. \tag{14.4}$$

For adiabatic motion the first law of thermodynamics is

$$p\rho^{-\gamma} = \text{constant}, \tag{14.5}$$

and expressed in the form of a differential equation, it is

$$\frac{dp}{dt} = c^2\frac{d\rho}{dt}, \tag{14.6}$$

where c is the speed of sound in the medium. The unperturbed (or equilibrium) state, characterized by the subscript zero, consists of the equations of *geostrophic* and *hydrostatic* equilibrium,

$$f_0 U_0 = -\frac{1}{\rho_0}\frac{\partial p_0}{\partial y}, \tag{14.7}$$

and

$$\frac{\partial p_0}{\partial z} = -g\rho_0, \tag{14.8}$$

respectively (with $g = |\mathbf{g}|$). The goal is to establish a dispersion relation from either a single ordinary differential equation for one of the field variables (the vertical component of velocity here, but any other will do) or for a set of them. Both approaches will be elucidated here. Eliminating the density perturbation from the system of equations (14.1), (14.4), and (14.6), we obtain the set [192]:

$$\rho_0 \frac{du}{dt} - \rho_0 f_0 v + \frac{\partial p}{\partial x} = 0; \tag{14.9}$$

$$\rho_0 \frac{dv}{dt} + \rho_0 f_0 u + \frac{\partial p}{\partial y} = 0; \tag{14.10}$$

$$\rho_0 \left(\frac{d^2}{dt^2} + N^2 \right) w + \frac{d}{dt} \left[\left(\frac{\partial}{\partial z} + \frac{g}{c^2} \right) p \right] = 0; \tag{14.11}$$

$$\frac{dp}{dt} - g\rho_0 w + c^2 \rho_0 \nabla \cdot \mathbf{u} = 0. \tag{14.12}$$

The buoyancy (or Brunt-Väisälä) frequency is N, where for an atmosphere in hydrostatic equilibrium,

$$N^2 = -g \left(\frac{g}{c^2} + \frac{1}{\rho_0} \frac{\partial \rho_0}{\partial z} \right). \tag{14.13}$$

Several of the coefficients in the above system of equations are clearly dependent on the equilibrium density ρ_0, and because the dispersion relation sought is a consequence of constant coefficient equations, the four dependent variables u, v, w, and p can be transformed to achieve this. In particular, for an *isothermal atmosphere*, N, c, and the density scale-height

$$H = -\frac{1}{\rho_0} \frac{\partial \rho_0}{\partial z}, \tag{14.14}$$

are all constant (and $\rho_0(z) \propto \exp(-z/H)$), so if we define

$$u(x, y, z, t) = \rho_0^{-1/2} U(x, y, z, t), \tag{14.15}$$

with corresponding expressions for the remaining variables, the set (14.9)–(14.12) becomes

$$\frac{dU}{dt} - f_0 V + \frac{\partial P}{\partial x} = 0; \tag{14.16}$$

$$\frac{dV}{dt} + f_0 U + \frac{\partial P}{\partial y} = 0; \tag{14.17}$$

$$\left(\frac{d^2}{dt^2} + N^2 \right) W + \frac{d}{dt} \left[\left(\frac{\partial}{\partial z} + E \right) P \right] = 0; \tag{14.18}$$

$$\frac{1}{c^2} \frac{dP}{dt} - EW + \nabla \cdot \mathbf{U} = 0, \tag{14.19}$$

where $\mathbf{U} = \langle U, V, W \rangle$, and E (or sometimes Γ) is known as the *Eckart coefficient*, defined by

$$E = \frac{1}{2\rho_0} \frac{d\rho_0}{dz} + \frac{g}{c^2} = -\frac{1}{2H} + \frac{g}{c^2} = \frac{1}{2g} \left(\frac{g^2}{c^2} - N^2 \right). \qquad (14.20)$$

It is a measure of the influence of both density stratification and compressibility on the inertial terms, whereas N is simply the buoyancy frequency for a fluid element in a stably stratified atmosphere. For an isothermal atmosphere, $N \approx 1.9 \times 10^{-2} \text{s}^{-1}$, and $E \approx 2.7 \times 10^{-5} \text{ m}^{-1}$.

If we a seek plane wave solutions for which all the field variables in the set (14.16)–(14.19) are proportional to $\exp[i(\mathbf{k} \cdot \mathbf{r} - \omega t)]$, where (with a trivial notational ambiguity) $\mathbf{k} = \langle k, l, m \rangle$, then in terms of constant amplitudes U_0, V_0, and so forth,

$$-i\omega U_0 - f_0 V_0 + ik P_0 = 0; \qquad (14.21)$$

$$-i\omega V_0 + f_0 U_0 + il P_0 = 0; \qquad (14.22)$$

$$\left(N^2 - \omega^2 \right) W_0 - i\omega \left(im + E \right) P_0 = 0; \qquad (14.23)$$

$$-\frac{i\omega}{c^2} P_0 + (im - E) W_0 + ik U_0 + il V_0 = 0. \qquad (14.24)$$

This system of algebraic equations has a nontrivial solution if the determinant of the coefficient matrix is nonzero, which implies that

$$m^2 = \kappa^2 \left(\frac{N^2 - \omega^2}{\omega^2 - f_0^2} \right) + \frac{\omega^2}{c^2} - \left(\frac{N^2}{c^2} + E^2 \right), \qquad (14.25)$$

where κ^2 is defined in (14.27).

This form is used in much of the literature on dynamical meteorology (e.g., [191], [192]). In the presence of a uniform zonal flow $\mathbf{U} = \langle \widehat{U}, 0, 0 \rangle$, we have $\omega \to \omega - k\widehat{U}$ in this equation.

The governing equation for the density-modified vertical velocity perturbation $W(z)$ is

$$\frac{d^2 W}{dz^2} + \left\{ \kappa^2 \left(\frac{N^2 - \omega^2}{\omega^2 - f_0^2} \right) + \left(\frac{\omega^2}{c^2} - \frac{1}{4H^2} \right) \right\} W = 0, \qquad (14.26)$$

where

$$\kappa^2 = k^2 + l^2. \qquad (14.27)$$

If $W(z) \propto \exp(imz)$, then the dispersion relation corresponding to (14.25) is

$$m^2 = \kappa^2 \left(\frac{N^2 - \omega^2}{\omega^2 - f_0^2} \right) + \left(\frac{\omega^2}{c^2} - \frac{1}{4H^2} \right), \qquad (14.28)$$

or equivalently,

$$\omega^4 - \omega^2 c^2 \left(\kappa^2 + M^2 + \frac{f_0^2}{c^2} \right) + c^2 \left(M^2 f_0^2 + N^2 \kappa^2 \right) = 0, \qquad (14.29)$$

where

$$M^2 = m^2 + \frac{1}{4H^2}. \tag{14.30}$$

As a quadratic equation in ω^2, (14.29) has the roots

$$\omega_{\pm}^2 = \frac{1}{2}c^2 \left(\kappa^2 + M^2 + \frac{f_0^2}{c^2} \right) \left\{ 1 \pm \left[1 - \frac{4c^2 \left(M^2 f_0^2 + N^2 \kappa^2 \right)}{c^2 \left(\kappa^2 + M^2 \right) + f_0^2} \right]^{1/2} \right\}. \tag{14.31}$$

The "+" modes represent gravity-modified acoustic waves propagating in opposite directions (as determined by the signs of ω), while the "−" modes represent compressibility-modified gravity waves, also propagating in opposite directions (see Figure (14.1)). If there is an unsheared zonal flow U, then ω is replaced by $\Omega \equiv \omega - k\widehat{U}$.

There are several special cases that can be examined based on equation (14.25) in particular, but first three basic categories of undamped waves should be noted. If $m^2 > 0$, the waves propagate vertically (as well as horizontally if $\kappa = \sqrt{k^2 + l^2} \neq 0$); if $m^2 = 0$, they propagate purely horizontally, and if $m^2 < 0$, they propagate horizontally; and decay exponentially in the vertical direction (retaining only the physically meaningful root $m = i\,|m|$). In this case they are variously classified as surface, trapped, or evanescent waves.

Note also from equations (14.28) and (14.25) that for $m = 0$,

$$\lim_{\kappa \to 0} \omega^2 = \frac{c^2}{4H^2} = N^2 + c^2 E^2 \equiv \varpi_a^2, \tag{14.32}$$

where $\varpi_a > N$ is the *acoustic cutoff frequency*.

Acoustic Waves

If we turn off the effects of gravity, density stratification, and rotation (via the Coriolis force), then the only restoring force that remains is compressibility, and we expect to recover the dispersion relation for acoustic waves propagating in a spherically symmetric medium. This is indeed the case; by respectively setting E, N, and f_0 to zero (and in the absence of a zonal flow), equation (14.25) simplifies to

$$K^2 \equiv \kappa^2 + m^2 = \frac{\omega^2}{c^2}. \tag{14.33}$$

Gravity Waves

In this case we filter out both acoustic and inertial waves by allowing $c \to \infty$ and setting $f_0 = 0$, whence

$$m^2 = \kappa^2 \left(\frac{N^2}{\omega^2} - 1 \right) - E^2, \tag{14.34}$$

that is,

$$\omega^2 = \frac{N^2 \kappa^2}{\kappa^2 + m^2 + E^2} = \frac{N^2 \kappa^2}{K^2 + E^2}. \tag{14.35}$$

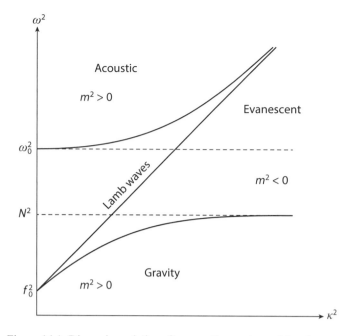

Figure 14.1 Dispersion relations for acoustic-gravity and Lamb waves.

Note that

$$\lim_{\kappa \to 0} \omega^2 = 0; \tag{14.36}$$

$$\lim_{\kappa \to \infty} \omega^2 = N^2. \tag{14.37}$$

The above results correspond to long and short wavelength limits, respectively.

Lamb Waves

For this mode, the vertical component of perturbation velocity $W_0 = 0$, and hence from equation (14.23), $im + E = 0$ (or $m = iE$). Since $E > 0$, these waves diminish exponentially with height; they are evanescent. Setting $W_0 = 0$ in the set (14.21)–(14.24) results in the dispersion relation for Lamb waves:

$$\omega^2 = f_0^2 + c^2 \kappa^2. \tag{14.38}$$

(See Figure 14.1.) Clearly,

$$\lim_{\kappa \to 0} \omega^2 = f_0^2, \tag{14.39}$$

and as $\kappa \to \infty$, $\omega^2 \sim c^2 \kappa^2$. Thus in the short wavelength limit they propagate horizontally as acoustic waves, because the Coriolis force has negligible influence on them.

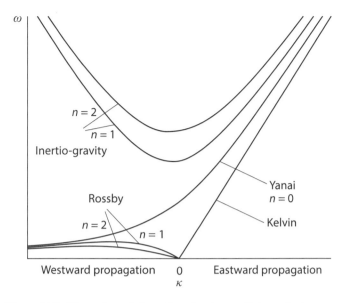

Figure 14.2 Dispersion relations for other types of atmospheric wave.

Rossby Waves

These waves have planetary length scales and so the *variation* of the Coriolis parameter f with latitude y becomes important (see Figure 14.2). If $f(y)$ is expressed as a linear Taylor polynomial, then the so-called β-plane approximation arises, that is,

$$f = f_0 + \beta y. \qquad (14.40)$$

It is customary to filter out the higher-frequency acoustic waves by letting $c^2 \to \infty$ and also neglecting the term $d^2 W/dt^2$ compared with $N^2 W$ in equations (14.19) and (14.18), respectively. For mid-latitude planetary-scale disturbances, the so-called geostrophic wind equations are approximately satisfied [190], namely,

$$U = -\frac{1}{f_0}\frac{\partial P}{\partial y} \equiv -\frac{\partial \eta}{\partial y}, \qquad (14.41)$$

$$V = \frac{1}{f_0}\frac{\partial P}{\partial x} \equiv \frac{\partial \eta}{\partial x}, \qquad (14.42)$$

where $\eta = P/f_0$. After some algebraic manipulation we arrive at the following coupled pair of partial differential equations, with ∇_h^2 referring to the "horizontal" Laplacian operator:

$$\frac{d}{dt}\left[\nabla_h^2 \eta + \frac{f_0^2}{N^2}\left(\frac{\partial^2 \eta}{\partial z^2} - E^2 \eta\right)\right] + \beta \frac{\partial \eta}{\partial x} = 0, \qquad (14.43)$$

and

$$N^2 W + f_0 \frac{d}{dt}\left(\frac{\partial \eta}{\partial z} + E\eta\right) = 0. \qquad (14.44)$$

Once more we seek plane wave solutions proportional to $\exp[i(\mathbf{k} \cdot \mathbf{r} - \omega t)]$ to obtain (in the presence of a uniform zonal flow $\widehat{U}$)

$$\Omega \equiv \omega - k\widehat{U} = -\frac{\beta k}{\kappa^2 + f_0^2 \left(m^2 + E^2\right)/N^2}, \tag{14.45}$$

or in terms of the vertical wavenumber m,

$$m^2 = -\left(\frac{N}{f_0}\right)^2 \left[\frac{\beta}{c_p - \widehat{U}} + \left(\frac{f_0 E}{N}\right)^2 + \kappa^2\right], \tag{14.46}$$

where $c_p = \omega/k$ is the x-direction phase speed of the wave relative to the mean flow. If the motion is purely horizontal, $m = 0$, so

$$c_p = \widehat{U} - \frac{\beta}{(f_0 E/N)^2 + \kappa^2}. \tag{14.47}$$

For stationary waves (forced by the topography), that is, such that $c_p = 0$, vertical propagation is possible if

$$m^2 = \left(\frac{N}{f_0}\right)^2 \left[\frac{\beta}{\widehat{U}} - \left(\frac{f_0 E}{N}\right)^2 - \kappa^2\right] > 0. \tag{14.48}$$

This forces the following inequality to be satisfied by $\widehat{U}$:

$$0 < \widehat{U} < \frac{\beta}{(f_0 E/N)^2 + \kappa^2} \equiv \widehat{U}_c, \tag{14.49}$$

and it means that Rossby waves cannot propagate vertically if the zonal wind is (i) easterly or (ii) is westerly and exceeds $\widehat{U}_c$.

Kelvin Waves

These are equatorial waves that propagate in the stratosphere in a zone of latitude $\pm 20°$ about the equator. They have very long periods (12–20 days) and one wavelength per circle of latitude! Their vertical wavelengths are in the range 10–12 km, and they travel eastward with phase speeds of about 30 m/s [190]. Because they are trapped close to the equator, the zero-order Coriolis parameter f_0 is neglected while retaining the β-plane approximation. By filtering out acoustic waves as before ($c^2 \to \infty$), the resulting system of equations is now

$$\frac{dU}{dt} - \beta y V + \frac{\partial P}{\partial x} = 0; \tag{14.50}$$

$$\frac{dV}{dt} + \beta y U + \frac{\partial P}{\partial y} = 0; \tag{14.51}$$

$$N^2 W + \frac{d}{dt}\left[\left(\frac{\partial}{\partial z} + E\right)P\right] = 0; \tag{14.52}$$

$$-EW + \nabla \cdot \mathbf{U} = 0. \tag{14.53}$$

Since the waves essentially propagate in an equatorial plane, we now seek plane wave solutions as before except that the meridional component of velocity V is zero, and we set $l = 0$ so $\kappa = k$. Furthermore, y is the distance (northward) from the equator, so we write the reduced set $\{U, W, P\}$ as follows:

$$\{U(x, y, z, t), W(x, y, z, t), P(x, y, z, t)\}$$

$$= \{u(y), w(y), p(y)\} \exp[i(kx + mz - \omega t)]. \tag{14.54}$$

From equations (14.50) and (14.51) we obtain

$$\frac{du}{dy} + \left(\frac{k\beta y}{\omega}\right) u = 0, \tag{14.55}$$

that is,

$$u(y) = u_0 \exp\left(-\frac{k\beta}{2\omega} y^2\right), \tag{14.56}$$

for some constant u_0, so it is clear that $|u(y)|$ vanishes exponentially away from the equator if $c_p > 0$ (or $c_p > \widehat{U}$ in the presence of a zonal flow), in which case they propagate from west to east (i.e., in the positive x direction). The e-folding equatorial decay distance is

$$y_e = \left|\frac{2\omega}{k\beta}\right|^{1/2}, \tag{14.57}$$

if $\widehat{U} = 0$. Otherwise,

$$y_e = \left|\frac{2\Omega}{k\beta}\right|^{1/2}. \tag{14.58}$$

Furthermore it follows that

$$p(y) = \frac{\omega}{k} u(y). \tag{14.59}$$

The dispersion relation is readily found from equations (14.50), (14.52), and (14.53) to be

$$\Omega^2 = \left(\frac{N^2 k^2}{m^2 + E^2}\right), \tag{14.60}$$

from which

$$c_p = \widehat{U} + \left(\frac{N^2}{m^2 + E^2}\right)^{1/2}. \tag{14.61}$$

Observations indicate in general that $m^2 \gg E^2$, in which case this expression reduces to that for internal gravity waves (for $k^2 \ll m^2$). (See Figure 14.2.)

Rossby-Gravity Waves

These waves have periods of 4–5 days and typically consist of 4 wavelengths per circle of latitude, with vertical wavelengths of about 6 km; they travel westward with phase speeds of approximately 20 m/s. Like Kelvin waves they are trapped in a zone about the equator, but unlike those waves, the velocity perturbation $v(y) \neq 0$, and it will be important to find the governing ordinary differential equation in terms of the independent variable y. To that end, applying the exponential expression $\exp[i(kx + mz - \omega t)]$ to the set of equations (14.50)–(14.53) yields the reduced form:

$$-i\omega u - \beta y v + ikp = 0; \tag{14.62}$$

$$-i\omega v + \beta y u + \frac{dp}{dy} = 0; \tag{14.63}$$

$$N^2 w + \omega mp - i\omega E p = 0; \tag{14.64}$$

$$iku + \frac{dv}{dy} + imw - Ew = 0. \tag{14.65}$$

SUBSET I: HORIZONTAL PROPAGATION ($m = 0$)

Setting $m = 0$ in equations (14.64) and (14.65) and eliminating all the dependent variables but v yields the following differential equation:

$$\frac{d^2 v}{dy^2} + \left\{ \left[\left(\frac{\omega E}{N} \right)^2 - k^2 - \frac{k\beta}{\omega} \right] - \left(\frac{\beta E}{N} y \right)^2 \right\} v = 0. \tag{14.66}$$

Exercise: Establish this result.

Observationally it is known that the waves are equatorially trapped, so it is natural to seek solutions of this equation that vanish in the limit $|y| \to \infty$. Fortunately, this equation is well known; it is identical in form to the time-independent Schrödinger equation for a harmonic oscillator potential. In standard form we may write this as

$$\frac{d^2 v}{dy^2} + \left(\mu^2 - \alpha^2 y^2 \right) v = 0, \tag{14.67}$$

and if we define

$$v(y) = \theta(y) \exp\left(-\frac{\alpha}{2} y^2 \right), \tag{14.68}$$

then $\theta(y)$ satisfies the following form of Hermite's differential equation:

$$\frac{d^2 \theta}{dy^2} - 2\alpha y \frac{d\theta}{dy} + \left(\mu^2 - \alpha \right) \theta = 0. \tag{14.69}$$

A standard expansion of θ as an ordinary power series shows that convergence is possible if and only if $\theta(y)$ is a polynomial function. The requirement for this is

$$\mu^2 = \alpha(2n + 1), \quad n = 0, 1, 2, \dots. \tag{14.70}$$

In terms of the original parameters this condition is

$$\left(\frac{N}{\beta E} \right) \left[\left(\frac{\omega E}{N} \right)^2 - k^2 - \frac{k\beta}{\omega} \right] = 2n + 1, \tag{14.71}$$

and in terms of the variable $\xi = (\beta E/N)^{1/2}\, y$, the solutions for $n = 0, 1$, and 2 are [190]

$$v(\xi) = v_0 \exp\left(-\xi^2/2\right); \tag{14.72}$$

$$v(\xi) = 2v_1 \xi \exp\left(-\xi^2/2\right); \tag{14.73}$$

$$v(\xi) = 2v_2 \left(2\xi^2 - 1\right) \exp\left(-\xi^2/2\right), \tag{14.74}$$

respectively, where the v_i, $i = 0, 1, 2$, are constants. The value of n signifies the number of nodes in the velocity profile, and in general.

ALTERNATIVE REPRESENTATIONS OF HERMITE'S EQUATION

If the independent variable in equation (14.69) is changed to $p = \alpha y^2$, then $\theta(p)$ satisfies the confluent hypergeometric equation

$$p\frac{d^2\theta}{dp^2} + \left(\frac{1}{2} - p\right)\frac{d\theta}{dp} + a\theta = 0, \tag{14.75}$$

with

$$a = \frac{1}{4} - \frac{\mu^2}{4\alpha}. \tag{14.76}$$

FUNDAMENTAL MODE

Equation (14.71) factors neatly for the fundamental mode $(n = 0)$, so that

$$\left(\frac{\omega E}{N} - k - \frac{\beta}{\omega}\right)\left(\frac{\omega E}{N} + k\right) = 0. \tag{14.77}$$

Although in principle there are three real roots for this equation, one can be ruled out immediately, because it corresponds to an inadmissable type of wave. Thus if the second factor is zero, we find

$$\frac{dv}{dy} - \left(\frac{k\beta y}{\omega}\right)v = 0, \tag{14.78}$$

or

$$v(y) = v_0 \exp\left(\frac{k\beta}{2\omega}y^2\right), \tag{14.79}$$

which is obviously unacceptable for physical reasons. From the remaining term, we have

$$\omega = \frac{kN}{2E}\left[1 \pm \left(1 + \frac{4\beta E}{Nk^2}\right)\right]^{1/2}. \tag{14.80}$$

The positive root corresponds to an eastward-propagating so-called equatorial *inertia-gravity wave* (meaning that restoring forces from both rotation and buoyancy are important) [190]. The negative root corresponds to waves of a more schizophrenic nature: for long length scales (zonal wavenumber $k \to 0$), the waves resemble inertia-gravity waves, whereas for scales of the order of 1,000 km, they behave more like inertial Rossby waves.

Exercise: Establish this result from the set (14.62)–(14.65).

SUBSET II: VERTICAL PROPAGATION

In Chapter 12 of [190] a dispersion relation for vertically propagating (linear) planetary waves is derived. The derivation involves meteorological terminology outside the scope of this chapter, but the requirement for vertical propagation is familiar: for waves with z behavior of the form $\exp(imz)$,

$$m^2 = -\left(\frac{N}{f_0}\right)^2 \left[\frac{k\beta}{\Omega} + \kappa^2\right] - \frac{1}{4H^2} > 0. \tag{14.81}$$

At first glance this may seem impossible to satisfy, but recall that the frequency $\Omega = \omega - k\widehat{U}$ (i.e., ω Doppler shifted by a zonal flow $\widehat{U}$) may be negative. For $m^2 > 0$ it is necessary that

$$\widehat{U} - c < \beta \left[\kappa^2 + \left(\frac{f_0}{4HN}\right)^2\right]^{-1}, \tag{14.82}$$

and for stationary waves in particular (i.e., $c = 0$)

$$\widehat{U} < \widehat{U}_c = \beta \left[\kappa^2 + \left(\frac{f_0}{4HN}\right)^2\right]^{-1}, \tag{14.83}$$

where $\widehat{U}_c$ is the Rossby critical velocity; the westerly winds must be weaker than this critical value, which depends on rotation, stratification, buoyancy, and the horizontal length scale. Since during the summer the mean zonal stratospheric winds are easterly, such stationary waves are trapped vertically except during the winter months.

YANAI WAVES

These are vertically propagating mixed Rossby-gravity waves corresponding to the $n = 0$ case in equation (14.72), and are sometimes called Yanai waves [191]. As may be seen from the dispersion curves in Figure 14.2, when propagating westward with short horizontal wavelengths they are closer in character to Rossby waves, whereas for short waves moving eastward their dispersion curves are asymptotic to those of Kelvin waves and $n = 1$ inertia-gravity waves as $\omega \to \infty$.

> A Yanai wave is an equatorially trapped planetary wave. Its meridional structure is a Gaussian centered at the equator in meridional velocity and has two opposite extrema off the equator in zonal velocity. Physically, it behaves like a short Rossby wave when the phase is propagating westward and like an inertia-gravity wave when the phase is propagating eastward. This dual behavior is reflected in the other name of the Yanai wave: a mixed Rossby-gravity wave. The most important Yanai waves observed in the ocean are found at the intraseasonal timescale. In the atmosphere, intraseasonal Yanai waves play an important role in the Quasi-Biennial Oscillation. [193]

14.2 A MATHEMATICAL MODEL OF LEE/MOUNTAIN WAVES OVER AN ISOLATED MOUNTAIN RIDGE

These waves are often referred to as terrain-generated gravity waves, and examples of such terrain include not only hills, ridges, and mountains but also canyons and valleys [194]. Indeed, any kind of significant change in topography can act as a catalyst for such waves. They are of considerable importance in the atmospheric sciences, because they can deposit energy and momentum in the middle and upper atmosphere and significantly affect global circulation. These waves are stationary relative to the ground, and so appear to be attached to the particular topographic feature inducing them, but they can grow as they move upward (and even "break"—this is a significant source of clear-air turbulence). Nevertheless, relative to the background flow they propagate upwind with the same phase speed as the wind (which we take to be in the positive x-direction), that is, $c = -U_0$, where U_0 is the component of the background flow directed over the terrain (e.g., perpendicular to the ridge line if that is the obstacle). We have already introduced the Doppler-shifted frequency $\Omega = \omega - kU_0$ so for a wave that is stationary relative to the ground, $\omega = 0$, and the above result (14.83) for c follows directly.

In this subsection the mathematical approach of both [195] and [194] is adopted and applied to an isolated ridge, but first it is important to distinguish between two types of terrain-generated gravity waves: *lee waves* and *mountain waves*. Lee waves extend downwind from the ridge, propagate horizontally, and are trapped in a duct bounded by the ground below and an upper surface from which they are reflected. They may extend many wavelengths downstream from the ridge and are frequently characterized by evenly spaced bands of clouds in the crests of the waves. From the point of view of upward energy and momentum transfer they are of little importance, unlike their sister waves, that is, mountain waves. (It should be pointed out that this distinction is not always made in the relevant literature; for example, [195] refers to waves induced by mountains as mountain *or* lee waves.) A more precise distinction is given by the American Meteorological Society definition of mountain waves:

> Mountain wave: An atmospheric gravity wave, formed when stable air flow passes over a mountain or mountain barrier. Mountain waves are often standing or nearly so, at least to the extent that upstream environmental conditions (and diurnal forcing) are stationary. Two divisions of mountain wave are recognized, vertically propagating and trapped lee waves. Vertically propagating mountain waves over a barrier may have horizontal wavelengths of many tens of kilometers or more, usually extend upward into the lower stratosphere, and in pure form, tilt upwind with height. They can accompany foehn, chinook, or bora wind conditions. They have the capability to concentrate momentum on the lee slopes, sometimes in structures resembling a hydraulic jump, leading to occasionally violent downslope windstorms. When sufficient moisture is present in the upstream flow, vertically propagating mountain waves produce interesting cloud forms, including altocumulus standing lenticular (ACSL) and other foehn clouds. Intense waves can present a significant hazard to aviation by producing severe or even extreme clear air turbulence. Trapped lee waves generally

have horizontal wavelengths of 5–35km. They occur within or beneath a layer of high static stability and moderate wind speeds at low levels of the troposphere (the lowest 1–5 km) lying beneath a layer of low stability and strong winds in the middle and upper troposphere. These conditions are often diagnosed using a vertical profile of the Scorer parameter, a sharp decrease in midtroposphere indicating conditions favorable to trapped lee wave formation [217].

14.2.1 Basic Equations and Solutions

If the ridge is sufficiently long and narrow that Coriolis force may be neglected (i.e., the Rossby parameter $f_0 \rightarrow \infty$) then we may consider steady-state ($\partial/\partial t = 0$) two-dimensional (x, z) flow over it, including buoyancy but neglecting density stratification. If the mean flow is $U(z)$, the governing equations may be written in a slightly different form [195]:

$$U \frac{\partial u}{\partial x} + w \frac{dU}{dz} + \frac{\partial P}{\partial x} = 0; \tag{14.84}$$

$$U \frac{\partial w}{\partial x} + \frac{\partial P}{\partial z} = r \equiv g \frac{\theta}{\theta_0}; \tag{14.85}$$

$$U \frac{\partial r}{\partial x} + N^2 w = 0; \tag{14.86}$$

$$\frac{\partial u}{\partial x} + \frac{\partial w}{\partial z} = 0. \tag{14.87}$$

The quantity

$$\theta \equiv T \left(\frac{P_0}{P} \right)^{(\gamma - 1)/\gamma}, \tag{14.88}$$

is called the *potential temperature*, where P_0 is a reference pressure (usually 1,000 millibars). Eliminating all the dependent variables in favor of the vertical component of the linear velocity perturbation, it can be shown that

$$\frac{\partial^2 w}{\partial x^2} + \frac{\partial^2 w}{\partial z^2} + S^2 w = 0, \tag{14.89}$$

where

$$S(z) = \left(\frac{N}{U} \right)^2 - \frac{1}{U} \frac{d^2 U}{dz^2} \tag{14.90}$$

is the so-called *Scorer parameter*, of which more anon.

Exercise: Derive (14.89), where S is defined by (14.90).

Before considering an isolated ridge, we examine the effect of an infinite periodic set of ridges(!) with surface corrugations given by

$$z = h(x) \equiv h_0 \cos kx, \tag{14.91}$$

and temporarily we set N and U to be constants (note that as defined, z can be negative; this can of course be remedied by adding a suitable constant). Because of the ground surface rigidity w must vanish there, and since

$$w(x; h(x)) = (U + u)\frac{dh}{dx}, \qquad (14.92)$$

and h is small, the linearized lower boundary condition becomes

$$w(x; 0) = U\frac{dh}{dx} = -Uh_0k \sin kx. \qquad (14.93)$$

Solutions of equation (14.89) can be sought in the form

$$w(x, z) = w_1(z) \cos kx + w_2(z) \sin kx, \qquad (14.94)$$

yielding the ordinary differential equation

$$\frac{d^2 w_i}{dz^2} + \left(S^2 - k^2\right) w_i = 0, \qquad i = 1, 2. \qquad (14.95)$$

For constant N and U the quantity $S^2 - k^2 = m^2$ is a constant, and the general solution of (14.95) may be written as

$$w_i(z) = A_i e^{\mu z} + B_i e^{-\mu z}, \quad k > S; \qquad (14.96)$$

$$w_i(z) = C_i \cos mz + D_i \sin mz, \quad k < S. \qquad (14.97)$$

The coefficients are determined by the upper and lower boundary conditions, and $\mu^2 = k^2 - S^2 = -m^2$. If there is no (partially) reflecting boundary at a finite altitude, the upper boundary condition must apply as $z \to \infty$, so if $k > S$, it is necessary that $A_i = 0$ for obvious physical reasons. To satisfy the lower condition (14.93), $B_1 = 0$ and $B_2 = -Uh_0k$. The complete solution for this case is therefore

$$w(z) = -Uh_0k e^{-\mu z} \sin kx. \qquad (14.98)$$

The waves are evanescent, because they decay exponentially with height.

For $k < S$ we require that the perturbation energy flux (which travels at the group velocity) is upward (since the ridge or mountain acts as an energy source for the waves). In this case it is necessary to examine the time-dependent version of (14.89). The corresponding dispersion relation is

$$\Omega^2 \left(m^2 + k^2\right) = N^2 k^2 + k\Omega U'', \qquad (14.99)$$

where as usual $\Omega = \omega - kU$. For constant flow this yields

$$\omega = kU \pm \frac{Nk}{\left(m^2 + k^2\right)^{1/2}}. \qquad (14.100)$$

$U > 0$ here, so for stationary waves, $\omega = 0$ implies that we take the negative root. The vertical component of the group velocity is therefore

$$\frac{\partial \omega}{\partial m} = \frac{Nkm}{\left(m^2 + k^2\right)^{3/2}} > 0 \qquad (14.101)$$

when $km > 0$. (Note that since the wave velocity and group velocity are orthogonal for these waves, if the latter is upward and downstream, the former is downward and upstream.) For $S > k$ the solution satisfying the lower boundary condition is [195]

$$w(z) = -Uh_0k \sin (kx + mz). \qquad (14.102)$$

14.2.2 An Isolated Ridge

One common choice is a bell-shaped ridge with height profile

$$h(x) = \frac{Hb^2}{x^2 + b^2};$$

another is a Gaussian-shaped ridge with profile

$$h(x) = He^{-x^2/b^2}.$$

Unlike the case for a series of ridges with constant periodicity, an isolated ridge must be represented by a continuous spectrum of wavenumbers, and in general such an integral superposition will allow for destructive interference except near the ridge, where they will combine to form a standing wave [194]. Here we shall work with the Gaussian profile, though both profiles give similar results in general. We define the Fourier x-transform pair for $h(x)$ to be

$$h(x) = \mathcal{F}_x^{-1}\left(\hat{h}(k)\right) = \frac{1}{2\pi} \int_{-\infty}^{\infty} \hat{h}(k)e^{ikx} dk,$$

$$\hat{h}(k) = \mathcal{F}_x(h(x)) = \int_{-\infty}^{\infty} h(x)e^{-ikx} dx = \pi^{1/2} Hbe^{-k^2b^2/4}.$$

Exercise: Prove this latter result.

We similarly define the Fourier transform pair for the linearized vertical wave velocity $w(x, z)$, so that its transform satisfies

$$\frac{d^2\hat{w}}{dz^2} + \left[\frac{N^2}{U^2} - k^2\right]\hat{w} = 0, \tag{14.103}$$

for constant N and U. From the condition

$$w(x, 0) = U\frac{dh}{dx}, \tag{14.104}$$

we find that

$$\hat{w}(k, 0) = ikU\hat{h}(k) = ikU\pi^{1/2} Hbe^{-k^2b^2/4}.$$

Furthermore, if we apply a radiation condition, namely that only upward-propagating wave perturbation energy is permitted, this corresponds to the vertical component of wave speed being negative, so

$$\hat{w}(k, z) = \hat{w}(k, 0)e^{-imz}, \quad z > 0$$

for propagating waves ($m^2 > 0$), where

$$m = \left(\frac{N^2}{U^2} - k^2\right)^{1/2} \in \mathbb{R},$$

and therefore

$$w(x, z) = \frac{1}{2\pi} \int_{-\infty}^{\infty} \hat{w}(k, 0)\, e^{i(kx - mz)} dk \tag{14.105}$$

$$= \frac{1}{\pi} \operatorname{Re} \int_{0}^{\infty} \hat{w}(k, 0)e^{i(kx - mz)} dk. \tag{14.106}$$

Because the waves are stationary with respect to the ridge, they are moving upstream with horizontal phase speed equal and opposite to the wind U (which is in the positive x-direction), so this means that $k < 0$. If we define $k = -\xi$, where $\kappa > 0$, then on using the even and odd nature, respectively, of the cosine and sine functions with respect to their arguments, we find after some algebraic manipulation that

$$w(x, z) = -\frac{UHb}{\pi^{1/2}} \int_0^\infty k e^{-k^2 b^2/4} [\sin(kx + mz)] \, dk, \qquad (14.107)$$

after replacing the dummy variable ξ by k.

When $k^2 > k_c^2 = N^2/U^2$ we write $m = -iq$, where

$$q = \left(k^2 - \frac{N^2}{U^2} \right)^{1/2},$$

and the negative root has been chosen so that $\hat{w}(k, z)$ is bounded as $z \to \infty$, so for $k^2 > k_c^2$,

$$\hat{w}(k, z) = \hat{w}(k, 0)e^{-qz}, \qquad q > 0, \quad z > 0.$$

This permits a useful decomposition of the wave field (14.107) into propagating and evanescent waves:

$$w(x, z) = w_p(x, z) + w_e(x, z) \qquad (14.108)$$

$$= -\frac{UHb}{\pi^{1/2}} \left[\int_0^{k_c} k e^{-k^2 b^2/4} \sin(kx + mz) \, dk \right.$$

$$\left. + \int_{k_c}^\infty k e^{-k^2 b^2/4} e^{-qz} \sin(kx) \, dk \right] \qquad (14.109)$$

respectively. Note that the vertical velocity field is directly proportional to the wind speed U, the height of the ridge H, and the "width" b (in fact the width at half-height for this ridge is $\approx 0.83b$). It is also strongly dependent on the dimensionless term $kb \exp\left[-(kb/2)^2\right]$. It is easily shown that the maximum occurs where $kb = \sqrt{2}$, which corresponds to a wavelength of $\lambda = \sqrt{2}\pi b \approx 4.4b$. If we take $x = 2.2b$, then $h/H \approx 0.008$, so the above wavelength is the width of the ridge at about 1 percent of its height. This means that the strongest or most excited wave is comparable to the width of the ridge. If the quantity $N^2/U^2 \equiv L^2 > k^2$, the wave is propagating ($m^2 > 0$). The wave structure over this isolated ridge is determined by the relative magnitude of L and b: using the maximum from the graph, we have $k = \sqrt{2}/b$, so

$$m^2 = L^2 - k^2 = L^2 - \frac{2}{b^2},$$

and $m^2 < 0$ when $Lb < \sqrt{2}$. Thus if $Lb < \sqrt{2}$, the waves produced will be evanescent (decaying exponentially away in z from the ridge). Since $Lb = Nb/U_0$, this inequality corresponds to combinations of narrow ridge, weak stratification, and

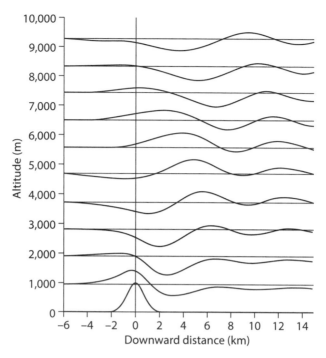

Figure 14.3 The wavefield over an isolated two-dimensional ridge with $b = 1$ km, $N = 10^{-2}$ s and $U_0 = 10$ m/s (after [194]).

strong winds. Conversely, waves for which $Lb > \sqrt{2}$ can be expected to be propagating; this implies combinations of wide ridge, strong stratification, and weak winds. A graph of a typical wavefield (for $b = 1$ km) is shown in Figure 14.3.

14.2.3 Trapped Lee Waves

Trapped lee waves assume the form of a series of waves running parallel to the ridges, and the crests of these waves often contain altocumulus, stratocumulus, wave clouds, or rotor clouds in parallel bands that can be very striking in satellite pictures. Because wave energy is trapped within the stable layer, these waves (and accompanying cloud bands) may dissipate only very slowly downwind, and they can continue downstream for many wavelengths spanning many tens of kilometers. Flow beneath the wave crests, occasionally made visible by rotor clouds, is often turbulent, thus presenting a significant hazard to low-level aviation. Vertically propagating mountain waves and trapped lee waves can coexist, and sometimes lee waves are incompletely trapped or "leaky," leading to a variety of complex rotor interactions. This complexity of rotor patterns often produces interesting variations in cloud forms. As mountain waves propagate upward, the rotor's amplitude can grow to the point that the rotor "breaks," that is, the rotor becomes convectively unstable and overturns. Wave breaking can have an important role in vertically redistributing horizontal atmospheric momentum, as it slows the atmosphere by turbulent transport of the earth's momentum upward. [217]

Recall from the definition of the Scorer parameter (14.90) that

$$S(z) = \left(\frac{N}{U}\right)^2 - \frac{1}{U}\frac{d^2U}{dz^2}. \tag{14.110}$$

It has been shown by [197] that waves trapped in the lee of isolated mountains or ridges (and generally confined to the lower troposphere [195]) occur only when $S'(z) < 0$. This can arise in various ways, other things being equal: an increase in U, a decrease in N, an increase in the curvature term U'', or indeed any appropriate combination of these effects. In [197] the relatively simple physical case (though still very complicated mathematically) of a two-layer atmosphere ("upper" and "lower") for which $S_U < S_L$, both being constants. The domain is here defined to be $z \in [-h, 0) \cup (0, \infty)$. It is interesting (and important) to note that S can be discontinuous without either of N or U being so. For trapping to occur the following condition must be satisfied for a lower layer of depth h:

$$\left(S_L^2 - k^2\right)^{1/2} \cot\left[h\left(S_L^2 - k^2\right)^{1/2}\right] = -\left(k^2 - S_U^2\right)^{1/2}. \tag{14.111}$$

To see how this result is obtained, we examine the basic model discussed in [196]. Based on the form of the solutions (14.96) and (14.97), the vertical perturbation *displacement* ξ for the evanescent solution in the upper layer can be characterized by

$$\xi = B_U e^{-\mu z}, \quad \mu = \left(k^2 - S_U^2\right)^{1/2}, \tag{14.112}$$

and for the propagating waves in the lower layer $[-h, 0)$,

$$\xi = B_L \sin m (z + h), \quad m = \left(S_L^2 - k^2\right)^{1/2}. \tag{14.113}$$

Thus $\xi(-h) = 0$, and by requiring continuity of both $\xi(z)$ and $\xi'(z)$ at $z = 0$, we obtain

$$mh \cot mh = -\mu h \tag{14.114}$$

(on multiplying both sides by h for ease of graphical representation). Thus for given values of the parameters S_U, S_L and h, the k values may be determined. A typical graph of the transcendental relation (14.114) is shown in Figure 14.4 (the vertical asymptotes at π and 2π are also indicated). Only the fourth quadrant is of interest because m and μ are real and positive quantities. The arcs for μ are circular because

$$\mu^2 + m^2 = S_L^2 - S_U^2 > 0. \tag{14.115}$$

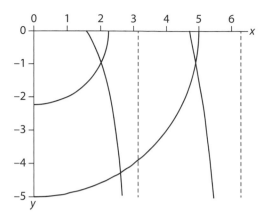

Figure 14.4 The right-hand side (circular areas) and left-hand side of (14.144) plotted as functions of $x = mh$. The dashed lines are the vertical asymptotes at π and 2π.

Note also that $\cot mh = 0$ for $mh = \pi/2, 3\pi/2$, etc. As see from Figure 14.4, the radius of the smaller circle lies in $(\pi/2, 3\pi/2)$ and that of the larger one is contained in $(3\pi/2, 5\pi/2)$. For the smaller circle there is a single real root of (14.114) and for the larger one there are two. This trend continues, so that there are n roots if

$$\left[\frac{(2n-1)\pi}{2}\right]^2 < \left(\mu^2 + m^2\right) h^2 < \left[\frac{(2n+1)\pi}{2}\right]^2, \qquad (14.116)$$

or

$$\left[\frac{(2n-1)\pi}{2h}\right]^2 < S_L^2 - S_U^2 < \left[\frac{(2n+1)\pi}{2h}\right]^2. \qquad (14.117)$$

Clearly no real solutions exist if $S_L^2 - S_U^2 < 0$; such solutions can only exist if

$$S_L > k > S_U, \qquad (14.118)$$

(see inequality (14.115)). Therefore the requirement for at least one such trapped wave to exist is that $n \geq 1$, or

$$S_L^2 - S_U^2 > \frac{\pi^2}{4h^2}. \qquad (14.119)$$

14.3 BILLOW CLOUDS, WIND SHEAR, AND HOWARD'S SEMICIRCLE THEOREM

Clear-air turbulence: A higher altitude (6–15 km) turbulence phenomenon occurring in cloud-free regions, associated with wind shear, particularly between the core of a jet stream and the surrounding air.

It is most common near upper air fronts and the tropopause, and can often affect an aircraft without warning. Clear-air turbulence also frequently occurs close to towering cumulus clouds (usually within 30 km), and near mountains. Airflow disrupted by mountains and other terrain can undulate in waves of turbulence for 1000 km or more. [218]

As already noted, most billow-type clouds and clear-air turbulence are believed to be the result of small-scale wave-like undulations that develop spontaneously when the vertical shear of a horizontal wind exceeds some critical value. The shear is a measure of how much the horizontal wind speed vector $\mathbf{U} = (U(z), 0, 0)$ varies as a function of altitude z, and this is usually defined by the derivative $U'(z)$ or its magnitude $|U'(z)|$. The process responsible for the development of these waves is called *shear* (or *Kelvin-Helmholtz*) *instability*. While growing, the waves must do work in exchanging denser fluid from below with less dense fluid from above. Generally, the more stable the stratification, the more work that must be done by the waves against the force of gravity. If wave amplification is to occur, the waves must extract kinetic energy from the undisturbed shear flow at a rate faster than they lose energy by doing work against gravity. This means that the critical shear $U'(z)$ required for wave amplification depends on the degree of density stratification; the greater the stratifications the larger the critical shear will be.

Both physical and mathematical arguments can be made to quantify this condition. For the former, suppose that two equal neighboring volumes of fluid, at heights z and $z + \delta z$ are interchanged. The work that must be done against gravity to effect this interchange (per unit volume) is

$$\delta W = -g\delta\rho\delta z,$$

where $\delta\rho$ is the (negative) density difference between the two heights in a stably stratified fluid (where the density decreases upward). The kinetic energy available to do this work (again, per unit volume) is the difference in kinetic energy between the flow at the two heights and the mean flow [226]:

$$\frac{1}{2}\rho\left[U^2 + (U + \delta U)^2 - 2\left\{\frac{(U + U + \delta U)}{2}\right\}^2\right] = \frac{1}{4}\rho(\delta U)^2.$$

A sufficient condition for stability is therefore

$$\frac{1}{4}\rho(\delta U)^2 < -g\delta\rho\delta z.$$

When the appropriate limiting procedures are carried out (U and ρ, it is assumed, being differentiable functions of z), an equivalent form of this condition is

$$\left(\frac{dU}{dz}\right)^2 < -4\left(\frac{g\rho'}{\rho}\right) \equiv 4g\beta.$$

In hydrodynamics, the *Richardson number* J is defined as

$$J = -\frac{g\rho'}{\rho U'^2} = \frac{g\beta}{U'^2},$$

and as can be seen, this number measures the ratio of the stabilizing effect of density stratification to the destabilizing effect of shear in a fluid. Thus, a sufficient condition for stability (and therefore no billows formed by shear instability) is that

$$J \geq \frac{1}{4},$$

everywhere in the fluid medium. Alternatively, a *necessary condition for instability* (and hence billow formation) is that $J < 1/4$ at least *somewhere* in the fluid. (Note that this last condition does not mean that instability will occur if $J < 1/4$; merely that if instability does occur, then J will be $< 1/4$ at least somewhere in the domain).

The mathematical proof of this *Richardson criterion* is interesting and will be presented after proving a very elegant theorem delineating bounds on complex eigenvalues, known as a *semicircle theorem*. As earlier, $U(z)$ is the speed of an undisturbed incompressible fluid with stratified density $\rho(z)$, $z \in [0, d]$, moving steadily and horizontally in the x-direction (though now we drop the subscript zero on these equilibrium quantities). We seek a Fourier mode solution for the linearized vertical displacement $\eta(x, z, t)$ as

$$\eta(x, z, t) = F(z) \exp[ik(x - ct)], \quad c = \frac{\omega}{k} \in \mathbb{C}.$$

The vertical component of velocity w is directly related to η and hence F via

$$w = \left(\frac{\partial}{\partial t} + U\frac{\partial}{\partial x}\right)\eta = ik[U(z) - c]F(z). \tag{14.120}$$

The function $F(z)$, which for $c_i = \mathrm{Im}(c) \neq 0$ is a complex function of the real variable $z \in [0, d]$, satisfies the following differential equation [219]:

$$\left[\rho(U - c)^2 F'\right]' + \rho\left[\beta g - k^2(U - c)^2\right]F = 0, \tag{14.121}$$

where $\beta(z) = -\rho'/\rho$ is a stratification parameter. The boundary conditions can be of the form

$$a_1 F(0) + b_1 F'(0) = 0, \tag{14.122}$$

$$a_2 F(d) + b_2 F'(d) = 0, \tag{14.123}$$

but without loss of generality as far as the final theorem is concerned we consider the simplest case of rigid boundaries ($F(0) = F(d) = 0$). Multiply equation (14.121) by $\bar{F}$, the complex conjugate of F, and integrate by parts over $[0, d]$ to obtain

$$g\int_0^d \beta\rho |F|^2 dz = \int_0^d \rho(U - c)^2\left[|F'|^2 + k^2|F|^2\right]dz. \tag{14.124}$$

The imaginary part of this equation is

$$2c_i \int_0^d \rho(U - c_r)\left[|F'|^2 + k^2|F|^2\right]dz = 0.$$

For unstable modes, $c_i \neq 0$, this equation implies that the real part of the phase speed lies in the range of the flow, that is,

$$a \equiv U_{\min} < c_r < U_{\max} \equiv b.$$

Correspondingly, the real part of equation (14.124) is

$$\int_0^d \left[(U - c_r)^2 - c_i^2\right]\phi \, dz = g\int_0^d \beta\rho |F|^2 dz,$$

where $\phi = \rho \left[|F'|^2 + k^2 |F|^2 \right] > 0$. It follows that

$$0 \geq \int_0^d (U - a)(U - b)\, \phi\, dz = \int_0^d \left[U^2 - (a + b)U + ab \right] \phi\, dz$$

$$= \left[c_r^2 + c_i^2 - (a + b)c_r + ab \right] \int_0^d \phi\, dz + g \int_0^d \beta\rho\, |F|^2\, dz$$

$$= \left\{ \left[c_r - \frac{1}{2}(a + b)^2 \right] + c_i^2 - \left[\frac{1}{2}(a - b)^2 \right] \right\} \int_0^d \phi\, dz + g \int_0^d \beta\rho\, |F|^2\, dz.$$

$$(14.125)$$

For a statically stable fluid $\beta > 0$, so it follows that

$$\left[c_r - \frac{1}{2}(a + b)^2 \right] + c_i^2 \leq \left[\frac{1}{2}(a - b)^2 \right]. \qquad (14.126)$$

This is *Howard's semicircle theorem* [219]; it states that the complex wave speed c for any unstable mode must lie inside the semicircle in the upper half-plane which has the range of U for diameter. It is interesting to note that although gravity-induced density stratification has been included in this formulation, the result is not optimal, because it holds for g or $\beta = 0$. In [220] an upper bound on the second integral in (14.125) was obtained in terms of ϕ, resulting in an elegant *semi-ellipse* theorem placing bounds on complex values of c. The effects of compressibility and magnetic fields have also been incorporated to modify and extend these results [221]–[225].

If we set $W = U - c$ and $G = W^{1/2} F$ in equation (14.121), the following equation results:

$$(\rho W G')' - \left\{ \frac{1}{2}(\rho U')' + k^2 \rho W + \frac{\rho}{W} \left(\frac{1}{4} U'^2 - g\beta \right) \right\} G = 0,$$

and corresponding to the boundary conditions invoked above, $G(0) = G(d) = 0$. Following a similar procedure using $\bar{G}$, the complex conjugate of G, we obtain

$$\int_0^d \rho W \left\{ \left(|G'|^2 + k^2 |G|^2 \right) + \frac{1}{2}(\rho U')' |G|^2 + \rho \left(\frac{1}{4} U'^2 - g\beta \right) \bar{W} \left| \frac{G}{W} \right|^2 \right\} dz = 0.$$

For $c_i \neq 0$ the imaginary part of this equation yields

$$\int_0^d \rho \left\{ |G'|^2 + k^2 |G|^2 \right\} dz + \int_0^d \rho \left(g\beta - \frac{1}{4} U'^2 \right) \left| \frac{G}{W} \right|^2 dz = 0, \qquad (14.127)$$

but this cannot be satisfied if

$$g\beta \geq \frac{1}{4} U'^2 \qquad (14.128)$$

everywhere in $[0, d]$, since for nontrivial motions all terms are positive on the left-hand side of the equation. This is a sufficient condition for the flow to be stable, and if $U' \neq 0$ anywhere in $[0, d]$, this condition can be expressed in terms of the Richardson number $J = g\beta / U'^2$ as simply $J \geq 1/4$ throughout the flow. The

Richardson number is a measure of the stabilizing effect of density stratification to the destabilizing effect of shear.

Using the fact that $|W|^{-2} \le c_i^{-2}$, an upper bound on the growth rate kc_i of unstable modes may readily be found. Thus from equation (14.127), we have

$$k^2 \int_0^d \rho \, |G|^2 \, dz = \int_0^d \rho \left(\frac{1}{4} U'^2 - g\beta \right) \left| \frac{G}{W} \right|^2 dz - \int_0^d \rho \, |G'|^2 \, dz \quad (14.129)$$

$$\le \frac{1}{c_i^2} \max_z \left(\frac{1}{4} U'^2 - g\beta \right) \int_0^d \rho \, |G|^2 \, dz, \quad (14.130)$$

that is,

$$k^2 c_i^2 \le \max_z \left(\frac{1}{4} U'^2 - g\beta \right). \quad (14.131)$$

14.4 THE TAYLOR-GOLDSTEIN EQUATION

With the usual x- and t- harmonic variation assumed in the Euler equations for internal gravity waves in an incompressible shear flow $U_0(z)$, the governing normal mode equation for the vertical velocity component $w(\omega, k; z)$ is found to be

$$\frac{d^2 w}{dz^2} + \frac{\rho'}{\rho} \frac{dw}{dz} + \left[\left(\frac{N}{c - U} \right)^2 + \frac{U''}{c - U} + \left(\frac{U'}{c - U} \right) \frac{\rho'}{\rho} - k^2 \right] w = 0, \quad (14.132)$$

where the square of the Brunt-Väisälä frequency is $N^2 = -\rho' g / \rho = \beta g$.

Exercise: Using equation (14.120), write $F(z)$ in terms of w, and expand equation (14.121) to derive equation (14.132).

As earlier, the second term in this equation represents the effects of changing atmospheric density on the wave amplitude [194], but if in particular the coefficient of w' is a constant (as it is for an exponentially decreasing density profile, in which case $\rho'/\rho = -1/H$), then setting

$$W = we^{-z/2H}, \quad (14.133)$$

leads to the Taylor-Goldstein equation

$$\frac{d^2 W}{dz^2} + \left[\left(\frac{N}{c - U} \right)^2 + \frac{U''}{c - U} - \frac{1}{H} \left(\frac{U'}{c - U} \right) - k^2 - \frac{1}{4H^2} \right] W = 0. \quad (14.134)$$

The first term inside the brackets is a measure of the buoyancy, the second is a curvature term, and the third is a shear term. Note that for this atmosphere, $N^2 = g/H$. In addition to the literature cited in the previous section, there is considerably more on the various stability and semicircle theorems that can be gleaned from this equation and its relatives (for a review, see [200]).

A Theorem about Internal Gravity Waves in the Absence of a Flow Field
Time-harmonic wave motion (with angular frequency ω and wavenumber $k \neq 0$) in a vertically stratified incompressible fluid of continuous density $\rho(z)$ (such as for internal gravity waves in an ocean or lake) is described by the following normal mode equation for a dependent variable $f(z)$ (such as the pressure perturbation or the vertical component of velocity):

$$\omega^2 \frac{d}{dz}\left(\rho(z)\frac{df}{dz}\right) - \left(\omega^2\rho(z) + g\rho'(z)\right)k^2 f = 0. \tag{14.135}$$

Consider waves in a channel of depth d with boundary conditions $f(0) = 0 = f(d)$; note that in general $f(z)$ is a complex function of the real variable z.

Prove that (i) ω^2 cannot be zero for $\rho'(z) \neq 0$ and nontrivial solutions, and (ii) $\omega^2 \in \mathbb{R}$, even if $\rho'(z)$ changes sign in $(0, d)$. **Hint**: Multiply (14.135) by the complex conjugate f and integrate over $[0, d]$.

Proof: (i) follows because the equation cannot be satisfied if $\omega^2 = 0$.

(ii) Use the hint to obtain

$$\omega^2 \int_0^d \bar{f}\frac{d}{dz}\left(\rho\frac{df}{dz}\right)dz - \int_0^d \left(\omega^2\rho + g\rho'\right)k^2 |f|^2 dz = 0. \tag{14.136}$$

Integrating by parts and using the boundary conditions yields

$$\omega^2 I_1 + gk^2 I_2 = 0, \tag{14.137}$$

where

$$I_1 = \int_0^d \left(\rho\,|f'|^2 + \rho k^2 |f|^2\right)dz > 0, \quad \text{and} \quad I_2 = \int_0^d \rho'\,|f|^2\,dz. \tag{14.138}$$

For nontrivial solutions for which $\rho'(z) \neq 0$, $I_1 > 0$ and I_2 is real, so $\omega^2 = -gk^2 I_2/I_1$ is real, and therefore ω is either purely real (if $I_2 < 0$) or purely imaginary (if $I_2 > 0$). If $\rho(z)$ changes sign in $(0, d)$, I_2 may also be of either sign (in which case $\omega^2 \in \mathbb{R}$ again), or even zero (in which case $\omega^2 = 0$). Note that if ρ is constant, $\omega^2 = 0$ as well.

PART III
Classical Scattering

Chapter Fifteen

The Classical Connection

In physics, action is an attribute of the dynamics of a physical system. It is a mathematical functional which takes the trajectory, also called path or history, of the system as its argument and has a real number as its result. Generally, the action takes different values for different paths. Action has the dimensions of [energy]·[time], and its SI unit is joule-second. This is the same unit as that of angular momentum.

... Empirical laws are frequently expressed as differential equations, which describe how physical quantities such as position and momentum change continuously with time. Given the initial and boundary conditions for the situation, the "solution" to these empirical equations is an implicit function describing the behavior of the system.

There is an alternative approach to finding equations of motion. Classical mechanics postulates that the path actually followed by a physical system is that for which the action is minimized, or, more generally, is stationary. In other words, the action satisfies a variational principle: the principle of stationary action. ... The action is defined by an integral, and the classical equations of motion of a system can be derived by minimizing the value of that integral. The action is typically represented as an integral over time, taken along the path of the system between the initial time and the final time of the development of the system.

[45]

15.1 LAGRANGIANS, ACTION, AND HAMILTONIANS

We revisit some earlier ideas to draw out the "alternative approach to finding equations of motion" mentioned in the quote. Thus let

$$\mathbf{F}(\mathbf{r}) = -\nabla V(\mathbf{r}) \tag{15.1}$$

be a conservative force acting on a particle of mass m with potential energy function $V(\mathbf{r})$. If $\dot{\mathbf{r}}$ denotes the velocity vector, the Lagrangian for the particle is simply

$$\mathcal{L}(\mathbf{r}, \dot{\mathbf{r}}) = \frac{1}{2}m\dot{\mathbf{r}}^2 - V(\mathbf{r}). \tag{15.2}$$

The *action* for the system is defined as

$$S_a(\mathbf{r}, t) = \int_{t_1}^{t} \mathcal{L}(\mathbf{r}, \dot{\mathbf{r}}, \tau) d\tau, \tag{15.3}$$

(where τ is a dummy variable for t). To find the equations of motion for the particle we use the Euler-Lagrange equations to determine where the action is stationary, that is,

$$\frac{d}{dt}\left(\frac{\partial \mathcal{L}}{\partial \dot{r}_i}\right) - \frac{\partial \mathcal{L}}{\partial r_i} = 0, \quad i = 1, 2, 3. \tag{15.4}$$

(See also (3.41).) Since

$$\frac{\partial \mathcal{L}}{\partial \dot{r}_i} = m\dot{r}_i, \quad \text{and} \quad \frac{\partial \mathcal{L}}{\partial r_i} = -\frac{\partial V}{\partial r_i 0}, \tag{15.5}$$

equations (15.4) simplify to

$$m\ddot{r}_i + \nabla V(r_i) = 0. \tag{15.6}$$

This is, of course, Newton's second law for the motion of a particle! More generally, if we denote the coordinates in a system with one degree of freedom as q and the momentum as p, the action integral (15.3) is

$$S_a(q, t) = \int_{t_0}^{t} \mathcal{L}(q, \dot{q}, \tau) d\tau. \tag{15.7}$$

Guided by the above example we can write the Lagrangian as [46]

$$\mathcal{L} = p\dot{q} - \mathcal{H}, \tag{15.8}$$

where

$$\mathcal{H}(p, q) = \frac{1}{2}m\dot{q}^2 + V(q) = \frac{p^2}{2m} + V(q) = \frac{1}{2}p\dot{q} + V(q), \tag{15.9}$$

is the *Hamiltonian* for the system. Then

$$S_a(q, t) = \int_{t_0}^{t} p\dot{q}\,d\tau - \int_{t_0}^{t} \mathcal{H}(p(t), q(t))d\tau = \int_{q(t_0)}^{q} p\,dq - \int_{t_0}^{t} \mathcal{H}(p(t), q(t))d\tau. \tag{15.10}$$

From the chain rule we have

$$S_a(q, t) = \int_{q(t_0)}^{q} \frac{\partial S_a}{\partial q}dq - \int_{t_0}^{t} \frac{\partial S_a}{\partial t}d\tau, \tag{15.11}$$

so formally we conclude that

$$p = \frac{\partial S_a}{\partial q}, \tag{15.12}$$

$$-\mathcal{H}(p, q) = -\mathcal{H}\left(\frac{\partial S_a}{\partial q}, q\right) = \frac{\partial S_a}{\partial t}. \tag{15.13}$$

Equation (15.13) is the famous *Hamilton-Jacobi equation*. If we rewrite this for now in terms of three-dimensional Cartesian coordinates $\mathbf{r} = \langle x, y, z \rangle$, then

$$\mathbf{p} = \nabla \mathcal{S}_a, \tag{15.14}$$

so

$$\frac{\partial \mathcal{S}_a}{\partial t} + \frac{1}{2m} |\nabla \mathcal{S}_a|^2 + V(\mathbf{r}) = 0. \tag{15.15}$$

For systems in which the energy is conserved, $\mathcal{H}$ is a constant of the motion and can be identified with the total energy E. In terms of Hamilton's characteristic function $\mathcal{W}$ [47] we thus infer that

$$\mathcal{S}_a(\mathbf{r}, t) = \mathcal{S}_a(\mathbf{r}, t_0) - E(t - t_0) \equiv \mathcal{W}(\mathbf{r}) - Et, \tag{15.16}$$

and

$$\frac{1}{2m} |\nabla \mathcal{S}_a|^2 + V(\mathbf{r}) - E = 0. \tag{15.17}$$

From this we deduce an important equation, a mathematical metaphor to be investigated shortly:

$$|\nabla \mathcal{S}_a|^2 = p^2 = |\nabla \mathcal{W}|^2 = 2m(E - V). \tag{15.18}$$

The equation

$$\mathcal{S}_a(\mathbf{r}, t) = \mathcal{W}(\mathbf{r}) - Et \tag{15.19}$$

indicates that the action $\mathcal{S}_a$ may be interpreted as the evolution of the surface $\mathcal{W}$ in time, or equivalently, that of the "wavefront" $\mathcal{S}_a(\mathbf{r}, t_0)$. We can go further and find the speed v of such wavefronts: in a time increment dt the distance moved by the wavefront normal to its surface is ds, and in the same increment the characteristic function has changed by an amount $dW = Edt$. But $dW = |\nabla W| ds$, so that

$$v = \frac{ds}{dt} = \frac{E}{|\nabla \mathcal{W}|}. \tag{15.20}$$

A connection (implicit in the result (15.18)) has already been made between the refractive index $n(\mathbf{r})$ of a medium and the potential energy function $V(\mathbf{r})$. For light traveling between points with position vectors $\mathbf{r}_0$ and $\mathbf{r}$ the elapsed time T is

$$T = \int_{t_0}^{t} d\tau = \int_{\mathbf{r}_0}^{\mathbf{r}} \frac{ds}{v} = \int_{\mathbf{r}_0}^{\mathbf{r}} \frac{n(\mathbf{r})}{c} ds, \tag{15.21}$$

where c is the constant speed of light in vacuo. As in Chapter 2 (see equation (2.4)), we may define the optical path length as

$$cT(t) = \int_{\mathbf{r}_0}^{\mathbf{r}} n(\mathbf{r}) ds = \int_{t_0}^{t} n(\mathbf{r}) |\dot{\mathbf{r}}| d\tau. \tag{15.22}$$

Fermat's principle states that the optical path length between two points is stationary. We may identify this integral with the action integral (15.7) with Lagrangian

$$\mathcal{L}(\mathbf{r}, \dot{\mathbf{r}}) = n(\mathbf{r}) |\dot{\mathbf{r}}|. \tag{15.23}$$

The components of momentum p_i are (see (15.4)) defined by

$$p_i = \frac{\partial \mathcal{L}}{\partial \dot{r}_i} = \frac{n\dot{r}_i}{|\dot{\mathbf{r}}|}, \quad i = 1, 2, 3. \tag{15.24}$$

From these equations it follows that (i)

$$p^2 = \sum_{i=1}^{3} p_i^2 = [n(\mathbf{r})]^2, \tag{15.25}$$

and (ii)

$$\sum_{i=1}^{3} p_i \dot{r}_i = n(\mathbf{r}) |\dot{\mathbf{r}}| = \mathcal{L} \equiv \mathcal{H} + \mathcal{L}, \tag{15.26}$$

from which we conclude that the Hamiltonian is identically zero, and so from (15.10) we find that

$$\mathcal{S}_a = \int_{t_0}^{t} p_i \dot{r}_i d\tau = \int_{q(t_0)}^{q} p_i dr_i. \tag{15.27}$$

This yields

$$p_i = \frac{\partial \mathcal{S}_a}{\partial r_i}, \frac{\partial \mathcal{S}_a}{\partial t} = 0, \tag{15.28}$$

so we again conclude that $\mathbf{p} = \nabla \mathcal{S}_a$, but in view of (15.25) we have once more the celebrated *eikonal equation* of geometrical optics, namely,

$$|\nabla \mathcal{S}_a(\mathbf{r})|^2 = [n(\mathbf{r})]^2. \tag{15.29}$$

Because of this we can make, from (15.18) the identification

$$n(\mathbf{r}) = [2m(E - V(\mathbf{r}))]^{1/2}. \tag{15.30}$$

Thus we see a close formal similarity between the Hamilton-Jacobi and the eikonal equations. From equation (15.25) the "momentum" of the "light ray" is therefore proportional to $\nabla \mathcal{S}_a(\mathbf{r})$, which means that the ray direction is orthogonal to the surfaces of constant $\mathcal{S}_a$.

15.2 THE CLASSICAL WAVE EQUATION

Following [47] (and also [48]) we can again derive the eikonal equation as the short-wavelength limit of geometrical optics, but in a little more detail than before. Consider the scalar wave equation

$$\nabla^2 \phi = \frac{n^2(\mathbf{r})}{c^2} \frac{\partial^2 \phi}{\partial t^2}, \tag{15.31}$$

where $n(\mathbf{r})$ is the refractive index of the medium. If we restrict ourselves to the case of time-harmonic waves such that

$$\phi(\mathbf{r}, t) = u(\mathbf{r})e^{-i\omega t}, \tag{15.32}$$

the resulting Helmholtz equation is

$$\nabla^2 u + k_0^2 n^2(\mathbf{r}) u \equiv \nabla^2 u + k^2(\mathbf{r}) u = 0, \tag{15.33}$$

where

$$k_0 = \omega/c = 2\pi/\lambda_0, \ k(\mathbf{r}) = k_0 n(\mathbf{r}). \tag{15.34}$$

Seeking a solution of the form

$$u(\mathbf{r}) = A(\mathbf{r}) e^{iS(\mathbf{r})}, \tag{15.35}$$

where u and A are real functions (note: for our purposes the phase function S is essentially the same as the action $\mathcal{S}_a$, though the latter was defined in terms of an integral of a Lagrangian). In the case of constant refractive index n, plane wave solutions exist, and A would be a constant amplitude while the phase term S would take the form $(\mathbf{k} \cdot \mathbf{r} - \omega t)$. Application of the gradient operator yields

$$\nabla^2 u = \left[\nabla^2 A + iA\nabla^2 S - A(\nabla S)^2 + 2i\nabla A \cdot \nabla S\right] e^{iS(\mathbf{r})}. \tag{15.36}$$

When this is substituted into (15.33) we find (on separating real and imaginary parts) that

$$\nabla^2 A + (k_0^2 n^2 - (\nabla S)^2) A = 0, \tag{15.37}$$

$$A\nabla^2 S + 2\nabla A \cdot \nabla S = 0. \tag{15.38}$$

A slight variation on the ansatz (15.35) can be useful: let $S(\mathbf{r}) = k_0 B(\mathbf{r})$. Then the system (15.37) and (15.38) takes the form

$$\nabla^2 A + (\nabla A)^2 + k_0^2(n^2 - (\nabla B)^2) = 0, \tag{15.39}$$

$$\nabla^2 B + 2\nabla A \cdot \nabla B = 0. \tag{15.40}$$

If the refractive index $n(\mathbf{r})$ is a slowly varying function in the sense that is does not change significantly over a wavelength λ (i.e. $\lambda \ll L$, where L is a typical length scale over which n does change significantly), then the free space wavenumber satisfies the condition $k_0 L \gg 1$. The dominant term in (15.39) is therefore the third one, from which we again encounter the eikonal equation, this time in the form

$$|\nabla B(\mathbf{r})|^2 = [n(\mathbf{r})]^2. \tag{15.41}$$

Returning to (15.38), it is readily seen that it may be written as

$$A^{-1}\nabla \cdot (A^2 \nabla S) = 0. \tag{15.42}$$

This expresses a conservation of flux for the vector $A^2 \nabla S$, where in view of (15.37) we may write $\nabla S = k_0 n = \mathbf{k}$, the wave vector in the eikonal approximation of geometrical optics. Furthermore, following [48], we can introduce the function

$$D(\mathbf{r}, \mathbf{k}) = \frac{c}{2k_0}\left[k^2 - (k_0 n)^2 - \frac{\nabla^2 A}{A}\right] = 0. \tag{15.43}$$

By virtue of

$$\nabla D \cdot d\mathbf{r} + \nabla_k D \cdot d\mathbf{k} = 0, \tag{15.44}$$

(where $\nabla_k = \langle \partial/\partial k_1, \partial/\partial k_2, \partial/\partial k_3 \rangle$) the following form of Hamilton's equations arises if we make the identifications

$$\frac{d\mathbf{r}}{dt} = \nabla_k D = \frac{c}{k_0}\mathbf{k}, \tag{15.45}$$

$$\frac{d\mathbf{k}}{dt} = -\nabla D = \frac{c}{2k_0}\nabla\left[(k_0 n)^2 + \frac{\nabla^2 A}{A}\right]. \tag{15.46}$$

In the terminology of [48] this generalization of the geometrical optics approximation retains the concept of rays traveling along "field lines" with $\mathbf{k} = \nabla S$ but with an additional feature: dependence on the amplitude distribution $A(\mathbf{r})$ of the ray tube. If the net flux of rays in the tube is zero, then $\nabla^2 S = 0$; from (15.38) this implies that $\nabla A \cdot \nabla S = 0$ and so the amplitude A is constant along each ray trajectory.

Recall from above that the action $S_a(\mathbf{r}, t) = W(\mathbf{r}) - Et$. From (15.41) W corresponds to B, so S_a must be proportional to the total phase of $\psi(\mathbf{r}, t)$, that is, to

$$S - \omega t = k_0 B - \omega t = k_0(B - ct) = 2\pi\left(\frac{B}{\lambda_0} - \nu t\right). \tag{15.47}$$

Formally then, the total energy E and wave frequency ν are proportional, whence we write $E = h\nu$–rather cheekily employing h as a constant of proportionality in a classical context. But $\lambda\nu = \upsilon$, the (phase) speed of the wave, and if the kinetic energy of the one-particle system is denoted by K, then

$$\upsilon = \frac{E}{|\nabla W|} = \frac{E}{(2mK)^{1/2}} = \frac{E}{p}, \tag{15.48}$$

since from (15.18) $p^2 = 2mK$. Therefore,

$$\lambda = \frac{\upsilon}{\nu} = \frac{h}{p} = \frac{h}{[2m(E - V)]^{1/2}}. \tag{15.49}$$

Returning to the Helmholtz equation (15.33) and noting that $k = 2\pi/\lambda$, we have

$$\nabla^2 u + \left(\frac{2\pi}{\lambda}\right)^2 u = 0. \tag{15.50}$$

Supposing that a similar type of equation arises in quantum mechanics for some wave function ψ and using (15.49) enables us to conjecture the existence of the equation

$$\nabla^2\psi + \frac{8\pi^2 m}{h^2}(E - V(\mathbf{r}))\psi = \nabla^2\psi + \frac{2m}{\hbar^2}(E - V(\mathbf{r}))\psi = 0, \tag{15.51}$$

where $\hbar = h/2\pi$. This is of course recognizable as the time-independent Schrödinger equation. We will use this as a springboard to the time-dependent Schrödinger equation and establish a connection to the classical equation of motion for a particle of mass m. Consider then

$$i\hbar\frac{\partial\Psi}{\partial t} = \mathcal{H}\Psi = \left[-\frac{\hbar^2}{2m}\nabla^2 + V(\mathbf{r})\right]\Psi; \tag{15.52}$$

now we generalize the ansatz (15.35) by writing

$$\Psi(\mathbf{r}, t) = A(\mathbf{r}, t)e^{iS(\mathbf{r}, t)/\hbar}. \tag{15.53}$$

(Note that this form implies there is an essential singularity at $\hbar = 0$. We shall return to this point later.) Applying this to (15.52) and separating real and imaginary parts as before, after a little rearrangement we obtain

$$\frac{\partial S}{\partial t} + \frac{1}{2m}(\nabla S)^2 - \frac{\hbar^2}{2m}\frac{\nabla^2 A}{A} + V(\mathbf{r}) = 0, \tag{15.54}$$

$$\frac{\partial A}{\partial t} + \frac{1}{m}\nabla A \cdot \nabla S + \frac{A}{2m}\nabla^2 S = 0. \tag{15.55}$$

Instead of the second equation being viewed as describing time-independent flux conservation, it is now in the form of a more general continuity equation for a probability density defined as

$$\rho = |\Psi|^2 = A^2. \tag{15.56}$$

Since

$$\nabla\Psi = \left[\frac{1}{A}\nabla A + \frac{i}{\hbar}\nabla S\right]\Psi, \tag{15.57}$$

we may define [49] a probability current density by

$$\mathbf{J} = \mathrm{Re}\left[\frac{-i\hbar}{m}\bar{\Psi}\nabla\Psi\right] = \frac{A^2}{m}\nabla S = \frac{\rho}{m}\nabla S, \tag{15.58}$$

so

$$\nabla \cdot \mathbf{J} = \frac{1}{m}\nabla \cdot (A^2\nabla S) = \frac{1}{m}\left[\nabla(A^2) \cdot \nabla S + A^2\nabla^2 S\right]. \tag{15.59}$$

Using (15.55) we have

$$\frac{\partial(A^2)}{\partial t} + \frac{1}{m}\nabla(A^2) \cdot \nabla S + \frac{A^2}{2m}\nabla^2 S = 0, \tag{15.60}$$

that is, we have established the continuity equation

$$\frac{\partial\rho}{\partial t} + \nabla \cdot \mathbf{J} = 0. \tag{15.61}$$

Thus far we have used only the imaginary part equation ((15.55)) of the above system. If we define

$$\mathbf{v} = \frac{\mathbf{J}}{\rho} = \frac{1}{m}\nabla S, \tag{15.62}$$

the real part (equation (15.54)) takes the form

$$\frac{\partial S}{\partial t} + \frac{1}{2}m\mathbf{v}^2 + V(\mathbf{r}) = \frac{\hbar^2}{2m}\frac{\nabla^2 A}{A}. \tag{15.63}$$

In the formal limit $\hbar \to 0$ the right-hand side is zero, and noting that

$$\nabla(\mathbf{v}^2) = 2(\mathbf{v} \cdot \nabla\mathbf{v}) + 2\mathbf{v} \times (\nabla \times \mathbf{v}) = 2(\mathbf{v} \cdot \nabla\mathbf{v}), \tag{15.64}$$

on using (15.62) it follows that

$$m\frac{d\mathbf{v}}{dt} \equiv m\left(\frac{\partial \mathbf{v}}{\partial t} + \mathbf{v}\cdot\nabla\mathbf{v}\right) = -\nabla V(\mathbf{r}). \qquad (15.65)$$

This is once more Newton's second law of motion for a particle of constant mass m moving in a field of potential V. In terms of the probability or "wave function" Ψ, this equation describes the motion of a fluid composed of noninteracting particles of mass m [50].

15.3 CLASSICAL SCATTERING: SCATTERING ANGLES AND CROSS SECTIONS

15.3.1 Overview

The Center-of-Mass (CoM) Frame of Reference

For particles of mass and position vectors m_1, $\mathbf{r}_1$ and m_2, $\mathbf{r}_2$, respectively, it is convenient to change from a laboratory frame of reference to a *center-of-mass* (CoM) coordinate system. This is done by introducing relative and CoM coordinates $\mathbf{r}$ and $\mathbf{R}$, where

$$\mathbf{r} = \mathbf{r}_1 - \mathbf{r}_2, \qquad (15.66)$$

$$\mathbf{R} = \frac{m_1\mathbf{r}_1 + m_2\mathbf{r}_2}{m_1 + m_2}. \qquad (15.67)$$

It turns out that what we need to do is transform the results into the CoM frame. This is simply a frame of reference in which the observer is traveling along with the center of mass of the system, so the CoM velocity $\mathbf{V} = \mathbf{0}$, where

$$\mathbf{V} = \dot{\mathbf{R}} = \frac{m_1\dot{\mathbf{r}}_1 + m_2\dot{\mathbf{r}}_2}{m_1 + m_2}. \qquad (15.68)$$

The total momentum in this frame is therefore zero. In this frame of reference the problem becomes a one-body problem for a fictitious particle of mass $m = m_1 m_2 / (m_1 + m_2)$ encountering a fixed center of scattering and thereby moving in a central potential $V(r) \equiv V(|\mathbf{r}_1 - \mathbf{r}_2|)$. This will be assumed in what follows.

Consider the problem of a particle of mass m moving under the influence of a central force (i.e., the force is always directed toward a fixed point). We will establish that the path of the particle lies in a plane and its angular momentum $J = |\mathbf{J}|$ is conserved. As will be noted from equation (16.7) in Chapter 16, the latter result can be written in polar coordinates $(r(\theta), \theta(t))$ as

$$J = mr^2\dot{\theta} = \text{constant}, \qquad (15.69)$$

where $\dot{\theta} = d\theta/dt$. It is well known (to those who know it!) that this implies *Kepler's second law*, namely, that the radius vector of the particle sweeps out equal areas in equal times. If $V(r)$ is the potential field in which the particle moves, then the total energy E of the particle may be written (writing dr/dt as $\dot{r}$)

$$E = \frac{1}{2}m(\dot{r}^2 + r^2\dot{\theta}^2) + V(r) = \frac{1}{2}m\dot{r}^2 + \frac{J^2}{2mr^2} + V(r), \qquad (15.70)$$

from which

$$\dot{r} = \left\{ \frac{2}{m}[E - V(r)] - \left(\frac{J}{mr}\right)^2 \right\}^{1/2}. \qquad (15.71)$$

This expression can be integrated to yield the time t explicitly, but noting that

$$\frac{d\theta}{dr} = \frac{\dot{\theta}}{\dot{r}}, \qquad (15.72)$$

we find

$$\theta = \int \frac{J}{r^2 \left\{2m[E - V(r)] - (J/r)^2\right\}^{1/2}} dr + \text{constant}, \qquad (15.73)$$

which is the general solution of the equation of the path $\theta(r)$. It also follows from (15.69) that θ varies monotonically with time, since $\dot{\theta}$ never changes sign. The quantity

$$\frac{J^2}{2mr^2} + V(r) = U(r) \qquad (15.74)$$

is called the *effective potential*, being the sum of the central potential $V(r)$ and the "centrifugal" potential. Values of energy for which $E = U(r)$ are intimately related to the limits of the motion, since at these values $\dot{r} = 0$, thus defining the *turning points* of the path. The radial motion of the particle ranges from $r = \infty$ to the largest root r_0 of $\dot{r} = 0$, thus defining r_0 as the *classical distance of closest approach* (or outermost radial turning point), and then from r_0 to $r = \infty$. The total (classical) deflection angle Θ (see Figure 15.1) is therefore

$$\Theta(J) = \pi - 2J \int_{r_0}^{\infty} \frac{dr}{r^2 \left\{2m[E - V(r)] - (J/r)^2\right\}^{1/2}}. \qquad (15.75)$$

The *impact parameter b* is defined as

$$b \equiv \frac{J}{mv} = \frac{J}{(2mE)^{1/2}}, \qquad (15.76)$$

where v is the initial speed of the particle, so $\Theta = \Theta(b)$ also. Thus equation (15.75) is equivalent to

$$\Theta(b) = \pi - 2b \int_{r_0}^{\infty} \frac{dr}{r^2 \left\{1 - V(r)/E - (b/r)^2\right\}^{1/2}}. \qquad (15.77)$$

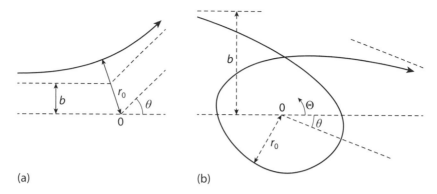

Figure 15.1 The deflection and scattering angles (See also Figure 1.1).

The deflection angle can be rewritten as

$$\Theta(E, J) = \pi - 2b \int_{r_0}^{\infty} \frac{dr}{r^2 \left[1 - V(r)/E - b^2/r^2\right]^{1/2}}. \tag{15.78}$$

Related to this quantity is the radial component of the *classical collision action* [101],

$$\Delta(E, J) = 2mv \left\{ \int_{r_0}^{\infty} \left[1 - V(r)/E - b^2/r^2\right]^{1/2} dr - \int_{b}^{\infty} \left[1 - b^2/r^2\right]^{1/2} dr \right\},$$

which is the difference between the radial momenta in the presence and absence of a potential. Another important quantity is the *collision delay time*, defined as

$$\tau(E, J) = 2 \left\{ \int_{r_0}^{\infty} \frac{dr}{\dot{r}} - \int_{b}^{\infty} \frac{dr}{\dot{r}} \right\}$$

$$= \frac{2}{v} \left\{ \int_{r_0}^{\infty} \left[1 - V(r)/E - b^2/r^2\right]^{-1/2} dr - \int_{b}^{\infty} \left[1 - b^2/r^2\right]^{-1/2} dr \right\}.$$

Again, this is a difference, now between the collision time in the presence and absence of a potential. In fact

$$\left(\frac{\partial \Delta}{\partial J}\right)_E = \Theta(E, J), \tag{15.79}$$

and

$$\left(\frac{\partial \Delta}{\partial E}\right)_J = \tau(E, J), \tag{15.80}$$

whence

$$d\Delta = \Theta(E, J)dJ + \tau(E, J)dE.$$

Exercise: Prove (15.79) and (15.80). These classical results foreshadow those based on the WKB(J) phase shift in semiclassical approaches (see, e.g., equations (22.194) and (22.196) in Chapter 22 and Appendix F).

For an entirely repulsive potential, $\Theta \in [0, \pi]$, but as pointed out in [26], it can take arbitrarily large negative values for an attractive one, since the particle may circle around the scattering center many times before finally emerging. The relationship between the deflection angle Θ and the *scattering angle* θ is

$$\Theta = \pm\theta - 2n\pi, \quad n = 0, 1, 2, \ldots,$$

where n is chosen such that $\theta \in [0, \pi]$, its physical range of variation (recall from Chapter 4 that the angle of minimum deviation for the (primary) rainbow ray was denoted by $D_{\min}$ or θ_R). For particles with impact parameters in the range $[b, b + \Delta b]$, the area of the associated annulus is $\Delta\sigma = 2\pi b \, |\Delta b|$; if these particles are scattered within an onion ring of area $\Delta S = 2\pi R^2 \sin\theta \, |\Delta\theta|$ (see Figure 15.2), or equivalently, within a solid angle $\Delta\Omega = 2\pi \sin\theta \, |\Delta\theta|$, then the *differential scattering cross section* is defined to be (in the appropriate limiting sense)

$$\frac{d\sigma}{d\Omega} = \frac{b \csc\theta}{|d\theta/db|} = \frac{b \csc\theta}{|d\Theta/db|}. \tag{15.81}$$

The distribution is independent of the azimuthal angle ϕ for a central force field. Note that for a repulsive potential in particular, $d\theta/db < 0$, because the larger the impact parameter is, the smaller will be the angle through which the particles are scattered in general. Note that the total scattering cross section is obtained by integrating (15.81) over all solid angles:

$$\sigma = \int \frac{d\sigma}{d\Omega} d\Omega = \int_0^{b_{\max}} 2\pi b \left| \frac{db}{d\theta} \right| d\theta = \pi b_{\max}^2, \tag{15.82}$$

where $b_{\max}$ is the maximal impact parameter, that is, if for $r > b_{\max}$ the potential vanishes (as it does for square well or barrier potentials), then particles with an impact parameter greater than $b_{\max}$ are not deflected at all (at least, classically; however, *tunneling* can occur, as will be discussed in Chapter 27). The scattering cross section is then just the geometrical cross section of the region; if the scattering potential has no such finite cutoff, then the classical scattering cross section is infinite. This is related to a sharp forward peaking of the differential cross section [51], [26] and is one of the classical singularities in $d\sigma/d\Omega$ discussed below. There may exist several trajectories that give rise to the same scattering angle, so (15.81) must be generalized to sum over all the impact parameters that lead to the same angle:

$$\frac{d\sigma}{d\Omega}(\theta) = \sum_j \frac{b_j(\theta) \csc\theta}{|d\theta/db_j|}, \tag{15.83}$$

where the derivative on the right side is equivalent to the Jacobian of the transformation relating θ and b. Some of these trajectories are illustrated in Figure 15.2 for

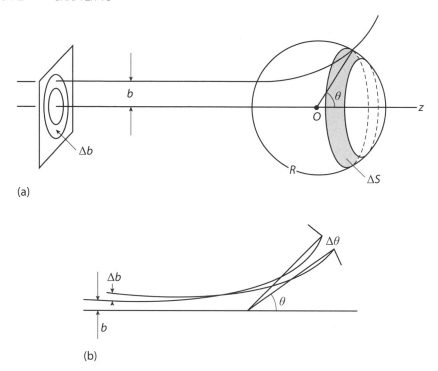

(a)

(b)

Figure 15.2 Terminology for the differential scattering cross section.

a typical potential type found in atomic and nuclear scattering problems, namely, an attractive outer region with a repulsive central core (also shown schematically in the figure, along with a typical deflection function $\Theta(b)$). There are three different types of singularities (or *caustics*—a caustic being an envelope of a family of rays, as noted in Chapter 6) associated with directions in which the differential cross section diverges. These are as follows.

(I) Rainbow scattering. This is a caustic arising when $\theta = \theta_R$, the rainbow angle; it occurs when

$$\left(\frac{d\theta}{db}\right)_{\theta_R} = 0, \tag{15.84}$$

so that the deflection function passes through an extremum. Its name arises from the analogy with the optical rainbow. In heavy-ion scattering, this caustic is referred to as Coulomb or nuclear rainbow scattering if the extremum is a maximum or minimum, respectively, and is manifested in terms of an angular distribution given by the Airy function (see Chapters 5–7). In view of (15.84) for θ sufficiently close to θ_R, we may write

$$\theta(b) \approx \theta_R + \frac{1}{2}\theta''(b_R)(b - b_R)^2, \tag{15.85}$$

where $b = b_R$ when $\theta = \theta_R$. Under these circumstances, the classical differential scattering cross section is

$$\frac{d\sigma}{d\Omega} \approx \left(\frac{b}{\sin\theta}\right)_R \left[\frac{2}{|\theta''(b_R)(\theta - \theta_R)|}\right]^{1/2}, \tag{15.86}$$

which diverges like $|\theta - \theta_R|^{-1/2}$.

(II) Glory scattering. This is a caustic caused by the vanishing of $\sin\theta$ in equation (15.81). In terms of the deflection angle Θ for nonzero impact parameter b_G,

$$\Theta(b_G) = n\pi \quad (n = 0, -1, -2, \ldots),$$

so that the differential cross section diverges like $(\sin\theta)^{-1}$ in the precisely forward direction (n even: forward glory scattering) or backward direction (n odd: backward glory scattering). Equation (15.75) implies that $\Theta = \pi$ can only occur when $b = 0$; this case does not constitute glory scattering.

(III) Forward peaking. As alluded to above, scattering potentials with tails extending asymptotically to zero at infinity will lead in general to a divergent forward differential cross section. This is because of the accumulation of small deflections for particles with arbitrarily large impact parameters, that is, from (15.83) in which some $b_j \to \infty$, $\theta \to 0$ while $|d\theta/db_j|$ remains bounded. Even for cutoff potentials, some forward anomaly is still to be expected.

(IV) Orbiting. This situation may occur when the effective potential $U(r) = U_J(r)$ for a given value of J has a relative maximum at $r = r_0$. In this case,

$$U_J(r) = E; \quad (dU_J/dr)_{r_0} = 0.$$

This is interpreted as the existence of an unstable circular orbit with radius r_0. Under these circumstances the integral (15.75) diverges logarithmically at its lower limit if $U_J''(r_0) \neq 0$ [73]. A simple calculation yields the result that $\Theta(J) \to -\infty$ as $r \to r_0$; this means that a particle with this energy and angular momentum will spiral indefinitely around the scattering center, taking an infinitely long time to reach the top of the barrier. In addition, the inverse function $b(\theta)$ becomes infinitely many-valued, so an infinite number of branches contribute to the differential cross section (15.83). If $U_J(r)$ has a local minimum for $r < r_0$, then particles with energy $E < U_J(r_0)$ can sustain oscillatory orbits in the potential well, and the unstable circular orbit functions as a separatrix between bounded and unbounded trajectories. In fact there are several interesting connections between the effective potential, tunneling, and the ray picture in rainbow and glory formation [25]. A brief discussion of this phenomenon in an optical context can be found in Chapter 28 (see also [73]).

15.3.2 The Classical Inverse Scattering Problem

This problem describes a particle of energy E scattered by a spherically symmetric scatterer; it is scattered through an angle θ, where $\theta = \theta(V(r), E, b)$, and b is the impact parameter. $V(r)$ can be uniquely determined from $\theta(b)$ for a single E if $V(r)$ is repulsive, that is, if $V(r) > 0$ and $V'(r) < 0$, then $V(0) > E$. Also, $V(r)$ can be determined for all r, but only for $r > r_{\min}(E)$, where $r_{\min}(E)$ is the distance of closest approach at energy E. Formally, this means that the largest root of $V(r_{\min}) = E$. The differential scattering cross section, a measured quantity in experiments (as opposed to $\theta(b)$) is for a repulsive potential

$$\frac{d\sigma}{d\Omega} = \left(\frac{-b}{\sin\theta}\right)\frac{db}{d\theta} = \frac{-1}{2\sin\theta}\frac{db^2}{d\theta} > 0,$$

that is,

$$b^2 = 2\int_\theta^\pi \frac{d\sigma(\theta')}{d\Omega'}\sin\theta'd\theta. \tag{15.87}$$

It follows that given $d\sigma/d\Omega$ at a single energy E, $V(r)$ is uniquely determined for $r > r_{\min}(E)$. In contrast, in quantum mechanics information is required for all energies E, and it is important to determine $V(r)$ for all r, though $d\sigma/d\theta$ is not required; only the phase shift for any one angular momentum is needed.

The classical deflection function is given by a variant of equation (15.75), namely,

$$\Theta(b) = \pi - 2\int_{r_0}^\infty \frac{bdr}{r^2\left[(1 - V/E) - b^2/r^2\right]^{1/2}}, \tag{15.88}$$

where $r_0 = r_{\min}(E)$.

The basic idea behind the inverse problem is to determine the function

$$W(r) = r^2\left(1 - \frac{V(r)}{E}\right), \tag{15.89}$$

from which the potential $V(r)$ can be determined. The approach of [101] is followed here, though it is related to the analysis of seismic rays in Chapter 9 (and the Luneberg lens discussed in Chapter 28). When $V(r) = 0$ it follows from equations (15.88) and (15.89) that, in terms of $W(r)$,

$$\int_{b^2}^\infty \frac{bdW}{W(W - b^2)^{1/2}} = \pi. \tag{15.90}$$

Eliminating π between equations (15.88) and (15.90), we find that

$$\Theta(b) = \int_{b^2}^\infty \frac{b(dY/dW)dW}{W(W - b^2)^{1/2}}, \tag{15.91}$$

where

$$Y(W) = \ln\left(\frac{W}{r^2}\right) = \ln\left[1 - \frac{V(r)}{E}\right] \tag{15.92}$$

(See also equation (22.201)) This implies that $Y \to 0$ as r (and hence $W) \to \infty$. Suppose that a typical point is $W_\tau = b_\tau^2$. If we multiply both sides of equation (15.91) by $\left(b^2 - W_\tau\right)^{-1/2}$ and integrate with respect to b over the interval $[b_\tau, \infty)$ then

$$\int_{b_\tau}^\infty \frac{\Theta(b)db}{\left(b^2 - W_\tau\right)^{1/2}} = \frac{1}{2}\int_{W_\tau}^\infty \left\{\int_{b^2}^\infty \frac{(dY/dW)dW}{\left[(b^2 - W_\tau)(W - b^2)\right]^{1/2}}\right\}db^2. \tag{15.93}$$

Formally reversing the order of integration on the right-hand side of this equation results in

$$\int_{b_\tau}^{\infty} \frac{\Theta(b)db}{(b^2 - W_\tau)^{1/2}} = \frac{1}{2} \int_{W_\tau}^{\infty} \left(\frac{dY}{dW}\right) \left\{ \int_{W_\tau}^{W} \frac{db^2}{[(b^2 - W_\tau)(W - b^2)]^{1/2}} \right\} dW$$

$$= \frac{\pi}{2} Y(W_\tau). \tag{15.94}$$

Use of equation (15.92) yields the result

$$r(W_\tau) = W_\tau^{1/2} \exp\left[\int_{W_\tau^{1/2}}^{\infty} \frac{\Theta(b)db}{(b^2 - W_\tau)^{1/2}}\right]. \tag{15.95}$$

This result can then be inverted (in principle) using equation (15.89) to obtain the potential $V(r)$. However, some caveats are in order with regard to the above procedure. It is necessary for dY/dW to be finite over the domain of integration in equation (15.94) for the reversal of integration order to be valid (in a very similar fashion to that discussed in Chapter 9; see also Chapter 28). This implies that $W(r)$ should be a monotonic function, that is, $r(W)$ should be single valued for all W. This in turn imposes the condition from (15.88) that the equation

$$1 - \frac{V(r)}{E} - \frac{W(r)}{r^2} = 0, \tag{15.96}$$

should possess a unique solution. A further problem is eliminated if the potential is known to be purely repulsive in character, for then $\Theta(b) = \theta(b)$, and the latter is uniquely determined by the differential scattering cross section, so inversion of $b(\theta)$ is well defined. If, however, the potential is attractive (or contains an attractive well), $b(\theta)$ is likely to be multivalued at angles less than the rainbow angle. It is then necessary to employ semiclassical considerations using the WKB(J) phase shift (again, see Chapter 22 and Appendix F) or a variant of the technique discussed above (see [101] for further details).

Chapter Sixteen

Gravitational Scattering

Gravitational scattering is the deflection of the paths of particles passing near a massive body. For small angle scattering, the change in velocity is given by

$$\Delta \mathbf{v} \equiv \int_{-\infty}^{\infty} \frac{\mathbf{F}}{m} dt, \tag{1}$$

where $\mathbf{F}$ is the gravitational force and m is the mass of the scattering particle. The perpendicular component of the speed change is then given by

$$\Delta v_{\perp} = \int_{-\infty}^{\infty} \frac{F \sin \theta}{m} dt, \tag{2}$$

where θ is the angle with respect to the initial unperturbed trajectory. Using

$$dt = \frac{dy}{v}, \tag{3}$$

where dy/dt is the perpendicular offset gives

$$\sin \theta \approx \frac{b}{r}, \tag{4}$$

where b is the so-called impact parameter and r is the distance from the central mass. [252]

We discuss in more detail some of the statements made in Chapter 15. In particular, the angular momentum vector for a particle of constant mass m and position vector r about the origin is

$$\mathbf{J} = \mathbf{r} \times m\dot{\mathbf{r}} = m\mathbf{r} \times \dot{\mathbf{r}}, \tag{16.1}$$

where $\dot{\mathbf{r}} = d\mathbf{r}/dt$. In what follows, we restrict ourselves to the study of central conservative forces (such as electrostatic forces, but primarily gravitational ones): these are forces that are (i) directed toward or away from a fixed point and (ii) can be expressed as the gradient of a potential function $V(\mathbf{r})$. Under these conditions, such a force $\mathbf{F}$ may be written

$$\mathbf{F} = -\frac{\mathbf{r}}{r} \frac{dV(r)}{dr} = F(r) \frac{\mathbf{r}}{r} = m\ddot{\mathbf{r}}, \tag{16.2}$$

where $r = |\mathbf{r}|$. Since $\mathbf{F}$ and $\mathbf{r}$ are parallel, $\mathbf{F} \times \mathbf{r} = 0$. But $\dot{\mathbf{J}}$, the time rate of change of $\mathbf{J}$, is

$$\dot{\mathbf{J}} = m\,(\dot{\mathbf{r}} \times \dot{\mathbf{r}} + \mathbf{r} \times \ddot{\mathbf{r}}) = m\mathbf{r} \times \ddot{\mathbf{r}} = \mathbf{r} \times \mathbf{F} = 0. \tag{16.3}$$

Therefore J is constant (in both magnitude and direction, of course). But $\mathbf{J}$ is orthogonal to the plane containing $\mathbf{r}$ and $\dot{\mathbf{r}}$, so these latter quantities must also lie in a plane. The two conservation laws of value to us are conservation of energy and angular momentum, namely, with $r = |\mathbf{r}|$,

$$\frac{1}{2}m\dot{\mathbf{r}}^2 + V(r) = E = \text{constant}, \tag{16.4}$$

and

$$m\mathbf{r} \times \dot{\mathbf{r}} = \text{a constant vector}, \tag{16.5}$$

respectively. In terms of polar coordinates (r, θ) in the plane, since $\mathbf{v} = \dot{\mathbf{r}} = \langle \dot{r}, r\dot{\theta}, 0 \rangle$, these equations take the form

$$\frac{1}{2}m\left(\dot{r}^2 + r^2\dot{\theta}^2\right) + V(r) = E, \tag{16.6}$$

and

$$mr^2\dot{\theta} = J. \tag{16.7}$$

Eliminating $\dot{\theta}$, we obtain

$$\frac{1}{2}m\dot{r}^2 + \frac{J^2}{2mr^2} + V(r) = E, \tag{16.8}$$

sometimes referred to as the radial energy equation. This conservation law contains an effective potential energy function

$$U(r) = \frac{J^2}{2mr^2} + V(r). \tag{16.9}$$

The additional term corresponds to a force of magnitude J^2/mr^3; this is the so-called centripetal force $mr\dot{\theta}^2$ (expressed in terms of J) required to maintain the particle in a closed bound orbit. Since $\dot{r}^2 \geq 0$, it follows that

$$\frac{J^2}{2mr^2} + V(r) \leq E. \tag{16.10}$$

Extrema of the radial distances permitted are given by the values of r for which equality holds in equation (16.10).

16.1 PLANETARY ORBITS: SCATTERING BY A GRAVITATIONAL FIELD

If we eliminate the time dependence from equations (16.6) and (16.7), we can obtain the polar equations of orbits in which the central field is a gravitational one [52]. To this end, let $u = r^{-1}$, so

$$\frac{du}{d\theta} = -r^{-2}\frac{dr}{d\theta}, \quad \text{and} \quad \dot{r} = \frac{dr}{d\theta}\dot{\theta} = -r^2\dot{\theta}\frac{du}{d\theta} = -\frac{J}{m}\frac{du}{d\theta}. \tag{16.11}$$

Using this result in equation (16.8), it follows that

$$\left(\frac{du}{d\theta}\right)^2 + u^2 = \frac{2m}{J^2}\left(E - V(u)\right). \tag{16.12}$$

In the case of an inverse square law for which, from (16.2),

$$\mathbf{F} = -\frac{\mathbf{r}}{r}\frac{dV(r)}{dr} = \frac{k}{r^2}\frac{\mathbf{r}}{r}, \tag{16.13}$$

we then have

$$V(r) = \frac{k}{r} = ku. \tag{16.14}$$

The constant k may be of either sign; positive for a repulsive "gravitational" force, and negative for an attractive one. The effective potential energy function is now

$$U(r) = \frac{J^2}{2mr^2} + \frac{k}{r}. \tag{16.15}$$

16.1.1 Repulsive Case: $k > 0$

Clearly $U(r)$ is a monotonically decreasing function on $(0, \infty)$ with range $(0, \infty)$. Therefore for a given value of $E > 0$, there is a unique positive solution of the equation $U(r) = E$, and this defines a minimum radius of approach to the scattering center. From equation (16.8),

$$\dot{r}^2 = \frac{2}{m}[E - U(r)], \tag{16.16}$$

so when $E < U(r)$ there are no real values for $\dot{r}$. Conversely, for $E > U(r)$, $\dot{r}$ is real and can be of either sign, depending on whether the particle is approaching or receding from the scattering center. As an example, consider a particle of charge q_1 moving in the field of a fixed point of charge q_2. Suppose that it is approaching with speed v and from such a great distance that its initial potential energy can be neglected. Furthermore, if it were to continue in a straight line, it would pass within a distance b of the charge q_2; b is the impact parameter. The angular momentum is $J = mvb$, the constant $k = q_1 q_2$, and asymptotically, $E = mv^2/2$, so at the point of closest approach $(r = r_c)$, $\dot{r} = 0$, implying from equation (16.8) that

$$v^2 = \frac{bv^2}{r_c^2} + \frac{2q_1 q_2}{mr_c},$$

from which the required positive root is

$$r_c = L + \left(L^2 + b^2\right)^{1/2}.$$

Here the parameter $L = q_1 q_2/mv^2$, so the distance of closest approach is slightly larger than b if $b \gg L$.

16.1.2 Attractive Case: $k < 0$

We define a convenient length scale, R, by $R = J^2/m |k|$, from which equation (16.15) can be expressed as

$$U(r) = \frac{|k|}{r} \left(\frac{R}{2r} - 1 \right). \tag{16.17}$$

Obviously $U(R/2) = 0$, and a unique minimum of U occurs at the point $(R, - |k| /(2R))$. Unlike the repulsive field, there are several cases of interest to consider here, classified by values and ranges of total energy E.

(i) $E = U_{min} = - |k| /(2R)$.

From (16.16), $\dot{r}$ is identically zero, corresponding to motion in a circle of radius R. The potential energy is $V = - |k| /R$, and the kinetic energy is $T \equiv E - V = |k| /(2R) \; (= mv^2/2$, thus defining the orbital speed v), which for this special orbit is half the magnitude of V.

(ii) $- |k| /(2R) < E < 0$.

This energy range confines the possible radial values to an interval $[r_{min}, r_{max}]$. As might be expected (and will be proven below), this orbit is elliptical, with $r = r_{min}$ and $r = r_{max}$ defining the perihelion and aphelion, respectively, for objects orbiting the sun. These terms are respectively named perigee and apogee for satellites orbiting the earth.

(iii) $E = 0$.

Because $E - U(r) > 0$ for $r > r_{min}$, the point of closest approach is defined by $r_{min} = R/2$; however r_{max} is unbounded (i.e., the particle can escape to infinity); since

$$T + V = 0, \quad \text{and} \quad \lim_{r \to \infty} V = 0^-,$$

it follows that the particle kinetic energy tends to zero at infinity; the particle only just makes it! Its orbit, as demonstrated below is parabolic.

(iv) $E > 0$.

As with (iii) above, $r \in [r_{min}, \infty)$, but this time $\lim_{r \to \infty} T > 0$, so the particle escapes to infinity with nonzero limiting speed, along, it will be shown, a hyperbolic orbit.

16.1.3 The Orbits

Returning to equation (16.12), we use the inverse square formulation for V, namely, $V = ku$, as before, with $R = J^2/(m |k|) = \pm J^2/(mk)$ for the repulsive and attractive cases, respectively. Then we have

$$\left(\frac{du}{d\theta} \right)^2 + u^2 \pm \frac{2u}{R} = \frac{2E}{R |k|}, \tag{16.18}$$

or, introducing the new dependent variable $z = Ru \pm 1$,

$$\left(\frac{dz}{d\theta} \right)^2 + z^2 = \frac{2ER}{|k|} + 1 = \text{constant} = e^2 > 0, \tag{16.19}$$

where e will be defined below as the eccentricity of the orbit. From this it follows that

$$\int \frac{dz}{\left(e^2 - z^2\right)^{1/2}} = \pm(\theta - K),$$

K being a constant of integration. If we take the negative sign, without loss of generality, and $K = \theta_0 + \pi/2$, then

$$z = e \cos\left(\theta - \theta_0\right). \tag{16.20}$$

In terms of r, for the repulsive case ($k > 0$), the equation of the orbit is

$$r\left[e \cos\left(\theta - \theta_0\right) - 1\right] = R, \tag{16.21}$$

and for the attractive case ($k < 0$),

$$r\left[e \cos\left(\theta - \theta_0\right) + 1\right] = R. \tag{16.22}$$

These are the polar equations of conic sections with focus at the origin of the coordinates. Note from equation (16.21) that e must exceed unity for the equation to be satisfied. From equation (16.19) this implies that $E > 0$, corresponding to a hyperbolic orbit as expected. The eccentricity determines the shape of the orbit; clearly for a circular orbit $e = 0$. For a hyperbolic orbit, the larger e is the greater the angle between the asymptotes of each branch of the hyperbola will be. Equation (16.22) requires that $0 \leq e < 1$ (and hence $E < 0$), corresponding to an elliptical orbit. In fact, in this case, for $k < 0$, $-k/(2R) < E < 0$, as has been noted in case (ii) above. For both repulsive and attractive orbits, the perihelion point (closest approach) is defined by the direction $\theta = \theta_0$, for which r is a minimum. This direction also has an important interpretation, which will be investigated below. Note also from equation (16.22) for elliptical orbits that $r = R$ when $\cos\left(\theta - \theta_0\right) = 0$ (i.e., when $\theta = \theta_0 \pm \pi/2$); this corresponds to the radial distance orthogonal to $\theta = \theta_0$. The results for elliptical orbits are also readily recognized when the orbit equation is expressed in Cartesian form, using the transformations $x = r \cos\theta$ and $y = r \sin\theta$. We align the x-axis along the direction $\theta = \theta_0$ so that the new $\theta_0 = 0$, and equation (16.22) in particular becomes

$$r\left[e \cos\theta + 1\right] = ex + r = R. \tag{16.23}$$

For $0 < e < 1$ this can be reconfigured into the Cartesian form using the intermediate result

$$x^2\left(1 - e^2\right) + 2eRx + y^2 = R^2, \tag{16.24}$$

or

$$\frac{(x + x_0)^2}{a^2} + \frac{y^2}{b^2} = 1, \tag{16.25}$$

where

$$x_0 = \frac{eR}{1 - e^2}, \quad a = \frac{R}{1 - e^2}, \quad \text{and} \quad b = \frac{R}{\left(1 - e^2\right)^{1/2}} = (aR)^{1/2}. \tag{16.26}$$

This is the equation of an ellipse centered at $(-x_0, 0)$ with semi-major axis a and semi-minor axis $b < a$. Clearly, $x_0 = ae$. Note that when $x = 0$ in equation (16.24), $y = \pm R$ (i.e., R is the semi-latus rectum for the ellipse). If $e = 0$, the equation of a circle is recovered. For $e = 1$, (16.24) reduces to

$$y^2 = 2R\left(\frac{R}{2} - x\right), \tag{16.27}$$

which is the equation of a parabola with vertex at $(R/2, 0)$, focus at the origin, directrix $x = R$, and y-intercepts at $(0, \pm R)$. Finally, if $e > 1$, equation (16.24) becomes

$$\frac{(x - x_0)^2}{a^2} - \frac{y^2}{b^2} = 1, \tag{16.28}$$

where now

$$x_0 = \frac{eR}{e^2 - 1}, \quad a = \frac{R}{e^2 - 1} \quad \text{and} \quad b = \frac{R}{\left(e^2 - 1\right)^{1/2}} = (aR)^{1/2}. \tag{16.29}$$

Equation (16.28) is the equation of a hyperbola centered at $(x_0, 0)$, that is, $\left(eR/(e^2 - 1), 0\right)$ or $(ae, 0)$. The asymptotes are defined by the equations

$$y = \pm\frac{b}{a}(x - x_0).$$

There is (literally) another far-reaching consequence of the fact that the angular momentum vector $\mathbf{J}$ is constant: Kepler's second law. The velocity vector $\dot{\mathbf{r}} = v = \langle v_r, v_\theta \rangle = \langle \dot{r}, r\dot{\theta} \rangle$, so the magnitude of $\mathbf{J}$ is, as noted in (16.7),

$$J = mrv_\theta = mr^2\dot{\theta}. \tag{16.30}$$

Geometrically, the area swept out in the angular interval $(\theta, \theta + d\theta)$ is $dA = r^2 d\theta/2$, so

$$\dot{A} = \frac{dA}{dt} = \frac{1}{2}r^2\dot{\theta} = \frac{J}{2m} = \text{constant}. \tag{16.31}$$

This is just Kepler's second law of planetary motion again; as the planetary particle moves around the sun (with the sun at one focus of its elliptical orbit), the radius vector sweeps out equal areas in equal times. Equivalently, the transverse component of the velocity vector $(r\dot{\theta})$ varies inversely with radial distance r. If we integrate the equation for $\dot{A}$ over one complete orbit of period T, we deduce that $A = JT/(2m)$, but since the orbit of a planet is elliptical, $A = \pi ab$, and $b^2 = aR$ (from (16.26)), it follows that

$$\left(\frac{T}{2\pi}\right)^2 = \frac{m^2 R}{J^2}a^3 = \frac{m}{|k|}a^3, \tag{16.32}$$

which is Kepler's third law of planetary motion: the square of the orbital period is proportional to the cube of the semi-major axis. The equation of motion for a particle of mass m moving under the influence of the force defined by equation (16.2) is

$$m\ddot{\mathbf{r}} = F(r)\frac{\mathbf{r}}{r}.$$

Taking the vector product of each side of this equation with $\mathbf{r}$ yields a result already noted, namely, $m\mathbf{r} \times \ddot{\mathbf{r}} = 0$. On integrating this, the constant angular momentum vector is

$$\mathbf{J} = \mathbf{r} \times m\dot{\mathbf{r}},$$

which is just equation (16.1). Now taking the vector product of the equation of motion with $\mathbf{J}$ eliminates any component of $m\ddot{\mathbf{r}}$ perpendicular to the plane of motion. Hence, rewriting

$$\mathbf{J} \times m\ddot{\mathbf{r}} = \mathbf{J} \times F(r)\frac{\mathbf{r}}{r}$$

as

$$\frac{d}{dt}(\mathbf{J} \times m\dot{\mathbf{r}}) = \mathbf{J} \times m\ddot{\mathbf{r}} = (\mathbf{r} \times m\dot{\mathbf{r}}) \times F(r)\frac{\mathbf{r}}{r} = \frac{mF(r)}{r}[(\mathbf{r} \times \dot{\mathbf{r}}) \times \mathbf{r}]$$

$$= \frac{mF(r)}{r}[(\mathbf{r} \cdot \mathbf{r})\dot{\mathbf{r}} - (\dot{\mathbf{r}} \cdot \mathbf{r})\mathbf{r}] = \frac{mF(r)}{r}[r^2\dot{\mathbf{r}} - r\dot{r}\mathbf{r}]. \quad (16.33)$$

This last result follows from the fact that $d(\mathbf{r} \cdot \mathbf{r})/dt = d(r^2)/dt$, or $\mathbf{r} \cdot \dot{\mathbf{r}} = r\dot{r}$. Therefore,

$$\frac{d}{dt}(\mathbf{J} \times \dot{\mathbf{r}}) = F(r)[r\dot{\mathbf{r}} - \dot{r}\mathbf{r}]. \quad (16.34)$$

This result is perfectly general, but for what central forces F, if any, can the right-hand side of this equation be a perfect differential, resulting in a trivial integration? That is, is it possible to write this equations as

$$\frac{d}{dt}(\mathbf{J} \times \dot{\mathbf{r}}) = \frac{d}{dt}[\beta(r)\mathbf{r}] \quad (16.35)$$

for some $\beta(r)$? For this to be the case, it follows that $\beta'(r) = -F(r)$, and $\beta(r) = rF(r)$, whence $\beta'/\beta = -r^{-1}$ and therefore $\beta(r) = \lambda r^{-1}$, λ being a constant, so that $F(r) = \lambda r^{-2}$, generating the inverse square force once again! Integration of equation (16.35) yields

$$\mathbf{J} \times \dot{\mathbf{r}} - \frac{\lambda}{r}\mathbf{r} = \boldsymbol{\kappa}, \quad (16.36)$$

where $\boldsymbol{\kappa}$ is a constant vector known as the *Lenz-Runge vector* [53]. Several properties associated with this vector follow immediately. They are:
(i) Since $\mathbf{J} = \mathbf{r} \times m\dot{\mathbf{r}}$,

$$\mathbf{r} \cdot \boldsymbol{\kappa} = \mathbf{r} \cdot \left(\mathbf{J} \times \dot{\mathbf{r}} - \frac{\lambda}{r}\mathbf{r}\right) = -\mathbf{J} \cdot (\mathbf{r} \times \dot{\mathbf{r}}) - \lambda r = -\frac{J^2}{m} - \lambda r, \quad (16.37)$$

Therefore, in terms of the angle θ between the vectors $\mathbf{r}$ and $\boldsymbol{\kappa}$,

$$r\kappa\cos\theta = -\frac{J^2}{m} - \lambda r,$$

or

$$r(\lambda + \kappa\cos\theta) = -\frac{J^2}{m}. \quad (16.38)$$

where $\kappa = |\boldsymbol{\kappa}|$. This is the polar form of the orbit under the inverse square force, similar in expression to equation (16.23). In the case of orbits, $R = J^2/(m\,|k|)$, so equation (16.38) has the equivalent form

$$r\left(1 - \frac{\kappa}{|k|}\cos\theta\right) = R,\tag{16.39}$$

for $k < 0$, and for $k > 0$,

$$r\left(1 + \frac{\kappa}{|k|}\cos\theta\right) = -R.\tag{16.40}$$

These equations are exactly equations (16.21) and (16.22) with $\theta_0 = -\pi$ and eccentricity

$$e = \frac{\kappa}{k} = \frac{\kappa}{Gm_1 m_2}$$

for a point of mass m_1 orbiting a point of mass m_2, G being the universal constant of gravitation.

(ii) We have

$$\boldsymbol{\kappa}\cdot\mathbf{J} = \left(\mathbf{J}\times\dot{\mathbf{r}} - \frac{\lambda}{r}\mathbf{r}\right)\cdot\mathbf{J} = 0.\tag{16.41}$$

Therefore since $\mathbf{J}$ is perpendicular to the plane of the orbit, $\boldsymbol{\kappa}$ lies in that plane, and the set $\{\mathbf{J}, \boldsymbol{\kappa}, \mathbf{J}\times\boldsymbol{\kappa}\}$ forms an orthogonal triad of constant basis vectors for the particle dynamics.

(iii) $\boldsymbol{\kappa}$ is a measure of the total energy E of the orbiting particle. Clearly,

$$\kappa^2 = \left(\mathbf{J}\times\dot{\mathbf{r}} - \frac{\lambda}{r}\mathbf{r}\right)\cdot\left(\mathbf{J}\times\dot{\mathbf{r}} - \frac{\lambda}{r}\mathbf{r}\right)$$

$$= (\mathbf{J}\times\dot{\mathbf{r}})\cdot(\mathbf{J}\times\dot{\mathbf{r}}) - \frac{2\lambda}{r}(\mathbf{J}\times\dot{\mathbf{r}})\cdot\mathbf{r} + \lambda^2$$

$$= |\mathbf{J}\times\dot{\mathbf{r}}|^2 + \frac{2\lambda}{r}\mathbf{J}\cdot(\dot{\mathbf{r}}\times\mathbf{r}) + \lambda^2$$

$$= J^2|\dot{\mathbf{r}}|^2\left(1 - \cos^2\theta\right) - \frac{2\lambda J^2}{mr} + \lambda^2$$

$$= J^2|\dot{\mathbf{r}}|^2 - (\mathbf{J}\cdot\dot{\mathbf{r}})^2 - \frac{2\lambda J^2}{mr} + \lambda^2.\tag{16.42}$$

But $\mathbf{J}\cdot\dot{\mathbf{r}} = 0$, and from equation (16.4),

$$\dot{\mathbf{r}}^2 = \frac{2}{m}(E - V(r)),$$

so for the inverse square force,

$$\kappa^2 = \frac{2J^2}{m}\left(E + \frac{\lambda}{r}\right) - \frac{2\lambda J^2}{mr} + \lambda^2$$

$$= \frac{2J^2 E}{m} + \lambda^2.\tag{16.43}$$

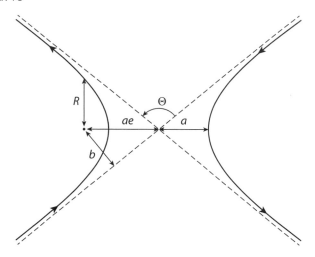

Figure 16.1 Geometry of the hyperbola.

In the hyperbolic case, an orbit can be either attractive or repulsive. The asymptotes of the trajectories are, from equation (16.28), given by $y = \pm (b/a)\, x$, so the angles θ for which $r \to \infty$ are such that $\tan\theta = (e^2 - 1)^{1/2}$ in the repulsive case, that is,

$$\theta = \pm \arccos\left(e^{-1}\right)$$

(see Figure 16.1).

In the attractive case

$$\theta = \pi \pm \arccos\left(e^{-1}\right).$$

The scattering angle Θ is defined as the angle through which the particle is deviated or deflected over its complete trajectory. For both cases it can be seen that

$$\Theta = \pi - 2\arccos\left(e^{-1}\right). \qquad (16.44)$$

There is a straightforward relationship between Θ, the impact parameter b, and the asymptotic speed v: we know that $E = mv^2/2$, and from (16.19),

$$e^2 - 1 = \frac{2ER}{|k|}, \qquad (16.45)$$

so

$$a = R/(e^2 - 1) = |k|\,/(2E) = |k|\,/(mv^2). \qquad (16.46)$$

But

$$\arccos\left(e^{-1}\right) = \left(\frac{\pi - \Theta}{2}\right), \quad \text{so} \quad e = \sec\left(\frac{\pi - \Theta}{2}\right),$$

and therefore,

$$B^2 = A^2\left(e^2 - 1\right) = A^2 \tan^2\left(\frac{\pi - \Theta}{2}\right) = A^2 \cot^2\frac{\Theta}{2}.$$

Hence the impact parameter is

$$b = \frac{|k|}{mv^2} \cot \frac{\Theta}{2}.$$ (16.47)

Note also that from (16.45) we can write

$$e = \left(\frac{2ER}{|k|} + 1\right)^{1/2} = \left(\frac{2J^2R}{mk^2} + 1\right)^{1/2}.$$ (16.48)

16.2 THE HAMILTON-JACOBI EQUATION FOR A CENTRAL POTENTIAL

In spherical coordinates the conjugate momenta are expressible in terms of the associated Lagrangian as

$$p_r = \frac{\partial \mathcal{L}}{\partial \dot{r}} = m\dot{r}; \quad p_\theta = \frac{\partial \mathcal{L}}{\partial \dot{\theta}} = mr^2\dot{\theta}; \quad p_\phi = \frac{\partial \mathcal{L}}{\partial \dot{\phi}} = mr^2\dot{\phi}\sin^2\theta,$$ (16.49)

where the following definition for $\mathcal{L}$ has been used:

$$\mathcal{L} = \frac{1}{2}m\left(\dot{r}^2 + r^2\dot{\theta}^2 + r^2\dot{\phi}^2\sin^2\theta\right) - V(r).$$ (16.50)

The Hamiltonian (or total energy) is therefore

$$\mathcal{H} = p_r\dot{r} + p_\theta\dot{\theta} + p_\phi\dot{\phi} - \mathcal{L} = \frac{1}{2m}\left(p_r^2 + \frac{p_\theta^2}{r^2} + \frac{p_\phi^2}{r^2\sin^2\theta}\right) + V(r).$$ (16.51)

The corresponding Hamilton-Jacobi equation is, from (15.13),

$$\frac{\partial \mathcal{S}_a}{\partial t} + \mathcal{H}\left(\frac{\partial \mathcal{S}_a}{\partial q}, q\right) = 0,$$ (16.52)

that is,

$$\frac{\partial \mathcal{S}_a}{\partial t} + \frac{1}{2m}\left(\mathcal{S}_r^2 + \frac{\mathcal{S}_\theta^2}{r^2} + \frac{\mathcal{S}_\phi^2}{r^2\sin^2\theta}\right) + V(r) = 0.$$ (16.53)

Although from earlier considerations we already know the t and ϕ dependence of the action $\mathcal{S}_a$, we shall proceed in a slightly more general manner by seeking a separable solution for $\mathcal{S}_a$ in the form

$$\mathcal{S}_a(t; r, \theta, \phi) = \mathcal{S}_1(t) + \mathcal{S}_2(r) + \mathcal{S}_3(\theta) + \mathcal{S}_4(\phi).$$ (16.54)

Therefore,

$$\frac{d\mathcal{S}_1}{dt} + \frac{1}{2m}\left[(\mathcal{S}_2')^2 + \frac{(\mathcal{S}_3')^2}{r^2} + \frac{(\mathcal{S}_4')^2}{r^2\sin^2\theta}\right] + V(r) = 0.$$ (16.55)

For central forces there are no explicit t or ϕ dependencies, so $d\mathcal{S}_1/dt$ (or $\mathcal{S}_1'$) and $\mathcal{S}_4'$ must be constant, or

$$\frac{d\mathcal{S}_1}{dt} = -E, \quad \frac{d\mathcal{S}_4}{d\phi} = F,$$ (16.56)

and so

$$S_a(t; r, \theta, \phi) = -Et + S_2(r) + S_3(\theta) + F\phi. \qquad (16.57)$$

This is actually a Legendre transform from the action S_a to its new decomposition. Clearly, we may now write the Hamilton-Jacobi equation as

$$\left(\frac{dS_3}{d\theta}\right)^2 + \left(\frac{F}{\sin\theta}\right)^2 = 2mr^2\left[E - V(r)\right] - r^2\left(\frac{dS_2}{dr}\right)^2 \equiv J^2, \qquad (16.58)$$

where J is a constant. This can be done because the left-hand side of (16.58) is a function of θ, while the right-hand side is a function of r. The resulting θ- and r-dependent equations are

$$\frac{dS_3}{d\theta} = \pm\left[J^2 - \left(\frac{F}{\sin\theta}\right)^2\right]^{1/2}, \quad \text{and} \qquad (16.59)$$

$$\frac{dS_2}{dr} = \pm\left[2m\left(E - V(r)\right) - \left(\frac{J}{r}\right)^2\right]^{1/2}. \qquad (16.60)$$

16.2.1 The Kepler Problem Revisited

For $V(r) = -K/r$,

$$\frac{dS_2}{dr} = \pm\left[2mE + \frac{2mK}{r} - \left(\frac{J}{r}\right)^2\right]^{1/2}. \qquad (16.61)$$

We know that the angular momentum J is conserved, that is, from (16.7),

$$J = mr^2\dot{\theta} = \text{constant}, \qquad (16.62)$$

which is Kepler's second law of planetary motion. We also know that $\dot{\phi} = 0$ and hence that the total energy E of the particle may be written as

$$E = \frac{1}{2}m(\dot{r}^2 + r^2\dot{\theta}^2) + V(r) = \frac{1}{2}m\dot{r}^2 + \frac{J^2}{2mr^2} + V(r), \qquad (16.63)$$

from which

$$\dot{r} = \left\{\frac{2}{m}[E - V(r)] - \left(\frac{J}{mr}\right)^2\right\}^{1/2}. \qquad (16.64)$$

This expression can be integrated to yield the time t explicitly,

$$t = \int \frac{dr}{\left\{(2/m)[E - V(r)] - (J/(mr))^2\right\}^{1/2}} + t_0, \qquad (16.65)$$

from which we find that (since $d\theta/dt = J/(mr^2)$)

$$\theta = J\int \frac{dr}{r^2\left\{(2m)[E - V(r)] - (J/r)^2\right\}^{1/2}} + \theta_0, \qquad (16.66)$$

and this is the general solution of the equation of the orbital path. For the Kepler problem

$$\theta(r) = J \int^r \frac{d\xi}{\xi \left(2m E \xi^2 + 2m K \xi - J^2\right)^{1/2}} + \theta_0. \tag{16.67}$$

After considerable effort this can be evaluated to yield

$$\theta(r) = \arccos \left[\frac{J^2 - m K r}{r \left(2m E J^2 + m^2 K^2\right)^{1/2}} \right] + \theta_0. \tag{16.68}$$

Simplifying this, we obtain

$$\left[m K + \left(2m E J^2 + m^2 K^2\right)^{1/2} \cos\left(\theta - \theta_0\right) \right] r = J^2. \tag{16.69}$$

If this is compared with the standard expression for conic sections in polar coordinates (choosing $\theta_0 = \pi$),

$$r\left(1 - e \cos\theta\right) = ed, \tag{16.70}$$

where e is the eccentricity and $x = d$ is the location of the vertical directrix in the Cartesian plane, it is seen that

$$e = \frac{\left(2m E J^2 + m^2 K^2\right)^{1/2}}{m K}; \quad d = \frac{m K J^2}{\left(2m E J^2 + m^2 K^2\right)^{1/2}}. \tag{16.71}$$

Exercise: It is useful to verify the following expected results: (i) if $E < 0$, the orbit is an ellipse ($e < 1$); (ii) if $E = 0$, it is a parabola ($e = 1$); and (iii) if $E > 0$, it is a hyperbola ($e > 1$).

16.2.2 Generalizations

If we consider the motion of a particle of mass m under the influence of a central force $F = m f(r) r$, the equations of motion take the form

$$\ddot{r} - r\dot{\theta}^2 = r f(r), \quad r^2 \dot{\theta} = h, \tag{16.72}$$

where h is a constant. Since we may write $\ddot{r}$ as $\dot{r} d(\dot{r})/dr$, this pair of equations implies that

$$\dot{r}^2 = -\frac{h^2}{r^2} + 2 \int^r \xi f(\xi) d\xi. \tag{16.73}$$

The speed v at any point in the orbit may be determined from

$$v^2 = \dot{r}^2 + \left(r\dot{\theta}\right)^2 = 2 \int^r \xi f(\xi) d\xi, \tag{16.74}$$

while the orbit itself can be determined in principle (if $f(r)$ is specified) from the differential equation

$$\left(\frac{dr}{d\theta}\right)^2 = \frac{\dot{r}^2}{\dot{\theta}^2} = -r^2 + 2\frac{r^4}{h^2} \int^r \xi f(\xi) d\xi. \tag{16.75}$$

Conversely, if we know the orbit, we can in principle determine the central force. Thus if $\theta = g(r)$, $\dot{\theta} = g'(r)\dot{r}$ so that

$$\dot{r} = \frac{h}{r^2 g'(r)}, \tag{16.76}$$

whence

$$\ddot{r} = -\frac{h\dot{r}\left[2g'(r) + rg''(r)\right]}{r^3 \left[g'(r)\right]^2} = -\frac{h^2 \left[2g'(r) + rg''(r)\right]}{r^5 \left[g'(r)\right]^3}. \tag{16.77}$$

The radial acceleration, and hence the force, is then obtained from the equation

$$\ddot{r} - r\dot{\theta}^2 = -\frac{h^2}{r^3}\left\{\frac{2}{r^2 \left[g'(r)\right]^2} + \frac{g''(r)}{r \left[g'(r)\right]^3} + 1\right\}. \tag{16.78}$$

Exercise: Show that if the orbit is a spiral given by $\theta = k/r$, the central force is attractive and of magnitude mh^2/r^3.

In closing this chapter, it will be usful to summarize some other classical scattering phenomena that will be reexamined from a less classical point of view in Part IV.

16.2.3 Hard Sphere Scattering

First, consider the scattering of a uniform beam of parallel particles by a hard sphere, that is, a perfectly elastic sphere, of radius R. The cross-sectional area presented by the target (or "seen" by a particle) is simply $\sigma = \pi R^2$. Suppose that a particle with speed v approaches the sphere parallel to the central axis with an impact parameter b (see Figure 16.2); if the angle of incidence is i, then

$$b = R \sin i. \tag{16.79}$$

For a hard sphere the law of reflection applies (essentially from the conservation of energy and angular momentum), and the particle is deflected through an angle $\theta = \pi - 2i$, whence

$$b = R \cos\frac{\theta}{2}. \tag{16.80}$$

Given an assemblage of such particles, those scattered through the angular interval $[\theta, \theta + d\theta]$ will be those with impact parameters $[b, b + db]$, where

$$db = -\frac{R}{2}\sin\frac{\theta}{2}d\theta. \tag{16.81}$$

For a small area $d\sigma$ of the incoming beam, where from Figure 16.2 (suitably generalized to include the azimuthal angle ϕ), $d\sigma = b\,|db|\,d\phi$, we have

$$d\sigma = \frac{R^2}{2}\sin\frac{\theta}{2}\cos\frac{\theta}{2}d\theta d\phi = \frac{R^2}{4}\sin\theta d\theta d\phi. \tag{16.82}$$

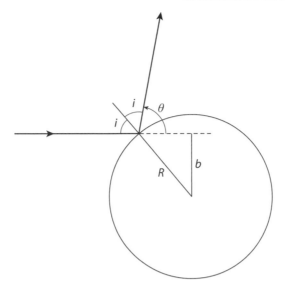

Figure 16.2 Hard sphere scattering.

Now the solid angle $d\Omega$ subtended at the origin by an area dA on the surface of a sphere of radius R is defined to be $d\Omega = R^{-2}dA$, and for such an area element $dA = R^2 \sin\theta d\theta d\phi$, so $d\Omega = \sin\theta d\theta d\phi$. Therefore for a hard sphere

$$\frac{d\sigma}{d\Omega} = \frac{1}{4}R^2. \tag{16.83}$$

This is the differential cross section for the sphere. The total cross section is found by integrating this result over all solid angles; since in this case (16.83) is independent of both θ and ϕ, it is equivalent to multiplying by 4π, yielding once more the cross sectional area of the sphere [54].

16.2.4 Rutherford Scattering

From expression (16.47) for the impact parameter b,

$$d\sigma = \frac{1}{2}\left(\frac{k}{mv^2}\right)^2 \cot\frac{\Theta}{2} \csc^2\frac{\Theta}{2} d\Theta d\phi. \tag{16.84}$$

Therefore,

$$\frac{d\sigma}{d\Omega} = \frac{1}{4}\left(\frac{k}{mv^2}\right)^2 \csc^4\frac{\Theta}{2}. \tag{16.85}$$

This is called the *Rutherford scattering cross section*. The total cross section is infinite.
Exercise: prove these statements.

As a postlude to the ray-theoretic optical scattering approach that was presented in Chapter 4, we consider the scattering of a particle of mass m and speed v by a transparent or "soft" sphere of radius a. It will be seen in Parts IV and V that such a sphere can be viewed in 'semiclassical' terms as a spherical potential well of radius a with depth V_0. This is a very useful concept connecting ray optics and potential scattering topics. Here the refractive index of the sphere can be defined using conservation of linear momentum, as

$$n = \left(1 + \frac{2V_0}{mv^2}\right)^{1/2} \equiv \frac{\sin i}{\sin r}, \tag{16.86}$$

where the angles i and r are the angles of incidence and refraction respectively [54].

Exercise: (i) If the impact parameter b is defined by

$$b = a \sin i,$$

show that

$$b = an \left[n^2 + 1 - 2n \cos \frac{\theta}{2}\right]^{-1/2} \sin \frac{\theta}{2}. \tag{16.87}$$

(ii) Sketch the graph of $b(\theta)/a$.

(iii) Show that the differential scattering cross section is

$$\frac{d\sigma}{d\Omega} = \frac{a^2 n^2 \left|n \cos (\theta/2) - 1\right| \left[n - \cos (\theta/2)\right]}{4 \cos (\theta/2) \left[n^2 + 1 - 2n \cos (\theta/2)\right]^2}. \tag{16.88}$$

(iv) Sketch the graph of

$$\frac{4}{a^2} \left(\frac{n-1}{n}\right)^2 \frac{d\sigma}{d\Omega}$$

as a function n, for $\theta = 0$.

(v) Show that the total scattering cross section is $\sigma = \pi a^2$.

Exercise: Evaluate the deflection angle $\Theta(J)$ for the Coulomb potential

$$V(r) = \frac{\alpha}{r}, \quad \alpha > 0. \tag{16.89}$$

(i) Show that

$$\Theta(J) = 2 \arctan \left[\frac{\alpha}{2J} \left(\frac{2m}{E}\right)^{1/2}\right]. \tag{16.90}$$

(ii) Also show that

$$\frac{d\sigma}{d\Omega} = \left(\frac{\alpha}{4E} \csc^2 \frac{\theta}{2} \right)^2 . \qquad (16.91)$$

(iii) Find $\tilde{\sigma}$, where

$$\tilde{\sigma} = \sigma(\theta \geq \theta_0), \quad \theta_0 > 0.$$

What happens as $\theta_0 \to 0$? Why do you think this is?

Chapter Seventeen

Scattering of Surface Gravity Waves by Islands, Reefs, and Barriers

Surface gravity waves propagating from the deep ocean to coastal regions may be substantially amplified by reflection, refraction, diffraction, and shoaling because of variation in water depth. Analytical solutions are particularly favorable in studies on wave scattering because of their high accuracy, low cost in labour, and timesaving; however, they are obtainable for only special topographies and simple governing equations. Regarding three-dimensional bathymetries, several analytical solutions to the long-wave equation have been determined for wave scattering by axi-symmetrical topographies, such as a circular cylindrical island mounted on a paraboloidal shoal, conical island and a circular paraboloidal shoal, circular cylindrical island mounted on a conical shoal, circular cylindrical island mounted on a general shoal, circular paraboloidal pit, truncated general shoal, circular general pit, circular island, circular general hump, dredge excavation pit, vertical cylinder containing an idealised scour pit, circular island consisting of combined topographies, and circular island mounted on a general shoal.

[261] [See the article for detailed references.]

In this chapter we consider long surface gravity waves (linear shallow water waves) such that (i) the depth $h(x, y)$ of the water is much greater than the vertical free surface displacement $\eta(x, y, t)$ (i.e., $\min(h) \gg \eta$) and (ii) the wavelength is much larger than the depth, (i.e., $\lambda \gg \max h$). Rather than manipulate the fluid equations from Chapters 11 or 13 into the required form we shall derive them directly for the present context. The linearized conservation laws of mass and momentum are, respectively,

$$\frac{\partial \eta}{\partial t} + \nabla \cdot (h\mathbf{u}) = 0, \tag{17.1}$$

and

$$\frac{\partial \mathbf{u}}{\partial t} + g\nabla \eta = 0. \tag{17.2}$$

Eliminating the velocity vector $\mathbf{u}$ yields the following linear hyperbolic partial differential equation with variable coefficients:

$$\frac{\partial^2 \eta}{\partial t^2} = g\nabla \cdot (h\nabla \eta). \tag{17.3}$$

17.1 TRAPPED WAVES

Now we consider $h = h(x)$ only, but retain y dependence for η and seek normal mode solutions of the form

$$\eta(x, y, t) = X(x)e^{i(\beta y - \omega t)}. \tag{17.4}$$

Equation (17.3) then simplifies to the Sturm-Liouville form:

$$\frac{d}{dx}\left(h(x)\frac{dX}{dx}\right) + \left(\frac{\omega^2}{g} - \beta^2 h(x)\right)X = 0. \tag{17.5}$$

We shall investigate this equation with a variety of mathematical tools, motivated in part by the approach of [43]. To begin, a phase-plane analysis of equation (17.5) is in order. To this end, we set

$$X'(x) = h^{-1}(x)Y(x), \tag{17.6}$$

so that (17.5) becomes

$$Y'(x) + \left(\frac{\omega^2}{g} - \beta^2 h(x)\right)X = 0. \tag{17.7}$$

Using the chain rule, we have the following first-order equation with x playing the role of a parameter. Solutions are represented by trajectories in the phase plane. Thus,

$$\frac{dY}{dX} = \frac{\left(\beta^2 h - \omega^2/g\right)hX}{Y}. \tag{17.8}$$

We make several further assumptions along the way:

$$\lim_{|x| \to \infty} h(x) = h_\infty; \tag{17.9}$$

$$h_0 \le h(x) \le h_\infty; \tag{17.10}$$

$$\beta^2 h_0 < \frac{\omega^2}{g} < \beta^2 h_\infty; \tag{17.11}$$

$$\frac{\omega^2}{g} = \beta^2 h(x_i), \quad i = 1, 2. \tag{17.12}$$

In what follows we shall refer to x_1 and x_2 as *turning points*. For $x \in (-\infty, x_1) \cup (x_2, \infty)$,

$$\beta^2 h_\infty > \frac{\omega^2}{g}, \tag{17.13}$$

(see Figure 17.1), so $X(x)$ is monotone in x. As $|x| \to \infty$,

$$h\left(\beta^2 h - \frac{\omega^2}{g}\right) \to h_\infty\left(\beta^2 h_\infty - \frac{\omega^2}{g}\right) \equiv h_\infty H^2 > 0, \tag{17.14}$$

so that

$$X(x) = O\left(e^{-H|x|}\right), \tag{17.15}$$

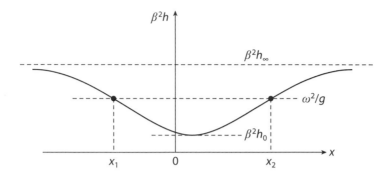

Figure 17.1 $\beta^2 h(x)$ for a submarine ridge.

(the positive exponent is rejected on physical grounds). For $x \in (x_1, x_2)$,

$$\beta^2 h(x) < \frac{\omega^2}{g}, \tag{17.16}$$

and $X(x)$ is oscillatory (such features will be discussed in more mathematical detail in Chapter 23). Such solutions correspond to *trapped waves*. From (17.8),

$$Y'(X) \to h_\infty H^2 XY^{-1} \quad \text{as} \quad |x| \to \infty. \tag{17.17}$$

Integrating this, we obtain

$$Y^2(X) = h_\infty H^2 X^2 + C,$$

where it is readily shown that $C = 0$, and that as $|x| \to \infty$ the trajectories emanate from or approach the origin along the asymptotic directions $Y = \pm h_\infty^{1/2} H X$ (Figure 17.2).

Exercise: Prove that $X(x)$ is monotone in x, given (17.13).

17.2 THE SCATTERING MATRIX $S(\alpha)$

We now consider the case of oblique incidence $(\beta^2 > 0)$ of propagating waves (i.e., waves such that $\beta^2 h_\infty < \omega^2/g$) on a two-dimensional submarine ridge. It is expedient to rearrange somewhat the governing equation by normalizing the depth function h as $h(x) = h_\infty a(x)$ and setting $\omega^2/g = (\alpha^2 + \beta^2) h_\infty$, where for the moment, $\alpha^2 > 0$. Then equation (17.5) has the form

$$\left(a(x)X'\right)' + \left[\alpha^2 + \beta^2 (1 - a)\right] X = 0, \tag{17.18}$$

where $a \leq 1$. The change in dependent variable, $X(x) = a^{-1/2}(x)\xi(x)$, leads to a well-known form—that of the one-dimensional time-independent Schrödinger equation:

$$\xi''(x) + [\lambda - V(x; \lambda)]\xi = 0. \tag{17.19}$$

There is, however, an important distinction between this equation and the standard problems of one-dimensional scattering presented in textbooks of quantum mechanics. In equation (17.19) the "potential function" V is *eigenvalue dependent,*

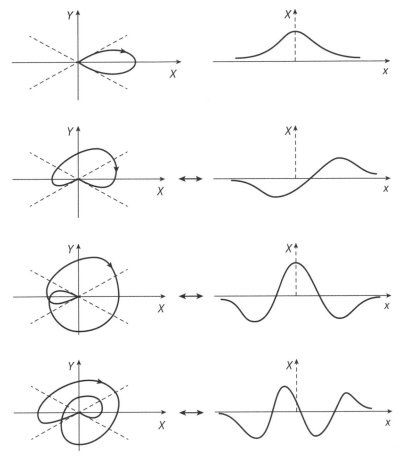

Figure 17.2 Some trajectories for equation (17.5) with corresponding modes. Dashed lines are the asymptotes $Y = \pm h_\infty^{1/2} H X$, and arrows indicate increasing values of x.

unlike in the many potential well/barrier type of problems for which $V = V(x)$ only. Specifically, here the "eigenvalue" is $\lambda = -\beta^2$, and

$$V(x;\lambda) = -\frac{1}{a(x)} \left(\alpha^2 + \beta^2 + \frac{3a'^2}{4a} - \frac{a'}{2a} - \frac{a''}{2} \right). \qquad (17.20)$$

It is worth emphasizing the point that in contrast to the standard potentials discussed in many books on scattering theory (e.g., [41]), *this potential is a function of the eigenvalue as well as the spatial variable.* A similar situation can arise in connection with the optics and acoustics of inhomogeneous media (see Appendix D for more details). It is sufficient to point out here that a Liouville transformation may be constructed to alleviate this problem (though, it creates another one elsewhere [127]).

We shall investigate this distinction further in Appendix D. For now, we shall merely note from equation (17.18) that $X(x)$ has an asymptotic form satisfying the equation $X'' + \alpha^2 X = 0$ as $|x| \to \infty$. Thus we write

$$X(x) \sim A_- e^{i\alpha x} + B_- e^{-i\alpha x}, \qquad x \to -\infty, \text{ and} \qquad (17.21)$$

$$X(x) \sim A_+ e^{i\alpha x} + B_+ e^{-i\alpha x}, \qquad x \to \infty. \qquad (17.22)$$

The constant coeffients will be shortly be related to so-called *transmission and reflection coefficients* T_i and R_i, $i = 1, 2$. Recall from equation (17.4) that there is an implicit $\exp(-i\omega t)$ dependence in these solutions, so the first term in each of equations (17.21) and (17.22) corresponds to a right-moving plane wave, and the second term to a left-moving one. The asymptotic solutions can be interpreted as a wave of amplitude A_- incident from negative infinity encountering the inhomogeneity $h(x)$ and being scattered (or decomposed) into a reflected wave of amplitude B_- and a transmitted wave of amplitude A_+. Correspondingly, a wave of amplitude B_+ incident from positive infinity is decomposed into a reflected wave of amplitude A_+ and a transmitted wave of amplitude B_-. These asymptotic forms arise because the depth function $h(x)$ (and hence $a(x)$) approaches a constant value as $|x| \to \infty$. Crudely speaking, the potential induced by the depth profile vanishes asymptotically (to high order, if necessary).

Next we define the *Jost solutions* $f_1(x; \alpha)$ and $f_2(x; \alpha)$ to be solutions to the right- and left-hand scattering problems, respectively, in the sense that

$$\lim_{x \to \infty} \left[e^{-i\alpha x} f_1(x; \alpha) \right] = 1, \quad \text{and} \quad \lim_{x \to -\infty} \left[e^{i\alpha x} f_2(x; \alpha) \right] = 1. \qquad (17.23)$$

More specifically we may write

$$f_1(x; \alpha) \sim \frac{1}{T_1} e^{i\alpha x} + \frac{R_1}{T_1} e^{-i\alpha x}, \qquad x \to -\infty, \qquad (17.24)$$

$$\sim e^{i\alpha x}, \qquad x \to \infty. \qquad (17.25)$$

Similarly, we have

$$f_2(x; \alpha) \sim \frac{1}{T_2} e^{-i\alpha x} + \frac{R_2}{T_2} e^{i\alpha x}, \qquad x \to \infty, \qquad (17.26)$$

$$\sim e^{-i\alpha x}, \qquad x \to -\infty. \qquad (17.27)$$

It is readily shown that the Wronskian

$$W\left[f_1(x; \alpha), f_2(x; \alpha) \right] = f_1 f_2' - f_1' f_2 = -\frac{2i\alpha}{T_1}, \qquad x \sim -\infty \qquad (17.28)$$

$$= -\frac{2i\alpha}{T_2}, \qquad x \sim \infty, \qquad (17.29)$$

and neither of these are zero in general. Note that these two asymptotic results together imply that $T_1 = T_2$, meaning that the transmission coefficients to the left and right are equal *even if* $h(x)$ is not symmetric about the origin. Furthermore, since f_1 and f_2 are linearly independent functions, (17.21) and (17.22) may be written generically as

$$X = C_1 f_1 + C_2 f_2. \qquad (17.30)$$

By comparing the asymptotic values of (17.21), (17.22), and (17.30), it is easy to obtain four equations for the coefficients of $\exp(\pm i\alpha x)$ as $x \to \pm\infty$. Eliminating C_1 and C_2, we find that

$$A_+ = T_1 A_- + R_2 B_+,\qquad(17.31)$$

$$B_- = R_1 A_- + T_2 B_+,\qquad(17.32)$$

or

$$\begin{pmatrix} A_+ \\ B_- \end{pmatrix} = \begin{pmatrix} T_1 & R_2 \\ R_1 & T_2 \end{pmatrix} \begin{pmatrix} A_- \\ B_+ \end{pmatrix} \equiv S(\alpha) \begin{pmatrix} A_- \\ B_+ \end{pmatrix},\qquad(17.33)$$

where the matrix $S(\alpha)$ is called the *scattering matrix or S*-matrix. Finally in this section note that the Jost solutions $f_1(x; \alpha)$ and $f_1(x; -\alpha)$ are also linearly independent, since from their asymptotic forms at $x \sim \infty$,

$$W[f_1(x; \alpha), f_1(x; -\alpha)] = -2i\alpha \neq 0.\qquad(17.34)$$

Therefore any solution (e.g., $f_2(x; \alpha)$) can be expressed as a linear combination of $f_1(x; \alpha)$ and $f_1(x; -\alpha)$. This will be of use in the next section. Note that the corresponding Wronskian for $x \sim -\infty$ is $-2i\alpha\left(1 - R_1^2\right)$, so the solutions are linearly dependent only if there is perfect reflection of waves from the left to the right.

17.3 TRAPPED MODES: IMAGINARY POLES OF $S(\alpha)$

From the behavior of the asymptotic solutions at $x \sim \infty$ we observe that

$$T_2 f_2(x; \alpha) = R_2 f_1(x; \alpha) + f_1(x; -\alpha),\qquad(17.35)$$

and on differentiating this with respect to x, we have a second equation:

$$T_2 f_2'(x; \alpha) = R_2 f_1'(x; \alpha) + f_1'(x; -\alpha).\qquad(17.36)$$

From these we obtain the following expressions for T_2 and R_2:

$$T_2 = \frac{-2i\alpha}{W[f_1(x; \alpha), f_2(x; \alpha)]},\qquad(17.37)$$

and

$$R_2 = \frac{W[f_1(x; -\alpha), f_2(x; \alpha)]}{W[f_1(x; \alpha), f_2(x; \alpha)]}.\qquad(17.38)$$

Clearly, any poles of T_2 correspond to the condition

$$W[f_1(x; \alpha), f_2(x; \alpha)] = 0,\qquad(17.39)$$

that is, when f_1 and f_2 become linearly dependent. We denote such poles (if they exist) by α_n, and they will also be the poles of R_2 and hence of $S(\alpha)$. Obviously $T_2^{-1} = 0$ and R_2/T_2 is finite, enabling us to infer that

$$f_2(x; \alpha) \sim \left(\frac{R_2}{T_2}\right)_{\alpha_n} \exp(i\alpha_n x), \qquad x \to \infty,\qquad(17.40)$$

$$\sim \exp(i\alpha_n x), \qquad x \to -\infty.\qquad(17.41)$$

Note from this result that if $\alpha_n \in \mathbb{R}$, then

$$f_2(x; \alpha) \notin \mathcal{L}_2(-\infty, \infty), \tag{17.42}$$

which we reject on physical grounds. Trapped modes decay exponentially as $|x| \to \infty$, and so for such modes $\alpha_n = \alpha_R + i\alpha_I$, where $\alpha_I > 0$. Returning to equation (17.18) with $\alpha_n \in \mathbb{C}$ and all the other quantities except X being real, we multiply that equation by $\bar{X}$, the complex conjugate of X, and the complex conjugate of (17.18) by X, and subtract one from the other. Then we arrive at the result

$$\left[a(x) \left(X\bar{X}' - \bar{X}X' \right) \right]' = 2i (\operatorname{Im}\alpha^2) |X|^2 . \tag{17.43}$$

At this point we identify X with f_2, integrate the expression (17.43) over $(-\infty, \infty)$, and apply the asymptotic exponential behavior to establish the important result:

$$\operatorname{Im}\alpha_n^2 \int_{-\infty}^{\infty} |f_2(x; \alpha_n)|^2 \, dx = 0. \tag{17.44}$$

Since the integral is positive definite for nontrivial f_2, it must be the case that $\operatorname{Im}\alpha_n^2 = 0$, and since $\operatorname{Im}\alpha_n = \alpha_I > 0$, it follows that $\operatorname{Re}\alpha_n = \alpha_R = 0$. Thus the poles are purely imaginary, and so the trapped modes decay monotonically as $|x| \to \infty$. Because only the square of α appears in (17.18), X can be considered a real function; on multiplying (17.18) by X and integrating by parts on $(-\infty, \infty)$ we obtain

$$\int_{-\infty}^{\infty} a(x) \left[X'(x) \right]^2 dx + \int_{-\infty}^{\infty} \left[\beta^2 a(x) - \frac{\omega^2}{gh_\infty} \right] X^2(x) dx = 0. \tag{17.45}$$

This implies that nontrivial nonconstant solutions can exist if and only if

$$\beta^2 a(x) - \frac{\omega^2}{gh_\infty} < 0 \tag{17.46}$$

somewhere in $(-\infty, \infty)$. But this is consistent with the condition (17.11) for trapped waves (since $a(x) \to 1^-$ as $|x| \to \infty$), so their existence has been confirmed.

17.4 PROPERTIES OF $S(\alpha)$ FOR $\alpha \in \mathbb{R}$

For real values of α it follows from (17.43) that

$$\left[a(x) \left(X\bar{X}' - \bar{X}X' \right) \right] = \text{constant}, \tag{17.47}$$

so from (17.21) and (17.22) we obtain the expressions

$$X\bar{X}' - \bar{X}X' = -2i\alpha \left(|A_-|^2 - |B_-|^2 \right), \quad x \to -\infty, \tag{17.48}$$

and

$$X\bar{X}' - \bar{X}X' = -2i\alpha \left(|A_+|^2 - |B_+|^2 \right), \quad x \to +\infty. \tag{17.49}$$

Note that the amplitudes of the outgoing waves are $[\longleftarrow B_-, A_+ \longrightarrow]$, and the amplitudes of the incoming waves are $[\longrightarrow A_-, B_+ \longleftarrow]$. Since $a(x) \to 1$ as

$|x| \to \infty$, it follows that these expressions must be equal, and hence the energies of the incoming and outgoing waves must be equal, because

$$|A_+|^2 + |B_-|^2 = |A_-|^2 + |B_+|^2 . \tag{17.50}$$

But now the scattering matrix $(S(\alpha))$ can "get into the act," since the conservation of energy statement (17.50) is equivalent to the expression

$$|A_+|^2 + |B_-|^2 = (A_+ B_-) \begin{pmatrix} \bar{A}_+ \\ \bar{B}_- \end{pmatrix} = (A_- B_+) S^T \bar{S} \begin{pmatrix} \bar{A}_- \\ \bar{B}_+ \end{pmatrix}, \tag{17.51}$$

where S^T is the transpose of S. From (17.50) this implies that

$$S^T \bar{S} = I = \begin{pmatrix} 1 & 0 \\ 0 & 1 \end{pmatrix}, \tag{17.52}$$

which means, as a matter of definition, that the S-matrix is *unitary*. This in turn implies from (17.33) that

$$\begin{pmatrix} T_1 & R_1 \\ R_2 & T_2 \end{pmatrix} \begin{pmatrix} \bar{T}_1 & \bar{R}_2 \\ \bar{R}_1 & \bar{T}_2 \end{pmatrix} = \begin{pmatrix} 1 & 0 \\ 0 & 1 \end{pmatrix}, \tag{17.53}$$

that is,

$$|T_1|^2 + |R_1|^2 = 1; \tag{17.54}$$

$$|T_2|^2 + |R_2|^2 = 1; \tag{17.55}$$

$$T_1 \bar{R}_2 + R_1 \bar{T}_2 = 0. \tag{17.56}$$

(A fourth equation is just the complex conjugate of the last one.) The first two results are again statements of energy conservation, and because (as noted above) $T_1 = T_2$, the third equation demonstates that $|R_1| = |R_2|$.

From equations (17.21) and (17.22) we note that

$$\bar{X}(x) \sim \bar{A}_- e^{-i\alpha x} + \bar{B}_- e^{i\alpha x}, \quad x \to -\infty, \quad \text{and} \tag{17.57}$$

$$\bar{X}(x) \sim \bar{A}_+ e^{-i\alpha x} + \bar{B}_+ e^{i\alpha x}, \quad x \to \infty, \tag{17.58}$$

so by comparing these two sets of equations, we make make the following replacements along with their complex conjugates: $A_- \to \bar{B}_-$ and $A_+ \to \bar{B}_+$. Hence (17.33) may be written as

$$\begin{pmatrix} \bar{B}_+ \\ \bar{A}_- \end{pmatrix} = S \begin{pmatrix} \bar{B}_- \\ \bar{A}_+ \end{pmatrix}, \tag{17.59}$$

or

$$\begin{pmatrix} B_+ \\ A_- \end{pmatrix} = \bar{S} \begin{pmatrix} B_- \\ A_+ \end{pmatrix}. \tag{17.60}$$

But there is more! Rewriting (17.21) and (17.22), we have

$$X(x) \sim B_- e^{i(-\alpha)x} + A_- e^{-i(-\alpha)x}, \quad x \to -\infty, \quad \text{and} \tag{17.61}$$

$$X(x) \sim B_+ e^{i(-\alpha)x} + A_+ e^{-i(-\alpha)x}, \quad x \to \infty. \tag{17.62}$$

Now we may associate the coefficients B_-, A_+ with incoming waves and A_-, B_+ with outgoing waves (i.e., $A_- \longleftrightarrow B_-$ and $A_+ \longleftrightarrow B_+$ in (17.33)), which becomes

$$\begin{pmatrix} B_+ \\ A_- \end{pmatrix} = S(-\alpha) \begin{pmatrix} B_- \\ A_+ \end{pmatrix}. \tag{17.63}$$

On comparing this equation with (17.60) it follows that

$$\bar{S}(\alpha) = S(-\alpha). \tag{17.64}$$

17.5 SUBMERGED CIRCULAR ISLANDS

In circular cylindrical geometry (17.3) becomes

$$\frac{\partial^2 \xi}{\partial t^2} = g\nabla_h \cdot [h(r)\nabla \xi] = gh(r)\left(\frac{\partial^2 \xi}{\partial r^2} + \frac{1}{r}\frac{\partial \xi}{\partial r} + \frac{1}{r^2}\frac{\partial^2 \xi}{\partial \theta^2}\right) + h'(r)\frac{\partial \xi}{\partial r}. \tag{17.65}$$

The solutions of this equation will be periodic with period 2π, so we seek modal solutions of the form

$$\xi(r, \theta, t) = R(r)\exp[i(n\theta - \omega t)], \quad n \in \mathbb{N}. \tag{17.66}$$

Hence

$$R''(r) + \left(\frac{1}{r} + \frac{h'}{h}\right)R'(r) + \left(\frac{\omega^2}{g} - \frac{n^2 h}{r^2}\right)R(r) = 0, \tag{17.67}$$

or, in Sturm-Liouville form,

$$\left(hr\,R'\right)' + \left(\frac{\omega^2}{g} - \frac{n^2 h}{r^2}\right)rR = 0. \tag{17.68}$$

As in the one-dimensional case, the radial solution behavior will be oscillatory or exponential accordingly as

$$\frac{\omega^2}{g} - \frac{n^2 h}{r^2} > 0 \tag{17.69}$$

or

$$\frac{\omega^2}{g} - \frac{n^2 h}{r^2} < 0, \tag{17.70}$$

respectively. There are some interesting differences in this problem compared with the one-dimensional Cartesian scattering problem. To investigate these consider first a submerged island with a monotonically increasing depth function h such that $0 < h(0) < h(r) < h(\infty)$. As noted from Figure 17.3, the solution will be oscillatory for $r > r_0$, where r_0 is the unique solution of

$$\frac{\omega^2}{g} - \frac{n^2 h(r_0)}{r_0^2} = 0. \tag{17.71}$$

This means that instead of waves being trapped in an interior region as we saw in the previous section (i.e., oscillatory there and monotonically decreasing outside

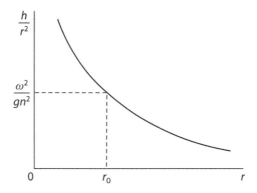

Figure 17.3 $h(r)/r^2$ for a submerged island.

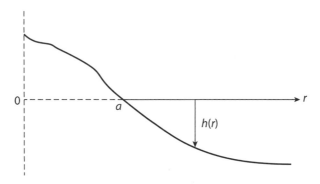

Figure 17.4 Typical $h(r)$ for island/shore.

that region), in the circular case the opposite is true; oscillations occur *outside* a central region. Consider next an island with a shore at $r = a$, that is, $h(r) = 0$ for $0 \le r \le a$ with (typically) $h'(r) > 0$ for $r > a$ (see Figure 17.4).

Now the function $r_0^{-2}h(r_0)$ possesses a single maximum for some $r > a$ and thereafter decreases to 0^+ as $r \to \infty$. Therefore if

$$\frac{\omega^2}{g} - \frac{n^2 h(r_i)}{r_i^2} = 0, \quad i = 1, 2; \quad r_1 < r_2, \tag{17.72}$$

and

$$\frac{\omega^2}{gn^2} < \max\left(\frac{h}{r^2}\right), \tag{17.73}$$

then the solutions are oscillatory in (a, r_1), exponentially attenuating in (r_1, r_2), and then oscillatory again for $r > r_2$ (see Figure 17.5).

The island acts as a barrier of finite thickness by trapping near the shore waves of (i) sufficiently low frequency or (ii) sufficiently large n. The smaller the frequency (or the larger the value of n), the wider the barrier becomes, and therefore the more effective it is at trapping energy nearer the shore line (corresponding to less energy leakage for large values of r). More specifically, a model of a stepped circular sill

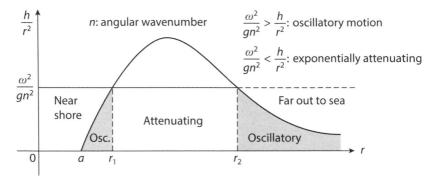

Figure 17.5 $h(r)/r^2$ for island/shore.

(like a pillbox) is mathematically very similar to a problem in optics for radially piecewise-constant media. Let

$$h(r) = h_1, \quad 0 < r < a, \tag{17.74}$$

$$h(r) = h_2, \quad r > a, \tag{17.75}$$

where $h_1 < h_2$. As seen from Figure 17.6, the function $r^{-2}h(r)$ possesses a discontinuity between two monotone decreasing profiles. The spike can give rise to interesting resonance phenomena, as it is known to do in optics. This will be discussed in considerable detail in Chapter 27.

17.6 EDGE WAVES ON A SLOPING BEACH

The Confluent Hypergeometric Equation; Laguerre Polynomials

The confluent hypergeometric equation with constants a and c may be written as

$$xy''(x) + (c - x)y'(x) - ay(x) = 0. \tag{17.76}$$

It has a regular singularity at $x = 0$ and an irregular point at infinity. One solution of this equation that is of interest in connection with edge waves is

$$y(x) = {}_1F_1(a, c; x) = 1 + \frac{a}{c}\frac{x}{1!} + \frac{a(a+1)}{c(c+1)}\frac{x^2}{2!} + \ldots, \tag{17.77}$$

convergent for all finite x, where $c \neq 0, -1, -2, \ldots$ Sometimes this solution is written as $M(a, c; x)$. It becomes a polynomial if a is zero or a negative integer. When $a = -n$ the equation is known as *Laguerre's equation*, and the solution above is expressed as a Laguerre polynomial, that is,

$$y(x) = {}_1F_1(-n, c; x) \equiv L_n(x). \tag{17.78}$$

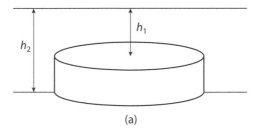

(a)

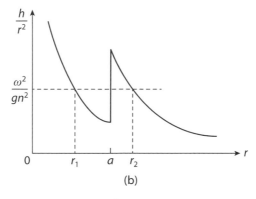

(b)

Figure 17.6 $h(r)/r^2$ for a stepped circular sill.

It is readily seen that

$$L_0(x) = 1; \quad L_1(x) = 1 - x, \quad L_2(x) = 1 - 2x + x^2/2, \quad (17.79)$$

and so forth. The set of functions $\{\varphi_n(x)\}$, where

$$\varphi_n(x) = e^{-x/2} L_n(x), \quad (17.80)$$

is orthonormal in $[0, \infty)$, that is,

$$\int_0^\infty e^{-x} L_m(x) L_n(x) dx = \delta_{mn}. \quad (17.81)$$

We consider a fixed rigid planar seafloor sloping down at an angle δ to the horizontal xz-plane $y = 0$, but because we wish to examine edge waves, it is necessary to study fully three-dimensional motions. In this geometry, y is in the vertical direction, and z is directed along the beach. The unit normal to the sloping bed is [36]

$$\mathbf{n} = \langle \sin \delta, \cos \delta, 0 \rangle, \quad (17.82)$$

and since the equation of the seafloor is $y = -x \tan \delta$, the condition for zero normal velocity on this rigid boundary is $\mathbf{n} \cdot \nabla \phi = 0$ (ϕ being the velocity potential) or

$$\sin \delta \frac{\partial \phi}{\partial x} + \cos \delta \frac{\partial \phi}{\partial y} = 0. \tag{17.83}$$

A solution of $\nabla^2 \phi = 0$ that describes wave motion parallel to the beach can take the form

$$\phi(\mathbf{r}, t) = g(z - ct) G(x, y); \tag{17.84}$$

hence with $\xi = z - ct$ the standard separation of variables approach yields the equations

$$\frac{d^2 g}{d\xi^2} - k^2 g = 0, \tag{17.85}$$

and

$$\frac{\partial^2 G}{\partial x^2} + \frac{\partial^2 G}{\partial y^2} - k^2 G = 0. \tag{17.86}$$

Therefore,

$$g(\xi(z, t)) = A \cos k\xi = A \cos k(z - ct) \tag{17.87}$$

for some constant A. (This equation can absorb the $\sin k\xi$ term by including a suitable phase factor.)

For obvious physical reasons, we may write a solution of the modified Helmholtz equation (17.86) as

$$G(x, y) = B \exp[-(c_1 x + c_2 y)]; \quad c_1 \geq 0, \quad c_2 \neq 0. \tag{17.88}$$

Using equation (17.86) and the boundary condition (17.83), it is readily shown that $c_1 = k \cos \delta$ and $c_2 = -k \sin \delta$, leading to the solution

$$\phi(\mathbf{r}, t) = D \exp(-kx \cos \delta + ky \sin \delta) \cos k(z - ct), \tag{17.89}$$

D being a constant. There are several things to note from this solution: (i) $\phi \to 0$ as $x \to \infty$, so the edge wave is confined to a small x neighborhood of the beach (hence the name), and (ii) as δ increases the x dependence of ϕ becomes less pronounced, and the motion extends farther out to sea. In fact, when $\delta = \pi/2$ the dispersion relation for deep water waves is recovered (recall that in linear theory the free surface corresponds to $y \approx 0$).

To see this, it is readily seen from the start of Chapter 10 that in the absence of surface tension, the velocity potential ϕ satisfies

$$\frac{\partial^2 \phi}{\partial t^2} + g \frac{\partial \phi}{\partial y} = 0 \tag{17.90}$$

on $y = 0$. On substituting solution (17.89) into this, the following dispersion relation arises for $D \neq 0$:

$$(gk \sin \delta - k^2 c^2) \exp(-kx \cos \delta) \cos k(z - ct) = 0. \tag{17.91}$$

This expression must hold for any and all x and z values, so that

$$c^2 = \frac{g \sin \delta}{k} \to \frac{g}{k}, \tag{17.92}$$

as $\delta \to \pi/2$.

17.6.1 One-Dimensional Edge Waves on a Constant Slope

We return to a one-dimensional version of equation (17.3) and consider normally incident waves on a gently sloping beach. The governing equation is then

$$\frac{\partial^2 \eta}{\partial t^2} = g \frac{\partial}{\partial x} \left(h(x) \frac{\partial \eta}{\partial x} \right). \tag{17.93}$$

When $h(x) = h_0 = $ constant, this reduces to the standard wave equation with wave speed $(gh_0)^{1/2}$, as is well known. Suppose that the seafloor slopes down linearly from the beach edge, now with *depth z* given by

$$z = -h(x) \equiv -sx, \quad s > 0. \tag{17.94}$$

Equation (17.93) reduces to

$$x\eta''(x) + \eta'(x) + \left(\frac{\omega^2}{sg} - \beta^2 x \right) \eta(x) = 0. \tag{17.95}$$

We make the following changes of dependent and independent variables:

$$\eta = e^{-r/2} f(r); \quad r = 2\beta x, \tag{17.96}$$

and then reduce equation (17.95) further to the form

$$r f''(r) + (1 - r) f'(r) + \frac{1}{2} \left(\frac{\omega^2}{\beta sg} - 1 \right) f(r) = 0. \tag{17.97}$$

Exercise: Derive equation (17.97) from (17.95).

Now it is apparent that nontrivial polynomial solutions to this equation satisfying $|f(0)| < \infty$ and $\lim_{r \to \infty} f(r) = 0$ will exist when

$$\frac{\omega^2}{\beta sg} = 2n + 1, \quad n = 0, 1, 2, \ldots, \tag{17.98}$$

so this becomes a Sturm-Liouville eigenvalue problem for $\omega(\beta)$ (or $\beta(\omega)$), one encountered above in connection with Laguerre polynomials. Because of the spatial exponential decay term present in the $\varphi_n(x)$, the eigenfunctions are appreciable only near the shore (Figure 17.7) and are referred to as edge waves; they may be associated with rip currents.

17.6.2 Wave Amplication by a Sloping Beach

We consider the next level of complexity (or simplicity!) by defining

$$h(x) = sx, \quad 0 \le x \le a, \tag{17.99}$$

$$h(x) = h_\infty, \quad x > a, \tag{17.100}$$

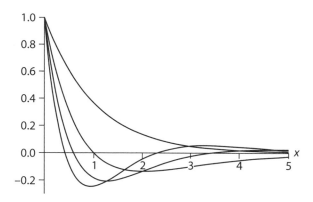

Figure 17.7 A generic form of the functions (17.80), $e^{-x} L_n(x)$ for $n = 0, 1, 2, 3$.

so that continuity of h means that $as = h_\infty$. The problem is simplified even further by considering the beach slope to be small, that is, $h_\infty \ll a$ so that the angle of dip, δ, satisfies

$$\tan \delta = \frac{h_\infty}{a} \approx \delta. \tag{17.101}$$

Then for $0 < x < a$, equation (17.93) is reduced to

$$\frac{\partial^2 \eta}{\partial t^2} = g\delta x \frac{\partial^2 \eta}{\partial x^2} + g\delta \frac{\partial \eta}{\partial x}. \tag{17.102}$$

Suppose that a wave incident from the open sea, possibly caused by tides or a distant storm, has the time-harmonic form

$$\eta(t) = D \cos (\omega t + \varepsilon), \tag{17.103}$$

D being a constant amplitude for $x > a$. We seek a "beach" solution for (17.102) of the form

$$\eta(x, t) = f(x) \cos (\omega t + \varepsilon), \tag{17.104}$$

so that $f(x)$ satisfies the equation

$$x f''(x) + f'(x) + \frac{\omega^2}{g\delta} f(x) = 0. \tag{17.105}$$

Making the change of independent variable $x = 2m^2$, this equation transforms to the form (with a slight abuse of notation)

$$m f''(m) + f'(m) + \frac{8\omega^2}{g\delta} m f(m) = 0 \tag{17.106}$$

after some algebra. This is Bessel's equation of order zero, and the solution bounded at the origin is

$$f(m) = A J_0(pm), \quad p = \left(\frac{8\omega^2}{g\delta} \right)^{1/2}, \tag{17.107}$$

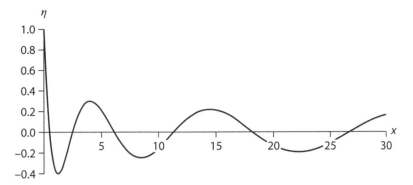

Figure 17.8 $\eta(x, t)$ at arbitrary time t.

or in terms of the original variable x, noting that $f(a) = D$,

$$\eta(x, t) = D \left[J_0 \left(2\omega \left[\frac{a}{g\delta} \right]^{1/2} \right) \right]^{-1} J_0 \left(2\omega \left[\frac{x}{g\delta} \right]^{1/2} \right) \cos(\omega t + \varepsilon). \quad (17.108)$$

As seen from Figure 17.8, the (linear) waves are amplified as they approach the shoreline. Of course, it is necessary to include the effect of nonlinearity for a full discussion of wave amplification.

However, it must be pointed out that such waves are in practice rarely, if ever observed because of the nonlinear wavebreaking that occurs as the coastline is approached, effectively destroying the results based on linear theory.[1]

[1] I am grateful to a reviewer for pointing this out.

Chapter Eighteen

Acoustic Scattering

Have you ever been in an auditorium of some kind, or a church, in which your view of a speaker is blocked by a pillar, but you can still hear what is being said? Why can your ears receive auditory signals, but your eyes cannot receive visual ones (excluding Superman, of course)? The reason for this is related to the wavelengths of the waves, being of orders 1 m and 10^{-6}–10^{-7} m respectively. The latter in effect scatters more like a particle, while the former is able to diffract ("bend") around an obstacle comparable in size to its wavelength. By the same token, therefore, we would expect that light waves can diffract around appropriately smaller obstacles, and indeed this is the case, as evidenced by softly colored rings of light around the moon (coronae) as thin cloud scuds past its face. Another diffraction-induced meteorological phenomenon is the green, purple-red, or blue iridescence occasionally visible in clouds.

But—and here is the point for this chapter—scattering (Rayleigh or otherwise) applies to more than just light; it works for sound as well under the right circumstances. In [147], the authors state that if one speaks into a large stand of fir trees from a distance, the echo may return raised by an octave! This is based on fascinating discussions in the literature about such natural echo chambers. It seems that in 1873 Lord Rayleigh studied this phenomenon; he stated that he heard an echo of a woman's voice that was raised by an octave, but that of a man did not (perhaps she walked into a tree). It transpires that the trees are the scattering centers, analogous to the molecules in the air for sunlight, but this requires that the wavelength of the incident sound should be large compared with the width of the trees. The first harmonic present in the sound, at twice the frequency of the fundamental (and hence half the wavelength) is scattered $(2)^4 = 16$ times more than the fundamental, and this may dominate the returning echo. Recall from Chapter 4 that a dimensional argument was used to elucidate the wavelength dependence of Rayleigh scattering.

Scattering of waves and/or particles is a ubiquitous phenomenon, though this might not be generally appreciated. The scattering of plane waves from spheres (the simplest situation to examine mathematically) is quite an old subject, but it is an extremely rich field, and it is still being developed and applied in fields as diverse as meteorology, elasticity, seismology, optics (as has already been noted), acoustics, biochemistry, medical physics, quantum mechanics (and the related areas of atomic, molecular and nuclear physics), and probably many others. The subject, seen both as direct and inverse problems, has enabled its practitioners to find out much about the world around us, doing so over large ranges of length and time scales. Indeed, almost all of what we see and much of what we hear, is respectively, a consequence of the scattering of electromagnetic and acoustic waves from various objects.

Historically, the names Lorenz, Mie, and Debye are associated with the theory of electromagnetic scattering from dielectric spheres (i.e., the spheres do not conduct electric current). The complete mathematical solution to the problem of electromagnetic wave scattering from a sphere was obtained by Lorenz in 1890 in terms of an infinite series of so-called partial waves (though Clebsch had also derived the result as early as 1863). As with so many instances in history, the solution is called by another name: the *Mie* or *Debye-Mie* solution, which will be discussed in some detail in Chapter 19. In principle this contains all the physics for any size of scattering sphere and all wavelengths. However, the number of terms required to be retained for a realistic physical understanding is $O(\beta)$, where β is the ratio of the scatterer's circumference to the wavelength. Since for molecular scattering of light in the atmosphere, $\beta \ll 1$, and (very simplistically speaking) only the first term is needed to understand blue sky problem; this is the Rayleigh scattering mentioned above. In contrast, $\beta \approx 5,000$ for the rainbow, wherein light is scattered from raindrops. Given the complexity of each term this is a nontrivial task, but more significantly, the physics can get swamped easily in a morass of computational details. Fortunately, due to work by some very eminent mathematical physicists and applied mathematicians (whose work has been summarized in [25], [148]), a valuable analytic technique (the *complex angular momentum* approach) has been developed to complement the computational work, although now it commands more mathematical than computational interest. It is important to note that the same kind of mathematics applies to the scattering of plane acoustic waves by spheres (which is the reason for mentioning it here), although that is a scalar problem as opposed to the vector problem arising in electromagnetic theory. The following places these comments in a broader historical context:[1]

Airy theory has been called the incomplete "complete" answer [see Chapter 28 for more details]. It did go beyond the models of the day in that it quantified the dependence upon the raindrop size of (i) the rainbow's angular width, (ii) its angular radius, and (iii) the spacing of the supernumerary bows. Also, unlike the models of Descartes and Newton, Airy's predicted a non-zero distribution of light intensity in Alexander's dark band (the darker region between the primary and secondary bows), and a finite intensity at the angle of minimum deviation (as noted above, the earlier theories predicted an infinite intensity there). However, spurred on by Maxwell's recognition that light is part of the electromagnetic spectrum, and the subsequent publication of his mathematical treatise on electromagnetic waves, several mathematical physicists sought a more complete theory of scattering, because it had been demonstrated by then that the Airy theory failed to predict precisely the angular position of many laboratory-generated rainbows. Among them were the German physicist Gustav Mie who published a paper in 1908 on the scattering of light by homogeneous spheres in a homogeneous medium, and Peter Debye who independently developed a similar theory for the scattering of electromagnetic waves by spheres. Mie's theory

---

[1] Partially cited at the start of Chapter 5.

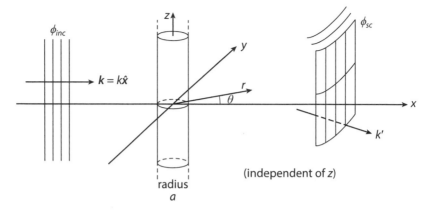

Figure 18.1 Cylindrical scattering geometry.

was intended to explain the colors exhibited by colloidally dispersed metal particles, whereas Debye's work, based on his 1908 thesis, dealt with the problem of light pressure on a spherical particle [149].

There are essentially three regimes appropriate to scattering phenomena: scattering of waves by objects which in size are (i) small, (ii) comparable with, and (iii) large compared to the wavelength of the impinging radiation. There may be considerable overlap of region (ii) with the others, depending on the problem of interest, but basically region (i) is responsible for why the sky is blue (and what a responsibility that is). At the other extreme, region (iii) is responsible for the rainbow, halo and glory phenomena we may on occasion observe. The wave-particle duality so fundamental in quantum mechanics is also relevant to region (ii), because the more subtle features exhibited by the phenomena involve both these aspects of description and explanation. [54]

18.1 SCATTERING BY A CYLINDER

Following the approach in [132], we will work here in terms of the velocity potential ϕ, which represents the total field and is decomposed into the incident plane wave and total scattered fields (see Figure 18.1). The latter is the result of acoustic wavefronts parallel to the z-axis of a cylinder of radius a being scattered by that cylinder (but also includes the incident field). Thus

$$\phi(\mathbf{r}) = \phi_i(\mathbf{r}) + \phi_s(\mathbf{r}), \tag{18.1}$$

where

$$\phi_i(\mathbf{r}) = e^{i\mathbf{k}\cdot\mathbf{r}}. \tag{18.2}$$

The scattered radiation (now dropping the subscript s) satisfies the scalar Helmholtz equation, which in this z-independent geometry is

$$(\nabla^2 + k^2)\phi = \left[\frac{1}{r}\frac{\partial}{\partial r}\left(r\frac{\partial}{\partial r}\right) + \frac{1}{r^2}\frac{\partial^2}{\partial\theta^2} + k^2\right]\phi = 0, \tag{18.3}$$

for $r > a$, subject to the rigid (or "hard") boundary condition

$$\left[\frac{\partial\phi}{\partial r}\right]_a = 0. \tag{18.4}$$

We seek solutions in terms of linear combinations of the functions

$$J_n(kr)e^{\pm in\theta}, \quad Y_n(kr)e^{\pm in\theta}, \quad n \in \mathbb{N}. \tag{18.5}$$

Since ϕ_i is bounded everywhere, it can be expanded in terms of Bessel functions of the first kind of order n:

$$\phi_i = e^{ikr\cos\theta} = \sum_{n=-\infty}^{\infty} i^n e^{in\theta} J_n(kr). \tag{18.6}$$

The outgoing radial function $R_n(r)$ for the nth angular term must be a linear combination of the incident wave and the outgoing scattered wave $H_n^{(1)}(kr) = J_n(kr) + iY_n(kr)$. Because equation (18.6) includes the term $i^n J_n(kr)$, we write (note that the field ϕ includes ϕ_i):

$$R_n(r) = i^n(J_n(kr) + BH_n^{(1)}(kr)). \tag{18.7}$$

Then

$$R_n'(a) = 0 \Rightarrow B = \frac{-J_n'(ka)}{H_n^{(1)'}(ka)}, \tag{18.8}$$

and therefore

$$\phi(r,\theta) = \sum_{n=-\infty}^{\infty} i^n e^{in\theta}\left(J_n(kr) - H_n^{(1)}(kr)\frac{J_n'(ka)}{H^{(1)'}(ka)}\right). \tag{18.9}$$

Hence

$$\phi_s(r,\theta) = \phi - \phi_i = -\sum_{n=-\infty}^{\infty} i^n e^{in\theta}\left(\frac{J_n'(ka)}{H^{(1)'}(ka)}H_n^{(1)}(kr)\right). \tag{18.10}$$

Again, we are primarily interested in $kr \gg 1$, the asymptotic region. We have that

$$i^n H_n^{(1)}(x) \sim \left(\frac{2}{\pi x}\right)^{1/2} e^{i(x-\pi/4)}, \quad x \to \infty, \tag{18.11}$$

so

$$\phi_s(r,\theta) \sim -\sum_{n=-\infty}^{\infty} e^{in\theta}\frac{J_n'(ka)}{H^{(1)'}(ka)}\left(\frac{2}{\pi kr}\right)^{1/2} e^{i(kr-\pi/4)} \sim \frac{f(\theta)}{r^{1/2}}e^{ikr}. \tag{18.12}$$

Hence the scattering amplitude $f(\theta)$ has, in cylindrical geometry, the Fourier representation

$$f(\theta) = -\sum_{n=-\infty}^{\infty} e^{i(n\theta - \pi/4)} \left(\frac{2}{\pi kr}\right)^{1/2} \frac{J_n'(ka)}{H_n^{(1)'}(ka)}. \tag{18.13}$$

18.2 TIME-AVERAGED ENERGY FLUX: A LITTLE BIT OF PHYSICS

From the common adages "work = force times distance" and "force = pressure times area," it follows dimensionally that the power per unit area, known as the *energy flux*, will involve pressure times (distance/time) or (in more than one spatial dimension) pressure times velocity, denoted below by $p\mathbf{v}$. Suppose now that we wish to find the time average of two real time-harmonic quantities:

$$\mathrm{Re}(\alpha e^{-i\omega t}) = \frac{1}{2}(\alpha e^{-i\omega t} + \bar{\alpha} e^{i\omega t}) \tag{18.14}$$

and

$$\mathrm{Re}(\beta e^{-i\omega t}) = \frac{1}{2}(\beta e^{-i\omega t} + \bar{\beta} e^{i\omega t}). \tag{18.15}$$

We can define the time average of a quantity $\Omega(t)$ (oscillating with angular frequency ω) by

$$\langle \Omega(t) \rangle = \frac{\omega}{2\pi} \int_0^{2\pi/\omega} \Omega(t)dt, \tag{18.16}$$

from which

$$\begin{aligned}
\left\langle \mathrm{Re}(\alpha e^{-i\omega t})\,\mathrm{Re}(\beta e^{-i\omega t}) \right\rangle &= \frac{1}{4}\left\langle \alpha\beta e^{-2i\omega t} + \bar{\alpha}\beta + \alpha\bar{\beta} + \bar{\alpha}\bar{\beta} e^{2i\omega t} \right\rangle \\
&= \frac{1}{4}\left\langle \bar{\alpha}\beta + \alpha\bar{\beta} \right\rangle = \frac{1}{2}\mathrm{Re}(\alpha\bar{\beta}) \\
&= \frac{1}{2}\mathrm{Re}(\bar{\alpha}\beta).
\end{aligned} \tag{18.17}$$

The time-averaged energy flux will therefore be

$$\mathbf{F} = \frac{1}{2}\mathrm{Re}(\bar{p}\mathbf{v}), \tag{18.18}$$

on using the second form. From the hydrodynamic equations laid out in Chapter 11, $p \propto \rho_0 \partial\phi/\partial t$, ρ_0 being the density of the medium, so the incident flux is

$$\mathbf{F} = \frac{1}{2}\mathrm{Re}(\nabla\phi_i \overline{[-i\omega\phi]}\rho_0), \tag{18.19}$$

and $\phi_i = e^{ikx}$ since $x = r\cos\theta$; therefore the associated flux is, in terms of the unit **x** vector,

$$\mathbf{F}_i = -\frac{1}{2}\omega\rho_0 k\hat{\mathbf{x}} = F_0\hat{\mathbf{x}}. \tag{18.20}$$

The scattered wave transports energy outward, and the time-averaged power crossing an area $d\mathbf{A} = r\,d\theta\,\hat{\mathbf{r}}$ is $\mathbf{F} \cdot d\mathbf{A}$, that is, $\mathrm{Re}(\bar{p}\mathbf{v} \cdot d\mathbf{A})/2$, so after some reduction, we have

$$\mathbf{F} \cdot d\mathbf{A} = -\frac{1}{2}|f(\theta)|^2 k\omega\rho_0 d\theta = F_0|f(\theta)|^2 d\theta \equiv F_0 d\sigma, \tag{18.21}$$

for a strip of width $d\sigma$ and unit length in z, thus defining the differential scattering "width" by

$$\frac{d\sigma}{d\theta} = |f(\theta)|^2. \tag{18.22}$$

Consider now $ka \ll 1$, corresponding to wavelengths that are large compared with the cylinder radius. We will use the small-argument expansions for Bessel functions of order $n = -1, 0$, and 1, because the other terms in equation (18.13) give $O(ka)^{|2n|}$ contributions that are negligible in this limit. Thus we have

$$f(\theta) \approx -\left(\frac{2}{\pi k}\right)^{1/2} e^{-i\pi/4} \left[e^{-i\theta} \frac{J'_{-1}(ka)}{H^{(1)'}_{-1}(ka)} + \frac{J'_0(ka)}{H^{(1)'}_0(ka)} + e^{i\theta} \frac{J'_1(ka)}{H^{(1)'}_1(ka)} \right]. \tag{18.23}$$

On simplifying the coefficients of the exponential terms in θ, the result for $ka \ll 1$ is

$$f(\theta) \approx \left(\frac{2}{\pi ki}\right)^{1/2} \left(\frac{\pi k^2 a^2}{4i}\right) (1 - 2\cos\theta). \tag{18.24}$$

Hence

$$\frac{d\sigma}{d\theta} = |f(\theta)|^2 \approx \frac{1}{8}\pi k^3 a^4 (1 - 2\cos\theta)^2 \tag{18.25}$$

in this limit. As seen in Figure 18.2, the polar diagram for $(1 - 2\cos\theta)^2$, and hence for the scattering width, is very anisotropic, with a forward intensity about 11 percent of the backward value. This polar diagram, by the way, represents an interesting exercise for students of calculus. Nodes are present at $\theta = \pm\pi/3$, with the forward lobe contained in $|\theta| < \pi/3$ and the backward lobe in $|\theta| > \pi/3$.

The total scattering cross section is given by

$$\sigma = \int_0^{2\pi} \frac{d\sigma}{d\theta}\,d\theta = \frac{3}{4}\pi^2 k^3 a^4 = 6\pi^5 a \left(\frac{a}{\lambda}\right)^3, \tag{18.26}$$

obviously proportional to a geometric width of $2a$ multiplying a factor $(a/\lambda)^3$ [132]. As anticipated in view of the above comments, the forward lobe contributes only a small amount to σ, since

$$2\int_0^{\pi/3} (1 - 2\cos\theta)^2\,d\theta = 2\pi - 3\sqrt{3},$$

which is a small percentage of the total area of the polar diagram.

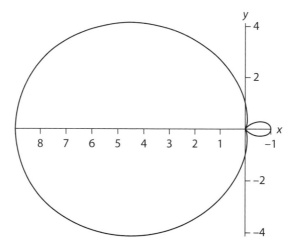

Figure 18.2 The polar graph of $(1 - 2\cos\theta)^2$.

A Calculation

Consider audible sound incident on a window screen ("What time is supper, Mom?"), and suppose that $a \approx 10^{-4}$ m, and the wavelength of the sound is of order 1 m, so that according to (18.26), the total scattering cross section is $\sigma \approx 10^{-13}$ m. The conclusion? Sound passes easily through such a screen, and the question will be readily heard (though it may be ignored). In the next section we examine acoustic scattering by a rigid (or impenetrable) sphere.

18.3 THE IMPENETRABLE SPHERE

18.3.1 Introduction: Spherically Symmetric Geometry

Let us use the notation of equation (18.1) above to cover several possible contexts. In spherical geometry the solution to the Helmholtz equation,

$$(\nabla^2 + k^2)\phi\,(\mathbf{r},k) = 0, \tag{18.27}$$

can be expressed generically via separation of variables (in terms of the spherical Hankel functions $h_l^{(1)}(kr)$ and $h_l^{(2)}(kr)$) as

$$\phi\,(\mathbf{r}, k) = \sum_{l,m} \left\{ A_{l,m} h_l^{(1)}(kr) + B_{l,m} h_l^{(2)}(kr) \right\} Y_l^m(\theta, \tilde{\phi}), \tag{18.28}$$

where the $Y_l^m(\theta, \tilde{\phi})$ are the spherical harmonic functions, an orthonormal set of basis functions on the surface of the unit sphere (and the azimuthal angle has been temporarily identified as $\tilde{\phi}$ to avoid confusion with the wave field $\phi(\mathbf{r}, k)$). It was noted earlier in the chapter that for an incident plane wave propagating in free space with wavevector $\mathbf{k}$ along an axial direction (i.e., x or z depending on the geometry used),

$$\phi_i(\mathbf{r}, k) = e^{i\mathbf{k}\cdot\mathbf{r}} = e^{ikr\cos\theta}. \tag{18.29}$$

Given that this is therefore independent of $\tilde{\phi}$ and cannot possess any singularities for finite $\mathbf{r}$, the representation (18.28) for $\phi_i(\mathbf{r}, k)$ reduces to the axially symmetric form involving the spherical Bessel functions $j_l(kr)$ and the Legendre polynomials $P_l(\cos\theta)$:

$$e^{ikr\cos\theta} = \sum_{l=0}^{\infty} C_l j_l(kr) P_l(\cos\theta). \tag{18.30}$$

As will be demonstrated shortly, the coefficients C_l are

$$C_l = (2l+1)i^l, \tag{18.31}$$

and the resulting expression,

$$e^{ikr\cos\theta} = \sum_{l=0}^{\infty} (2l+1)i^l j_l(kr) P_l(\cos\theta), \tag{18.32}$$

is known as *Bauer's expansion*. It is a so-called partial-wave expansion for a plane wave, and appears complicated, because we are trying to expand it in terms of azimuthally symmetric eigenfunctions!

Returning to the general form (18.28), it similarly follows that

$$\phi(\mathbf{r}, k) = \phi(r, \theta; k) = \sum_{l=0}^{\infty} (2l+1)i^l \left\{ A_l h_l^{(1)}(kr) + B_l h_l^{(2)}(kr) \right\} P_l(\cos\theta) \tag{18.33}$$

$$\equiv \sum_{l=0}^{\infty} (2l+1)i^l \phi_l(r) P_l(\cos\theta), \tag{18.34}$$

where we define $\phi_l(r)$ for use below.

For a scalar wave field, such as acoustic waves with ϕ being the velocity potential, the condition of impenetrability is equivalent to the Dirichlet boundary condition $\phi(a, \theta; k) = 0$. This corresponds to an acoustically "soft" sphere, in contrast to the cylindrical geometric case discussed above; as there, the boundary condition for a "hard" sphere would be of the Neumann type, namely $\phi'(a, \theta; k) = 0$. In the latter case no waves enter the sphere, because the radial component of velocity vanishes at the surface. Applying the boundary condition $\phi(a, \theta; k) = 0$ yields the following relation between A_l and B_l:

$$\frac{A_l}{B_l} = -\frac{h_l^{(2)}(\beta)}{h_l^{(1)}(\beta)} \equiv S_l(\beta), \tag{18.35}$$

where $\beta = ka = 2\pi a/\lambda$ is called the *size parameter*; clearly, it is the ratio of the circumference of the scatterer to the wavelength of the incident radiation. For an impenetrable sphere of radius a, without loss of generality [54],

$$\phi(r, \theta; k) = \sum_{l=0}^{\infty} (2l + 1)i^l \left\{ A_l h_l^{(1)}(kr) + B_l h_l^{(2)}(kr) \right\} P_l(\cos \theta). \qquad (18.36)$$

Therefore,

$$\phi(r, \theta; k) = \sum_{l=0}^{\infty} (2l + 1)i^l \left[h_l^{(1)}(kr) + S_l(\beta) h_l^{(2)}(kr) \right] P_l(\cos \theta). \qquad (18.37)$$

In this expression, $B_l = 1$ because as $r \to \infty$, the solution $\phi(r, \theta; k) \to \exp(ikr \cos \theta)$, a unit-amplitude plane wave, as expected. Noting that

$$h_l^{(2)}(kr) = \overline{h_l^{(1)}(kr)}, \qquad (18.38)$$

(the overbar denoting complex conjugation), we see from (18.35) that $|S_l(\beta)| = 1$, so $S_l(\beta)$ is just a phase factor, which we write as

$$S_l(\beta) = e^{2i\delta_l(\beta)}. \qquad (18.39)$$

$\delta_l(\beta)$ is a real *phase shift* and in general is an element in the S-matrix.

18.3.2 The Scattering Amplitude Revisited

We have previously defined the scattering amplitude $f(k, \theta)$ by the condition that in the far zone, $\phi(r, \theta)$ behaves like the sum of the incident plus the scattered waves:

$$\lim_{kr \to \infty} \phi(r, \theta) = e^{ikr} + f(k, \theta) \frac{e^{ikr}}{r}. \qquad (18.40)$$

Since for large enough r this will approximate a plane wave, using the asymptotic form of equation (18.37), we seek the relationship

$$\sum_{l=0}^{\infty} (2l + 1)i^l P_l(\cos \theta) \frac{\sin(kr - l\pi/2)}{kr} + \frac{e^{ikr}}{r} f(k, \theta) + O(r^{-2})$$

$$= \sum_{l=0}^{\infty} a_l \frac{\sin(kr - l\pi/2 + \delta_l(\beta))}{br} P_l(\cos \theta) + O(r^{-2}). \qquad (18.41)$$

Therefore

$$\sum_{l=0}^{\infty} \frac{(2l + 1)i^l}{kr} \left\{ \frac{\exp[i(kr - l\pi/2)] - \exp[-i(kr - l\pi/2)]}{2i} \right\} P_l(\cos \theta) + \frac{e^{ikr}}{r} f(k, \theta)$$

$$= \sum_{l=0}^{\infty} \frac{a_l}{kr} \left\{ \frac{\exp[i(kr - l\pi/2 + \delta_l)] - \exp[-i(kr - l\pi/2 + \delta_l)]}{2i} \right\} P_l(\cos \theta). \qquad (18.42)$$

Comparing coefficients of the exponential terms, we find that

$$a_l = C_l e^{i\delta_l} = (2l + 1)i^l e^{i\delta_l}, \qquad (18.43)$$

and so

$$2ikf(\theta) + \sum_{l=0}^{\infty}(2l+1)i^l e^{-il\pi/2} P_l(\cos\theta) = \sum_{l=0}^{\infty} a_l e^{i(\delta_l - l\pi/2)} P_l(\cos\theta), \quad (18.44)$$

which is valid for all θ. Therefore

$$2ikf(\theta) = \sum_{l=0}^{\infty}(2l+1)i^l e^{-il\pi/2} P_l(\cos\theta)\left(e^{2i\delta_l} - 1\right), \quad (18.45)$$

and

$$f(\theta) = f(\theta, k) = \frac{1}{2ik}\sum_{l=0}^{\infty}(2l+1)\left(e^{2i\delta_l} - 1\right) P_l(\cos\theta) \quad (18.46)$$

$$= \frac{1}{2ik}\sum_{l=0}^{\infty}(2l+1)\left(S_l(\beta) - 1\right) P_l(\cos\theta) \quad (18.47)$$

$$= \frac{1}{k}\sum_{l=0}^{\infty}(2l+1)e^{i\delta_l}\sin(\delta_l) P_l(\cos\theta). \quad (18.48)$$

Now

$$\frac{d\sigma}{d\Omega} = |f(\theta, k)|^2, \quad (18.49)$$

so therefore

$$\sigma = 2\pi \int_0^\pi \frac{d\sigma}{d\Omega}\sin\theta d\theta = 2\pi \int_0^\pi |f(\theta, k)|^2 \sin\theta d\theta. \quad (18.50)$$

Consider next the obvious identity $|f(\theta)|^2 = f(\theta)\overline{f(\theta)}$. Using the form of (18.48),

$$\frac{1}{k^2}\left\{\sum_{l=0}^{\infty}(2l+1)e^{i\delta_l}(\sin\delta_l)\,P_l(\cos\theta)\right\} \times \left\{\sum_{l=0}^{\infty}(2l'+1)e^{-i\delta_{l'}}(\sin\delta_{l'})\,P_{l'}(\cos\theta)\right\},$$

will yield the products

$$(2l'+1)(e^{i(\delta_l - \delta_{l'})}\sin\delta_l \sin\delta_{l'}\left[P_l(\cos\theta) P_{l'}(\cos\theta)\right]. \quad (18.51)$$

The orthogonality result

$$\int_{-1}^{1} P_l(x) P_{l'}(x) dx = \frac{2}{2l+1}\delta_{ll'}, \quad (18.52)$$

where $x = \cos(\theta)$ will permit the reduction

$$2\pi \int_0^\pi |f(\theta)|^2 d\theta = \frac{4\pi}{k^2}\sum_{l=0}^{\infty}(2l+1)\sin^2\delta_l = \sigma = \sum_{l=0}^{\infty}\sigma_l, \quad (18.53)$$

which is a sum over l only. σ_l denotes the lth partial cross section, and this result is related to the *optical theorem* below.

18.3.3 The Optical Theorem

Since

$$f(\theta) = \frac{1}{2ik} \sum_{l=0}^{\infty} (2l+1) \left(e^{2i\delta_l} - 1\right) P_l(\cos\theta), \tag{18.54}$$

it follows that

$$f(0) = \frac{1}{2ik} \sum_{l=0}^{\infty} (2l+1) \left(e^{2i\delta_l} - 1\right), \tag{18.55}$$

since $P_l(1) = 1$. Then

$$f(0) - \overline{f(0)} = \frac{1}{ik} \sum_{l=0}^{\infty} (2l+1) (\cos 2\delta_l - 1) = -\frac{2}{ik} \sum_{l=0}^{\infty} (2l+1) \sin^2 \delta_l. \tag{18.56}$$

Therefore

$$2i \operatorname{Im} f(0) = -\frac{\sigma k^2}{2\pi i k}, \tag{18.57}$$

or

$$\sigma = \frac{4\pi}{k} \operatorname{Im} f(0). \tag{18.58}$$

This is known as the optical theorem, relating the forward scattering amplitude to the total cross section of the scatterer. $f(0)$ is the scattering amplitude with an angle of zero, that is, the amplitude of the wave scattered to the center of a distant screen.

18.3.4 The Sommerfeld Radiation Condition

From (18.34) we have the following representation for the radial component of the Helmholtz equation:

$$\phi_l(r) = A_l h_l^{(1)}(kr) + B_l h_l^{(2)}(kr), \tag{18.59}$$

where in general $B_l \neq 0$ for radiation problems in which $k > 0$. Note that if an outgoing spherical wave has the radial form (where we recall the time-harmonic behavior $\exp(-i\omega t)$)

$$f(r) = \frac{e^{ikr}}{r}, \tag{18.60}$$

(see below), then it is easily seen that

$$r\frac{\partial f}{\partial r} - ikrf = -f.$$

Now, as $r \to \infty$, $|f| \to 0$, but neither term on the left-hand side tends to zero individually. Therefore, $f(r)$ must satisfy the *Sommerfeld radiation condition*

$$\lim_{r \to \infty} \left(\frac{\partial f}{\partial r} - ikrf \right) = 0. \tag{18.61}$$

Note that the incoming radial wave given by

$$g(r) = \frac{e^{-ikr}}{r}, \tag{18.62}$$

does *not* satisfy this condition, and so (18.61) may be used to eliminate such converging solutions. This, then, is an alternative form for the second (asymptotic) boundary condition.

Question/Exercise: Why should an outgoing spherical wave have the form (18.60)? To see this, show that the wave equation for spherically symmetric waves,

$$\frac{\partial^2 \phi}{\partial t^2} = \frac{c^2}{r^2} \frac{\partial}{\partial r} \left(r^2 \frac{\partial \phi}{\partial r} \right), \tag{18.63}$$

can be rewritten as

$$\frac{\partial^2 g}{\partial t^2} = c^2 \frac{\partial^2 g}{\partial r^2}, \tag{18.64}$$

where $g(r, t) = r\phi(r, t)$. The D'Alembert solution for the simpler form (18.64) therefore implies that

$$\phi(r, t) = \frac{1}{r} [F(r - ct) + G(r + ct)], \tag{18.65}$$

for some functions F and G. Clearly, to preserve only waves outgoing from the source, $G \equiv 0$, from which the particular choice of $F(r) = e^{ikr}$ yields (18.60).

18.4 RIGID SPHERE: SMALL ka APPROXIMATION

We revisit the scattering of sound from a solid sphere of radius a, but this time applied to a hard sphere (i.e., one with a rigid boundary). For $r > a$, if $\widetilde{\phi}(r, t)$ is the velocity potential of motion (so the velocity field $\mathbf{v} = -\nabla \widetilde{\phi}$), then its governing wave equation is (for $r > a$)

$$\nabla^2 \widetilde{\phi} - \frac{1}{c^2} \frac{\partial^2 \widetilde{\phi}}{\partial t^2} = 0. \tag{18.66}$$

At $r = a$, velocity perturbation should have no radial component, so the Neumann condition applies. This means that

$$\left[\frac{\partial \widetilde{\phi}}{\partial r} \right]_a = 0. \tag{18.67}$$

As before, if

$$\tilde{\phi}(r, t) = \phi(r)e^{-i\omega t}, \tag{18.68}$$

then

$$\nabla^2 \phi + k^2 \phi = 0, \quad r > a, \tag{18.69}$$

with

$$\left[\frac{\partial \phi}{\partial r}\right]_a = 0.$$

As in the case of scattering from a rigid cylinder, we seek a solution of the form

$$\phi = \phi_i + \phi_s, \tag{18.70}$$

where

$$\phi_i = \exp\left(ikr\cos\theta\right), \tag{18.71}$$

and ϕ_s must satisfy an outgoing radiation condition. From equations (18.37) and (18.39) we see that

$$\left[\frac{d}{dr}\left(h_l^{(1)}(kr) + e^{2i\delta_l}h_l^{(2)}(kr)\right)\right]_a = 0, \tag{18.72}$$

or

$$k\left[h_l^{(1)'}(ka) + e^{2i\delta_l}h_l^{(2)'}(ka)\right] = 0. \tag{18.73}$$

Since for $x \in \mathbb{R}$, $h_l^{(2)}(x) = \overline{h_l^{(1)}(x)}$, we can write

$$h_l^{(1)'}(ka) = |h_l^{(1)'}(ka)|e^{i\gamma_l}, \text{ and } h_l^{(2)'}(ka) = |h_l^{(1)'}(ka)|e^{-i\gamma_l}, \tag{18.74}$$

for some phase term γ_l, which can be determined from

$$h_l^{'(1)} = j_l' + in_l' = |h_l^{(1)'}(ka)|e^{i\gamma_l}, \tag{18.75}$$

so that

$$\cos\gamma_l = \frac{j_l'(ka)}{\left([j_l'(ka)]^2 + [n_l'(ka)]^2\right)^{1/2}}. \tag{18.76}$$

Therefore equation (18.73) becomes

$$e^{2i\delta_l} + e^{2i\gamma_l} = 0 \Rightarrow e^{2i\delta_l} = e^{2i(\gamma_l + n\pi/2)}, \quad n = 1, 3, 5, \dots, \tag{18.77}$$

or for $n = 1$ in particular,

$$\delta_l = \gamma_l + \frac{\pi}{2}. \tag{18.78}$$

From (18.53) it therefore follows that

$$\sigma = \frac{4\pi}{k^2}\sum_{l=0}^{\infty}(2l+1)\cos^2\gamma_l = \frac{4\pi}{k^2}\sum_{l=0}^{\infty}\frac{(2l+1)\left[j_l'(ka)\right]^2}{\left[j_l'(ka)\right]^2 + \left[n_l'(ka)\right]^2}. \tag{18.79}$$

When $x \equiv ka \ll 1$, this series converges rapidly. We find (after considerable algebra) that for the two lowest l values,

$$\cos \gamma_0 = \frac{j_0'(ka)}{\left(\left[j_0'(ka) \right]^2 + \left[n_0'(ka) \right]^2 \right)^{1/2}} \approx -\frac{x^3}{3}, \tag{18.80}$$

$$\cos \gamma_1 \approx \frac{x^3}{6}. \tag{18.81}$$

For higher l values, $j_l' = O(x^{l-1})$, and $n_l' = O(x^{-(l+2)})$, so when $l \geq 2$,

$$\cos \gamma_l = O(x^{2l+1}), \tag{18.82}$$

and for $x \ll 1$ each successive l value gives rise to a term $O(x^2)$ smaller than the preceding one and can be neglected. Therefore we have

$$\sigma = \frac{4\pi}{k^2} \left\{ \cos^2 \gamma_0 + 3 \cos^2 \gamma_1 + \sum_{l=2}^{\infty} (2l+1) \cos^2 \gamma_l \right\}$$

$$\approx \frac{4\pi}{k^2} \left(\frac{x^6}{9} + \frac{x^6}{12} \right) = \frac{7}{9} \pi k^4 a^6. \tag{18.83}$$

Note that (18.83) can be variously written as

$$\sigma \approx \frac{7}{9} \pi \beta^4 a^2 = \frac{112}{9} \frac{\pi^5 a^6}{\lambda^4},$$

because $\beta = ka = 2\pi a \lambda^{-1}$. This is the scalar version of Rayleigh's inverse fourth-power law of scattering for wavelengths large compared to the size of the scatterer. Note also that $\sigma \propto a^6$, as will also be demonstrated for the electromagnetic case (Chapter 19)

18.5 ACOUSTIC RADIATION FROM A RIGID PULSATING SPHERE

Let us write the radius at any time as

$$R(t) = a + h \sin \omega t, \quad h \ll a. \tag{18.84}$$

For a pulsating sphere in air, the velocity potential $\tilde{\phi}(r, \theta, \phi; t)$, defined by $v = -\nabla \tilde{\phi}$, satisfies the wave equation (18.66). At radius $R(t)$ the radial component of the air velocity must equal the corresponding value at the surface of the sphere, so modifying the notation slightly in this case (for obvious reasons),

$$v_R = -\left[\frac{\partial \tilde{\phi}}{\partial r} \right]_R = \frac{dR(t)}{dt} = A\omega \cos \omega t. \tag{18.85}$$

The potential must therefore vary harmonically at $r = a$, so in general we consider complex solutions of the form

$$\tilde{\phi}(r, \theta, \phi; t) = \phi(r, \theta, \phi)e^{-i\omega t}, \tag{18.86}$$

(the real part of this expression is taken at the end of the calculation). Hence we define the complex radius by

$$R(t) = a + he^{-i\omega t}, \tag{18.87}$$

where $h \ll c/\omega$ for subsonic pulsations. Therefore

$$\left[\frac{\partial \phi}{\partial r}\right]_R = i\omega h. \tag{18.88}$$

Spherical symmetry implies that $\phi = \phi(r)$ only, so ϕ satisfies

$$\frac{d^2\phi}{dr^2} + \frac{2}{r}\frac{d\phi}{dr} + k^2\phi = 0, \tag{18.89}$$

where $k = \omega/c$ as usual. Equation (18.89) is the spherical Bessel equation of order zero, with linearly independent solutions

$$j_0(x) = \left(\frac{\pi}{2x}\right)^{1/2} J_{1/2}(x) = \frac{\sin x}{x}, \tag{18.90}$$

$$n_0(x) = \left(\frac{\pi}{2x}\right)^{1/2} N_0(x) = \frac{-\cos x}{x}, \tag{18.91}$$

in terms of Bessel functions of the first and second kinds of order zero, respectively ($x = kr$). In terms of complex constants A and B the spatial component of the solution is

$$\phi(r) = \phi_0(r) = A j_0(kr) + B n_0(kr), \tag{18.92}$$

where $r > 0$ implies $B \neq 0$ in general. But so far we have only the boundary condition (18.88). Thus

$$\phi_0(r) = \frac{1}{kr}\left(A\sin kr - B\cos kr\right), \tag{18.93}$$

and so

$$\phi(r, t) = \left(\frac{A - iB}{2ikr}\right) e^{i(kr - \omega t)} - \left(\frac{A + iB}{2ikr}\right) e^{-i(kr + \omega t)}$$

$$\equiv \frac{D}{kr} \exp\left[\frac{i\omega}{c}(r - ct)\right] - \frac{C}{kr} \exp\left[-\frac{i\omega}{c}(r + ct)\right]. \tag{18.94}$$

The first term represents an outgoing spherical wave, and the second an incoming spherical wave, which must be excluded since we require only waves that diverge from their source. This is again a radiation condition and represents (in essence) the second boundary condition. Hence,

$$C = A + iB = 0 \implies \phi(r, t) = \frac{D}{kr} \exp\left[\frac{i\omega}{c}(r - ct)\right]. \tag{18.95}$$

We now take a short physical digression. The power flux vector $\mathbf{P}$ is defined by $\mathbf{P} = p\mathbf{v}$, which is the amount of energy passing per second through a unit area normal to it. The Euler equation [34] linearized about the equilibrium state $p = p_0$, $\rho = \rho_0$, and $\mathbf{u} = \mathbf{0}$ is

$$\rho_0 \frac{\partial \mathbf{v}}{\partial t} + \nabla p_1 = 0, \tag{18.96}$$

where

$$p = p_0 + p_1, \ |p_1| \ll p_0; \ \rho = \rho_0 + \rho_1, \ |\rho_1| \ll \rho_0; \ \mathbf{u} = \mathbf{0} + \mathbf{v}. \tag{18.97}$$

Now $\upsilon = -\nabla\phi$ (ignoring the exponential time factor), so equation (18.96) implies, for constant density ρ_0, that

$$\nabla\left(-\rho_0 \frac{\partial\phi}{\partial t} + p_1\right) = 0. \tag{18.98}$$

Therefore to within an additive constant

$$p = p_0 + \rho_0 \frac{\partial\phi}{\partial t}, \tag{18.99}$$

so that

$$\mathbf{P} = -\left(p_0 + \rho_0 \frac{\partial\phi}{\partial t}\right)\nabla\phi. \tag{18.100}$$

But these are complex solutions, so ϕ must be replaced by its real part, hence

$$\mathbf{P} = -p_0 \operatorname{Re}(\nabla\phi) - \rho_0 \operatorname{Re}\left(\frac{\partial\phi}{\partial t}\right) \operatorname{Re}(\nabla\phi). \tag{18.101}$$

When a quantity such as ϕ varies harmonically in time, the most useful measure is the *average power* (over one cycle or period), as noted earlier. The first term will not contribute to $\langle\mathbf{P}\rangle$ so

$$\langle\mathbf{P}\rangle = \left\langle -\rho_0 \operatorname{Re}\left(\frac{\partial\phi}{\partial t}\right) \operatorname{Re}(\nabla\phi)\right\rangle. \tag{18.102}$$

From equation (18.95),

$$\phi(r, t) = \frac{\tilde{D}}{r}e^{i\Theta}, \tag{18.103}$$

where $\tilde{D} = k^{-1}D$, $\Theta = kr - \omega t$, and therefore,

$$\frac{\partial\phi}{\partial t} = -\frac{i\omega\tilde{D}}{r}e^{i\Theta}. \tag{18.104}$$

Also,

$$\nabla\phi = \frac{\tilde{D}}{r^2}e^{i\Theta}(ikr - 1)\mathbf{r}_0, \tag{18.105}$$

where $\mathbf{r}_0 = \mathbf{r}/|\mathbf{r}|$ is the unit radial vector. After a great deal of algebra, equation (18.102) yields the result

$$\langle\mathbf{P}\rangle = \frac{\rho_0\omega k}{2r^2}|\tilde{D}|^2\mathbf{r}_0. \tag{18.106}$$

From the boundary condition at the surface of the sphere it is readily shown that the constant

$$A = \frac{i\omega h k a^2}{ka + i} e^{-ika}, \tag{18.107}$$

so that

$$|\tilde{D}|^2 = \frac{\omega^2 h^2 a^4}{k^2 a^2 + 1}, \tag{18.108}$$

and so in terms of the wavelength λ,

$$\langle \mathbf{P} \rangle = \frac{\rho_0 h^2 \omega^4 a^4}{2c \left(1 + 4\pi^2 a^2 \lambda^{-2}\right)} \frac{\mathbf{r}_0}{r^2}. \tag{18.109}$$

If $ka = 2\pi a/\lambda \ll 1$, then with $\omega = kc = 2\pi c/\lambda$,

$$\langle \mathbf{P} \rangle \approx \frac{\rho_0 h^2 a^4}{2c} \left(\frac{2\pi c}{\lambda}\right)^4 \frac{\mathbf{r}_0}{r^2} = \frac{\rho_0 h^2 c^3}{2} \left(\frac{2\pi a}{\lambda}\right)^4 \frac{\mathbf{r}_0}{r^2}. \tag{18.110}$$

Hence,

$$|\langle \mathbf{P} \rangle| \propto \lambda^{-4}, \tag{18.111}$$

which is once more the acoustic counterpart of Rayleigh scattering! The total power radiated is the surface integral of $|\langle \mathbf{P} \rangle|$ over a sphere of arbitrary radius $R > a$, is

$$P_{\text{tot}} = \int_0^{2\pi} d\phi \int_0^{\pi} |\langle \mathbf{P} \rangle|_R R^2 \sin\theta d\theta \approx \frac{2\pi \rho_0 h^2 \omega^4 a^4}{2c} \int_0^{\pi} \sin\theta d\theta$$

$$= \frac{2\pi \rho_0 h^2 \omega^4 a^4}{c}, \tag{18.112}$$

and, in this same limit $ka \ll 1$. Not surprisingly, $P_{\text{tot}} \propto \lambda^{-4}$, as before.

18.6 THE SOUND OF MOUNTAIN STREAMS

The sound of mountain streams arises in part from the radial oscillations and bursting of small air bubbles [133]. Given the previous analysis for an oscillating sphere, we will consider the former (i.e., the oscillations of small air bubbles). Let the center of a spherical bubble be at the origin of coordinates, with radius $R(t)$, oscillating about a fixed radius a. This will be the basis for the discussion that follows, after a physical argument to determine the velocity field. Both the air inside the bubble and the water outside it oscillate, but the momentum of the oscillating air mass is negligible compared with that of the water since the air density is $\approx 10^{-3}$ times that of the water. We consider next two concentric spherical bubble surfaces of radius R_1 and R_2, and assume that spherical symmetry is maintained at all times, so that the velocity field is radial (i.e., $\mathbf{v} = \langle v(r), 0, 0 \rangle$). The total mass of the water enclosed by two concentric spheres is constant, and since water is (essentially) incompressible, $\nabla \cdot \mathbf{v} = 0$. Therefore the volume (and mass) inflow per unit time and outflow per unit time are equal:

$$4\pi R_1^2 v(R_1) = 4\pi R_2^2 v(R_2),$$

and hence

$$\frac{v(R_1)}{v(R_2)} = \frac{R_2^2}{R_1^2},$$
(18.113)

or more generally,

$$v(r) \propto r^{-2}.$$
(18.114)

Obviously this follows directly from the divergence-free condition above:

$$\frac{1}{r^2} \frac{d}{dr} \left[r^2 v(r) \right] = 0.$$
(18.115)

Thus, at the bubble surface, $r = R(t)$, $v(R) = v_R$, and for $r \geq R$,

$$v(r) = \frac{R^2}{r^2} v_R.$$
(18.116)

Now we consider the total kinetic energy of the water as a result of this one oscillating bubble, given by,

$$E_k = \int_R^\infty 4\pi r^2 \left[\frac{1}{2} \rho v^2(r) \right] dr,$$
(18.117)

where the term in square brackets is the kinetic energy per unit volume. Thus, by (18.116)

$$E_k = 2\pi\rho \int_R^\infty r^2 v^2(r) dr = 2\pi\rho \int_R^\infty r^2 \left(\frac{R^2}{r^2} v_R \right)^2 dr$$

$$= 2\pi\rho R^4 v_R^2 \int_R^\infty \frac{dr}{r^2} = 2\pi\rho v_R^2 R^3.$$
(18.118)

In terms of the momentum mv associated with the bubble oscillation, we can write

$$mv = \frac{d}{dv} \left(\frac{1}{2} mv^2 \right) = \frac{d}{dv_R} (E_k) = 4\pi\rho R^3 v_R.$$
(18.119)

The force exerted by the bubble on the surrounding liquid is equal to the rate of change of momentum by Newton's second law of motion. Thus, if $p(R)$ is the pressure inside the bubble when the radius is R and $p(a) = p_a$, then since force = pressure × area, we have

$$4\pi\rho \frac{d}{dt} \left(R^3 v_R \right) = 4\pi R^2 (p - p_a).$$
(18.120)

But how does the pressure vary with the radius? We shall invoke the ideal gas law and assume that the oscillation is fast enough for it to be an adiabatic process. In terms of the volume of the undisturbed bubble,

$$pV^\gamma = p_a V_a^\gamma,$$
(18.121)

where $\gamma \approx 1.4$ for air. It is also assumed that the oscillations about the mean radius a are small in the sense that

$$R(t) = a + \delta R(t),$$
(18.122)

where $|\delta_R| / a \ll 1$. Therefore [133]

$$\frac{p}{p_a} = \left(\frac{V_a}{V}\right)^\gamma = \left(\frac{a}{a + \delta R}\right)^{3\gamma} = \left(1 + \frac{\delta R}{a}\right)^{-3\gamma} \approx 1 - \frac{3\gamma \delta R}{a}, \qquad (18.123)$$

so that

$$p - p_a = \left(\frac{p}{p_a} - 1\right) p_a \approx -3\gamma p_a \frac{\delta R}{R_0} \propto \delta R. \qquad (18.124)$$

Now the left-hand side of (18.120) is

$$4\pi\rho \left(3R^2 \frac{dR}{dt} v_R + R^3 \frac{dv_R}{dt} \right),$$

and since we are looking for small amplitude oscillations, we seek, without loss of generality, perturbations of the form

$$\delta R = A \sin \omega t, \quad A \ll R_0. \qquad (18.125)$$

Since $v_R = dR/dt$, this implies that for small oscillations the term

$$3R^2 \frac{dR}{dt} v_R \propto \left(\frac{dR}{dt}\right)^2 = \left(\frac{d\delta R}{dt}\right)^2 \propto A^2, \qquad (18.126)$$

whereas to this level of approximation,

$$R^3 \frac{dv_R}{dt} \propto A, \qquad (18.127)$$

so this term is retained and the previous one neglected by comparison. Folding all these approximations into (18.120) yields (not surprisingly) the equation of simple harmonic motion in the form

$$\frac{d^2}{dt^2}(\delta R) + \omega^2 \delta R = 0, \qquad (18.128)$$

where

$$\omega^2 = \frac{3\gamma p_a}{\rho R_0^2}. \qquad (18.129)$$

The period of this oscillation is

$$T = v^{-1} = \frac{2\pi}{\omega} = 2\pi \left(\frac{3\gamma p_a}{\rho a^2}\right)^{-1/2}, \qquad (18.130)$$

hence

$$\delta R = A \sin \left[\left(\frac{3\gamma p_a}{\rho R_0^2}\right)^{-1/2} t \right]. \qquad (18.131)$$

If we put in the following numerical values

$$\gamma = 1.4; \quad p_a = 101 \text{ kPa} \approx 10^5 \text{ N/m}; \quad \rho = 10^3 \text{ kg/m}; \quad a = 10^{-3} \text{ m},$$

we obtain $v \approx 3,300$ Hz. Note that this is not unreasonable, since the approximate acoustic frequency range for humans is $20 \lesssim v \lesssim 20{,}000$ Hz.

18.6.1 Bubble Collapse

Now we examine the collapse—or bursting—of a bubble with a more rigorous eye (or ear?). For the sake of completeness we reiterate the physical details above and consider an infinite incompressible liquid volume of density ρ containing a bubble of radius R_0, in equilibrium with its surroundings under a uniform pressure P. This equilibrium is then perturbed, and the bubble undergoes radial oscillations. Under these ideal circumstances, the pressure p inside the bubble and its volume V are related via the adiabatic gas law,

$$pV^\gamma = \text{constant},$$

or equivalently, $p \propto R^{-3\gamma}$, where again $\gamma > 1$ is the ratio of specific heats for the gas in the bubble. Since the oscillations are radial, the velocity $\mathbf{v}$ of the surface will be radial only, and a function of time and the radius r only; that is, in spherical polar coordinates, $\mathbf{v}(r, t) = \langle v(r, t), 0, 0 \rangle$. The equation of continuity for an incompressible fluid is, as before,

$$\nabla \cdot \mathbf{v} = 0, \quad \text{or} \quad \frac{1}{r^2} \frac{d}{dr}(vr^2) = 0, \tag{18.132}$$

so that $v = f_1(t)r^{-2}$, where $f_1(t)$ is (at this stage) an arbitrary function of time. For a vector $\mathbf{v}$ the velocity potential ϕ is defined by $\mathbf{v} = \nabla\phi$ (implying that $\nabla^2\phi = 0$), so here

$$f_1(t)r^{-2} = \frac{d\phi}{dr}, \text{ so that } \phi = -f_1(t)r^{-1}. \tag{18.133}$$

Note also that

$$\lim_{r \to \infty} v = 0, \tag{18.134}$$

as should be expected. Since

$$\nabla \times \mathbf{v} = \mathbf{0}, \tag{18.135}$$

the motion is irrotational. In the absence of body forces (such as gravity) Bernoulli's equation for this problem can be written as

$$\frac{p}{\rho} + \frac{v^2}{2} - \frac{1}{r}\frac{df_1}{dt} = f_2(t), \tag{18.136}$$

where $f_2(t)$ is another arbitrary function of time. Hence

$$\frac{p}{\rho} = \frac{1}{r}\frac{df_1}{dt} + f_2(t) - \frac{f_1^2}{2r^4}, \tag{18.137}$$

from which it follows that as $r \to \infty$, $f_2(t) = P/\rho = \text{constant}$, since the pressure field tends in this limit to the constant pressure P at infinity. Imposing the condition of continuity of pressure *and* normal velocity at the surface of the bubble ($r = R$), and using the fact that

$$PR_0^{3\gamma} = pR^{3\gamma}, \tag{18.138}$$

we find that $f_1(t) = R^2 \dot{R}$, and hence the Bernoulli equation takes the form

$$\frac{d}{dt}(R^2 \dot{R}) - \frac{1}{2}R\dot{R}^2 = \frac{PR}{\rho}\left\{\left(\frac{R_0}{R}\right)^{3\gamma} - 1\right\}. \tag{18.139}$$

The left-hand side can be arranged into the form

$$\frac{d}{dR}\left(\frac{1}{2}R^3\dot{R}^2\right) = \frac{P}{\rho}(R_0^{3\gamma}R^{2-3\gamma} - R^2). \tag{18.140}$$

Integrating this equation with respect to R yields the first-order nonlinear equation for the time evolution of the bubble radius:

$$\frac{1}{2}R^3\dot{R}^2 = \frac{P}{\rho}\int_{R_{min}}^{R}(R_0^{3\gamma}R^{2-3\gamma} - R^2)d\bar{R}$$

$$= \frac{P}{\rho}\left\{\left[\frac{R_0^{3\gamma}(R^{3(1-\gamma)} - R_{min}^{3(1-\gamma)})}{3(1-\gamma)} - \frac{1}{3}(R^3 - R_{min}^3)\right]\right\}. \tag{18.141}$$

The left-hand side of this equation is in fact proportional to the kinetic energy of the resulting oscillations. Suppose now that the bubble oscillates between maximum and minimum radii $R_{max} = R_1$ and $R_{min} = R_2$, respectively. If the above integration is carried out between these limits for the left-hand side, then since $\dot{R}$ vanishes at these two extremes, it follows that the following relationship between R_0, R_1, and R_2 holds:

$$R_0^{3\gamma}(R_2^{3(1-\gamma)} - R_1^{3(1-\gamma)}) = (R_1^3 - R_2^3)(\gamma - 1). \tag{18.142}$$

If the oscillations are large in the sense that $R_2 \ll R_1$, then the equation simplifies to the approximate relation

$$\frac{R_1}{R_2} \approx (\gamma - 1)^{-1/3}\left(\frac{R_0}{R_2}\right)^{3\gamma}. \tag{18.143}$$

The left-hand side of equation (18.139) can be rewritten, and if the inertial terms dominate the right-hand side of the equation, then we find as a limiting case that

$$R\ddot{R} + \frac{3}{2}\dot{R}^2 = 0, \tag{18.144}$$

which can be solved by first defining $w = \dot{R}^2$ to obtain

$$\dot{w}R + 3\dot{R}w = 0, \tag{18.145}$$

with general solution

$$w = \text{constant} \times R^{-3}. \tag{18.146}$$

This can be integrated directly to give

$$R(t) = (Dt + E)^{2/5}. \tag{18.147}$$

If it takes the bubble a time t_c to collapse from an initial radius of R_0, then imposing these "boundary conditions in time" yields the interesting result

$$R(t) = R_0\left[\frac{t_c - t}{t_c}\right]^{2/5}. \tag{18.148}$$

Noting that $\dot{R} \propto (t_c - t)^{-3/5}$, it is clear that, in this (subsequently unrealistic) limiting case the bubble-wall velocity becomes divergent as the bubble nears collapse. The resulting sound pressure field will also (theoretically) diverge, though in practice this does not occur: how many of us have been nearly deafened by the collapse of a bubble? (Bursting balloons at a child's party are, however, another matter.) At this point the neglected terms (together with viscous and surface tension terms) kick in to modify the evolutionary demise of the bubble. The bubble collapse is eventually short circuited by heating of the gas inside the bubble (due to adiabatic compression) and emission of sound, and even disintegration of the bubble for this or other reasons. Nevertheless, this solution does give an indication of the level of *cavitation* damage that can be generated by ship propellors, for example. Indeed, Lord Rayleigh was approached by the Royal Navy during World War I concerning such damage to the propellors of surface ships and submarines. He showed that collapsing bubbles were to blame: fast rotating propellors cause the pressure near the blades to drop below the water vapor pressure, so bubbles form and subsequently collapse quite violently.

18.6.2 Playing with Mathematical Bubbles

Let's play some more with this fascinating problem, following Rayleigh's analysis. From the continuity equation we know that $vr^2 = f_1(t)$, so that if at any given time the velocity at the bubble boundary $r = R$ is v_R, and that at $r > R$ it is v, it again follows that

$$v(r) = R^2 v_R / r^2. \tag{18.149}$$

Furthermore, the total kinetic energy of motion in the fluid is

$$K = \frac{1}{2} \int_R^\infty v^2(r) dm = 2\pi\rho \int_R^\infty v^2 r^2 dr = 2\pi\rho R^4 v_R^2 \int_R^\infty r^{-2} dr = 2\pi\rho R^3 v_R^2. \tag{18.150}$$

The work done in reducing the radius from R_0 to R is

$$W = -P \int_{R_0}^R 4\pi r^2 dr = \frac{4}{3}\pi P \left[R_0^3 - R^3 \right]. \tag{18.151}$$

Equating K and E, we obtain

$$v_R^2 = \dot{R}^2 = \frac{2P}{3\rho} \left(\frac{R_0^3}{R^3} - 1 \right), \tag{18.152}$$

from which an explicit value for the time of collapse from radius R_0 to R, is found to be

$$t = -\left(\frac{3\rho}{2P} \right)^{1/2} \int_{R_0}^R \left(\frac{\tilde{R}^3}{R_0^3 - \tilde{R}^3} \right)^{1/2} d\tilde{R} = R_0 \left(\frac{\rho}{6P} \right)^{1/2} \int_b^1 \tilde{b}^{-1/6} (1 - \tilde{b})^{-1/2} d\tilde{b}, \tag{18.153}$$

where b (and dummy variable $\tilde{b}$) is equal to $(R/R_0)^3$. The time for complete collapse is therefore

$$t_c = R_0 \left(\frac{\rho}{6P} \right)^{1/2} \int_0^1 \tilde{b}^{-1/6} (1 - \tilde{b})^{-1/2} d\tilde{b}. \tag{18.154}$$

Using the definition of the *Beta function*, that is,

$$B(m + 1, n + 1) = \int_0^1 \tilde{b}^m (1 - \tilde{b})^n d\tilde{b} = \frac{\Gamma(m + 1)\Gamma(n + 1)}{\Gamma(m + n + 2)}, \qquad (18.155)$$

we see that

$$t_c = R_0 \left(\frac{\rho}{6P}\right)^{1/2} \frac{\Gamma(5/6)\Gamma(1/2)}{\Gamma(4/3)} \approx 0.915 R_0 \left(\frac{\rho}{P}\right)^{1/2}. \qquad (18.156)$$

Thus the time to collapse varies as the size of the bubble, as the square root of the fluid density outside the bubble, and as the inverse square root of the inside pressure. The mathematics of this argument is nicely complemented by the following dimensional one.

We can also use dimensional analysis to find the dependence of t_c on R_0, P, and ρ, but it provides no information on the constant of proportionality. Suppose that

$$t_c \propto R_0^a P^b \rho^c. \qquad (18.157)$$

If we equate the dimensions $[M]$, $[L]$, and $[T]$ of mass, length, and time, respectively, we have that

$$[T][M]^0[L]^0 = [L]^a \{[M][L]^{-1}[T]^{-2}\}^b \{[M][L]^{-3}\}^c$$
$$= [T]^{-2b}[M]^{b+c}[L]^{a-b-3c}, \qquad (18.158)$$

or $b = -1/2$, $c = 1/2$ and $a = 1$, so as expected from (18.156),

$$t_c \propto R_0 \left(\frac{\rho}{P}\right)^{1/2}. \qquad (18.159)$$

Question: What assumptions about the functional dependence for t_c are implicit in relation (18.157)?

It is also interesting to note that when this method is applied to find the time dependence of the radius r of the shock front from an atomic explosion in the atmosphere, that is also proportional to $t^{2/5}$.

Chapter Nineteen

Electromagnetic Scattering: The Mie Solution

In a delightful book by Robert Ehrlich entitled "The Cosmological Milkshake" [227] there appears a cartoon showing a child sitting on a bench with her father; the caption to this picture reads 'The budding urban scientist asks "Daddy, why is the sky brown?"' Pollution notwithstanding, sunlight entering the earth's atmosphere is scattered by the molecules in the air, which are small compared with wavelengths of light. The electric field of the incident sunlight causes electrons in the molecules to oscillate, and re-radiate the light; this is what is meant here by the word 'scattering.' The degree of scattering is inversely proportional to the fourth power of the wavelength of the light; blue light being of shorter wavelength than red, it is scattered the most, and consequently we see blue sky except when we look in the direction of the sun at sunrise or sunset, when the long path through which the light passes depletes the blue light, leaving a predominance of the longer wavelength red light. This phenomenon, coherent scattering, is often referred to as *Rayleigh scattering*, named for Lord Rayleigh, who developed the theoretical basis for this type of scattering. Sunset colors are also determined by the amount of dust (or aerosols in general) in the atmosphere; after major volcanic eruptions they can be really spectacular, as are the sunrises, and can occasionally give rise to so-called 'blue moons' (but only once in a blue moon).

By now the alert reader may be thinking: *so why isn't the sky violet, since that has a shorter wavelength than blue light?* The reasons depend on both external and internal factors; firstly, sunlight is not uniformly intense at all wavelengths (otherwise it would be pure white before entering the atmosphere). It has a peak intensity somewhere in the green part of the visible spectrum, so the entering intensity of violet light is considerably less than that of blue. The other reason is physiological in origin: our eyes are less sensitive to violets than blues (and indeed greens). The scattering of sunlight by molecules and dust is much more complex than described here, of course; there are subtle dependences of color and intensity as a function of angle from the sun, polarization of the light, dust particles are typically *not* small compared to the the wavelengths of light, and so Rayleigh scattering is not occurring.

[122]

In what follows a very brief and simplistic account of light scattering is presented, but to more fully appreciate both the underlying physics and profound beauty of light scattering in the world around us, one can do no better than read [124] and [125], and then, for the more mathematically ambitious [56].

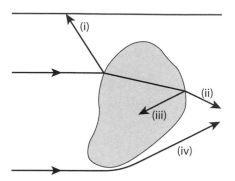

Figure 19.1 Light incident on a large particle.

For electromagnetic waves in the visible spectrum, the amount of scattering depends on the range of particle sizes present compared with the wavelengths of visible light. The latter range is approximately 400–700 nm. A convenient measure of relative size is the radius-to-wavelength ratio R/λ. When this is of the order of at least about 10, the particles are considered to be large, and it is convenient to regard light in terms of rays. This is the domain of geometrical optics, and as illustrated in Figure 19.1, incident light may undergo some or all of the indicated processes: (i) reflection at the surface, (ii) refraction into and out of the particle, (iii) internal reflection, and (iv) edge diffraction. So light rays can be partially reflected from the surface of the particles, refracted on passing through its interior, or diffracted (bent) around the edges. The second and third mechanisms are exhibited in the phenomenon of the rainbow; light is refracted and reflected by raindrops to produce this beautiful colored arc in the sky. Less familiar is the third important mechanism—diffraction—a consequence of the wavelike properties of light. This is responsible for some of the more subtle rainbow features, as mentioned in Chapter 4 and elsewhere—pale fringes below the top of the primary bow, and as already noted, iridescence in clouds near the sun.

Depending on the size of the particle, the amount of light scattered by diffraction can be as much as that by the other two mechanisms. Some of the refracted light may be absorbed by the particles; if so, this will affect the color of the outgoing radiation. An extreme example of this is black smoke—in this case most of the incident radiation is absorbed. When little or no absorption occurs, large particles scatter light pretty much in the forward direction, so the observer looking toward the light source—the sun, usually—will see a general whitish color. When the particles are not large the light ray approach of geometrical optics is inadequate to describe the scattering processes; the wave nature of light, as mentioned above, must be taken into account. Particles for which $R/\lambda \approx 1$ scatter light in a wider band away from the incident direction (resulting in the sky appearing hazy), but it is smaller particles (for which $R/\lambda \ll 1$) that are better able to scatter light multidirectionally. If the aerosols are smaller than about 0.1 microns ($0.1\,\mu$), the light is scattered much more uniformly in all directions; as much backward as forward and not much less off to the sides. Furthermore, the amount of scattering is

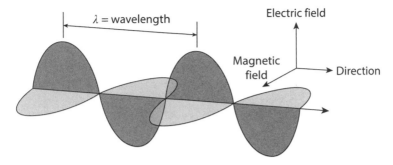

Figure 19.2 Cartoon of electromagnetic wave orthogonal triad.

very sensitive to the wavelength of the incident light; as seen in Section 4.3.3, the intensity of scattering is $\propto \lambda^{-4}$. This means that light of shorter wavelength, such as blue or violet, is scattered much more than the longer wavelength red light. Only when we look in the direction of the setting sun, for example, do we see the red light predominating—most of the blue has been scattered out of the line of sight. Think for a moment of cigarette smoke curling upward from an ash tray; typically it is bluish in color—a consequence of the smoke particles being smaller than the wavelengths of light. It is the blue light that is scattered more, and this is what we see. This is an example of Rayleigh scattering, the same phenomenon that makes the sky blue. Rayleigh scattering arises because of wavelength-dependent molecular scattering.

> If the smoke is exhaled, it appears to be whiter, because moisture from the air in the lungs has coated the smoke particles. This increases their effective size, and the wavelength dependence of the scattering is altered. The light is now scattered more uniformly than before, and hence the smoke appears whiter.

To understand the phenomenon of scattering from a more analytic point of view we need to recall some basic physics. An electromagnetic wave has, not surprisingly, both an electric and magnetic field that are functions of time and space as it propagates. The direction of propagation and the directions of these fields form a mutually orthogonal triad (Figure 19.2).

When an electromagnetic field encounters an electron bound to a molecule, the electron is accelerated by the electric field of the wave. It's a type of "chicken and egg" situation, because an accelerated electron will also radiate electromagnetic energy in the form of waves in all directions (to some extent), and this is the scattered radiation that we have been discussing. Consider Figure 19.3, which illustrates such a situation for a small particle with $R/\lambda \ll 1$ as a snapshot in time.

As shown, the wave propagates with speed c in the x direction, with the electric field in the z direction (it is said to be polarized in that direction; this is a qualification we shall address below). This field varies periodically with frequency $\nu = c/\lambda$,

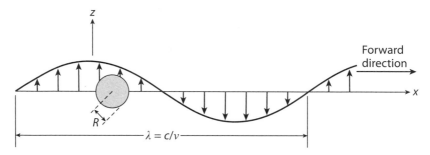

Figure 19.3 Electromagnetic wave incident on a spherical particle of radius $R \ll \lambda$.

and its fluctuations affect the electrons it encounters. To a much lesser extent, the more massive nuclei are also affected, but this will be ignored here. We shall denote the incident electric field by

$$E(x_0, t) = E_0 \sin \omega t, \tag{19.1}$$

where x_0 is the location of this particular electron on the x-axis, E_0 is the amplitude of the wave, and $\omega = 2\pi\nu$ is the *angular frequency* of the wave. As the wave passes, electrons will be accelerated back and forth in the z direction, and they will in turn radiate electromagnetic waves—this radiation is the scattered light. An important consequence of our assumption that the particle is small is this: since $R \ll \lambda$, the electric field is almost uniform throughout the particle, so every electron (with charge e) experiences close to the same force (eE) accelerating it, proportional to its displacement s from its former position of equilibrium in the absence of the wave. The force will always be such as to move the electron back toward that position, so it can be incorporated in Newton's second law of motion as follows:

$$ma \equiv m\frac{d^2s}{dt^2} = eE_0 \sin \omega t - As, \tag{19.2}$$

m being the electron mass, a being its acceleration, and A being a constant of proportionality. This is recognizable as an inhomogeneous second-order differential equation with constant coefficients. It is in fact the equation of forced simple harmonic motion.

Exercise: Show that the solution to equation (19.2) satisfying the simplest initial conditions $s(0) = 0$ and $s'(0) = 0$ is given by

$$s(t) = \frac{eE_0}{A - m\omega^2} \left\{ \sin \omega t - \left(\frac{m}{A}\right)^{1/2} \omega \sin \left[\left(\frac{A}{m}\right)^{1/2} t\right] \right\}, \tag{19.3}$$

provided that $A \neq m\omega^2$. In the event that these quantities are equal—a case known as resonance, which will not be pursued here—the solution can be found directly from the original differential equation or by applying L'Hopital's rule to the solution (19.3).

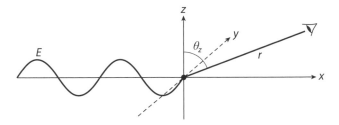

Figure 19.4 Scattering from a particle at the origin.

Exercise: Show that the solution for $s(t)$ when resonance *does* occur is given by

$$s(t) = \frac{eE_0}{2A}\left\{\sin\left[\left(\frac{A}{m}\right)^{1/2}t\right] - \left(\frac{A}{m}\right)^{1/2}t\cos\left[\left(\frac{A}{m}\right)^{1/2}t\right]\right\}. \qquad (19.4)$$

Returning to equation (19.3), we note because of the tight binding of electrons in most aerosols, the term $(m/A)^{1/2}\omega \ll 1$ so the second term can usually be neglected. Then the electron acceleration can be approximated by the expression

$$a = -\frac{\omega^2 eE_0}{A}\sin\omega t. \qquad (19.5)$$

The amplitude of this acceleration is proportional to $\omega^2 E_0$. Recall that the electric field (and hence acceleration of the electron) is perpendicular to the direction of the incident wave. Reversing this, we anticipate that the electric field of the scattered wave will be proportional to the perpendicular component of the acceleration at large distances from the particle, that is, to $a\sin\theta_z$, where θ_z is the angle between the direction of the scattered light and the z-axis (see Figure 19.4), which is the direction of the incident electric field.

We also expect that the total electric field is proportional to the number of electrons present in the particle, and therefore to its volume V. The intensity I of this scattered light is proportional to the square of the electric field, and allowing for the usual inverse square fall-off with distance r, consistent with energy conservation, we arrive at the proportionality relation

$$I_s^{xz} \propto I_0\frac{\omega^4 V^2}{r^2}\sin^2\theta_z \propto I_0\frac{V^2}{\lambda^4 r^2}\sin^2\theta_z \equiv I_0\frac{V^2}{\lambda^4 r^2}\cos^2\theta_x, \qquad (19.6)$$

where $I_0 = E_0^2$. The scattered wave makes an angle $\theta_x = \pi/2 - \theta_z$ with the forward direction (the x-axis). We have therefore obtained from these simple arguments a fundamental result for Rayleigh scattering, namely that the intensity of scattered light is inversely proportional to the fourth power of the wavelength (and recall that this was accomplished in several other ways in Chapter 18; see also Section 4.3.3). More accurately, the probability that a photon of sunlight will be scattered from its original direction by an air molecule is $\propto \lambda^{-4}$. In any case, a higher proportion of blue light ($\lambda \approx 0.4\mu$) than red light ($\lambda \approx 0.7\mu$) is scattered; in fact about $(7/4)^4 \approx 9$ times more.

There are some other things to note from this formula. The scattered radiation is proportional to V^2; this dependence results in a reduction of intensity in the forward

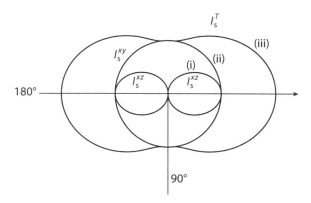

Figure 19.5 Angular variation of polarization intensities.

direction (because of loss of intensity from the incident beam) and corresponds to a dependence on the sixth power of radius R. This can be contrasted with intensity reduction arising from absorption within the particle. The latter will be proportional to the number of absorbing molecules present, and therefore to V. Therefore, in relative terms the importance of scattering (as opposed to absorption) for intensity reduction in the forward direction will be greater for larger particles (while still satisfying the requirement that $R/\lambda \ll 1$). Also it is clear that the intensity of scattered light depends on the angle θ_z, but recall that this is for the special case of polarized light. For scattering in the xy-plane there is no angular dependence because of symmetry about the z-axis, and so for this case,

$$I_s^{xy} \propto I_0 \frac{V^2}{\lambda^4 r^2}. \tag{19.7}$$

Sunlight is *unpolarized*, because the electric field in light from the sun vibrates in all possible directions. We can use the result (19.6) and generalize it to this case as follows. We define the *scattering plane* to be the plane containing both the forward and scattered directions. An unpolarized incident wave can be written as the arithmetic mean of two independent linearly polarized components, one being parallel to and the other being perpendicular to the scattering plane. Essentially, to do this we average the scattered intensity by replacing the term $\sin^2 \theta_z$ by $(\sin^2 \theta_z + \sin^2 \theta_y)/2$, where θ_y is the angle the direction of observation makes with the y-axis. By including the corresponding angle with the x-axis, we note that the squares of the direction cosines sum to one:

$$\cos^2 \theta_x + \cos^2 \theta_y + \cos^2 \theta_z = 1. \tag{19.8}$$

This is equivalent to

$$\sin^2 \theta_y + \sin^2 \theta_z = 1 + \cos^2 \theta_x. \tag{19.9}$$

Therefore in total

$$I_s^T \propto I_0 \frac{V^2}{\lambda^4 r^2} \left(1 + \cos^2 \theta_x\right). \tag{19.10}$$

for unpolarized sunlight. This total scattering intensity is just the sum of the two polarization intensities. This is illustrated in Figure 19.5.

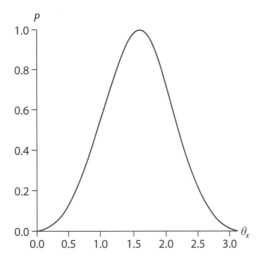

Figure 19.6 Degree of polarization from (19.11).

The perpendicular component of polarization (I_s^{xz}) is the central figure, the middle figure is the parallel component (I_s^{xy}), independent of angle because this is scattering in the xy-plane (perpendicular to the electric field vector), and the outer figure (I_s^{T}) is the sum of these two. Therefore the scattered light is most polarized (and least bright) perpendicular to the incoming light, and least polarized (and most bright) in the forward and backward directions [203].

The angular dependence of the polarization for unpolarized incident light is conveniently expressed in terms of the quotient

$$P = \frac{I_s^{xy} - I_s^{xz}}{I_s^{xy} + I_s^{xz}} = \frac{1 - \cos^2 \theta_x}{1 + \cos^2 \theta_x}. \tag{19.11}$$

A graph of P is shown in Figure 19.6 (in Cartesian form with θ_x in radians), illustrating afresh that (for ideal molecular scatterers) the light from the incoming radiation is 100 percent polarized at $90°$, and completely unpolarized in the forward ($0°$) and backward ($180°$) directions. (In practice this will not be the case, because molecules are not perfect scatterers [103].)

For particles larger than about 0.05μ the Rayleigh scattering argument is inappropriate, because the processes of reflection, refraction, and diffraction cannot be neglected. It turns out that most of the light is scattered near the forward direction, but for the case of particles whose size is comparable with the wavelengths of light, the situation is quite complex both mathematically and physically (as will be seen). The exact solution to this problem for spherical particles is often referred to as the *Mie solution*, from a 1908 paper by Gustav Mie; Debye also solved the problem independently in 1909, but the solution had already been published in 1863 by Clebsch, and again independently by Lorenz in 1890 and 1898. The rediscovery of existing results is a rather common theme in the history of science and mathematics. It is to the establishment of the Debye solution that we now turn.

19.1 MAXWELL'S EQUATIONS OF ELECTROMAGNETIC THEORY

There are several different (but entirely equivalent) forms of Maxwell's equations. In what follows, the fundamental vectors $\mathbf{E}$ and $\mathbf{B}$ are the electric and magnetic induction vectors, respectively. In addition, there are three other vectors that must be included for a complete description of the effect of the electromagnetic field on material objects: the electric displacement $\mathbf{D}$, the magnetic field $\mathbf{H}$, and the electric current density $\mathbf{j}$. Gaussian units will be used here. For continuous media, Maxwell's equations consist of a pair of vector equations and a pair of scalar ones, supplemented by the so-called constitutive relations, which describe the behavior of materials in the presence of the electromagnetic field [13]:

$$\nabla \times \mathbf{H} - \frac{1}{c}\frac{\partial \mathbf{D}}{\partial t} = \frac{4\pi}{c}\mathbf{j}; \qquad (19.12)$$

$$\nabla \times \mathbf{E} + \frac{1}{c}\frac{\partial \mathbf{B}}{\partial t} = \mathbf{0}; \qquad (19.13)$$

$$\nabla \cdot \mathbf{D} = 4\pi\rho; \qquad (19.14)$$

$$\nabla \cdot \mathbf{B} = 0; \qquad (19.15)$$

$$\mathbf{j} = \sigma\mathbf{E}; \qquad (19.16)$$

$$\mathbf{D} = \varepsilon\mathbf{E}; \qquad (19.17)$$

$$\mathbf{B} = \mu\mathbf{H}. \qquad (19.18)$$

The last three equations are valid under various assumptions, (e.g., that the bodies are at rest and the medium is isotropic), otherwise σ (electrical conductivity), ε (dielectric constant or permittivity), and μ (magnetic permeability) are rank-2 tensors. They are constant in vacuo and independent of time here, but in inhomogeneous media they are in general spatially dependent. In the absence of currents and charges ($\mathbf{j} = \mathbf{0}$, $\rho = 0$), the first four equations may be written for constant ε and μ as

$$\nabla \times \mathbf{H} = \frac{\varepsilon}{c}\frac{\partial \mathbf{E}}{\partial t}; \qquad (19.19)$$

$$\nabla \times \mathbf{E} = -\frac{\mu}{c}\frac{\partial \mathbf{H}}{\partial t}; \qquad (19.20)$$

$$\nabla \cdot (\varepsilon\mathbf{E}) = 0; \qquad (19.21)$$

$$\nabla \cdot (\mu\mathbf{H}) = 0. \qquad (19.22)$$

In a vacuum, the permittivity and permeability are denoted by ε_0 and μ_0, respectively. The speed of light in vacuo is $c = (\varepsilon_0\mu_0)^{-1/2}$. In a medium it is $(\varepsilon\mu)^{-1/2}$, and consequently the refractive index n of a medium, which is defined as the ratio of the speed of light in vacuo to that in the medium, is defined by

$$n = \left(\frac{\varepsilon \mu}{\varepsilon_0 \mu_0} \right)^{1/2}. \tag{19.23}$$

A further definition is useful in terms of the wave frequency v_0:

$$c = \omega_0 / k_0; \quad \text{where } \omega_0 = 2\pi v_0 \text{ and } k_0 = 2\pi / \lambda_0.$$

Here ω_0 is the angular frequency of electromagnetic waves in vacuo, k_0 is the constant wavenumber, and λ_0 is the wavelength. In general, for *any* wave of individual angular frequency ω and wavenumber k, in any medium, the wave speed (or phase speed) is $c = \omega / k$. Generally, the zero subscript will not be used further; the appropriate speed c will be determined from the context.

19.2 THE VECTOR HELMHOLTZ EQUATION FOR ELECTROMAGNETIC WAVES

The desired expansion of a plane wave in spherical harmonics was not achieved without difficulty. This is undoubtedly the result of the unwillingness of a plane wave to wear a guise in which it feels uncomfortable; expanding a plane wave in spherical wave functions is somewhat like trying to force a square peg into a round hole. However, the reader who has painstakingly followed the derivation ..., and thereby acquired virtue through suffering, may derive some comfort from the knowledge that it is relatively clear sailing from here on. [56]

For constant magnetic permeability μ and electric permeability ε, equations (19.21) and (19.22) reduce to

$$\nabla \cdot \mathbf{E} = 0; \tag{19.24}$$

$$\nabla \cdot \mathbf{H} = 0. \tag{19.25}$$

For monochromatic plane waves with time-harmonic dependence of the form $\exp(-i\omega t)$ equations (19.20) and (19.19), respectively, become

$$\nabla \times \mathbf{E} = i \frac{\omega \mu}{c} \mathbf{H} \equiv k_2 \mathbf{H}; \tag{19.26}$$

$$\nabla \times \mathbf{H} = -i \frac{\omega \varepsilon}{c} \mathbf{E} \equiv -k_1 \mathbf{E}. \tag{19.27}$$

Note that

$$-k_1 k_2 \equiv k^2 = \frac{\omega^2}{c^2} \varepsilon \mu = \frac{\omega^2}{c^2} n^2. \tag{19.28}$$

The wavenumber k is sometimes known as the propagation constant. If the medium is a conducting one ($\sigma \neq 0$), then the refractive index is complex, so that

$$n^2 = \mu \left(\varepsilon + i \frac{4\pi \sigma}{\omega} \right) = \left(\varepsilon + i \frac{4\pi \sigma}{\omega} \right), \tag{19.29}$$

for a nonmagnetic medium.

Using a standard vector identity and the solenoidal nature of **E** and **H**, it follows that both fields for a homogeneous medium (i.e., one with constant k) satisfy the vector Helmholtz equation [55], [56]:

$$\nabla^2 \mathbf{M} + k^2 \mathbf{M} = \mathbf{0}, \tag{19.30}$$

where $k^2 = \omega^2 \varepsilon \mu = \omega^2 / c^2$, and $c = (\varepsilon \mu)^{1/2}$ is the speed of light in the medium. Let

$$\mathbf{M} = \nabla \times (\mathbf{c}\psi), \tag{19.31}$$

where (for now) **c** is a constant vector. Then

$$\mathbf{M} = \psi \nabla \times \mathbf{c} + (\nabla \psi) \times \mathbf{c} = \nabla \psi \times \mathbf{c} = -\mathbf{c} \times \nabla \psi. \tag{19.32}$$

This implies that

$$\nabla \cdot \mathbf{M} = 0. \tag{19.33}$$

Also

$$\nabla^2 \mathbf{M} = \nabla^2 [\nabla \times (\mathbf{c}\psi)] = \nabla \times [\nabla^2 (\mathbf{c}\psi)] = \nabla \times [\mathbf{c}\nabla^2 \psi], \tag{19.34}$$

and

$$k^2 \mathbf{M} = k^2 \nabla \times (\mathbf{c}\psi) = \nabla \times (\mathbf{c}\kappa^2 \psi). \tag{19.35}$$

On combining these equations, we arrive at an important result:

$$\nabla^2 \mathbf{M} + k^2 \mathbf{M} = \nabla \times [\mathbf{c}(\nabla^2 \psi + k^2 \psi)]. \tag{19.36}$$

Why is this so important? Well, if ψ satisfies the *scalar* Helmholtz equation,

$$\nabla^2 \psi + k^2 \psi = 0, \tag{19.37}$$

then **M** satisfies the *vector* Helmholtz equation (19.30). Next, we consider the vector field

$$\mathbf{N} \equiv k^{-1} \nabla \times \mathbf{M}. \tag{19.38}$$

Then

$$\nabla^2 \mathbf{N} = k^{-1} \nabla^2 (\nabla \times \mathbf{M}) = k^{-1} \nabla \times (\nabla^2 \mathbf{M})$$
$$= k^{-1} \nabla \times (-k^2 \mathbf{M}) = -k\nabla \times \mathbf{M} = -k^2 \mathbf{N}. \tag{19.39}$$

Therefore

$$\nabla^2 \mathbf{N} + k^2 \mathbf{N} = \mathbf{0}. \tag{19.40}$$

Also we note that

$$\nabla \times \mathbf{N} = k^{-1} \nabla \times (\nabla \times \mathbf{M}) = k^{-1} [\nabla(\nabla \cdot \mathbf{M}) - \nabla^2 \mathbf{M}]$$
$$= -k^{-1} \nabla^2 \mathbf{M} = k\mathbf{M}, \tag{19.41}$$

and hence

$$\mathbf{M} = k^{-1} \nabla \times \mathbf{N}. \tag{19.42}$$

These relationships make the vector fields $\mathbf{M}$ and $\mathbf{N}$ ideal solutions for $\mathbf{E}$ and $\mathbf{H}$, given the form of Maxwell's equations. Suppose further, that (as is often the case) we require vector solutions in spherical coordinates that are orthogonal to the radius vector $\mathbf{r}$. Then it follows that

$$0 = \mathbf{M} \cdot \mathbf{r} = (\nabla\psi \times \mathbf{c}) \cdot \mathbf{r} = \mathbf{r} \cdot (\nabla\psi \times \mathbf{c}) = (\nabla\psi) \cdot (\mathbf{c} \times \mathbf{r}) \implies \mathbf{c} \times \mathbf{r} = \mathbf{0},$$
(19.43)

so that the choice $\mathbf{c} = \mathbf{r}$ will suffice for this purpose ($\mathbf{c}$ of course no longer being a constant). If we then write

$$\mathbf{M} = \nabla \times (\mathbf{r}\psi) \equiv \nabla\psi \times \mathbf{r} = -\mathbf{r} \times \nabla\psi,$$
(19.44)

where $\mathbf{r}$ is the radius vector in a spherical coordinate system, and $\psi = \psi(\mathbf{r})$ is an arbitrary scalar function of position, then it follows from another vector identity that

$$\nabla \cdot \mathbf{M} = 0.$$
(19.45)

(This was previously established for the constant vector $\mathbf{c}$.) After a considerable amount of vector algebra (**Exercise**), it can be shown that

$$\nabla^2\mathbf{M} = \nabla \times \nabla \times (\mathbf{r} \times \nabla\psi) = \nabla \times (\mathbf{r}\nabla^2\psi),$$

and

$$k^2\mathbf{M} = k^2\left[\nabla \times (\mathbf{r}\psi)\right] = \nabla \times (\mathbf{r}k^2\psi),$$

so that

$$\nabla^2\mathbf{M} + k^2\mathbf{M} = \nabla \times \left[\mathbf{r}(\nabla^2\psi + k^2\psi)\right].$$

Hence a *sufficient* condition for

$$\nabla^2\mathbf{M} + k^2\mathbf{M} = \mathbf{0}$$
(19.46)

is that

$$\nabla^2\psi + k^2\psi = 0.$$
(19.47)

Clearly this is not the most general result, but it is quite adequate for the purposes of this section. If $\mathbf{r}(\nabla^2\psi + k^2\psi)$ can be written as the gradient of a scalar function of position, and this gradient is in the radial direction, then $\mathbf{M}$ is a more general solution of the vector Helmholtz equation. A similar problem to this arises occurs elasticity theory in connection with *Lamé's theorem* (see Appendix E).

Returning to the result (19.46), if ψ satisfies the scalar Helmholtz equation, then $\mathbf{M}$ satisfies the corresponding vector one. A second such solution follows if we define

$$\mathbf{N} = k^{-1}\left(\nabla \times \mathbf{M}\right),$$
(19.48)

so that

$$\nabla^2\mathbf{N} = k^{-1}\nabla^2(\nabla \times \mathbf{M}) = k^{-1}\nabla \times (\nabla^2\mathbf{M}) = -k(\nabla \times \mathbf{M}) = -k^2\mathbf{N}.$$
(19.49)

Remember that k is assumed to be constant for a given material (this will be relaxed in appendices B–D). Note also that

$$\nabla \times \mathbf{N} = k^{-1} \nabla \times (\nabla \times \mathbf{M}) = -k^{-1} \nabla^2 \mathbf{M} = k\mathbf{M}, \qquad (19.50)$$

that is,

$$\mathbf{M} = k^{-1} (\nabla \times \mathbf{N}), \qquad (19.51)$$

which recovers (19.42). As is frequently the case (and especially so in electromagnetic theory), we require vector solutions in spherical coordinate systems that are orthogonal to $\mathbf{r}$ (as noted above). In the present case,

$$\mathbf{M} \cdot \mathbf{r} = (\nabla \psi \times \mathbf{r}) \cdot \mathbf{r} = 0, \qquad (19.52)$$

on using the triple scalar product. Now although the vectors $\mathbf{A}$ and $\nabla \times \mathbf{A}$ are generally not orthogonal [134], it is the case that if $\mathbf{A}$ has no components along any two of three mutually orthogonal directions, then $\mathbf{A} \cdot (\nabla \times \mathbf{A}) = 0$ at all points. This is the situation here, so $\mathbf{M} \cdot \mathbf{N} = 0$ also.

Proof That a Vector Field A Is Defined Uniquely by Its Curl and Divergence

Suppose $\nabla \times \mathbf{A} = \mathbf{F}_1(\mathbf{r})$ and $\nabla \cdot \mathbf{A} = F_2(\mathbf{r})$, where the scalar function F_2 and the vector function $\mathbf{F}_1$ are prescribed and defined everywhere such that $\lim_{|\mathbf{r}| \to \infty} |\mathbf{A}| = 0$. Suppose also that $\mathbf{A}_1$ and $\mathbf{A}_2$ both satisfy the above equations. Then we define $\mathbf{A}_- = \mathbf{A}_1 - \mathbf{A}_2$. Clearly, $\nabla \times \mathbf{A}_- = \mathbf{0}$ and $\nabla \cdot \mathbf{A}_- = 0$. Thus $\mathbf{A}_- = \nabla u$, and hence $\nabla^2 u = 0$ in some region V (which may be infinite). Suppose further that $U = U_s$ on the bounding surface S (if V is finite), or satisfies $\lim_{|\mathbf{r}| \to \infty} |u| = 0$ (if V is all space). We shall assume the latter here without loss of generality; also that $U \in \mathbb{C}^2(V)$. We use Green's first integral identity:

$$\iiint_V \{u\nabla^2 v + (\nabla u) \cdot (\nabla v)\} dV = \iint_S [u(\nabla v) \cdot \mathbf{n}] dS \to 0 \text{ as } |\mathbf{r}| \to \infty. \qquad (19.53)$$

Let $u = v = U$, so

$$\iiint_V \{U\nabla^2 U + |\nabla U|^2\} dV \Rightarrow \iiint_V |\nabla U|^2 dV = 0, \qquad (19.54)$$

and since the integrand is continuous and non-negative everywhere, $|\nabla U|^2 = 0 \Rightarrow U$ is constant in V. But since U vanishes as $|\mathbf{r}| \to \infty$ (or on S if V is finite), $U = 0$ as required, and so $\mathbf{A}_- = \mathbf{0} \Rightarrow \mathbf{A}_1 = \mathbf{A}_2$ as required. (See also Appendix E.)

19.3 THE LORENTZ-MIE SOLUTION

Our point here is not that the exact Mie theory describes the natural rainbow inadequately, but rather that the approximate Airy theory can describe it quite well. Thus the supposedly outmoded Airy theory generates a more natural-looking map of real rainbow colors than Mie theory does, even though Airy theory makes substantial errors in describing the scattering of monochromatic light by isolated small drops. As in many hierarchies of scientific models, the virtues of a simpler theory can, under the right circumstances, outweigh its vices. [169]

Airy theory has been called the incomplete "complete" answer (see the concluding Chapter 28 for some historical details). It did go beyond the models of the day in that it quantified the dependence on raindrop size of (i) the rainbow's angular width, (ii) its angular radius, and (iii) the spacing of the supernumeraries. Also, unlike the models of Descartes and Newton, Airy's predicted a nonzero distribution of light intensity in Alexander's dark band, and a finite intensity at the angle of minimum deviation (as noted in Section 4.2.3, the earlier theories predicted an infinite intensity there). However, spurred on by Maxwell's recognition that light is part of the electromagnetic spectrum, and the subsequent publication of his mathematical treatise on electromagnetic waves, several mathematical physicists sought a more complete theory of scattering, because it had been demonstrated by then that the Airy theory failed to predict precisely the angular position of many laboratory-generated rainbows. Among them were the German physicist Gustav Mie, who published a paper in 1908 on the scattering of light by homogeneous spheres in a homogeneous medium, and Peter Debye, who independently developed a similar theory for the scattering of electromagnetic waves by spheres. Mie's theory was intended to explain the colors exhibited by colloidally dispersed metal particles, whereas Debye's work, based on his 1908 thesis, dealt with the problem of light pressure on a spherical particle. The resulting body of knowledge is usually referred to as Mie theory, and typical computations based on it are formidable compared with those based on Airy theory, unless the drop size is sufficiently small. A similar (but scalar) formulation arises in the scattering of sound waves by an impenetrable sphere, studied by Lord Rayleigh and others in the nineteenth century. In fact, Ludvig Lorenz, a Danish theorist, preceded Mie by about 15 years in the treatment of the scattering of electromagnetic waves by spheres. His contributions to electromagnetic scattering theory and optics are rather overlooked, probably because his work was published in Danish (in 1890). Further details of his research, including his contributions to applied mathematics, may be found in [150] (the article immediately following this one is about Gustav Mie). Clebsch also provided essentially the same solution in 1863, but in connection with elastic waves (see the detailed exposition of Mie and his work [204]). In fact, the work of Clebsch was largely ignored, as noted in an excellent historical account of the subject:

In October 30, 1861, A. Clebsch (1833–1872) [submitted] a 68-page memoir in which he developed the mathematical theory required to solve (by the method of separation of variables) the class of boundary-value problems in which a wave propagating in an elastic medium impinges upon a spherical surface. This paper could have become the cornerstone upon which future generations of scientists could base their theoretical studies. Scores of writers since Clebsch have sought to increase our understanding of this class of problems. However, the mathematical ingenuity of this master craftsman was doomed to lie buried within the pages of one of the leading mathematical journals of the middle of the nineteenth century while later writers rediscovered the results which were to be found in Clebsch's paper.

The same fate was to befall the equally great memoir upon the reflection and refraction of light by a transparent sphere, which was published in 1890 by L. Lorenz (1829–1891). Few problems that have captured the attention of mathematical physicists can claim such an illustrious history as that in the papers published on the relatively simple boundary-value problem of a spherical obstacle which is illuminated by a plane electromagnetic wave. The history has been enriched because of the many diverse fields of scientific endeavor in which the solution of this class of problems has found application. [86]

The Mie solution is based on the solution of Maxwell's equations of electromagnetic theory for a monochromatic plane wave from infinity incident on a homogeneous isotropic sphere of radius a. The surrounding medium is transparent (as the sphere may be), homogeneous, and isotropic. The incident wave induces forced oscillations of both free and bound charges in synchrony with the applied field, and this induces a secondary electric and magnetic field, each of which has components inside and outside the sphere. Of crucial importance in the theory are what are termed the scattering amplitudes for the two independent polarizations; these amplitudes can be expressed as an infinite sum called a *partial wave expansion*. Each term (or partial wave) in the expansion is defined in terms of combinations of Legendre functions of the first kind, Riccati-Bessel functions, and Riccati-Hankel functions.

It is obviously of interest to determine under what conditions such an infinite set of terms can be truncated, and what the resulting error may be by so doing. It transpires that the number of terms that must be retained is of the same order of magnitude as the size parameter β (i.e., up to several thousand for the rainbow problem). In contrast, the "why is the sky blue?" scattering problem—Rayleigh scattering—requires only one term, because the scatterers are molecules that are much smaller than a wavelength of light, so the simplest truncation—retaining only the first term—is perfectly adequate. Although in principle the rainbow problem can be "solved" with enough computer time and resources, numerical solutions by themselves offer little or no insight into the physics of the phenomenon.

However, there was a significant mathematical development in the early twentieth century that eventually had a profound impact on the study of scalar and vector scattering. What is now known as the *Watson transform* was originally introduced in 1918 by G. N. Watson in connection with the diffraction of radio waves around the earth (and subsequently modified by several mathematical physicists in studies of the rainbow problem). It is a method for transforming the slowly converging partial-wave series into a rapidly convergent expression involving an integral in the complex angular-momentum plane. This allows the above transformation to effectively redistribute the contributions to the partial wave series into a few points in the complex plane—specifically, poles (called *Regge poles* in elementary particle physics) and saddle points. Such a decomposition means that instead of identifying angular momentum with certain discrete real numbers, it is now permitted to move continuously through complex values. However, despite this modification, the poles and saddle points have profound physical interpretations in the rainbow problem [86].

In what follows reference will be made to the intensity functions i_1, i_2, the Mie coefficients a_n, b_n (or a_l, b_l depending on context, so as to avoid confusion with the refractive index n), and the angular functions π_n, τ_n. The former pair are proportional to the square of the magnitude of two incoherent, plane-polarized components scattered by a single particle; they are related to the scattering amplitudes S_1 and S_2. The function i_1 is associated with the electric oscillations perpendicular to the plane of scattering (sometimes called horizontally polarized) and i_2 is associated with the electric oscillations parallel to the plane of scattering (or vertically polarized). The scattered wave is composed of many partial waves, the amplitudes of which depend on a_n and b_n. Physically, these may be interpreted as the nth electrical and magnetic partial waves, respectively. The intensity functions i_1, i_2 are represented in the Mie solution as a spherical wave composed of two sets of partial waves: electric (a_n) and magnetic (b_n). The first set is that part of the solution for which the radial component of the magnetic vector in the incident wave is zero; in the second set the corresponding radial component of the electric vector is zero. A given partial wave can be thought of as coming from an electric or a magnetic multipole field, the first wave coming from a dipole field, the second from a quadrupole, and so on. The angular functions $\pi_n(\theta)$ and $\tau_n(\theta)$ are, as their names imply, independent of the size parameter (β) and refractive index.

For a point P located a distance r from the origin of coordinates, at polar angle θ and azimuthal angle ϕ the scattered intensities I_θ and I_ϕ are, respectively,

$$I_\theta = \left(\frac{i_2}{kr}\right)^2 \cos^2 \phi, \tag{19.55}$$

and

$$I_\phi = \left(\frac{i_1}{kr}\right)^2 \sin^2 \phi, \tag{19.56}$$

where $i_j = |S_j|^2$, $j = 1, 2$, and the amplitude functions S_j are given by

$$S_1(\beta, \theta) = \sum_{n=1}^{\infty} \frac{2n+1}{n(n+1)} [a_n(\beta)\pi_n(\cos\theta) + b_n(\beta)\tau_n(\cos\theta)], \tag{19.57}$$

and

$$S_2(\beta, \theta) = \sum_{n=1}^{\infty} \frac{2n+1}{n(n+1)} [a_n(\beta)\tau_n(\cos\theta) + b_n(\beta)\pi_n(\cos\theta)], \tag{19.58}$$

$\beta = ka$ being the size parameter; n here is the order of the induced electric or magnetic multipole. To avoid confusion, subscripts l will be used below when the refractive index n appears explicitly. The Legendre functions $\pi_n(\cos\theta)$ and

$\tau_n(\cos\theta)$ are defined in terms of the associated Legendre functions of the first kind, $P_n^1(\cos\theta)$, as

$$\pi_n(\cos\theta) = \frac{P_n^1(\cos\theta)}{\sin\theta}, \tag{19.59}$$

and

$$\tau_n(\cos\theta) = \frac{d}{d\theta}P_n^1(\cos\theta). \tag{19.60}$$

For completeness, note the recursion relations

$$(n-1)\pi_n = [(2n+1)\cos\theta]\,\pi_{n-1} - n\pi_{n-2}, \tag{19.61}$$

$$\tau_n = (n\cos\theta)\pi_n - (n+1)\pi_{n-1}. \tag{19.62}$$

Exercise: For future reference, show that $\pi_1 = 1$ and $\tau_1 = \cos\theta$.

19.3.1 Construction of the Solution

In this section we utilize the approaches of [56], [57] in obtaining the Mie solution, while returning to the scalar Helmholtz equation in spherical polar coordinates, that is,

$$\nabla^2\psi + k^2\psi \equiv \frac{1}{r^2}\frac{\partial}{\partial r}\left(r^2\frac{\partial\psi}{\partial r}\right) + \frac{1}{r^2\sin\theta}\frac{\partial}{\partial\theta}\left(\sin\theta\frac{\partial\psi}{\partial\theta}\right)$$

$$+\frac{1}{r^2\sin^2\theta}\frac{\partial^2\psi}{\partial\phi^2} + k^2\psi = 0. \tag{19.63}$$

As pointed out by Arfken [44], the restriction that k^2 be a constant is unnecessarily restrictive: it can be generalized. In fact if

$$k^2 \to k^2 + f(r) + \frac{g(\theta)}{r^2} + \frac{h(\phi)}{r^2\sin^2\theta}$$

for the arbitary functions $f(r)$, $g(\theta)$, and $h(\phi)$ (where k^2 is still a constant), the scalar Helmholtz equation is still separable in spherical polar coordinates.
   Exercise: Show that

$$\nabla^2\psi(r,\theta,\phi) + \left[k^2 + f(r) + \frac{g(\theta)}{r^2} + \frac{h(\phi)}{r^2\sin^2\theta}\right]\psi(r,\theta,\phi) = 0$$

is separable in spherical polar coordinates.

The standard separation of variables in the form

$$\psi(r,\theta,\phi) = R(r)\Theta(\theta)\Phi(\phi), \tag{19.64}$$

yields the following ordinary differential equations:

$$\frac{1}{r^2}\frac{d}{dr}\left(r^2\frac{dR}{dr}\right) + \left(k^2 - \frac{l(l+1)}{r^2}\right)R = 0;$$ (19.65)

$$\frac{1}{\sin\theta}\frac{d}{d\theta}\left(\sin\theta\frac{d\Theta}{d\theta}\right) + \left(l(l+1) - \frac{m^2}{\sin^2\theta}\right)\Theta = 0;$$ (19.66)

$$\frac{d^2\Phi}{d\phi^2} + m^2\Phi = 0,$$

where l and m are separation constants. With the introduction of the variable $\xi = \cos\theta$, equation (19.66) becomes

$$\frac{d}{d\xi}\left[(1-\xi^2)\frac{d\Theta(\xi(\theta))}{d\xi}\right] + \left[l(l+1) - \frac{m^2}{1-\xi^2}\right]\Theta(\xi(\theta)) = 0.$$ (19.67)

The requirement that ψ be single valued imposes conditions on Θ and Φ. The basis solutions for each of the separated equations are as follows:

$$R(r) = \{j_l(kr), y_l(kr)\},$$ (19.68)

$$\Theta(\theta) = \{P_l^m(\cos\theta)\},$$ (19.69)

$$\Phi(\phi) = \{\cos(m\phi), \sin(m\phi)\}.$$ (19.70)

In these solution sets, the j_l and y_l functions (denoted by z_l in general) are, respectively, spherical Bessel functions of the first and second kind and of order l, and P_l^m is an associated Legendre function of the first kind of degree l and order m (where $l = m, m+1, \ldots$). The requirement that ψ be a single-valued function of the azimuthal angle ϕ restricts m at this stage to be an integer or zero (when $m = 0$ the functions reduce to the Legendre polynomials P_n). The P_l^m functions form an orthogonal set, that is, if $\xi = \cos\theta$, then in terms of the Kronecker delta,

$$\int_{-1}^{1} P_l^m(\xi)\,P_{l'}^m(\xi)\,d\xi = \frac{2}{2l+1}\frac{(l+m)!}{(l-m)!}\delta_{ll'},$$ (19.71)

The various components of the vectors $\mathbf{M}$ and $\mathbf{N}$ (see equations (19.30) and (19.38)) may now be determined from their definitions in the previous subsection. Given the modal definition (where the function z_l at this stage can mean either j_l or y_l),

$$\psi(r, \theta, \phi) = \exp(im\phi)P_l^m(\cos\theta)z_l(kr)$$ (19.72)

(the real and imaginary parts of subsequent quantities will be taken where appropriate), it follows that

$$\mathbf{M} = \nabla\psi \times \mathbf{r} = \langle 0, M_\theta, M_\phi \rangle;$$ (19.73)

where

$$M_\theta = \frac{im}{\sin\theta}e^{im\phi}P_l^m(\cos\theta)z_l(kr);$$ (19.74)

$$M_\phi = -\frac{dP_l^m(\cos\theta)}{d\theta}e^{im\phi}z_l(kr).$$ (19.75)

Similarly,

$$\mathbf{N} = k^{-1}\nabla \times \mathbf{M} = \langle N_r, N_\theta, N_\phi \rangle = \frac{1}{kr^2 \sin\theta}\langle A, B, C \rangle, \tag{19.76}$$

where

$$A = r\left[\frac{\partial}{\partial\theta}\left(M_\phi \sin\theta\right) - \frac{\partial}{\partial\phi}\left(M_\theta\right)\right]; \tag{19.77}$$

$$B = -r\sin\theta\frac{\partial}{\partial r}\left(rM_\phi\right); \tag{19.78}$$

$$C = r\sin\theta\frac{\partial}{\partial r}\left(rM_\theta\right). \tag{19.79}$$

We evaluate the various components as follows:

$$\frac{\partial}{\partial\theta}\left(M_\phi \sin\theta\right) = -\left(\cos\theta\frac{d\,P_l^m(\cos\theta)}{d\theta} + \sin\theta\frac{d^2\,P_l^m(\cos\theta)}{d\theta^2}\right)e^{im\phi}z_l(kr); \tag{19.80}$$

$$\frac{\partial}{\partial\phi}\left(M_\theta\right) = -\frac{m^2}{\sin\theta}e^{im\phi}\,P_l^m(\cos\theta)z_l(kr); \tag{19.81}$$

$$\frac{\partial}{\partial r}\left(rM_\phi\right) = -\frac{d\,P_l^m(\cos\theta)}{d\theta}e^{im\phi}\frac{d\,(rz_l(kr))}{dr}; \tag{19.82}$$

$$\frac{\partial}{\partial r}\left(rM_\theta\right) = \frac{im}{\sin\theta}e^{im\phi}\,P_l^m(\cos\theta)\frac{d\,(rz_l(kr))}{dr}. \tag{19.83}$$

Therefore

$$N_r = -\frac{e^{im\phi}z_l(kr)}{kr\sin\theta}\left[\sin\theta\frac{d^2\,P_l^m(\cos\theta)}{d\theta^2} + \cos\theta\frac{d\,P_l^m(\cos\theta)}{d\theta} - \frac{m^2}{\sin\theta}P_l^m(\cos\theta)\right]. \tag{19.84}$$

But the $P_l^m(\cos\theta)$ satisfy the differential equation (19.66), so that

$$N_r = \frac{e^{im\phi}z_l(kr)}{kr\sin\theta}l\,(l+1)\,P_l^m(\cos\theta). \tag{19.85}$$

Also

$$N_\theta = \frac{e^{im\phi}}{kr}\left[\frac{d\,P_l^m(\cos\theta)}{d\theta}\right]\left[\frac{d\,(rz_l(kr))}{dr}\right], \tag{19.86}$$

and

$$N_\phi = \frac{ime^{im\phi}}{kr\sin\theta}P_l^m(\cos\theta)\left[\frac{d\,(rz_l(kr))}{dr}\right]. \tag{19.87}$$

On taking the real and imaginary parts of the vectors $\mathbf{M}$ and $\mathbf{N}$, respectively, we find that

$$\operatorname{Re} \mathbf{M} \equiv \mathbf{M}_r(m, l; k)$$

$$= -\frac{m}{\sin \theta} \sin(m\phi) P_l^m(\cos \theta) z_l(kr) \mathbf{e}_\theta$$

$$- \cos(m\phi) \frac{d P_l^m(\cos \theta)}{d\theta} e^{im\phi} z_l(kr) \mathbf{e}_\phi;$$

$$\operatorname{Im} \mathbf{M} \equiv \mathbf{M}_i(m, l; k)$$

$$= \frac{m}{\sin \theta} \cos(m\phi) P_l^m(\cos \theta) z_l(kr) \mathbf{e}_\theta$$

$$- \sin(m\phi) \frac{d P_l^m(\cos \theta)}{d\theta} z_l(kr) \mathbf{e}_\phi; \qquad (19.88)$$

$$\operatorname{Re} \mathbf{N} \equiv \mathbf{N}_r(m, l; k)$$

$$= \frac{\cos(m\phi)}{kr \sin \theta} l(l+1) P_l^m(\cos \theta) z_l(kr) \mathbf{e}_r$$

$$+ \frac{\cos(m\phi)}{kr} \left[\frac{d P_l^m(\cos \theta)}{d\theta} \right] \left[\frac{d(rz_l(kr))}{dr} \right] \mathbf{e}_\theta$$

$$- \frac{m \sin(m\phi)}{kr \sin \theta} P_l^m(\cos \theta) \left[\frac{d(rz_l(kr))}{dr} \right]; \qquad (19.89)$$

$$\operatorname{Im} \mathbf{N} \equiv \mathbf{N}_i(m, l; k)$$

$$= \frac{\sin(m\phi)}{kr \sin \theta} l(l+1) P_l^m(\cos \theta) z_l(kr) \mathbf{e}_r$$

$$+ \frac{\sin(m\phi)}{kr} \left[\frac{d P_l^m(\cos \theta)}{d\theta} \right] \left[\frac{d(rz_l(kr))}{dr} \right] \mathbf{e}_\theta$$

$$+ \frac{m \cos(m\phi)}{kr \sin \theta} P_l^m(\cos \theta) \left[\frac{d(rz_l(kr))}{dr} \right]. \qquad (19.90)$$

In spherical coordinates, an x-polarized wave traversing the z direction is given by

$$\mathbf{E}_i = E_0 \exp(ikr \cos \theta) \mathbf{e}_x,$$

where $\quad \mathbf{e}_x = (\sin \theta \cos \phi) \mathbf{e}_r + (\cos \theta \cos \phi) \mathbf{e}_\theta - (\sin \phi) \mathbf{e}_\phi. \qquad (19.91)$

Exercise: Derive (19.91).

The dependence of this expression on the azimuthal angle ϕ limits the separation constant m to $m = 1$. Assuming that the vectors $\mathbf{M}$ and $\mathbf{N}$ form a complete set, we can express $\mathbf{E}_i$ in terms of them. As we have seen, the coefficients of the infinite vector spherical harmonic expansion of the plane wave can be obtained by using the orthogonality of the basis functions (now rejecting harmonics that involve the y_l functions, since they diverge at the origin). Thus we write

$$\exp(ikr\cos\theta)\mathbf{e}_x = \sum_{l=1}^{\infty}(a_l\mathbf{M}_i + b_l\mathbf{N}_r). \tag{19.92}$$

But why is this particular expansion chosen instead of some other combination of the four vectors? The answer follows from comparing the ϕ dependence in the $\mathbf{e}_\theta$ and $\mathbf{e}_\phi$ components in (19.91) with the corresponding components of $\mathbf{M}_i$ and $\mathbf{N}_r$. To find the coefficients a_n and b_n we need to evaluate the integral

$$\int_0^{\pi}\int_0^{2\pi}(\mathbf{e}_x\cdot\mathbf{M}_i)\exp(ikr\cos\theta)\sin\theta d\theta d\phi. \tag{19.93}$$

Using Bauer's formula (18.32)

$$\exp(ikr\cos\theta) = \sum_{l=0}^{\infty}i^l(2l+1)j_l(kr)P_l(\cos\theta), \tag{19.94}$$

and the orthogonality relations discussed earlier, we can show that

$$\int_0^{\pi}\int_0^{2\pi}(\mathbf{e}_x\cdot\mathbf{M}_i)\exp(ikr\cos\theta)\sin\theta d\theta d\phi = 2\pi i^l l(l+1)[j_l(kr)]^2, \tag{19.95}$$

whence

$$a_l = \frac{2l+1}{l(l+1)}i^l. \tag{19.96}$$

Exercise: Derive (19.96) from (19.95).

Similarly,

$$\int_0^{\pi}\int_0^{2\pi}(\mathbf{e}_x\cdot\mathbf{N}_r)\exp(ikr\cos\theta)\sin\theta d\theta d\phi$$
$$= -2\pi i^{l+1}\frac{l(l+1)}{2l+1}\left\{(l+1)[j_{l-1}(kr)]^2 + l[j_{l+1}(kr)]^2\right\}, \tag{19.97}$$

from which

$$b_l = -\frac{2l+1}{l(l+1)}i^{l+1}. \tag{19.98}$$

Hence

$$\mathbf{e}_x e^{ikz} = \sum_{l=1}^{\infty}i^l\frac{2l+1}{l(l+1)}(\mathbf{M}_i - i\mathbf{N}_r). \tag{19.99}$$

This is the expansion for an incoming plane wave with electric vector polarized in the x direction. A similar argument for a wave polarized in the y direction yields the result

$$\mathbf{e}_y e^{ikz} = -\sum_{l=1}^{\infty}i^l\frac{2l+1}{l(l+1)}(\mathbf{M}_r + i\mathbf{N}_i). \tag{19.100}$$

The incident electric and magnetic fields (suppressing the time-harmonic dependence) are therefore (in the notation of [57]),

$$\mathbf{E}_i = E_0 \sum_{l=1}^{\infty} i^l \frac{2l+1}{l(l+1)} \left(\mathbf{M}_i^{(1)} - i\mathbf{N}_r^{(1)} \right), \tag{19.101}$$

and

$$\mathbf{H}_i = -\frac{k_2 E_0}{\omega} \sum_{l=1}^{\infty} i^l \frac{2l+1}{l(l+1)} \left(\mathbf{M}_r^{(1)} + i\mathbf{N}_i^{(1)} \right), \tag{19.102}$$

where in a nonmagnetic and medium (assumed here) $\mu = 1$, and the propagation constants k_1 and k_2 are defined above by equations (19.26) and (19.27).

In the above equation for $\mathbf{H}_i$ and in subsequent equations, the subscripts (1) and (2) on constants and parameters refer to the spherical scatterer and the external homogeneous medium, respectively. The superscripts (1) and (3), respectively, denote radial dependence given by spherical Bessel functions of the first and third kinds (the latter, spherical Hankel functions of the first kind, $h_l^{(1)}(k_2 r)$, corresponding to outwardly propagating waves). The corresponding scattered $(r > a)$ and interior (or transmitted) fields $(r < a)$ are given respectively by

$$\mathbf{E}_s = E_0 \sum_{l=1}^{\infty} i^l \frac{2l+1}{l(l+1)} \left(a_l^s \mathbf{M}_i^{(3)} - i b_l^s \mathbf{N}_r^{(3)} \right) \tag{19.103}$$

$$\mathbf{H}_s = -\frac{k_2 E_0}{\omega} \sum_{l=1}^{\infty} i^l \frac{2l+1}{l(l+1)} \left(b_l^s \mathbf{M}_r^{(3)} + i a_l^s \mathbf{N}_i^{(3)} \right), \quad r > a; \tag{19.104}$$

and

$$\mathbf{E}_t = E_0 \sum_{l=1}^{\infty} i^l \frac{2l+1}{l(l+1)} \left(a_l^t \mathbf{M}_i^{(1)} - i b_l^t \mathbf{N}_r^{(1)} \right), \tag{19.105}$$

$$\mathbf{H}_t = -\frac{k_1 E_0}{\omega} \sum_{l=1}^{\infty} i^l \frac{2l+1}{l(l+1)} \left(b_l^t \mathbf{M}_r^{(1)} + i a_l^t \mathbf{N}_i^{(1)} \right), \quad r < a. \tag{19.106}$$

Next we solve for the coefficients of the transmitted and scattered fields. To accomplish this, we must impose boundary conditions on the fields at the surface of the sphere:

$$(\mathbf{E}_i + \mathbf{E}_s - \mathbf{E}_t) \times \mathbf{e}_r = (\mathbf{H}_i + \mathbf{H}_s - \mathbf{H}_t) \times \mathbf{e}_r = 0. \tag{19.107}$$

These conditions require the field components tangential to the boundary to be continuous and provide us with four independent equations when written in component form, from which we can find (for each l), the four unknown coefficients a_l^s, b_l^s, a_l^t, and b_l^t. In what follows the prime denotes a derivative, and $\beta = ka$ is the size parameter. In terms of the functions

$$\psi_l(z) \equiv z j_l(z), \quad \zeta_l^{(1)}(z) \equiv z h_l^{(1)}(z), \tag{19.108}$$

these four equations can be solved (in particular) for the coefficients of the external (i.e., scattered) field to yield [54]

$$a_l^s(\beta) = \frac{\psi_l(\beta)\psi_l'(n\beta) - n\psi_l(n\beta)\psi_l'(\beta)}{\zeta_l^{(1)}(\beta)\psi_l'(n\beta) - n\psi_l(n\beta)\zeta_l^{(1)\prime}(\beta)}, \qquad (19.109)$$

$$b_l^s(\beta) = \frac{\psi_l(n\beta)\,\psi_l'(\beta) - n\psi_l(\beta)\psi_l'(n\beta)}{\zeta_l^{(1)\prime}(\beta)\psi_l(n\beta) - n\psi_l'(n\beta)\zeta_l^{(1)}(\beta)}. \qquad (19.110)$$

This means (in principle at least) that the problem of scattering has been solved. Of course, there is a great deal of physics implicit in these formulas, and much of the research into electromagnetic scattering has been concerned with the numerical "unfolding" of the physics [26], [54]. For an alternative approach to deriving these coefficients and several detailed applications of these results, the reader is encouraged to consult chapter 14 of [13].

Exercise: Show that the coefficients a_l^t and b_l^t for the transmitted field are

$$a_l^t(\beta) = \frac{-in}{\zeta_l^{(1)}(\beta)\psi_l'(n\beta) - n\psi_l(n\beta)\zeta_l^{(1)\prime}(\beta)}, \qquad (19.111)$$

$$b_l^t(\beta) = \frac{in}{\zeta_l^{(1)\prime}(\beta)\psi_l(n\beta) - n\psi_l'(n\beta)\zeta_l^{(1)}(\beta)}. \qquad (19.112)$$

19.3.2 The Rayleigh Scattering Limit: A Condensed Derivation

In contrast to the geometrical optics limit of $\beta \gg 1$, the Rayleigh scattering limit arises for long wavelengths (or equivalently, small spheres), that is, $\beta = ka = 2\pi a/\lambda \ll 1$. Additionally, $n\beta \ll 1$, though that is trivially satisfied for $n = O(1)$. In this same limit the coefficients $a_l^s(\beta)$ and $b_l^s(\beta)$ simplify to the forms [54]

$$a_l^s(\beta) = -i\beta^{2l+1}\frac{(2l+1)(l+1)(n^2-1)}{[(2l+1)!!]^2\,(n^2 l+l+1)}\left[1 + O(\beta^2)\right], \qquad (19.113)$$

$$b_l^s(\beta) = -\frac{i}{2}\frac{\beta^{2l+3}(l+1)n^2-1}{[(2l+1)!!]^2\,(2l+3)}\left[1 + O(\beta^2)\right]. \qquad (19.114)$$

Note that in the limit of small β the a_l^s terms dominate the b_l^s ones, and indeed, it is only necessary here to consider $l = 1$, that is, a_1^s. Substituting these values into equations (19.57) and (19.58) gives

$$S_1(\beta, \theta) \approx \frac{3}{2}\left(a_1^s\pi_1 + b_1^s\tau_1\right) \approx -i\beta^3\left(\frac{n^2-1}{n^2+2}\right), \qquad (19.115)$$

$$S_2(\beta, \theta) \approx \frac{3}{2}\left(b_1^s\pi_1 + a_1^s\tau_1\right) \approx -i\beta^3\left(\frac{n^2-1}{n^2+2}\right)\cos\theta. \qquad (19.116)$$

Following [54]

$$\frac{d\sigma}{d\Omega} = \frac{1}{k^2}\left(|S_2(\beta,\theta)|^2\cos^2\phi + |S_1(\beta,\theta)|^2\sin^2\phi\right)$$

$$\approx k^4 a^6 \left|\frac{n^2-1}{n^2+2}\right|^2 \left(\cos^2\theta\cos^2\phi + \sin^2\phi\right),$$

from which the total cross section, by means of the appropriate angular integrations, is found to be

$$\sigma(k,a,n) \approx \frac{6\pi}{k^2}|a_1^s|^2 \approx \frac{8\pi}{3}k^4 a^6 \left|\frac{n^2-1}{n^2+2}\right|^2 = \frac{128\,\pi^5 a^6}{3\,\lambda^4}, \qquad (19.117)$$

once more illustrating the inverse-fourth power dependence on wavelength in this limit.

The Radiation Produced by an Accelerating Charge

Now what happens if a charge starts out at rest, and then is suddenly accelerated to some constant velocity? The field should initially be that of a stationary charge: observers have no way of knowing that it will suddenly start moving. Even after it starts moving, distant observers will take time to realize this: information about the sudden change in motion cannot reach them any faster than the maximum speed c allowed by relativity. (This speed is commonly called "the speed of light", though it is actually the maximum speed of light, or of any other physical particle or wave.) Meanwhile, once the charge reaches a uniform velocity, observers close to it should simply see the ordinary field of a moving charge: the fact that it used to be "stationary" is not permanently imprinted on the charge.... Close in we have the field of a moving charge, and farther out we have the field of a stationary charge. Between these two regions is a spherical shell of stretched field lines connecting the two fields. This shell carries the information about the charge's sudden surge of acceleration: it expands at speed c, but has a constant thickness equal to $c\Delta t$, where Δt is the duration of the acceleration. The stretched field lines in this shell are what we call electromagnetic radiation. Two properties are immediately obvious. ... The fields in electromagnetic radiation are not radial, but transverse (i.e. perpendicular to the radius) [and] far from the source, the field lines of the radiation are much more tightly packed than the "backgound" of the stationary or uniformly moving source. [229]

It is further pointed out in [229] that the field lines will be perpendicular to the surface of a sphere of radius r enclosing the source, so that

$$E_r = \frac{Q}{4\pi\varepsilon_0} \times \frac{1}{r^2} \qquad (19.118)$$

where Q is the charge, and $1/(4\pi\varepsilon_0)$ is Coulomb's electric constant. In contrast, in the radiation shell, the field lines are predominantly transverse,

so that

$$E_\perp \approx \frac{Q}{4\pi\varepsilon_0} \times \frac{a\perp}{c^2} \times \frac{1}{r^2}, \qquad (19.119)$$

where $a\perp$ is the component of the charge's acceleration perpendicular to the radial line.

19.3.3 The Radiation Field Generated by a Hertzian Dipole

With the above physical description as background we now examine a special case of electromagnetic wave generation using a theoretical construct called a *Hertzian dipole*. Consider once more the scalar wave equation for a field $\psi(\mathbf{r}, t)$:

$$\nabla^2\psi = \frac{1}{c^2}\frac{\partial^2\psi}{\partial t^2}. \qquad (19.120)$$

If now the vector $\mathbf{c} = \hat{\mathbf{a}}$ is a unit vector of fixed direction, then $\hat{\mathbf{a}}\psi$ is a solution of the corresponding *vector* wave equation. But $\hat{\mathbf{a}}\psi$ does not, in general, satisfy the divergence condition

$$\nabla \cdot \mathbf{E} = 0. \qquad (19.121)$$

However, as already noted, a vector field of the form

$$\mathbf{E} = \nabla \times (\hat{\mathbf{a}}\psi), \qquad (19.122)$$

a solenoidal or divergence-free vector field, satisfies both of the above equations. Another such class is

$$\mathbf{E} = \nabla \times \left[\nabla \times (\hat{\mathbf{a}}\psi)\right]. \qquad (19.123)$$

(Higher-order multiple curls are redundant for time-harmonic fields.) A spherically symmetric solution $\psi(r, t)$ ($r = |r|$) corresponding to outward-traveling waves is

$$\psi(r, t) = Ar^{-1}\exp\left[i(\kappa r - \omega t)\right], \quad r \neq 0, \qquad (19.124)$$

for some constant A. In what follows we write $\kappa r - \omega t$ as Φ. Note that the wavenumber is now defined to be $\kappa = \omega/c$ to avoid confusion with the unit vector $\hat{\mathbf{k}}$ along the polar axis ($\theta = 0$) in a spherical coordinate system. If we choose $\hat{\mathbf{a}} = \hat{\mathbf{k}}$ and substitute this into equation (19.123), suppressing the t dependence for notational simplicity, we obtain

$$\mathbf{E} = \nabla(\nabla \cdot \mathbf{V}) + \hat{\mathbf{k}}\kappa^2\psi(r), \qquad (19.125)$$

where $\mathbf{V} = \hat{\mathbf{k}}\psi(r)$. Using the decomposition

$$\hat{\mathbf{k}} = \hat{\mathbf{r}}\cos\theta - \hat{\boldsymbol{\theta}}\sin\theta, \qquad (19.126)$$

it is easily shown that

$$\nabla \cdot \mathbf{V} = \psi'(r)\cos\theta, \qquad (19.127)$$

so from (19.125),

$$\mathbf{E} = \hat{\mathbf{r}} \frac{\partial}{\partial r} (\psi' \cos \theta) + \hat{\boldsymbol{\theta}} \frac{\partial}{r \partial \theta} (\psi' \cos \theta) + (\hat{\mathbf{r}} \cos \theta - \hat{\boldsymbol{\theta}} \sin \theta) \kappa^2 \psi \quad (19.128)$$

$$= \left\langle (\psi'' + \kappa^2 \psi) \cos \theta, - \left(\frac{\psi'}{r} + \kappa^2 \psi \right) \sin \theta, 0 \right\rangle \equiv \langle E_r, E_\theta, 0 \rangle. \quad (19.129)$$

From the form (19.124) this means that

$$E_r = \frac{2A}{r^3} e^{i\Phi} \cos \theta (1 - i\kappa r), \quad (19.130)$$

$$E_\theta = \frac{A}{r^3} e^{i\Phi} \sin \theta (1 - i\kappa r - \kappa^2 r^2), \quad (19.131)$$

$$E_\phi = 0. \quad (19.132)$$

The associated magnetic field is given by

$$\mathbf{B} = -\frac{i}{\omega} \nabla \times \mathbf{E} = -\frac{i}{\omega r} \langle 0, 0, B_\phi \rangle, \quad (19.133)$$

where

$$B_\phi = \frac{\partial (r E_\theta)}{\partial r} - \frac{\partial E_r}{\partial \theta} = -\kappa^2 r \psi' \sin \theta = -\frac{A}{c r^3} e^{i\Phi} \sin \theta (i\kappa r + \kappa^2 r^2). \quad (19.134)$$

Thus, from equations (19.129) and (19.134) we see that the vector $\mathbf{E}$ lies in the plane of the incident wave, and $\mathbf{B}$ cycles around the dipole. At large distances, $\kappa r \gg 1$ and terms $O(r^{-1})$ dominate in these expressions (as noted in the above box), so

$$\mathbf{E} \sim -\hat{\boldsymbol{\theta}} \frac{\kappa^2 A}{r} e^{i\Phi} \sin \theta = -\hat{\boldsymbol{\theta}} \frac{A\omega^2}{rc^2} e^{i\Phi} \sin \theta \equiv -\hat{\boldsymbol{\theta}} \frac{A_0 \omega^2}{rc^2} e^{i\omega t} \sin \theta, \quad (19.135)$$

(by subsuming $e^{i\kappa r}$ into the constant A_0 and taking the real part of the time-harmonic term), and

$$\mathbf{B} \sim -\hat{\boldsymbol{\phi}} \frac{\kappa^2 A}{cr} e^{i\Phi} \sin \theta = -\hat{\boldsymbol{\phi}} \frac{A\omega^2}{rc^2} e^{i\Phi} \sin \theta. \quad (19.136)$$

For a charge of magnitude e undergoing simple harmonic motion of amplitude A (and hence real acceleration a) we may write

$$- |A_0| \omega^2 \cos \omega t = a, \quad (19.137)$$

$$|\mathbf{E}| = \frac{ea \sin \theta}{rc^2} = \frac{e |A_0| \omega^2}{rc^2} \cos \omega t \sin \theta. \quad (19.138)$$

Furthermore, we may interpret θ as the angle between the direction of acceleration and the vector from the charge (at large distances) to the observer. The electric energy density in an electromagnetic wave is

$$W_E = \frac{1}{2} \varepsilon_0 |\mathbf{E}|^2, \quad (19.139)$$

and the magnetic energy density is

$$W_B = \frac{1}{2\mu_0} |\mathbf{B}|^2 . \tag{19.140}$$

Since these are equal, the total energy can be written in particular as $W = \varepsilon_0 |\mathbf{E}|^2$.

Then the mean rate at which energy is emitted is the total amount of electromagnetic energy contained in a sphere of radius $r = c$, that is [205],

$$\mathcal{W} = \int_0^c \int_0^\pi \int_0^{2\pi} \varepsilon_0 |\mathbf{E}|^2 r^2 \sin\theta d\phi d\theta dr$$

$$= \left(\frac{e\bar{a}}{c^2}\right)^2 (2\pi c\varepsilon_0) \int_0^\pi \sin^3\theta d\theta = \frac{8}{3} \frac{\pi\varepsilon_0 e^2 \bar{a}^2}{c^3}; \tag{19.141}$$

in this expression $\bar{a}$ is the mean square acceleration in the dipole:

$$\bar{a} = \frac{\omega}{2\pi} \left\{ |A_0|^2 \omega^4 \int_0^{2\pi/\omega} \cos^2 \omega t dt \right\} = \frac{1}{2} |A_0|^2 \omega^4. \tag{19.142}$$

Therefore

$$\mathcal{W} = \frac{4\pi |A_0|^2 \omega^4 \varepsilon_0 e^2}{3c^3}. \tag{19.143}$$

We write this in terms of the wavelength λ, where $\omega = 2\pi c/\lambda$ so that

$$\mathcal{W} = \frac{64\pi^4 |A_0|^2 e^2 c\varepsilon_0}{3\lambda^4}. \tag{19.144}$$

Thus, based on this "toy" model of radiation, we have arrived at yet another form of Rayleigh scattering: given a set of oscillators with the same amplitude and charge but different frequencies, they will emit radiation in inverse proportion to the fourth power of the wavelength.

Chapter Twenty

Diffraction of Plane Electromagnetic Waves by a Cylinder

Radio signals may also undergo diffraction. It is found that when signals encounter an obstacle they tend to travel around them. This can mean that a signal may be received from a transmitter even though it may be "shaded" by a large object between them. This is particularly noticeable on some long wave broadcast transmissions. For example the BBC long wave transmitter on 198 kHz is audible in the Scottish glens where other transmissions could not be heard. As a result the long wave transmissions can be heard in many more places than transmissions on VHF FM. To understand how this happens it is necessary to look at Huygen's Principle. This states that each point on a spherical wave front can be considered as a source of a secondary wave front. Even though there will be a shadow zone immediately behind the obstacle, the signal will diffract around the obstacle and start to fill the void. It is found that diffraction is more pronounced when the obstacle becomes sharper and more like a "knife edge." For a radio signal a mountain ridge may provide a sufficiently sharp edge. A more rounded hill will not produce such a marked effect. It is also found that low frequency signals diffract more markedly than higher frequency ones. It is for this reason that signals on the long wave band are able to provide coverage even in hilly or mountainous terrain where signals at VHF and higher would not.

[253]

We tend to think of shadow boundaries as being sharp, but even a moment's study of the shadow of a tree, for example, shows that boundaries are quite blurry, especially as one looks farther along the shadow. This blurriness is due, of course, to the increasing width of the penumbral shadow, which occurs because the sun is not a point source of light. (Note that if we observe a partial solar eclipse, we are in the penumbral shadow). With such macrosopic phenomena, readily discussed using geometrical optics, the effects diffraction are generally negligible. But from physical or wave optics we know that no sharp shadows can exist; electromagnetic (or indeed, other) radiation can enter the region of geometrical shadow. Although the vector wave equations for both polarizations must be considered for many optical situations, polarization is not as significant in diffraction problems, and we shall devote our attention here to the electric polarization.

20.1 ELECTRIC POLARIZATION

We again consider a normally incident plane wave incident on the cylinder (parallel to the z-axis of the cylinder). In what follows, θ is the polar angle when viewed along the axis of the cylinder. From equation (18.6) for the incident field, we have (not surprisingly) a similar expression to that for acoustic wave scattering, namely,

$$\phi_i = e^{ikx} = e^{ikr\cos\theta} = \sum_{n=-\infty}^{\infty} i^n e^{in\theta} J_n(kr), \tag{20.1}$$

where as noted already, the total field is given by the sum of the incident field and the scattered field. We follow the succinct approach adopted in [5] for much of this section. This expansion is used when the implicit time-harmonic field varies as $\exp(-i\omega t)$, because then the scattered radiation is outgoing, as it should be. Using the condition that the total field must be zero on the boundary of the cylinder, we have that

$$\phi(r,\theta) = \sum_{n=-\infty}^{\infty} \frac{i^n e^{in\theta}}{H_n^{(1)}(ka)} \left[J_n(kr)H_n^{(1)}(ka) - H_n^{(1)}(kr)J_n(ka) \right]. \tag{20.2}$$

Note that in some of the mathematical literature in physics and engineering (e.g., [5], [55]), the opposite time-harmonic convention is used, namely, $\exp(i\omega t)$, in which case the sign of k must be changed in expression (20.1) for ϕ_i to ensure that the field is incident from the negative x direction (we adopt that convention in this section). Consequently, the Hankel functions of the first kind in equation (20.2) must be replaced by those of the second kind. Then (20.2) becomes

$$\phi(r,\theta) = \sum_{n=-\infty}^{\infty} \frac{i^{-n} e^{in\theta}}{H_n^{(2)}(ka)} \left[J_n(kr)H_n^{(2)}(ka) - H_n^{(2)}(kr)J_n(ka) \right]. \tag{20.3}$$

It is known, however, that these eigenfunction expansions are not very useful for high-frequency waves because of their poor convergence when $ka \gg 1$. The answer is to use the Watson transform [26]. Consider the function $f(\xi)$, analytic in the vicinity of the real ξ-axis. Suppose that C is a contour that is deformed such that all poles of $f(\xi)$ are contained in C. Then [5], [58],

$$\sum_{n=-\infty}^{\infty} (-1)^n f(n) = -\frac{i}{2} \oint_C \frac{f(\xi)}{\sin \pi \xi} d\xi. \tag{20.4}$$

In particular, the integral

$$I = \frac{i}{2} \oint_C \frac{\exp\left[i\xi\,(\theta + \pi)\right]}{\sin \pi \xi} f(\xi) d\xi, \tag{20.5}$$

has an infinite set of first-order poles along the real ξ-axis, where $\xi = n$, $n = 0, \pm 1, \pm 2, \dots$. If C encloses all the poles (Figure 20.1), then by summing the residues, we find

$$I = \sum_{n=-\infty}^{\infty} e^{in\theta} f(n). \tag{20.6}$$

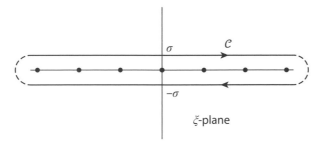

Figure 20.1 Contour for the integral (20.5).

Comparing the forms (20.3) and (20.6) gives

$$f(\xi) = \frac{i^{-n}}{H_\xi^{(2)}}(ka)\left[J_\xi(kr)H_\xi^{(2)}(ka) - H_\xi^{(2)}(kr)J_\xi(ka) \right]$$

$$= \frac{i^{-n}}{2H_\xi^{(2)}(ka)}\left[H_\xi^{(1)}(kr)H_\xi^{(2)}(ka) - H_\xi^{(2)}(kr)H_\xi^{(1)}(ka) \right], \qquad (20.7)$$

because

$$2J_\xi(x) = H_\xi^{(1)}(x) + H_\xi^{(2)}(x). \qquad (20.8)$$

Furthermore, the analytic continuation formulas

$$H_{-\xi}^{(1)}(x) = e^{i\pi\xi}H_\xi^{(1)}(x); \qquad H_{-\xi}^{(2)}(x) = e^{-i\pi\xi}H_\xi^{(2)}(x), \qquad (20.9)$$

imply that

$$f(-\xi) = f(\xi)e^{2i\pi\xi}. \qquad (20.10)$$

Using the contour shown in Figure 20.1, we can write the integral (20.5) as

$$I = \frac{i}{2}\left\{ \int_{\infty-i\sigma}^{-\infty-i\sigma} + \int_{-\infty+i\sigma}^{\infty+i\sigma} \right\} \frac{\exp\left[i\xi\,(\theta + \pi)\right]}{\sin\pi\xi} f(\xi)d\xi$$

$$= i\int_{\infty-i\sigma}^{-\infty-i\sigma} i^{2\xi} f(\xi)\frac{\cos\xi\theta}{\sin\xi\pi}d\xi. \qquad (20.11)$$

The final integral has been derived by setting $\xi \to -\xi$ in the preceding second integral and using the condition (20.10). Since $\sigma > 0$ the path of integration lies below the real axis and must therefore be closed in the lower-half ξ-plane, and this requires knowing where any poles ξ_n of $f(\xi)$ are located in that region. From equation (20.7) these are just the zeros of $H_{\xi_n}^{(2)}(ka)$, and for large values of ka can be found from the asymptotic form of the Hankel function. In contrast, if $\xi_n \ll ka$, there are no solutions to $H_{\xi_n}^{(2)}(ka) = 0$, since then

$$H_{\xi_n}^{(2)}(ka) \sim \left(\frac{2}{\pi ka}\right)^{1/2} \exp\left[-i\left(ka - \frac{\xi_n\pi}{2} - \frac{\pi}{4}\right)\right], \qquad (20.12)$$

for all $\arg \xi$. But if $\xi \simeq ka$, the following asymptotic form for $H_{\xi_n}^{(2)}(ka)$ can be used:

$$H_{\xi_n}^{(2)}(ka) \sim 2 \left(\frac{2}{ka}\right)^{1/3} e^{i\pi/3} \, \mathrm{Ai} \left(\varpi e^{-2\pi i/3}\right), \tag{20.13}$$

where

$$\varpi = (\xi_n - ka) \left(\frac{2}{ka}\right)^{1/3}. \tag{20.14}$$

If we denote the zeros of Ai (x) by $-\gamma_n$, then

$$\xi_n = ka + \gamma_n \left(\frac{ka}{2}\right)^{1/3} e^{-i\pi/3}. \tag{20.15}$$

For large values of ξ_n (i.e., $|\xi_n| \gg ka$), two other asymptotic results for $H_{\xi_n}^{(2)}(ka)$ are required. These are

$$H_\xi^{(2)}(ka) \sim i \left(\frac{2}{\pi \xi}\right)^{1/2} \left(\frac{2\xi}{eka}\right)^\xi, \qquad |\arg \xi| \le \frac{\pi}{2}, \tag{20.16}$$

and

$$H_\xi^{(2)}(ka) \sim -\left(\frac{2}{\pi \xi}\right)^{1/2} \left(\frac{2\xi}{eka}\right)^\xi, \qquad \frac{\pi}{2} < |\arg \xi| < \frac{3\pi}{2}. \tag{20.17}$$

With a large circle of radius $R \to \infty$ in mind, we set $\xi = R \exp(i\varphi)$ in these to obtain

$$\left|H_\xi^{(2)}(ka)\right| \sim \left(\frac{2}{\pi R}\right)^{1/2} \exp\left\{ R \left[\cos \varphi \ln \left(\frac{2R}{eka}\right) - \varphi \sin \varphi\right]\right\}, \qquad |\varphi| \le \frac{\pi}{2}, \tag{20.18}$$

and

$$\left|H_\xi^{(2)}(ka)\right| \sim \left(\frac{2}{\pi R}\right)^{1/2} \exp\left\{ -R \left[\cos \varphi \ln \left(\frac{2R}{eka}\right) + \varphi \sin \varphi\right]\right\},$$
$$\frac{\pi}{2} < |\varphi| < \frac{3\pi}{2}. \tag{20.19}$$

Examination of the exponents indicates that there are no zeros for $\varphi \in (\pi/2, 3\pi/2)$, and in the left half-plane the zeros are located very close to the imaginary axis. In view of equation (20.15), Figure 20.2 is a sketch of possible pole locations in the fourth quadrant. From equation (20.7) it therefore follows that

$$\lim_{|\xi| \to \infty} f(\xi) = 0, \qquad |\varphi| \le \frac{\pi}{2}, \tag{20.20}$$

apart from poles near the imaginary axis, so by closing the contour $\mathcal{C}$ in the integral (20.11) by a semicircle at infinity, no additional contributions are collected. The task is then to evaluate the residues at the poles. From the summation (20.6) we find the residue series to be

$$\phi(r, \theta) = \pi \sum_{n=1}^{\infty} i^{\xi_n} \frac{H_{\xi_n}^{(2)}(kr) H_{\xi_n}^{(1)}(ka) \cos \xi_n \theta}{\left[\partial H_\xi^{(2)}(ka)/\partial \xi\right]_{\xi=\xi_n} \sin \xi_n \pi}. \tag{20.21}$$

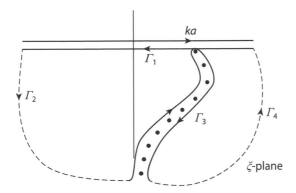

Figure 20.2 Sketch of $f(\xi)$ pole distribution.

Now if $\xi_n \pi = \alpha + i\beta$, $\alpha, \beta \in \mathbb{R}$, then from (20.15) for large ka, β is large and negative, so that to a good approximation

$$\sin \xi_n \pi \approx \frac{e^{i\xi_n \pi}}{2i}. \tag{20.22}$$

In view of this,

$$i^{\xi_n} \frac{\cos \xi_n \theta}{\sin \xi_n \pi} \approx i \left\{ \exp\left[-i\xi_n \left(\frac{\pi}{2} - \theta\right)\right] + \exp\left[-i\xi_n \left(\frac{\pi}{2} + \theta\right)\right] \right\}, \tag{20.23}$$

and so

$$\phi(r, \theta) = i\pi \sum_{n=1}^{\infty} \frac{H_{\xi_n}^{(2)}(kr) H_{\xi_n}^{(1)}(ka)}{\left[\partial H_\xi^{(2)}(ka)/\partial \xi\right]_{\xi=\xi_n}}$$

$$\times \left\{ \exp\left[-i\xi_n \left(\frac{\pi}{2} - \theta\right)\right] + \exp\left[-i\xi_n \left(\frac{\pi}{2} + \theta\right)\right] \right\}. \tag{20.24}$$

Of particular interest here is the evaluation of terms in (20.24) when the order and argument are large and comparable, so $\xi \simeq ka$. In view of the more general functional form of (20.13), that is,

$$H_\xi^{(2)}(x) \sim 2 \left(\frac{2}{x}\right)^{1/3} e^{i\pi/3} \operatorname{Ai}\left(\varpi e^{-2\pi i/3}\right); \quad \varpi = (\xi - x) \left(\frac{2}{x}\right)^{1/3}, \quad \xi \approx x, \tag{20.25}$$

it follows that

$$\left[\frac{\partial H_\xi^{(2)}(x)}{\partial \xi}\right]_{\xi=\xi_n} = -\left[H_\xi^{(2)\prime}(x)\right]_{\xi_n} \approx 2 \left(\frac{2}{x}\right)^{1/3} \left[\operatorname{Ai}(x)'\right]_{-\gamma_n} e^{-i\pi/3}. \tag{20.26}$$

The Wronskian of the two types of Hankel functions can be used to determine the remaining information needed to evaluate the total electric field; since

$$W\left[H_\xi^{(1)}(x), H_\xi^{(2)}(x)\right] = H_\xi^{(1)}(x) H_\xi^{(2)\prime}(x) - H_\xi^{(1)\prime}(x) H_\xi^{(2)}(x) = -\frac{4i}{\pi x}, \tag{20.27}$$

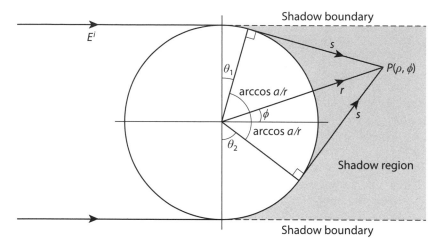

Figure 20.3 Shadow boundary details.

and

$$H^{(2)}_{\xi_n}(ka) = 0, \tag{20.28}$$

then

$$H^{(1)}_{\xi_n}(ka) = -\frac{4i}{\pi ka \, H^{(2)'}_{\xi_n}(ka)}. \tag{20.29}$$

Now we can rewrite (20.24) as

$$\phi(r, \theta) \approx \frac{1}{2} \left(\frac{ka}{2}\right)^{1/3} e^{2\pi i/3} \sum_{n=1}^{N} \frac{H^{(2)}_{\xi_n}(kr)}{\left\{[\mathrm{Ai}\,(x)']_{-\gamma_n}\right\}^2}$$
$$\times \left\{\exp\left[-i\xi_n\left(\frac{\pi}{2} - \theta\right)\right] + \exp\left[-i\xi_n\left(\frac{\pi}{2} + \theta\right)\right]\right\}. \tag{20.30}$$

The upper limit of the sum has been truncated to some integer N, because the approximations used for $\xi_n \simeq ka$, and therefore the terms are dependent on the validity of the results (20.26) and (20.15). This result may be further reduced for points outside the cylinder such that $kr > \xi_n$. We use the asymptotic result, valid for $\xi < x$,

$$H^{(2)}_{\xi}(x) \sim \left\{\frac{2}{\pi \left(x^2 - \xi^2\right)^{1/2}}\right\}^{1/2} \exp\left\{-i\left[\left(x^2 - \xi^2\right)^{1/2} - \xi \, \mathrm{arcsec}\left(\frac{x}{\xi}\right) - \frac{\pi}{4}\right]\right\}. \tag{20.31}$$

From Figure 20.3 it is seen that because $s = \left(r^2 - a^2\right)^{1/2}$, and $\xi_n \lesssim ka$, we have

$$(kr)^2 - \xi_n^2 \approx (ks)^2, \tag{20.32}$$

which is valid for at most the first N terms of the residue series. Then

$$\phi(r, \theta) \approx \left(\frac{ka}{2}\right)^{1/3} \frac{\exp\left[-i\left(\pi/12 + ks\right)\right]}{(2\pi ks)^{1/2}} \sum_{n=1}^{N} \frac{\{A + B\}}{\left\{\left[\text{Ai}(x)'\right]_{-\gamma_n}\right\}^2}, \tag{20.33}$$

where

$$A = \exp\left[-i\xi_n \left(\frac{\pi}{2} - \theta - \arccos(a/r)\right)\right]; \tag{20.34}$$

$$B = \exp\left[-i\xi_n \left(\frac{\pi}{2} + \theta - \arccos(a/r)\right)\right], \tag{20.35}$$

and once more, ξ_n is given by

$$\xi_n \approx ka + \gamma_n \left(\frac{ka}{2}\right)^{1/3} e^{-i\pi/3}. \tag{20.36}$$

If we define

$$\delta = \frac{\pi}{2} - \arccos\left(\frac{a}{r}\right), \tag{20.37}$$

then by examining the exponents in the above approximation (20.33), recalling from (20.36) that $\text{Im}\,\xi_n < 0$, we see that the series for $\phi(r, \theta)$ will converge rapidly, provided that $|\theta| < \delta$ or is the angular region

$$-\frac{\pi}{2} + \arccos\left(\frac{a}{r}\right) < \theta < \frac{\pi}{2} - \arccos\left(\frac{a}{r}\right). \tag{20.38}$$

As may be seen from Figure 20.3 this corresponds to field points in the geometrical optics shadow region, and the first several terms of the residue series should be sufficient to determine the field in this region to any desired degree of precision. Regarding the illuminated region, the integral (20.11) together with the approximation (20.22) can be recast as

$$I = -\int_{\infty - i\sigma}^{-\infty - i\sigma} f(\xi) \left[e^{i\xi\theta} + e^{-i\xi\theta}\right] d\xi. \tag{20.39}$$

Because $\text{Im}\,\xi = -\sigma$ along the path of integration, the first term in the integrand will be dominant for $\theta \in (0, \pi)$, and the second term will dominate for $\theta \in (-\pi, 0)$. These contributions can be combined by writing the integral as

$$I = \int_{-\infty - i\sigma}^{\infty - i\sigma} f(\xi) e^{i\xi|\theta|} d\xi, \tag{20.40}$$

or in full, from (20.7),

$$\phi(r, \theta) = \frac{1}{2} \int_{-\infty - i\sigma}^{\infty - i\sigma} \left\{H_\xi^{(1)}(kr) - \frac{H_\xi^{(1)}(ka)}{H_\xi^{(2)}(ka)} H_\xi^{(2)}(kr)\right\} \exp\left[i\xi\left(|\theta| - \frac{\pi}{2}\right)\right] d\xi. \tag{20.41}$$

Now $|\xi|$ is large everywhere on the path of integration, with a possible exception in the vicinity of $\text{Re}\,\xi = 0$, but if σ is large enough to accommodate the pole closest to the Re ξ-axis, this can be resolved. Then we can use the asymptotic forms for the Hankel functions to deduce that

$$\frac{H_\xi^{(1)}(ka)}{H_\xi^{(2)}(ka)} \sim -1, \qquad |\xi| > ka, \tag{20.42}$$

and

$$\frac{H_\xi^{(1)}(ka)}{H_\xi^{(2)}(ka)} \sim \exp\left\{2i\left[(k^2a^2 - \xi^2)^{1/2} - \xi \arccos\left(\frac{\xi}{ka}\right) - \frac{\pi}{4}\right]\right\}, \qquad |\xi| < ka. \tag{20.43}$$

As $kr \to \infty$ the $H_\xi^{(2)}(kr)$ term will dominate the $H_\xi^{(1)}(kr)$ one, because Im $\xi < 0$ and $|\text{Im}\,\xi|$ is large. Then in a more succinct form

$$\phi(r, \theta) = \int_{-\infty-i\sigma}^{\infty-i\sigma} h(\xi)e^{ig(\xi)}d\xi, \tag{20.44}$$

with two different expressions for each of $h(\xi)$ and $g(\xi)$, depending on whether $|\xi| > ka$ or $|\xi| < ka$. For $|\xi| > ka$,

$$h = h_+(\xi) = \left[2\pi\left(k^2r^2 - \xi^2\right)^{1/2}\right]^{-1/2}, \tag{20.45}$$

$$g = g_+(\xi) = -\left(k^2r^2 - \xi^2\right)^{1/2} + \xi\left[\arccos\left(\frac{\xi}{kr}\right) + |\theta| - \frac{\pi}{2}\right] + \frac{\pi}{4}. \tag{20.46}$$

For $|\xi| < ka$,

$$h = h_-(\xi) = -\left[2\pi\left(k^2r^2 - \xi^2\right)^{1/2}\right]^{-1/2}, \tag{20.47}$$

$$g = g_-(\xi) = -\left(k^2r^2 - \xi^2\right)^{1/2} + 2\left(k^2a^2 - \xi^2\right)^{1/2}$$
$$+ \xi\left[\arccos\left(\frac{\xi}{kr}\right) - 2\arccos\left(\frac{\xi}{ka}\right) + |\theta| - \frac{\pi}{2}\right] - \frac{\pi}{4}. \tag{20.48}$$

And *now* the method of stationary phase can be employed to evaluate the leading terms of the field in the illuminated region! The stationary points are given by $g'_\pm(\xi) = 0$, that is, after some cancellation of terms,

$$g'_+(\xi) = \arccos\left(\frac{\xi}{kr}\right) + |\theta| - \frac{\pi}{2} = 0 \Rightarrow \xi_0^+ = kr\sin|\theta|, \tag{20.49}$$

and some trigonometric and algebraic manipulation,

$$g'_-(\xi) = \arccos\left(\frac{\xi}{kr}\right) - 2\arccos\left(\frac{\xi}{ka}\right) + |\theta| - \frac{\pi}{2} = 0$$

$$\Rightarrow \xi_0^- = \pm ka\cos\frac{|\theta|}{2}, \qquad \pm ka\sin\frac{|\theta|}{2}, \tag{20.50}$$

to leading order in the limit $kr \to \infty$. These points are located on the real axis, and so the contour can be deformed to pass through them while avoiding any singularities in the complex ξ-plane (see Figure 20.4).

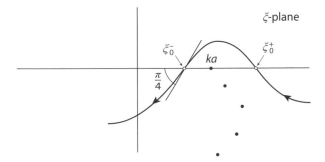

Figure 20.4 Stationary phase contour deformation.

Exercise: Establish the result (20.50).

Now the second derivatives of $g_\pm(\xi)$ are required to obtain the leading terms in the stationary phase expansion,

$$g_+''(\xi) = -\left(k^2 r^2 - \xi^2\right)^{-1/2} \Rightarrow g_+''(\xi_0^+) = -(kr\cos\theta)^{-1}, \qquad (20.51)$$

and

$$g_-''(\xi) = -(k^2 r^2 - \xi^2)^{-1/2} + 2\left(k^2 a^2 - \xi^2\right)^{-1/2} \Rightarrow g_+''(\xi_0^-) = -(kr\cos\theta)^{-1}, \qquad (20.52)$$

so from the leading contributions we have

$$\phi(r,\theta) \sim \sum_{\xi_0} \left(\frac{2\pi}{|g_\pm''(\xi_0)|}\right)^{1/2} h(\xi_0)\exp\left\{i\left[g_\pm(\xi_0) + \frac{\pi}{4}\operatorname{sgn}\left(g_\pm''(\xi_0)\right)\right]\right\}. \qquad (20.53)$$

After considerable algebra we find, for $kr \to \infty$ and $|\theta| \in (0, \pi)$,

$$\phi(r,\theta) \sim \exp\left(-ikr\cos\theta\right) - \left(\frac{a}{2r}\sin\frac{\theta}{2}\right)^{1/2}\exp\left[-ik\left(r - 2a\sin\frac{\theta}{2}\right)\right]. \qquad (20.54)$$

The first term is, as expected, the incident field, while the second term is related to the reflected field (see Figure 20.3). The analysis for the magnetic polarization is carried out in a very similar manner. For each polarization there are, however, major technical issues in the vicinity of the shadow boundary, since the residue series is very slowly convergent there, and the stationary phase solution for the illuminated region is not useful. Though beyond the scope of this book, some quite sophisticated mathematics is necessary to obtain a solution for this transition region where $\xi \approx ka$. This region is important, because as seen from the analysis for the illuminated region, a significant contribution to the integral arises for large kr from this region of the ξ-plane.

Exercise: Derive (20.54).

20.2 MORE ABOUT CLASSICAL DIFFRACTION

20.2.1 Huygen's Principle

Simply put, Huygen's principle states that "every point on a wavefront can be considered to be a source of a spherical wave (or Huygen's wavelet), and the envelope of the wavelets gives the wavefront at a later time." Crudely speaking, the Huygen's Principle includes interference effects in the ray propagating, and allows interference to happen while the wave propagates. Huygen's Principle is usually applied to diffraction problems in the following simple way. Suppose we know the (scalar) optical field $E(x, y)$ over some aperture in a plane Σ. The basic idea is to consider each point in the aperture to be a source of Huygen's wavelets. The total field at each observation point P is just the (integral) sum of all the wavelets. [231]

20.2.2 The Kirchhoff-Huygens Diffraction Integral

Having jumped straight into the mathematics in the previous section, perhaps it is time to take a breath, so to speak, and consider a different but not unrelated approach. As noted above, polarization is not as significant in diffraction problems, so the scalar wave equation will be all that is needed (and it serves as a template of sorts for any components of the $\mathbf{E}$ or $\mathbf{H}$ vectors); indeed, for monochromatic radiation the scalar Helmholtz equation is all that is necessary here.

Suppose then, that (following [13]) we wish to determine the field at a point P arising from a superposition of secondary waves emanating from a surface between a point source of radiation P_0 and P. We state Green's theorem for two scalar functions $\psi(\mathbf{r})$ and $\psi_0(\mathbf{r})$ defined in a volume V bounded by a closed surface S. We also require that these functions have continuous first- and second-order partial derivatives in and on S. If $\mathbf{n}$ is the *inward* unit normal vector (following [13]) and $\partial/\partial n$ denotes differentiation along that inward normal, then

$$\oint_S \left(\psi \frac{\partial \psi_0}{\partial n} - \psi_0 \frac{\partial \psi}{\partial n} \right) dS = - \int_V \left(\psi \nabla^2 \psi_0 - \psi_0 \nabla^2 \psi \right) dV. \tag{20.55}$$

If $\psi(\mathbf{r})$ and $\psi_0(\mathbf{r})$ both satisfy the Helmholtz equation

$$\nabla^2 \phi + k^2 \phi = 0, \tag{20.56}$$

then

$$\psi_0(\mathbf{r}) = \frac{e^{iks}}{s}, \tag{20.57}$$

where s is the distance from the point P to the point $\mathbf{r} = \langle x, y, z \rangle$. Note that this is the Green's function for the equation

$$\nabla^2 \psi_0 + k^2 \psi_0 = -\delta(s). \tag{20.58}$$

Clearly the integral over V vanishes in (20.55), but in this case V must consist of a volume from which the point P has been excluded by a small spherical surface S'

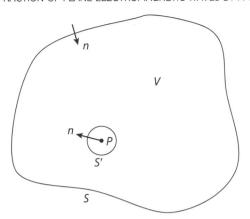

Figure 20.5 Region of integration for (20.62).

of radius ε. Then

$$\oint_S \left[\psi \frac{\partial}{\partial n} \left(\frac{e^{iks}}{s} \right) - \left(\frac{e^{iks}}{s} \right) \frac{\partial \psi}{\partial n} \right] dS \qquad (20.59)$$

$$= -\oint_{S'} \left[\psi \left(\frac{e^{iks}}{s} \right) \left(ik - \frac{1}{s} \right) - \left(\frac{e^{iks}}{s} \right) \frac{\partial \psi}{\partial n} \right] dS' \qquad (20.60)$$

$$= -\oint_\Omega \left[\psi \left(\frac{e^{ik\varepsilon}}{\varepsilon} \right) \left(ik - \frac{1}{\varepsilon} \right) - \left(\frac{e^{ik\varepsilon}}{\varepsilon} \right) \frac{\partial \psi}{\partial n} \right] \varepsilon^2 d\Omega, \qquad (20.61)$$

where $d\Omega$ is an element of solid angle. Since ε is arbitrarily small, we may evaluate the last integral in the limit $\varepsilon \to 0$, yielding the result $4\pi \psi(P_0)$, and hence the *Helmholtz-Kirchhoff integral theorem*

$$\psi(P) = \frac{1}{4\pi} \oint_S \left[\psi \frac{\partial}{\partial n} \left(\frac{e^{iks}}{s} \right) - \left(\frac{e^{iks}}{s} \right) \frac{\partial \psi}{\partial n} \right] dS. \qquad (20.62)$$

It means that if the field and its normal derivative are known on an arbitrary surface S enclosing a point P, then the field $\psi(P)$ is known.

Using this result in connection with Figure 20.5, equation (20.62) becomes, with $S = S_1 \cup S_2 \cup S_3$, where surfaces $S_1 \equiv A$, $S_2 \equiv B$, and $S_3 \equiv C$,

$$\psi(P) = \frac{1}{4\pi} \left(\oint_{S_1} + \oint_{S_2} + \oint_{S_3} \right) \left[\psi \frac{\partial}{\partial n} \left(\frac{e^{iks}}{s} \right) - \left(\frac{e^{iks}}{s} \right) \frac{\partial \psi}{\partial n} \right] dS. \qquad (20.63)$$

Physically reasonable boundary conditions (see [13] for details) are that on the aperture S_1,

$$\psi = \psi_i \equiv \frac{A e^{ikr}}{r}, \qquad \frac{\partial \psi}{\partial n} = \frac{\partial \psi_i}{\partial n} = \frac{A e^{ikr}}{r} \left(ik - \frac{1}{r} \right) \mathbf{n} \cdot \hat{\mathbf{r}}, \qquad (20.64)$$

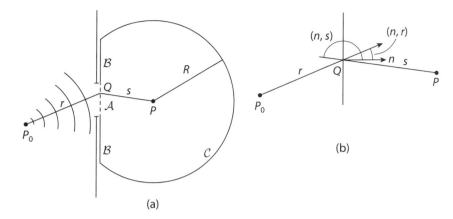

Figure 20.6 Notation for Fresnel-Kirchhoff diffraction.

(see Figure 20.6) C being a constant amplitude, and on the screen S_2,

$$\psi = 0, \qquad \frac{\partial \psi}{\partial n} = 0, \tag{20.65}$$

where S_1, S_2, and S_3 are an aperture, screen, and (distant) spherical surface, respectively. These are known as Kirchhoff's boundary conditions, and they imply, respectively, that the field and its normal derivative on S_1 (except near the edges) are identical to what they would be without the screen, and that the screen itself is perfectly opaque so that the field and its normal derivative are zero there.

But what of the contribution from the spherical surface portion S_3? It transpires that a similar argument can be used, resulting in the expression

$$\psi(P) = \frac{C}{4\pi} \oint_{S_3} \left[\left(ik - \frac{1}{r} \right) \mathbf{n} \cdot \hat{\mathbf{r}} + \left(ik - \frac{1}{s} \right) \mathbf{n} \cdot \hat{\mathbf{s}} \right] \frac{e^{ik(r+s)}}{rs} \, dS. \tag{20.66}$$

In effect, the surface S_3 is considered to be sufficiently far from P that ψ and its normal derivative are zero there; this is a form of the Sommerfeld radiation condition for spherical waves.

Note from Figure 20.6b that, in terms of unit vectors, $(n, r) \equiv \mathbf{n} \cdot \hat{\mathbf{r}} = \cos\theta$ and $(n, s) = \mathbf{n} \cdot \hat{\mathbf{s}} = \cos\theta_0$. When the source and observation point are large distances from the surface $S = S_1 \cup S_2 \cup S_3$ in comparison to a wavelength of the radiation (i.e., $r, s \gg 2\pi/k$), we have

$$\psi(P) \approx \frac{iC}{\lambda} \oint_{S_3} \left(\frac{\cos\theta + \cos\theta_0}{2} \right) \frac{e^{ik(r+s)}}{rs} \, dS. \tag{20.67}$$

This is known as the *Kirchhoff-Fresnel diffraction formula* and is essentially a mathematical statement of Huygen's principle. (Note that some treatments of this result have the sign of k reversed in this integral; an example of this is found in the next section.) The first term in the integrand is usually named the *obliquity factor*. If we now replace the open surface S_3 by a portion of the wavefront incident on the circular aperture A, and (consistent with the above assumptions) the radius of curvature of the wavefront is sufficiently large, then the surface of integration is a spherical cap bounded by the aperture opening. Then s is constant, and $\theta_0 \approx 0$, and the formula (20.67) reduces to

$$\psi(P) \approx \frac{ikC}{4\pi}(1 + \cos\theta)\frac{e^{ikr_0}}{r_0}\int_A \frac{e^{iks}}{s}dA. \tag{20.68}$$

The exponential term outside the integral represents the complex amplitude of the wave incident on the aperture (where r_0 is the radius of the wavefront) and from this primary wave each aperture element dA generates a secondary spherical wave; the integral of these elements is the total field at the point P. Note that in the forward direction $\theta = 0$, so the obliquity factor is one, whereas in the backward direction it is *zero*, meaning that the original wavefront generates no backward wave [231].

20.2.3 Derivation of the Generalized Airy Diffraction Pattern

To derive a more general form for the diffraction pattern of particular interest in meteorological optics (for example), Grandy [54] used a Green's function argument, sketched out as follows. The free space Green's function for the scalar Helmholtz equation

$$\left(\nabla^2 + k^2\right)G(\mathbf{r}, \mathbf{r}') = -4\pi\delta(\mathbf{r} - \mathbf{r}'), \tag{20.69}$$

is

$$G(\mathbf{r}, \mathbf{r}') = \frac{e^{ikR}}{R}, \qquad R = |\mathbf{r} - \mathbf{r}'|. \tag{20.70}$$

In the asymptotic region $G(\mathbf{r}, \mathbf{r}')$ can be shown to take the form

$$\frac{e^{ikR}}{R} \sim \frac{e^{ikr}}{r}e^{-i\mathbf{k}'\cdot\mathbf{r}'} = \frac{e^{ikr}}{r}e^{-kr'\cos\theta}, \qquad \mathbf{k}' = k\hat{\mathbf{r}}, \tag{20.71}$$

where θ is the scattering angle. This can be seen by examining the component of the vector $\mathbf{r} - \mathbf{r}'$ in the direction of unit radial vector $\hat{\mathbf{r}} = \mathbf{r}/r$. In spherical polar coordinates $\hat{\mathbf{r}}$ is expressible in terms of the corresponding unit vectors in a three-dimensional Cartesian system as

$$\hat{\mathbf{r}} = \langle\sin\theta\cos\phi, \sin\theta\sin\phi, \cos\theta\rangle \tag{20.72}$$

After some arrangement the scattered field may be written as [54]

$$\psi_s(\mathbf{r}) \approx ikC\frac{e^{ikr}}{4\pi r}(1 + \cos\theta)\int_A \exp\left[-ik\sin\theta\left(x'\cos\phi + y'\sin\phi\right)\right]dx'dy'. \tag{20.73}$$

Since $x' = r'\cos\phi'$ and $y' = r'\sin\phi'$, this can be written in polar coordinates as

$$\psi_s(\mathbf{r}) \approx \frac{ikCe^{ikr}}{4\pi r}(1 + \cos\theta)\int_0^{2\pi}d\phi'\int_0^a r'\exp\left[-ik\sin\theta\cos\left(\phi - \phi'\right)\right]dr'. \tag{20.74}$$

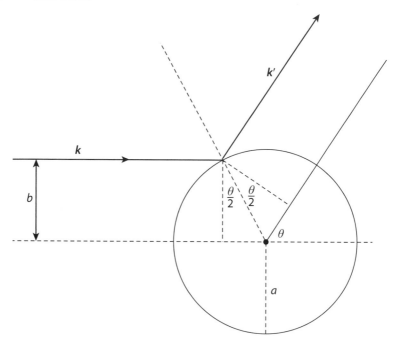

Figure 20.7 Path difference details for specular reflection.

Some integral representations of Bessel functions can be employed to simplify this result. First, the angular integration is of the form [28]

$$\frac{1}{\pi} \int_0^{\pi} \exp\left(i\alpha \cos\rho\right) d\rho = J_0(\alpha),\qquad(20.75)$$

and second, we have, also from [28],

$$\int_0^a p J_0(p\beta) dp = \frac{a}{\beta} J_1(a\beta),\qquad(20.76)$$

so

$$\psi_s(\mathbf{r}) \approx \frac{1}{2} iaC \left(\frac{1+\cos\theta}{\sin\theta}\right) J_1(ka \sin\theta) \frac{e^{ikr}}{r}.\qquad(20.77)$$

This is the contribution from the so-called shadow-forming waves [54]. There is an additional term corresponding to the reflection of the incident wave at the surface of the sphere. A point where the wave is specularly reflected in the direction $\mathbf{k}'$ is a saddle point of the integrand for the "lit" portion of the sphere, meaning that a point of stationary phase associates a geometrical optics ray with a saddle point, so this contribution can be approximated geometrically. As can be seen from Figure 20.7, there is a phase difference between the reflected ray and a ray that would have been reflected in the same direction by a sphere of vanishingly small radius. This is clearly an amount $-2k \sin(\theta/2)$. According to [54], this phase difference must be multiplied by a factor $a/2$ (which is the area-to-perimeter ratio for a *disk* of radius a), because the geometric cross section of the sphere is πa^2.

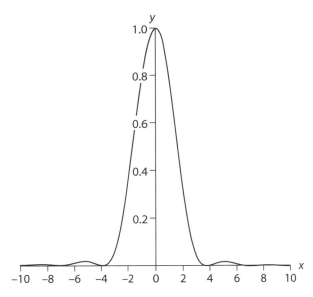

Figure 20.8 The Airy diffraction pattern.

The asymptotic form of the scattered wave field is, in terms of the scattering amplitude $f(k, \theta)$,

$$\psi_s(\mathbf{r}) \sim C f(k, \theta) \frac{e^{ikr}}{r}, \tag{20.78}$$

so on combining the field from the geometric reflection term with that from the approximation (20.77), we have the expression for the modulus of the scattering amplitude (based only on classical considerations):

$$|f(k, \theta)| \approx \left| \frac{1}{2} ia \left\{ i \exp \left[-2ika \sin \left(\frac{\theta}{2} \right) \right] + \left(\frac{1 + \cos \theta}{\sin \theta} \right) J_1 (ka \sin \theta) \right\} \right|, \tag{20.79}$$

hence the classical differential scattering cross section is

$$\frac{d\sigma}{d\Omega} \approx \frac{a^2}{4} \left[1 + \cot^2 \left(\frac{\theta}{2} \right) J_1^2 (ka \sin \theta) \right]. \tag{20.80}$$

These terms respectively represent (in the short-wavelength limit) the contribution from geometrical optics and that from diffraction. We may also write the second term in (20.79) as

$$|f_d(k, \theta)| \approx \left| \frac{a}{2} \left(\frac{1 + \cos \theta}{\sin \theta} \right) J_1 (ka \sin \theta) \right| \tag{20.81}$$

$$= \left| \frac{ka^2}{2} (1 + \cos \theta) \frac{J_1 (ka \sin \theta)}{ka \sin \theta} \right| \tag{20.82}$$

$$\approx ka^2 \left| \frac{J_1 (ka \sin \theta)}{ka \sin \theta} \right|, \quad \theta \ll 1. \tag{20.83}$$

The square of the quantity $2J_1(x)/x$ is shown in Figure 20.8 (normalized to unity). It exhibits the so-called *Airy diffraction pattern*, with characteristic forward diffraction peak containing about 84 percent of the total intensity. This is relevant to the coronal rings often seen around a full moon when viewed through a low thin cloud laden with water vapor (discussed in Chapter 4).

PART IV
Semiclassical Scattering

Chapter Twenty-One

The Classical-to-Semiclassical Connection

21.1 INTRODUCTION: CLASSICAL AND SEMICLASSICAL DOMAINS

Geometrical optics and wave (or physical) optics are two very different but complementary approaches to describing many optical phenomena. However, a broad middle ground, exists between them: the semiclassical regime. Thus there are essentially three domains in which scattering phenomena may be described: the scattering of waves by objects which in size are (i) small, (ii) comparable with, and (iii) large compared to the wavelength of the incident (plane wave) radiation. There may be considerable overlap of region (ii) with the others, depending on the problem of interest, but basically, the wave-theoretic principles in region (i) tell us why the sky is blue (among many other things!). At the other extreme, the classical domain (iii) enables us in particular to be able to describe the basic features of the rainbow in terms of ray optics. The wave-particle duality so fundamental in quantum mechanics is relevant to region (ii), because the more subtle features exhibited by such phenomena involve both these aspects of description and explanation. Indeed, it is useful to relate (somewhat loosely) the regimes (i)—(iii) to three domains, as stated in [54]:

1. The classical domain: geometrical optics; particle and particle/ray-like trajectories;
2. The wave domain: physical optics, acoustic and electromagnetic waves, quantum mechanics; and
3. The semiclassical domain: the vast intermediate region between the above two, containing many interesting physical phenomena.

Geometrical optics is associated with real rays, but their analytic continuation to complex values of some associated parameters enables the concept of complex rays to be used, often in connection with surface or evanescent rays traveling along a boundary while penetrating the less dense medium in an exponentially damped manner. However, complex rays can also be used to describe the phenomenon of diffraction: the penetration of light into regions that are forbidden to the real rays of geometrical optics, so there are several different contexts in which this term can be used. In fact, the primary bow light/shadow transition region is associated physically with the confluence of a pair of geometrical rays and their transformation into complex rays; mathematically this corresponds to a pair of real saddle points merging into a complex saddle point. For the primary bow the two exiting (supernumerary) rays coalesce at the rainbow angle θ_R when they are incident on the sphere surface at the Descartes angle i_c, and the subsequent vanishing of these rays is associated with the complex ray on the shadow side of the rainbow. In contrast, rays that graze the sphere and just miss grazing it may "tunnel" into the interior, or more accurately, both of these regions together form an "edge region" that gives rise to the tunneling ray. This phenomenon is well known in quantum mechanics; specifically, tunneling through a classically forbidden potential barrier. Because it occurs in the edge region of semiclassical scattering, it permits grazing rays (and those just outside the sphere) to interact with it and contribute to the radiation field. It has been shown in a series of very elegant but technical papers [83], [84] that scattering of scalar waves by a transparent sphere is in many respects isomorphic to the problem of scattering of particles by a spherical potential well. In quantum mechanics, as will be shown in Chapter 22, the bound states of a potential well correspond to poles in the elements of a certain matrix (the scattering matrix) on the negative real energy axis, whereas *resonances* of the well correspond to poles that are just below the positive real energy axis of the second Riemann sheet associated with those matrix elements. The closer these poles are to the real axis, the more the resonances behave like very long-lived bound states, or "almost bound" states of the system. In simplistic terms, if a particle with a resonance energy is shot at the well from far enough away, it is captured by the well for a considerable time and acts like a bound particle, but eventually it escapes from the well (this is also a crude description of the mechanism for α -decay from a nucleus, though that is a decay phenomenon, not a scattering one). The reciprocal of the half-width of the resonance is a measure of the lifetime of the resonance particle in the well.

21.2 INTRODUCTION: THE SEMICLASSICAL FORMULATION

The energy E of a photon of frequency v is $E = hv$, where h is Planck's constant. h is approximately equal to $6.6260755 \times 10^{-34}$ J-s. The associated momentum is defined to be $p = h/\lambda$, thus implicitly defining a wavelength λ, and when applied to a particle, its associated wavelength λ is referred to as the de Broglie wavelength. These are often referred to as "matter waves".

In a sense, what is called the primitive semiclassical approach is the "geometric mean" of classical and quantum mechanical descriptions of phenomena; while retaining the concept of particle trajectories and their individual contributions, there is nevertheless an associated de Broglie wavelength for each particle, so that interference and diffraction effects enter the picture. The latter do so via the transition from geometrical optics to wave optics. In fact the term "simple" semiclassical has been used to describe the scattering at angles well separated from the classical rainbow and glory singularities [101]. This allows for the use of a variant of the saddle-point method in approximating a sum of terms (the stationary phase approximation). It relies on replacing the sum by an integral if the number of contributing terms is suffcently large. Then a quadratic expansion for the phase of the resulting integrand about its stationary value is performed. This is possible because any rapid variation of phase with l must lead to destructive interference. The differential scattering cross section is related to the quantum scattering amplitude $f(k, \theta)$ and this in turn is expressible as a partial wave expansion. The formal relationship between this and the classical differential cross section is established using the WKB(J) approximation, and the principle of stationary phase is used to evaluate asymptotically a certain phase integral. Moreover, a point of stationary phase can be identified with a classical trajectory, but if more than one such point is present (provided they are well separated and of first order), the corresponding expression for $|f(k, \theta)|^2$ will contain interference terms. This is a distinguishing feature of the simple semiclassical formulation and has significant implications for the four effects (rainbow scattering, glory scattering, forward peaking and orbiting) noted in Chapter 16. The infinite intensities (incorrectly!) predicted by geometrical optics at focal points, lines, and caustics in general are breeding grounds for diffraction effects, as are light/shadow boundaries for which geometrical optics predicts finite discontinuities in intensity [102]. Such effects are most significant when the wavelength is comparable with (or larger than) the typical length scale for variation of the physical property of interest (this is the domain (c) mentioned above). Thus a scattering object with a sharp boundary (relative to one wavelength) can give rise to diffractive scattering phenomena.

Therefore under circumstances appropriate to the four critical effects noted above, the simple semiclassical approximation breaks down, and these diffraction effects cannot be ignored. Although the angular ranges in which such critical effects become significant get narrower as the wavelength decreases, the differential cross section can oscillate very rapidly and become very large within them. There are so-called *transitional asymptotic approximations* to the scattering amplitude in these critical angular domains, but they have very narrow domains of validity and do not match smoothly with neighboring noncritical angular domains. It has therefore been of considerable importance to seek *uniform asymptotic approximations* that by definition do not suffer from these failings. Fortunately, the problem of plane wave scattering by a homogeneous sphere exhibits all of the critical scattering effects (and it can be solved exactly, in principle) and is therefore an ideal laboratory in which to test both the efficacy and accuracy of the various approximations. Furthermore, it has relevance to both quantum mechanics (as a square well or barrier problem) and optics (Mie scattering). Indeed, it also serves as a model for the scattering of acoustic and elastic waves, and (as previously noted) was studied in the early twentieth century as a model for the diffraction of radio waves around the surface of the earth. More technical details of the simple semiclassical approach are provided in Appendix F.

It may seem overly pedantic, but to remind the reader that in what follows we are dealing with aspects of the semiclassical approximation, representing a type of transition zone between the classical and fully quantum mechanical approaches, I shall use the dependent variables Ψ or ψ respectively in the wave equation or Helmholtz equation instead of (in the latter case) the earlier variable ϕ. Although we cannot legitimately regard either Ψ or ψ as wavefunctions (though they may be eigenfunctions), this choice is intended to remind us that contextually at least, we have moved closer in some sense to the quantum mechanical concept of probability waves or distributions.

21.2.1 The Total Scattering Cross Section

Recall from Chapter 15 that in scattering theory the total scattering cross section σ is a measure of the probability that an interaction of some type occurs; the larger the cross section, the greater will be the probability that an interaction happens when a particle is incident on a target. As we have seen, both classical and quantum mechanical scattering phenomena can be characterized by σ. For particles of mass m and energy $E > 0$, scattering occurs as a result of a potential field $V(\mathbf{r})$, $r = |\mathbf{r}|$ being the radial distance from the center of the scatterer. Suppose we are interested in the scattering of particles (classical or quantum) or waves from a spherically symmetric potential $V(r)$, and suppose also that the incoming plane wave is incident along the direction of the z-axis and is scattered as an outgoing wave. Sufficiently close to the scattering object, the incident and scattered waves interfere with each other, and each is distorted by the other. At large values of r, it is assumed that the potential has weakened sufficiently that the scattered field and incident field are essentially decoupled. In broad terms, the radiation field ψ consists of a linear combination of the incident plane wave and an outwardly propagating spherical wave. This is commonly expressed in the form of (1.13), that is,

$$\psi(r, z, \theta) \sim e^{ikz} + \frac{f(\theta)}{r} e^{ikr}, \tag{21.1}$$

in terms of wavenumber k and the scattering amplitude $f(\theta)$, which gives the directional dependence of the scattered field. The origin of the spherical coordinate system (r, θ, ϕ) used here is the center of $V(r)$, but there is no azimuthal ϕ dependence because of the cylindrical symmetry imposed on the spherically symmetric potential by the incident field direction. The r^{-1} dependence provides the necessary inverse-square energy (or intensity) fall-off for an outgoing spherical wave. From standard classical or quantum mechanical considerations we can identify the *current density* as

$$\mathbf{j} = \frac{i}{2} \left(\psi \nabla \bar{\psi} - \bar{\psi} \nabla \psi \right). \tag{21.2}$$

The flux of particles (or radiation) passing through an area $d\mathbf{S}$ for a given (θ, ϕ) is

$$\mathbf{j} \cdot d\mathbf{S} = \frac{i}{2} \left(\psi \nabla \bar{\psi} - \bar{\psi} \nabla \psi \right) \cdot d\mathbf{S}. \tag{21.3}$$

Substituting the first term in (21.1) into this expression yields, for the incoming flux of particles,

$$(\mathbf{j} \cdot d\mathbf{S})_i = k \, dA, \tag{21.4}$$

where dA is the area differential with normal direction parallel to the z-axis. For the second term in (21.1) we find

$$(\mathbf{j} \cdot d\mathbf{S})_{\text{scatt}} = k \frac{|f(\theta)|^2}{r^2} dS = k |f(\theta)|^2 d\Omega, \tag{21.5}$$

where $d\Omega$ is an element of solid angle. The *differential cross section* is defined by

$$\frac{d\sigma}{d\Omega} = \frac{\text{scattered flux/unit solid angle}}{\text{incident flux/unit area}} = |f(\theta)|^2. \tag{21.6}$$

The *total cross section* σ is obtained by integrating the differential cross section over all scattering angles:

$$\sigma = \int |f(\theta)|^2 d\Omega = \int_0^{2\pi} d\phi \int_0^\pi |f(\theta)|^2 \sin\theta d\theta = 2\pi \int_0^\pi |f(\theta)|^2 \sin\theta d\theta. \tag{21.7}$$

21.2.2 Classical Wave Connections

For the case of scalar plane wave scattering by a spherically symmetric scatterer, the form of the inhomogeneous Helmholtz equation and the time-independent Schrödinger wave equation are similar. Therefore we can often treat classical and quantum problem by analogous methods. In particular, for acoustic wave propagation in radially inhomogeneous media, the governing equation is

$$\rho\nabla \cdot (\rho^{-1}\nabla\Psi(\mathbf{r}, t)) - \frac{1}{c^2(r)}\frac{\partial^2\Psi(\mathbf{r}, t)}{\partial t^2} = 0. \tag{21.8}$$

Here $\Psi(\mathbf{r}, t)$ is the acoustic pressure perturbation, and $\rho(r)$ is the density of the medium (see [129]). If the density is constant (as assumed here), equation (21.8) reduces to the standard spherically symmetric wave equation. Consider a monochromatic time-harmonic dependence of $\Psi(\mathbf{r}, t)$ such that

$$\Psi(\mathbf{r}, t) = \psi(\mathbf{r})e^{-i\omega t}. \tag{21.9}$$

Consequently the spatial part of $\Psi(\mathbf{r}, t)$ satisfies the inhomogeneous Helmholtz equation

$$\nabla^2\psi(\mathbf{r}) + k_0^2\psi(\mathbf{r}) = V(r)\psi(\mathbf{r}), \tag{21.10}$$

where $r = |\mathbf{r}|$. In this equation $k_0 = \omega/c_0$ is the "free space" wavenumber, where c_0 is the constant wave speed of the acoustic wave outside the scatterer. In electromagnetic terminology c_0 corresponds to an outer region where the refractive index $n = 1$. The potential in this case is

$$V(r) = k_0^2 - k^2(r) \equiv k_0^2(1 - n^2(r)) = k_0^2\left(1 - \frac{c_0^2}{c^2(r)}\right) = k_0^2 - \frac{\omega^2}{c^2(r)}. \tag{21.11}$$

Thus $k(r) = \omega/c(r)$ is the wavenumber in the scattering medium. Equation (21.10) corresponds to the canonical Helmholtz form of the time-independent Schrödinger equation, where $k_0^2 = 2mE/\hbar^2$, $V(r) = 2mU(r)/\hbar^2$, $\hbar = h/2\pi$, h being Planck's constant, and m is the mass of a particle of total energy E moving in a potential $U(r)$, that is,

$$\nabla^2\psi(\mathbf{r}) + [k_0^2 - V(r)]\psi(\mathbf{r}) = 0. \tag{21.12}$$

In what follows k_0 will be written as k, and $k(r)$ will be used for $kn(r)$. Note from definition (21.11) that $V(r) \leq 0$ for $n^2(r) \geq 1$, so the quantum mechanical formulation corresponds to scattering by a spherical potential well. Therefore, the wave function can be interpreted both as the spatial part of the acoustic pressure (classically) and as the Schrödinger wave function (quantum mechanically). The basis for this section is the application of the Jost function formulation of quantum scattering theory [64] to the scattering of a scalar plane wave by a medium with spherically symmetric inhomogeneities. By the method of separation of variables, the solution to (21.12) can be written in a spherical coordinate system as

$$\psi(r) = \sum_{l=0}^{\infty} \sum_{m=-l}^{+l} A_{lm} R_l(r) Y_{lm}(\theta, \phi), \tag{21.13}$$

where $R_l(r) = u_l(r)/r$, $Y_{lm}(\theta, \phi)$ is a spherical harmonic, and l and m are angular momentum parameters. Then it follows that $u_l(r)$ satisfies the partial-wave radial equation

$$\frac{d^2 u_l(r)}{dr^2} + \left[k^2 - \frac{l(l+1)}{r^2} - V(r) \right] u_l(r) = 0. \tag{21.14}$$

21.3 THE SCALAR WAVE EQUATION

21.3.1 Separation of Variables

For convenience some of the material in Chapter 19 regarding vector wave fields is repeated here with a slightly different formulation for scalar waves. In a source-free region the wave equation for $\tilde{\psi}(\mathbf{r}, t)$ satisfies the free-space wave equation (c being constant here):

$$\nabla^2 \tilde{\psi} - \frac{1}{c^2} \frac{\partial^2 \tilde{\psi}}{\partial t^2} = 0. \tag{21.15}$$

If

$$\tilde{\psi}(\mathbf{r}, t) = \int_{-\infty}^{\infty} \psi(\mathbf{r}, \omega) e^{-i\omega t} d\omega, \tag{21.16}$$

then $\psi(\mathbf{r}, \omega)$, written now as $\psi(\mathbf{r}, k) = \psi$ satisfies the Helmholtz equation

$$\left(\nabla^2 + k^2 \right) \psi(\mathbf{r}, k) = 0, \qquad k = \omega/c. \tag{21.17}$$

k is the wavenumber in free space, and as mentioned above, it can be generalized easily to the case of a medium with refractive index $n(\mathbf{r})$. In a spherical coordinate system (r, θ, ϕ), equation (21.17) takes the form

$$\left\{ \frac{1}{r^2} \frac{\partial}{\partial r} \left(r^2 \frac{\partial}{\partial r} \right) + \frac{1}{r^2 \sin\theta} \frac{\partial}{\partial \theta} \left(\sin\theta \frac{\partial}{\partial \theta} \right) + \frac{1}{r^2 \sin^2\theta} \frac{\partial^2}{\partial \phi^2} + k^2 \right\} \psi(r, \theta, \phi) = 0. \tag{21.18}$$

Formally seeking a separable solution as a product of radially dependent functions and spherical harmonics $\psi(r, \theta, \phi) = R(r)Y(\theta, \phi)$ yields the equations

$$\frac{d}{dr}\left(r^2 \frac{dR}{dr}\right) + \left(k^2 r^2 - \lambda\right) R = 0, \tag{21.19}$$

$$\frac{1}{\sin\theta}\frac{\partial}{\partial\theta}\left(\sin\theta \frac{\partial Y}{\partial\theta}\right) + \frac{1}{\sin^2\theta}\frac{\partial^2 Y}{\partial\phi^2} + \lambda Y = 0. \tag{21.20}$$

where, λ is a separation constant. Formally, therefore, if $Y(\theta, \phi)$ is finite for $\theta \in [0, \pi)$, $\phi \in [0, 2\pi)$, then

$$\psi(\mathbf{r}, k) = \sum_\lambda c_\lambda R_\lambda(r) Y_\lambda(\theta, \phi). \tag{21.21}$$

To find acceptable values of λ we proceed to decompose the spherical harmonics as $Y(\theta, \phi) = \Theta(\theta)\Phi(\phi)$ to obtain

$$\frac{d}{d\theta}\left(\sin\theta \frac{d\Theta}{d\theta}\right) + \left(\frac{\mu}{\sin\theta} - \lambda\sin\theta\right)\Theta = 0, \tag{21.22}$$

$$\frac{d^2\Phi}{d\phi^2} - \mu\Phi = 0. \tag{21.23}$$

Imposing the requirement of periodicity in ϕ, $\mu = -m^2$, $m = \pm 1, \pm 2, \ldots$, and $m = 0$ if ψ is independent of ϕ (a case of great importance, as we shall see!), equation (21.22) takes the form

$$\frac{d}{d\theta}\left(\sin\theta \frac{d\Theta}{d\theta}\right) - \left(\frac{m^2}{\sin\theta} + \lambda\sin\theta\right)\Theta = 0. \tag{21.24}$$

The change of independent variable $x = \cos\theta$ converts (21.24) to the *associated Legendre equation*:

$$\frac{d}{dx}\left((1 - x^2)\frac{dy}{dx}\right) - \left(\frac{m^2}{1 - x^2} + \lambda\right)y = 0. \tag{21.25}$$

The dependent variable $y(x) = \Theta(\arccos x) = \Theta(\theta)$. It becomes an eigenvalue equation for λ if the solution is required to be finite at the regular singular points $x = \pm 1$. When $m = 0$ this equation reduces to the simpler eigenvalue form of Legendre's equation:

$$\frac{d}{dx}\left((1 - x^2)\frac{dy}{dx}\right) - \lambda y = 0. \tag{21.26}$$

Since this case is one that will be needed later, we state the following theorem.

Theorem: Equation (21.26) possesses a bounded solution (to within a scalar multiple) in $[-1, 1]$ if and only if $\lambda = l(l + 1)$, $l = 0, 1, 2, \ldots$. (It will come as no surprise that this solution is the Legendre polynomial $P_n(x)$.)

Exercise: Prove this result.

21.3.2 Bauer's Expansion Again

In equation (18.32) reference was made to Bauer's expansion (or formula), and the quote at the beginning of Section 19.2 gives the reader some idea of why this expansion is so unnatural and yet so important! Consider a unit-amplitude plane wave propagating in the z direction in a spherical coordinate system. This direction of propagation immedially imposes a symmetry-breaking change on the system, it now being cylindrically symmetrical and therefore independent of the azimuthal angle ϕ (corresponding to $m = 0$ in the associated Legendre equation). The plane wave is therefore represented by [54], [62]

$$\psi(r) = e^{i\mathbf{k}\cdot\mathbf{r}} = e^{ikr\cos\theta}, \tag{21.27}$$

where $\mathbf{k}$ is, as usual, the wave vector. Since this expression is not singular at $r = 0$, it must be representable in a spherical coordinate system in the following way:

$$e^{ikr\cos\theta} = \sum_{l=0}^{\infty} C_l \, j_l(kr) \, P_l(\cos\theta). \tag{21.28}$$

The constant coefficients C_l are to be determined. A simple way to accomplish this is to multiply both sides of this posited equation by $P_{l'}(\cos\theta)\sin\theta$ and integrate over $[0, \pi]$. We use the orthogonality relation satisfied by Legendre polynomials, namely,

$$\int_{-1}^{1} P_l(x) P_{l'}(x)dx = \frac{2}{2l+1}\delta_{ll'}. \tag{21.29}$$

With $x = \cos\theta$ (and dropping the prime on l), on inverting the expansion (21.28), it follows that

$$C_l \, j_l(kr) = \left(l + \frac{1}{2}\right) \int_{-1}^{1} e^{ikrx} P_l(x)dx. \tag{21.30}$$

At this point we can take a short route or a long one. The former makes use of the fact that

$$j_l(\alpha) = \frac{(-i)^l}{2} \int_{-1}^{1} e^{i\alpha x} P_l(x)dx. \tag{21.31}$$

Hence

$$C_l = (2l+1)\,i^l, \tag{21.32}$$

and we have re-established Bauer's expansion, namely,

$$e^{ikr\cos\theta} = \sum_{l=0}^{\infty}(2l+1)i^l \, j_l(kr) \, P_l(\cos\theta). \tag{21.33}$$

The long route, which is also well worth examining, makes use of the asymptotic form of the spherical Bessel function $j_l(kr)$. Integrating expression (21.30) by parts, we obtain

$$C_l \, j_l(kr) = \frac{(2l+1)}{2ikr}\left\{\left[e^{ikrx} P_l(x)\right]_{-1}^{1} - \int_{-1}^{1} e^{ikrx}\frac{dP_l}{dx}dx\right\}. \tag{21.34}$$

The integral successively generates terms of order r^{-n}, $n \geq 2$, that become negligibly small in the asymptotic region, so we may write

$$C_l j_l(kr) \sim \frac{(2l+1)}{2ikr} \left[e^{ikrx} P_l(x) \right]_{-1}^{1}. \tag{21.35}$$

Noting that $P_l(1) = 1$ and $P_l(-1) = (-1)^l$, (21.32) follows after some algebra, using

$$j_l(kr) = \frac{\sin(kr - l\pi/2)}{kr} + O((kr)^{-2}) \text{ as } kr \to \infty. \tag{21.36}$$

The details of this derivation of (21.35) are left as an exercise.

21.4 THE RADIAL EQUATION: FURTHER DETAILS

Consider first the following equation

$$\frac{d}{dr}\left(r^2 \frac{dR_l}{dr}\right) + \left(k^2 r^2 - l(l+1)\right) R_l = 0. \tag{21.37}$$

This equation has solutions in terms of the by now familiar spherical Bessel functions defined in terms of the ordinary Bessel functions J_l and Y_l, where

$$j_l(kr) = \left(\frac{\pi}{2kr}\right)^{1/2} J_{l+1/2}(kr), \text{ and } y_l(kr) = \left(\frac{\pi}{2kr}\right)^{1/2} Y_{l+1/2}(kr), \tag{21.38}$$

Now we add a potential $V(r)$ to the mix:

$$\frac{1}{r^2}\frac{d}{dr}\left(r^2 \frac{dR_l}{dr}\right) + \left(k^2 - V(r) - \frac{l(l+1)}{r^2}\right) R_l = 0. \tag{21.39}$$

Let us note a few properties of the equation and its solutions. The change of dependent variable $u_l(r) = r R_l(r)$ reduces this equation to the simpler form

$$\frac{d^2 u_l(r)}{dr^2} + \left[k^2 - V(r) - \frac{l(l+1)}{r^2}\right] u_l(r) = 0, \quad r > 0. \tag{21.40}$$

From the indicial equation, the solutions can be seen to have the following behavior near the origin (provided that $V(r)$ possesses at most a simple pole there):

$$u_l(r) \sim r^{l+1} + O(r^{l+2}), \text{ or } u_l(r) \sim r^{-l} + O(r^{-l+1}). \tag{21.41}$$

Clearly, the second independent solution can be rejected on physical grounds. Heuristically, we anticipate that the last two terms in the bracket in (21.40) will become negligible as $r \to \infty$, so any such solution will take the asymptotic form

$$u_l(r) \sim A \sin(kr + \alpha), \tag{21.42}$$

for constants A and α. The circumstances under which this can occur can be inferred by setting $u_l(r) = U(r)e^{ikr}$, where $U(r)$ now satisfies the equation

$$\frac{d^2 U}{dr^2} + 2ik\frac{dU}{dr} - \left[V(r) + \frac{l(l+1)}{r^2}\right] U = 0. \tag{21.43}$$

By the assumption that the asymptotic form (21.42) for $u_l(r)$ is oscillatory, U must be slowly varying for large enough r, so $|U''(r)| \ll |kU'(r)|$ leads to the approximate result

$$2ik \ln U \approx \int^r \left[V(\xi) + \frac{l(l+1)}{\xi^2} \right] d\xi. \tag{21.44}$$

The integral approaches a constant asymptotically if and only if $V(r) \sim r^{-(1+\varepsilon)}$, $\varepsilon > 0$. This excludes the Coulomb field, of course, for which a separate treatment is necessary. For the specified functional form of V, the oscillatory solution (21.42) for $u_l(r)$ is therefore valid. Returning to equation (21.39) for R_l, we are then able to define the solution that is finite at the origin as having the asymptotic form

$$R_l(r) \sim (kr)^{-1} \sin \left(kr - \frac{l\pi}{2} + \delta_l(k) \right), \tag{21.45}$$

where $\delta_l(k)$ is the phase shift, and the multiple of $\pi/2$ is chosen to be consistent with the asymptotic form of the solution in the absence of a potential $V(r)$. There are some further conclusions to be drawn regarding the nature of δ_l. It follows from the discussion above that, to within a factor of k^{-1},

$$U(r) \sim \sin \left(kr - \frac{l\pi}{2} + \delta_l(k) \right). \tag{21.46}$$

If

$$F(r; k) = k^2 - V(r) - \frac{l(l+1)}{r^2}, \tag{21.47}$$

and we again assume that $V(r)$ has no pole higher than order r^{-1}, it follows that for small enough values of r, $F(r) < 0$ but for large enough values, $F(r) > 0$. F therefore has at least one zero in $(0, \infty)$. For simplicity consider the case where there is a unique zero at $r = r(l, k) \equiv r_l$. The following argument can be made: for small values of r the solution $u_l(r)$ in equation (21.40) is of the form Kr^{l+1} (where we assume for the moment that $K > 0$). This implies that $u_l(r) > 0$ and $u_l'(r) > 0$ for small r, and hence, from (21.40), $u_l''(r) > 0$ also. As r increases away from zero, u_l will start to decrease when u_l' changes sign, and it does this beyond the first zero of u_l''. But from the above definition of F this zero must be at $r = r_l$. The proof is similar for $K < 0$. (**Exercise:** Show this.) Finally, for $r > r_l$, $u_l(r)$ is oscillatory. From a classical perspective, a particle of mass m and energy E under the influence of a central force, with angular momentum J about the center of force (be it electrostatic or gravitational in nature) must satisfy the law of energy conservation,

$$E = \frac{1}{2}mv^2 + V(r), \tag{21.48}$$

where v is the speed of the particle at the *point of closest approach* (where the radial component of velocity is zero). Furthermore, invoking the conservation of angular momentum,

$$J = mvr.$$

On eliminating υ, we find that

$$E - V(r) - \frac{J^2}{2mr^2} = 0. \tag{21.49}$$

The left-hand side of this expression has been encountered on several earlier occasions (Chapters 15 and 16). If we identify the energy E with k^2 and the angular momentum J with $[l\,(l+1)]^{1/2}$, then equation (21.48) is formally equivalent to $F(r;k) = 0$, so we see that classically r_l is the closest distance of approach of a particle with angular momentum J. Instead of considering what happens for small r, we now examine some consequences of the angular momentum parameter being large. Specifically, this requires $V(r_l)$ to be small enough that r_l approximately satisfies $k^2 r_l^2 = l(l+1)$. The corresponding differential equation satisfied by a different u_l, $\tilde{u}_l$, say, is

$$\frac{d^2\tilde{u}_l(r)}{dr^2} + \left[k^2 - \frac{l\,(l+1)}{r^2} \right] \tilde{u}_l(r) = 0, \tag{21.50}$$

which possesses the Riccati-Bessel functions $S_l(kr)$ and $C_l(kr)$ as linearly independent solutions. In terms of (i) spherical Bessel functions of the first and second kind of order $l\,(j_l(kr),\ y_l(kr))$ or (ii) Bessel functions of the first and second kind of order $l\,(J_l(kr),\ Y_l(kr))$, they are defined as

$$S_l(kr) = kr\,j_l(kr) = \left(\frac{\pi kr}{2} \right)^{1/2} J_{l+1/2}(kr); \tag{21.51}$$

$$C_l(kr) = -kr\,y_l(kr) = - \left(\frac{\pi kr}{2} \right)^{1/2} Y_{l+1/2}(kr). \tag{21.52}$$

Only the S_l function is bounded at the origin, and asymptotically, therefore, $\tilde{u}_l(r) \sim \sin(kr - l\pi/2)$ as may be readily shown. The arguments invoked above for the behavior of $R_l(r)$ also apply for $\tilde{u}_l(r)$: it decreases exponentially as $r \to 0$ for $r < r_l$. We can now use the solution $\tilde{u}_l(r)$ to construct a perturbation-based solution for $u_l(r)$ as follows. Let

$$u_l(r) = \tilde{u}_l(r) + \phi(r), \tag{21.53}$$

where, following [62], we assume that the product $V(r)\phi(r)$ can be neglected when compared with either quantity alone. From (21.40) it follows that

$$\frac{d^2\phi(r)}{dr^2} + \left[k^2 - \frac{l\,(l+1)}{r^2} \right] \phi(r) \approx V(r)\tilde{u}_l(r). \tag{21.54}$$

We regard this approximation as exact from this point forward. Making the further choice for $\phi(r) = \tilde{u}_l(r)\xi(r)$, the above equation can be rearranged as

$$\xi''\tilde{u}_l + 2\xi'\tilde{u}_l' = V\tilde{u}_l. \tag{21.55}$$

Multiplying both sides by $\tilde{u}_l$ and integrating, it follows that, for some constant c,

$$\xi'\tilde{u}_l^2 = \int_c^r V(\mu)\tilde{u}_l^2(\mu)\,d\mu. \tag{21.56}$$

Since $\tilde{u}_l(r) = S_l(kr)$, and therefore $\tilde{u}_l(0) = 0$, c must be zero. Hence

$$\frac{d\xi}{dr} = [\tilde{u}_l(r)]^{-2} \int_0^r V(\mu)\tilde{u}_l^2(\mu)\,d\mu, \tag{21.57}$$

and so for sufficiently large values of r

$$\frac{d\xi}{dr} \sim \csc^2\left(kr - \frac{l\pi}{2}\right) \int_0^\infty V(r)\tilde{u}_l^2(r)dr \equiv \csc^2\left(kr - \frac{l\pi}{2}\right) I_l. \tag{21.58}$$

The integral in (21.58) converges for the stated conditions on $V(r)$, since $\tilde{u}_l(r)$ is bounded at infinity. The integral I_l is small for two reasons: (i) $V(r)$ is small for $r > r_l$, and (ii) $\tilde{u}_l$ is small for $r < r_l$. Furthermore, for constant α it follows that

$$\xi \sim -\left[\cot\left(kr - \frac{l\pi}{2}\right) + \alpha\right]k^{-1}I_l, \tag{21.59}$$

whence

$$u_l(r) = \tilde{u}_l(1 + \xi) \sim \sin\left(kr - \frac{l\pi}{2}\right)$$
$$-\left[\cos\left(kr - \frac{l\pi}{2}\right) + \alpha\sin\left(kr - \frac{l\pi}{2}\right)\right]k^{-1}I_l. \tag{21.60}$$

Note that if $k^{-1}I_l \equiv -\delta_l$, then on neglecting terms of order δ_l^2 and higher, it can be seen that

$$u_l(r) \sim \sin\left(kr - \frac{l\pi}{2} + \delta_l\right), \tag{21.61}$$

where

$$\delta_l \approx -\frac{\pi}{2}\int_0^\infty V(r)\left[J_{l+1/2}(kr)\right]^2 rdr. \tag{21.62}$$

This, by the way, is the *first Born approximation*. In [62] it is stated that this result is valid if the right-hand side is small, showing that δ_l is small under the conditions stated. The result (21.62) is valid for large values of l and may be used to test the validity of the formula for the scattering amplitude $f(\theta)$. It is utilized in Appendix D in connection with the scattering of electromagnetic waves from a weak scatterer.

Exercise: Comment on this argument.

21.5 SOME EXAMPLES

21.5.1 Scattering by a One-Dimensional Potential Barrier

The standard form of the nonrelativistic one-dimensional time-independent Schrödinger equation is

$$-\frac{\hbar^2}{2m}\frac{d^2\Psi}{dx^2} + V(x)\Psi = E\Psi. \tag{21.63}$$

In this equation, $\hbar$ is the reduced Planck's constant $h/(2\pi)$, m is the mass of the particle (the rest mass of an electron, for example, is approximately $9.1093897 \times 10^{-31}$ kg), and E is its energy. As mentioned above $\Psi(x)$ is actually *not* the probability wave function for the particle (though we shall freely adopt this pseudonym); because the particle has momentum, Ψ is in fact a momentum eigenfunction, and $V(x)$ is the potential that it encounters. Obviously, by everyday standards, the quantity $\hbar^2/(2m)$ is very tiny, so we shall rewrite equation in more general terms as

$$\varepsilon^2 \frac{d^2\Psi}{dx^2} + (E - V(x))\,\Psi = 0. \tag{21.64}$$

We shall consider (for now) $V(x)$ to be a rectangular potential barrier equal to $V_0 > 0$ (a constant) for $x \in (0, a)$ and zero elsewhere. Furthermore, we shall set the incoming wave (from $-\infty$) to have unit amplitude, and outside the barrier, the reflected wave to have amplitude B and the transmitted wave to have amplitude C. Thus

$$\Psi(x) = e^{ikx} + Be^{-ikx}, \qquad x \le 0; \tag{21.65}$$

$$= Fe^{iKx} + Ge^{-iKx}, \qquad 0 < x \le a; \tag{21.66}$$

$$= Ce^{ikx}, \qquad x > a. \tag{21.67}$$

The quantities k and K are, respectively, the wavenumbers outside and inside the barrier. For $E > V_0$,

$$k = \frac{E^{1/2}}{\varepsilon}; \qquad K = \frac{(E - V_0)^{1/2}}{\varepsilon}. \tag{21.68}$$

Because we can always choose variables such that the coefficient of Ψ'' is one, we shall set $\varepsilon = 1$ here. By requiring $\Psi(x)$ and $\Psi'(x)$ to be continuous at $x = 0, a$, consider the following two cases:

(i) $E > V_0$

Exercise: Show by eliminating the coefficients F and G that

$$C = \frac{4kKe^{i(k-K)a}}{(K+k)^2 - (K-k)^2 e^{-2iKa}}, \tag{21.69}$$

and hence that the energy transmission coefficient

$$T = |C|^2 = \left[1 + \frac{\left(K^2 - k^2\right)^2 \sin^2 Ka}{4k^2 K^2} \right]^{-1} = \left[1 + \frac{V_0^2 \sin^2 Ka}{4E(E - V_0)} \right]^{-1}, \tag{21.70}$$

where

$$k = E^{1/2}, \text{ and } K = (E - V_0)^{1/2}. \tag{21.71}$$

Further show that

$$|B|^2 + |C|^2 = 1. \tag{21.72}$$

What does this mean physically?

Sketch graphs of both T and the energy reflection coefficient $R = |B|^2$ versus E/V_0. What happens when $Ka = n\pi$, $n = 1, 2, 3, ...$?
(ii) $0 < E < V_0$

Exercise: Show that

$$T = |C|^2 = \left[1 + \frac{V_0^2 \sinh^2 \beta a}{4E(V_0 - E)}\right]^{-1}, \tag{21.73}$$

where

$$\beta = (V_0 - E)^{1/2}. \tag{21.74}$$

Show further that, for $\beta a \gg 1$,

$$T \approx \frac{16E(V_0 - E)}{V_0^2} \exp\left[-2(V_0 - E)^{1/2}\right]. \tag{21.75}$$

Exercise: In the case $V_0 < 0$, the barrier becomes a well [60]. How does this change the expressions for T and R?
Hints: The mathematics is not hard, but the algebra is quite tedious at times. In part (i) you can use the fact that, by the conservation of energy, $T + R = 1$ (or you can calculate R algebraically if you wish). In part (ii), just use the fact that (in part (i)) $K \to i\beta$.

21.5.2 The Radially Symmetric Problem: Phase Shifts and the Potential Well

The Case $l = 0$

From (21.14) we have the time-independent radial Schrödinger equation for a spherically symmetric potential (where $k^2 = E$):

$$\frac{d^2 u_l(r)}{dr^2} + \left[k^2 - \frac{l(l+1)}{r^2} - V(r)\right] u_l(r) = 0. \tag{21.76}$$

As a special case, let $l = 0$ and suppose that $V(r)$ corresponds to a spherical potential well of depth V_0 and radius a, as considered before. If we require that $u_0(0) = 0$ so that wave function can penetrate the well, then clearly

$$u_0(r) = A \sin \eta r, \quad 0 \le r < a; \tag{21.77}$$

$$= B \sin [kr + \delta_0(\beta)], \quad r > a, \tag{21.78}$$

where to avoid confusion with the exercises in the previous section 21.5.1, we define $\eta = (k^2 + V_0)^{1/2}$ and $\beta = ka$. Continuity of u_0 and u_0' at $r = a$ (or equivalently of the logarithmic derivative u_0'/u_0 there) implies that

$$\eta \tan(ka + \delta_0) = k \tan \eta a. \tag{21.79}$$

Solving for the phase shift, we obtain

$$\delta_0(\beta) = \arctan\left(\frac{k}{\eta}\tan\eta a\right) - ka$$

$$= \arctan\left[\frac{k}{(k^2 + V_0)^{1/2}}\tan\left(k^2 + V_0\right)^{1/2}a\right] - ka. \tag{21.80}$$

The Case $l \neq 0$

In the more general case, there are three somewhat fuzzy regions of interest regarding (21.76), assuming that $V(r)$ is negligible for some $r > r_0$. These are region I: $V(r)$ is not negligible, so $u_l = u_l(V(r), r)$; region II: $V(r)$ is negligible, but $l(l+1)/r^2$ is not; and region III: both $l(l+1)/r^2$ and $V(r)$ are negligible. In region II, solutions of (21.76) are expressible in terms of Riccati-Bessel functions in the form (to be shown)

$$u_l(r) = D_l kr[j_l(kr) - K_l n_l(kr)], \tag{21.81}$$

whereas in region III, asymptotically,

$$u_l(r) \sim D_l \sin\left(kr - \frac{l\pi}{2} + \delta_l\right). \tag{21.82}$$

In general $u_l(r)$ must be consistent with the physical requirement that $|u_l(0)| < \infty$. Thus if (21.81) is valid for sufficiently small kr, then $u_l(r) \propto kr j_l(kr)$. Then the exact solution (whether known or not) must revert to the form (21.81) with $K_l \to 0$ as $kr \to 0$ but with $K_l \neq 0$ as $kr \to \infty$. Now from (21.1) we can write

$$\psi = \psi_k(r, \theta) \sim e^{ikr\cos\theta} + \frac{f(\theta)}{r}e^{ikr} \tag{21.83}$$

as $r \to \infty$, but we also know from Bauer's expansion (21.33) that

$$e^{ikr\cos\theta} = \sum_{l=0}^{\infty}(2l+1)i^l j_l(kr)P_l(\cos\theta), \tag{21.84}$$

and

$$f(\theta) = \frac{1}{2ik}\sum_{l=0}^{\infty}(2l+1)(e^{2i\delta_l} - 1)P_l(\cos\theta). \tag{21.85}$$

This is called the Faxen-Holtzmark formula, and it was stated in a slightly different form in equation (1.14).

To find the coefficient K_l we evaluate the lth term in the expansion of (21.83) with the corresponding asymptotic form of (21.81), noting that

$$j_l(x) \propto x^{-1}\sin\left(x - \frac{l\pi}{2}\right); \quad n_l(x) \propto -x^{-1}\cos\left(x - \frac{l\pi}{2}\right). \tag{21.86}$$

By comparing in turn the coefficients of $\exp(ikr)$ and $\exp(-ikr)$ in these two forms, we eventually arrive at

$$(2l+1)e^{2i\delta_l} = (-i)^l D_l(1 + iK_l), \tag{21.87}$$

and

$$D_l = -\frac{(2l+1)i^l i^l}{i^l[-1+iK_l]} = \frac{(2l+1)i^l}{1-iK_l}. \tag{21.88}$$

Solving for K_l, we obtain

$$K_l = \frac{e^{i\delta_l} - e^{-i\delta_l}}{i(e^{i\delta_l} + e^{-i\delta_l})} = \frac{2i \sin \delta_l}{i(2 \cos \delta_l)} = \tan \delta_l. \tag{21.89}$$

Thus in region II, solutions of (21.76) take the form

$$u_l(r) = D_l kr \sec \delta_l [j_l(kr) \cos \delta_l - n_l(kr) \sin \delta_l],$$
$$\equiv \tilde{D}_l kr[j_l(kr) \cos \delta_l - n_l(kr) \sin \delta_l]. \tag{21.90}$$

Application to a Perfectly Rigid (Impenetrable) Sphere of Radius a

The condition $u_l(a) = 0$ means that

$$\tan \delta_l = \frac{j_l(ka)}{n_l(ka)}. \tag{21.91}$$

(This case also corresponds to scattering from an acoustically "soft" sphere.) Again, let us examine the $l = 0$ or s-wave case. Using

$$j_0(x) = \frac{\sin x}{x}, \quad n_0(x) = -\frac{\cos x}{x},$$

it follows that

$$\tan \delta_0 = -\tan ka, \tag{21.92}$$

that is,

$$\delta_0 = -ka + n\pi, \quad n = 0, \pm 1, \pm 2, \dots. \tag{21.93}$$

From the result (18.53) the $l = 0$ total scattering cross section is

$$\sigma_0 = \frac{4\pi}{k^2} \sin^2 \delta_0 = \frac{4\pi}{k^2} \sin^2 ka = 4\pi a^2 \left(\frac{\sin ka}{ka}\right)^2 \to 4\pi a^2 \text{ as } ka \to 0, \tag{21.94}$$

and we have taken $n = 0$ in equation (21.93) without loss of generality. The scattering is spherically symmetric in this limit, and the total scattering cross section is *four times* the cross-sectional area, equal to the surface area of the sphere in fact! The reason is that in this long wavelength limit the waves (to put it crudely) "feel" their way around the sphere, as opposed to classical particles that only encounter the sphere via its head-on cross section. It is in fact a diffraction-related problem.

A More General Finite-Range Potential

Consider next $V(r) = 0$ for $r > a$ but arbitrary in $[0, a]$. Applying the logarithmic derivative $L_l(r)$ at the boundary for the radial solution of equation (21.76), where $R_l(kr) = r^{-1} u_l(kr)$, we have that

$$L_l(a) \equiv \lim_{r \to a^+} \frac{R'_l(r)}{R_l(r)} = \lim_{r \to a^-} \frac{R'_l(r)}{R_l(r)}. \tag{21.95}$$

Since it follows from (21.90) that

$$R_l(r) \propto (j_l \cos \delta_l - n_l \sin \delta_l), \qquad (21.96)$$

then

$$L_l(a) = \left[\frac{k(j_l' \cos \delta_l - n_l' \sin \delta_l)}{j_l \cos \delta_l - n_l \sin \delta_l} \right]_a, \qquad (21.97)$$

where the prime refers to the derivative with respect to the argument (in this case kr, evaluated at $r = a$). Solving for the phase shift yields the result

$$\tan \delta_l = \frac{kj_l'(ka) - j_l(ka)L_l}{kn_l'(ka) - n_l(ka)L_l}. \qquad (21.98)$$

An alternative representation follows if we use the dimensionless version of (21.98) in the form

$$L_l = \frac{a R_l'(a)}{R_l(a)}, \qquad (21.99)$$

with $\beta = ka$, that is,

$$\tan \delta_l(\beta) = \frac{\beta j_l'(\beta) - j_l(\beta)L_l}{\beta n_l'(\beta) - n_l(\beta)L_l}. \qquad (21.100)$$

THE PENETRABLE SQUARE-WELL POTENTIAL REVISITED

Now we reuse the standard form for $V(r)$ and apply the condition $u_l(0) = 0$, which forces the solution $u_0(kr) = rj_l(\eta r)$ for $r \le a$, so if $\hat{\beta} = \eta a$, (inside the well) we now have

$$L_l = \frac{\hat{\beta} j_l'(\hat{\beta})}{j_l(\hat{\beta})}. \qquad (21.101)$$

Again for $l = 0$ and using the definitions above, it is readily found that

$$L_0 = \hat{\beta} \cot \hat{\beta} - 1. \qquad (21.102)$$

For $r > a$, we have $\eta \to k$ and $\hat{\beta} \to \beta$ in (21.100) (but *not* in (21.102)), so after some reduction,

$$\tan \delta_0 = \frac{\beta j_0'(\beta) - j_0(\beta)L_0}{\beta n_0'(\beta) - n_0(\beta)L_0} = \frac{\beta \cos \beta - \hat{\beta} \sin \beta \cot \hat{\beta}}{\beta \sin \beta + \hat{\beta} \cos \beta \cot \hat{\beta}}. \qquad (21.103)$$

Furthermore, from the fact that

$$f_l(k) = \frac{e^{i\delta_l}}{k} \sin \delta_l, \qquad (21.104)$$

we have

$$\frac{d\sigma_0}{d\Omega} = |f_0(k)|^2 = \frac{\sin^2 \delta_0}{k^2} = \frac{\tan^2 \delta_0}{k^2 (1 + \tan^2 \delta_0)}, \qquad (21.105)$$

and we can substitute for $\tan \delta_0$ to obtain the rather complicated expression

$$\sin^2 \delta_0 = \frac{\beta^2 \cos^2 \beta - \beta \hat{\beta} \sin 2\beta \cot \hat{\beta} + \hat{\beta}^2 \sin^2 \beta \cot^2 \hat{\beta}}{\beta^2 + \hat{\beta}^2 \cot^2 \hat{\beta}}. \qquad (21.106)$$

In the low-energy limit $k \to 0$, we have $\eta \to (V_0)^{1/2}$ so $\beta \to \beta_0$, say, and therefore in this limit

$$\frac{d\sigma_0(\theta)}{d\Omega} \approx \frac{a^2(1 - \beta_0 \cot \beta_0)^2}{\beta_0^2 \cot^2 \beta_0} = a^2 \left(\frac{\tan \beta_0}{\beta_0} - 1 \right)^2. \tag{21.107}$$

This result is independent of θ, so

$$\sigma_0 = 4\pi a^2 \left(\frac{\tan \beta_0}{\beta_0} - 1 \right)^2, \tag{21.108}$$

as $k \to 0$.

THE PHASE SHIFTS FOR LARGE l: TEDIOUS DETAILS

When l is sufficiently large, L_l will differ little from its value when $V(r) = 0$, so $\hat{\beta} \approx \beta$ and we write

$$L_l = \beta \left[\frac{j_l'(\beta)}{j_l(\beta)} + \varepsilon_l \right], \tag{21.109}$$

where

$$|\varepsilon_l| \ll \left| \frac{j_l'(\beta)}{j_l(\beta)} \right|, \tag{21.110}$$

since then δ_l will be small. Then equation (21.100) becomes, after some algebra

$$\tan \delta_l = \frac{\varepsilon_l j_l^2}{\varepsilon_l n_l j_l - W[j_l, n_l]}, \tag{21.111}$$

and since the Wronskian is

$$W[j_l(\beta), n_l(\beta)] = \beta^{-2}, \tag{21.112}$$

(21.111) reduces to the form

$$\tan \delta_l = \frac{\varepsilon_l(\beta)\beta^2 j_l^2(\beta)}{\varepsilon_l(\beta)\beta^2 n_l(\beta) j_l(\beta) - 1}. \tag{21.113}$$

Now as $\beta \to 0$,

$$j_l(\beta) \to \frac{\beta^l}{(2l+1)!!}; \quad n_l(\beta) \to \frac{(2l-1)!!}{\beta^{l+1}}, \tag{21.114}$$

where for even values of n the double factorial is defined by

$$n!! = \prod_{k=1}^{n/2} (2k) = n(n-2)... 2, \tag{21.115}$$

and for odd values of n,

$$n!! = \prod_{k=1}^{(n+1)/2} (2k-1) = n(n-2)... 1. \tag{21.116}$$

Therefore

$$j_l'(\beta) \to \frac{l\beta^{l-1}}{(2l+1)!!}; \quad \frac{j_l'(\beta)}{j_l(\beta)} \to \frac{l}{\beta}, \tag{21.117}$$

and so $|\varepsilon_l| \ll l\beta^{-1}$. Furthermore [60],

$$\tan \delta_l = \frac{\varepsilon_l \beta^2 j_l^2(\beta)}{\varepsilon_l \beta^2 j_l(\beta)n_l(\beta) - 1} \approx -\frac{\varepsilon_l \beta^{2l+2}}{[(2l+1)!!]^2}. \tag{21.118}$$

Now

$$(2l+1)! = (2l+1)!!2^l (l!)^2, \tag{21.119}$$

as is readily verified, so that

$$\tan \delta_l \approx -\frac{\varepsilon_l \beta^{2l+2} 2^{2l} (l!)^2}{[(2l+1)!]^2} \approx \delta_l, \tag{21.120}$$

since $\beta \to 0$, and obviously

$$|\tan \delta_l| \approx |\delta_l|. \tag{21.121}$$

From (21.120), (21.121)

$$\ln |\delta_l| = \ln |\varepsilon_l| + 2\ln(\ln 2) + 2\ln(l!) + (2l+2)\ln \beta - 2\ln[(2l+1)!]. \tag{21.122}$$

The leading term in the asymptotic expansion of $\ln s!$ can be found using Stirling's formula in the form:

$$\ln s! \approx \frac{1}{2}\ln(2\pi) + \left(\frac{1}{2} + s\right)\ln s - s, \tag{21.123}$$

Neglecting constants and retaining only terms greater than $O(\ln l)$, we find eventually that

$$\ln |\delta_l| \approx \ln |\varepsilon_l| + 2l[\ln \beta + 1 - \ln 2] - 2l(\ln l), \tag{21.124}$$

and therefore

$$|\delta_l| \approx |\varepsilon_l|\beta^{2l} \left(\frac{e}{2l}\right)^{2l}, \tag{21.125}$$

and this tends to zero faster than exponentially, so the series converges rapidly for large l.

It's time to move on to pastures new!

Chapter Twenty-Two

The WKB(J) Approximation Revisited

The WKB approximation is a "semiclassical calculation" in quantum mechanics in which the wave function is assumed an exponential function with amplitude and phase that slowly varies compared to the de Broglie wavelength λ, and is then semiclassically expanded. While Wentzel, Kramers and Brillouin developed this approach in 1926, earlier in 1923, a mathematician, Harold Jeffreys, had already developed a more general method of approximating linear, second-order differential equations. ... Jeffreys is rarely given his proper credit.

[254]

In this paper, we look at specific cases of integrals and differential equations. Solutions of such equations commonly lack closed forms, and so asymptotic methods are often useful for determining their leading-order behavior. We will begin by developing two common methods to study integral asymptotics. The first, Laplace's Method, was presented by Pierre-Simon Laplace in 1774 and handles integrals of real-valued functions. The second extends this method to handle to complex integrals and was used first by Riemann in 1863; it is known as the Method of Steepest Descent. We then look at a specific differential equation containing a small parameter. The WKB Method will be used to find approximate solutions. The credit for derivation of the WKB Method is usually given toWentzel, Kramers, and Brillouin, who developed it in 1926. Although it is sometimes referred to by other names when credit is given to those who formulated the method previously, we refer to it as the WKB Method in this paper. Its primary usefulness is seen in physics to derive approximate solutions to Schrödinger's Equation.

[255]

We are now in a position to revisit this approximation in light of earlier chapters (especially Chapter 21) that dealt with potential wells and barriers from a broad perspective. We have examined square wells and barriers and now proceed to the somewhat more complicated case of a triangular barrier (primarily because of its intimate connection with Airy functions).

22.1 THE CONNECTION FORMULAS REVISITED: AN ALTERNATIVE APPROACH

We start in a general way by recapitulating the basic details. Given the Airy differential equation in the form

$$\frac{d^2\psi}{dy^2} - y\psi = 0, \tag{22.1}$$

we have the following linearly independent solutions for real y (with $\xi = 2y^{3/2}/3$) [28]:

$$\text{Ai}(y) = \frac{1}{2}\pi^{-1/2}y^{-1/4}e^{-\xi}\left[1 + O(\xi^{-1})\right]; \tag{22.2}$$

$$\text{Bi}(y) = \pi^{-1/2}y^{-1/4}e^{\xi}\left[1 + O(\xi^{-1})\right]. \tag{22.3}$$

Note also that [66]

$$\text{Ai}(-y) = \pi^{-1/2}y^{-1/4}\left[\sin(\xi + \pi/4) + O(\xi^{-1})\right]; \tag{22.4}$$

$$\text{Bi}(-y) = \pi^{-1/2}y^{-1/4}\left[\cos(\xi + \pi/4) + O(\xi^{-1})\right]. \tag{22.5}$$

If instead the argument is a complex number, z say, then there are restrictions on $\arg z$; $|\arg z| < \pi$ for (22.2), and $|\arg z| < 2\pi/3$ for (22.4).

The time-independent Schrödinger equation is, in the suppressed notation used here (i.e., with $\varepsilon = 1$),

$$\frac{d^2\psi}{dx^2} + (E - V)\psi = 0. \tag{22.6}$$

We have previously defined the function $p(x)$ by

$$p(x) = E - V(x), \tag{22.7}$$

Expanding about a (simple) zero of $p(x)$ at $x = x_0$, we have

$$p(x) = p(x_0) + p'(x_0)(x - x_0) + O[(x - x_0)^2] \tag{22.8}$$

$$\approx p'(x_0)(x - x_0) = -V'(x_0)(x - x_0). \tag{22.9}$$

Hence, if we set $x_0 = 0$, without loss of generality,

$$V(x) \approx E + V'(0)x. \tag{22.10}$$

Then equation (22.6) is approximated by

$$\frac{d^2\psi}{dx^2} - V'(0)x\psi = 0, \tag{22.11}$$

so if

$$y = \left[V'(0)\right]^{1/3}x \equiv ax, \tag{22.12}$$

we recover the original form (22.1). As would be expected, the "patching wave function" (see Section 7.1.2) linking the regions $x < 0$ and $x > 0$ is the linear combination

$$\psi_p(x) = \alpha \, \text{Ai}(ax) + \beta \, \text{Bi}(ax). \tag{22.13}$$

In terms of the function

$$p^{1/2}(x) = [E - V(x)]^{1/2} \approx [-V'(0)x]^{1/2} = a^{3/2}(-x)^{1/2} \tag{22.14}$$

$$= \pm i a^{3/2} |x|^{1/2} = \pm i a^{3/2} x^{1/2} \quad \text{for } x > 0, \tag{22.15}$$

$$\int_0^x |p(\xi)|^{1/2} \, d\xi \approx a^{3/2} \int_0^x |\xi|^{1/2} \, d\xi = \frac{2}{3} (ax)^{3/2}, \tag{22.16}$$

and therefore the WKB(J) 'wave function' is

$$\psi(x) \approx A a^{-3/4} x^{-1/4} \exp\left[-\frac{2}{3}(ax)^{3/2}\right], \tag{22.17}$$

for some constant A. The large-argument asymptotic forms of the Airy functions in the region $x > 0$ (region II in Figure 7.2, Chapter 7) imply that the patching wavefunction is

$$\psi_p(x) \approx \frac{\alpha}{2} \pi^{-1/2}(ax)^{-1/4} \exp\left[-\frac{2}{3}(ax)^{3/2}\right] + \beta \pi^{-1/2}(ax)^{-1/4} \exp\left[\frac{2}{3}(ax)^{3/2}\right]. \tag{22.18}$$

Comparing the expressions (22.17) and (22.18) indicates that

$$\alpha = \left(\frac{4\pi}{a}\right)^{1/2} A, \tag{22.19}$$

and $\beta = 0$.

Next we repeat the procedure for the overlapping region II in $x < 0$. Then

$$\int_x^0 p(\xi) d\xi \approx \frac{2}{3}(-ax)^{3/2}, \tag{22.20}$$

so that the WKB(J) wavefunction is

$$\psi(x) \approx a^{-3/4}(-x)^{-1/4} \left\{ B \exp\left[\frac{2i}{3}(-ax)^{3/2}\right] + C \exp\left[-\frac{2i}{3}(-ax)^{3/2}\right] \right\}. \tag{22.21}$$

Now we employ the asymptotic forms of the Airy functions for large negative y. From equation (22.13) (with $\beta = 0$) we have

$$\psi_p(x) \approx \alpha \pi^{-1/2}(-ax)^{-1/4} \sin\left[\frac{2}{3}(-ax)^{3/2} + \frac{\pi}{4}\right] \tag{22.22}$$

$$= \frac{\alpha}{2i} \pi^{-1/2}(-ax)^{-1/4} \left\{ \exp\left(i\left[\frac{2}{3}(-ax)^{3/2} + \frac{\pi}{4}\right]\right) \right.$$

$$\left. - \exp\left(-i\left[\frac{2}{3}(-ax)^{3/2} + \frac{\pi}{4}\right]\right) \right\}. \tag{22.23}$$

Comparing the forms (22.21) and (22.23), it follows that

$$\frac{\alpha}{2i}\pi^{-1/2}e^{i\pi/4} = B\alpha^{-1/2}; \quad \frac{-\alpha}{2i}\pi^{-1/2}e^{-i\pi/4} = C\alpha^{-1/2}. \tag{22.24}$$

Using (22.19), we find the coefficients to be

$$B = -ie^{i\pi/4}A; \quad C = ie^{-i\pi/4}A. \tag{22.25}$$

Hence the WKB(J) wavefunction (for $V'(0) > 0$ in particular) is

$$\psi(x) \approx [p(x)]^{-1/4}\left\{B\exp\left(i\int_x^0 p(\xi)d\xi\right) + C\exp\left(-i\int_x^0 p(\xi)d\xi\right)\right\}, \quad x < 0, \tag{22.26}$$

and

$$\psi(x) \approx A\,|p(x)|^{-1/4}\exp\left(-\int_0^x p(\xi)d\xi\right), \quad x > 0. \tag{22.27}$$

Therefore in terms of A alone, and shifting the origin for a turning point at $x = x_0$, this becomes

$$\psi(x) \approx 2A\,[p(x)]^{-1/4}\sin\left(\int_x^{x_0} p(\xi)d\xi + \frac{\pi}{4}\right), \quad x < x_0, \tag{22.28}$$

and

$$\psi(x) \approx A\,|p(x)|^{-1/4}\exp\left(-\int_{x_0}^x p(\xi)d\xi\right), \quad x > x_0. \tag{22.29}$$

An alternative representation, based on [135] can be found in [136]. In this case, with $f(x) \equiv E - V(x) = p^2(x)$,

$$f(x) < 0, x > x_0; \quad f(x) > 0, \quad x < x_0, \tag{22.30}$$

and the connection formulas are summarized as

$$(f)^{-1/4}\exp\left[\pm i\left(L + \frac{\pi}{4}\right)\right] \longleftrightarrow (-f)^{-1/4}\left[e^M \pm \frac{i}{2}e^{-M}\right], \tag{22.31}$$

where

$$M = \int_{x_0}^x (-f)^{1/2}\,dx; \quad L = \int_x^{x_0} f^{1/2}dx. \tag{22.32}$$

This connects neatly (to use a phrase) with the corresponding result (7.57) in Chapter 7.

22.2 TUNNELING: A PHYSICAL DISCUSSION

Using the square well/barrier scattering problems discussed in Chapter 21 as a guide, suppose that waves are scattered from a one-dimensional barrier with compact support. Sufficiently far to the left (such that $V(x) \approx 0$) we may write the wavefunction as a combination of complex exponentials:

$$\psi_<(x) \approx Ae^{ikx} + Be^{-ikx}, \tag{22.33}$$

where $k = E^{1/2}$. Sufficiently far to the right the transmitted wave is

$$\psi_>(x) \approx C e^{ikx}, \tag{22.34}$$

and the transmission probability may be defined as

$$T = \left| \frac{C}{A} \right|^2. \tag{22.35}$$

But in the so-called nonclassical region for which $E < V(x)$, the (bounded) WKB(J) approximation is, for a sufficiently high and wide barrier,

$$\psi(x) \approx D |E - V(x)|^{-1/4} \exp\left(-\int_{x_1}^{x_2} |E - V(x)|^{1/2}\, dx \right), \tag{22.36}$$

where x_1 and x_2 are the local turning points (i.e. where $V(x_1) = E = V(x_1)$). By appropriate matching, the transmission coefficient is found to be

$$T \approx \exp\left(-2 \int_{x_1}^{x_2} [V(x) - E]^{1/2}\, dx \right). \tag{22.37}$$

Exercise: Derive equation (22.37).

22.3 A TRIANGULAR BARRIER

Let us set $\lambda = 1$ without loss of generality in equation (7.68) so that, as above,

$$\psi''(x) - p(x)\, \psi(x) = 0, \tag{22.38}$$

where

$$p(x) \equiv V(x) - E \equiv V(x) - \alpha^2. \tag{22.39}$$

Suppose that the potential $V(x)$ is a symmetrical triangular barrier such that

$$\begin{align}
V(x) &= 0, \quad (x \le -V_0/a); \tag{22.40}\\
&= V_0 + ax, \quad (-V_0/a < x < 0); \tag{22.41}\\
&= V_0 - ax, \quad (0 < x < V_0/a); \tag{22.42}\\
&= 0, \quad (x > V_0/a). \tag{22.43}
\end{align}$$

Then in the four resulting regimes, (22.38) takes the various forms

$$\begin{align}
\psi'' + \alpha^2 \psi &= 0, \quad (|x| \ge V_0/a); \tag{22.44}\\
\psi'' - (\beta^2 + ax)\, \psi &= 0, \quad (-V_0/a < x < 0); \tag{22.45}\\
\psi'' - (\beta^2 - ax)\, \psi &= 0, \quad (-V_0/a < x < 0), \tag{22.46}
\end{align}$$

where $\beta^2 = V_0 - \alpha^2$. The new ξ substitution,

$$\xi = a^{1/3}\left(x + \frac{\beta^2}{a} \right), \tag{22.47}$$

reduces equation (22.45) to a convenient form, namely, Airy's differential equation:

$$\frac{d^2\psi}{d\xi^2} - \xi\psi = 0, \tag{22.48}$$

with general solution

$$\psi\left(\xi(x)\right) = C_1 \operatorname{Ai}(\xi) + C_2 \operatorname{Bi}(\xi), \tag{22.49}$$

in terms of Airy integrals, this time in the form

$$\operatorname{Ai}(\xi) = \frac{1}{\pi} \int_0^\infty \cos\left(\xi s + \frac{1}{3}s^3\right) ds, \tag{22.50}$$

$$\operatorname{Bi}(\xi) = \frac{1}{\pi} \int_0^\infty \left\{ \exp\left(\xi s - \frac{1}{3}s^3\right) + \sin\left(\xi s + \frac{1}{3}s^3\right) \right\} ds. \tag{22.51}$$

A similar substitution

$$\mu = a^{1/3}\left(-x + \frac{\beta^2}{a}\right), \tag{22.52}$$

reduces equation (22.46) to

$$\frac{d^2\psi}{d\mu^2} - \mu\psi = 0. \tag{22.53}$$

The solutions may then be written as [136]

$$\psi = e^{i\alpha x} + A e^{-i\alpha x}, \quad (x \le -V_0/a); \tag{22.54}$$

$$= C\operatorname{Ai}(\xi) + D\operatorname{Bi}(\xi), \quad (-V_0/a < x < 0); \tag{22.55}$$

$$= E\operatorname{Ai}(\mu) + F\operatorname{Bi}(\mu), \quad (0 < x < V_0/a); \tag{22.56}$$

$$= B e^{i\alpha x}, \quad (x > V_0/a). \tag{22.57}$$

This system has a unique solution of the set of constants $\{A, B, C, D, E, F\}$ if we require continuity of ψ and its derivative at $x = 0$ and $x = \pm V_0/a$, cumbersome though they can be. After some algebra the transmission coefficient T is found to be [66]

$$T = |B|^2 = \left\{ 1 + \pi^2 \left(\operatorname{Ai}[\operatorname{Ai}]' + \operatorname{Bi}[\operatorname{Bi}]'\right)^2 \right\}^{-1}, \tag{22.58}$$

the arguments being $(\beta^3/a)^{2/3}$ and $-(\alpha^3/a)^{2/3}$. For tall and thin barriers (i.e., for $a \ll \beta^3$ and $a \ll \alpha^3$) the asymptotic expansions

$$\operatorname{Ai}(y) \sim \frac{1}{2\pi^{1/2}} y^{-1/4} \exp\left(-\frac{2}{3}y^{3/2}\right), \tag{22.59}$$

and

$$\operatorname{Bi}(y) \sim \frac{1}{\pi^{1/2}} y^{1/4} \exp\left(\frac{2}{3}y^{3/2}\right), \tag{22.60}$$

may be used to give

$$T \approx \exp\left(-\frac{8\beta^3}{3a}\right). \tag{22.61}$$

If w is the width of the barrier at the height of penetration E, then by simple proportion $w = 2a^{-1}\beta^2$, and (22.61) becomes

$$T \approx \exp\left(-\frac{4}{3}\beta w\right). \tag{22.62}$$

Use of a General WKB(J) Result for the Triangular Barrier

Using the connection formulas discussed above, it may be shown in general that

$$T = \left(e^I - \frac{1}{4}e^{-I}\right)^{-2}, \tag{22.63}$$

where in reference to Figure 22.1,

$$I = \int_{x_1}^{x_2} [V(x) - E]^{1/2}\,dx, \tag{22.64}$$

so that if I is large (as would be the case for a high and wide barrier) we recover the approximation (22.37) namely,

$$T \approx \exp\left\{-2\int_{x_1}^{x_2} [V(x) - E]^{1/2}\,dx\right\}. \tag{22.65}$$

Hence for the triangular barrier

$$T \approx \exp\left[-2\left\{\int_{-w/2}^{0}\left[a\left(x + \frac{w}{2}\right)\right]^{1/2}dx + \int_{0}^{w/2}\left[a\left(\frac{w}{2} - x\right)\right]^{1/2}dx\right\}\right] \tag{22.66}$$

$$= \exp\left(-\frac{4}{3}\beta w\right) \tag{22.67}$$

as before. For completeness, the corresponding solution for a rectangular barrier is shown in Figure 22.2. Compare this with the exact solution for this problem, discussed in the previous Chapter 21.

Exercise: Derive (22.63).

Note also that for a spherically symmetric potential the transmission coefficient includes the centripetal potential; thus (22.37) becomes

$$T \approx \exp\left\{-2\int_{r_1}^{r_2}\left[V(r) + \frac{l(l+1)}{r^2} - E\right]^{1/2}dx\right\}. \tag{22.68}$$

22.4 MORE NUTS AND BOLTS

In this section a somewhat different, rather circuitous but informative approach to the WKB(J) formulation is attempted (I was told as a child that "there are many ways to skin a cat." On reflection, that is a rather ghastly phrase and a questionable one!).

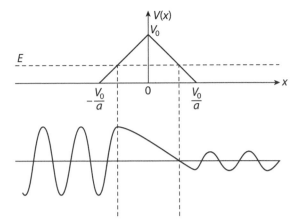

Figure 22.1 The triangular barrier and corresponding WKB(J) solution.

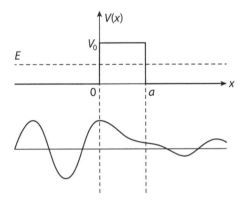

Figure 22.2 The rectangular barrier and corresponding WKB(J) solution.

Given the definition (21.47),

$$F(r) = k^2 - V(r) - \frac{l(l+1)}{r^2}, \qquad (22.69)$$

equation (21.40) can be written as

$$\frac{d^2 u_l(r)}{dr^2} + F(r)u_l(r) = 0, \quad r > 0. \qquad (22.70)$$

We already know that the asymptotic form of the solution for which $u_l(0) = 0$ is

$$u_l(r) \sim \sin\left(kr - \frac{l\pi}{2} + \delta_l\right), \qquad (22.71)$$

provided that $V(r) \to 0$ faster than r^{-2} for large r. If the potential is slowly-varying such that F changes little over a distance $|F|^{1/2}$, then in the spirit of the WKB(J) approach for $F > 0$, we can regard the quantity $2\pi / F^{1/2}$ as a local wavelength and according seek a solution of the form

$$u_l(r) = B(r) \exp\left(\pm i \int F^{1/2}(r) dr\right).$$ (22.72)

This ansatz results in the following differential equation(s) for B:

$$B'' \pm 2i B' F^{1/2} \pm \frac{i}{2} F^{-1/2} F' B = 0,$$ (22.73)

or

$$\frac{B''}{F^{1/2}} \pm 2i B' \pm i \frac{F'}{2F} B = 0.$$ (22.74)

A solution for B would be much simpler to obtain if the term B'' could be ignored, so it seems worthwhile to see under what circumstances this could be done (with apologies to my pure mathematician colleagues!). It has been said that applied mathematicians change equations they cannot solve into ones they can. Very sensible...

Now $2\pi B''/F^{1/2}$ is approximately the change in B', $\Delta B'$ say, over a local wavelength, and since F by assumption varies little over the same distance, it is very reasonable to assume the same for B and B' [66]. Then $\Delta B' \ll B'$ and therefore an approximate solution for B is

$$B(r) = B_0 F^{-1/4},$$ (22.75)

so that

$$u_l(r) \approx B_0 F^{-1/4} \exp\left(\pm i \int F^{1/2}(r) dr\right).$$ (22.76)

If $F < 0$, then under these same assumptions

$$u_l(r) \approx B_0 G^{-1/4} \exp\left(\pm \int G^{1/2}(r) dr\right),$$ (22.77)

where $G = -F$. As noted in Chapter 21, there may be a unique value $r = r_0$ such that $F(r_0) = 0$, because $F < 0$ for small r and $F > 0$ for large r. Clearly, in a small interval around $r = r_0$ the above assumptions are invalid and the solutions (22.76) and (22.77) are unbounded, so some kind of matching is required between solutions that are zero at the origin and bounded at infinity. To attempt this, we use what is called a *Langer transformation* (to be discussed in more detail in Section 22.9.1) and make the following changes of dependent and independent variables:

$$u_l(r) = r^{1/2} p_l(r); \quad r = e^{\mu}.$$ (22.78)

Then (22.70) takes the form

$$\frac{d^2 p_l(\mu)}{d\mu^2} + \left[\{k^2 - V(\mu)\} e^{2\mu} - \left(l + \frac{1}{2}\right)^2 \right] p_l(\mu) = 0,$$ (22.79)

so defining

$$F_\mu(\mu) = [k^2 - V(\mu)] e^{2\mu} - \left(l + \frac{1}{2}\right)^2,$$ (22.80)

equation (22.79) is simply

$$\frac{d^2 p_l(\mu)}{d\mu^2} + F_\mu(\mu) p_l(\mu) = 0. \tag{22.81}$$

Since $r \to 0^+ \Rightarrow \mu \to -\infty$, the condition $u_l(0) = 0$ is satisfied if $p_l(-\infty)$ is bounded. Now $F(r_0) = 0$ implies that $F_\mu(\mu) = 0$ for some $\mu = \mu_0$, so if we replace $F_\mu(\mu)$ by its linear approximation

$$F_\mu(\mu) \approx b^2 (\mu - \mu_0) \tag{22.82}$$

near μ_0, (22.81) is approximated by the equation

$$\frac{d^2 p_l}{d\mu^2} + b^2 (\mu - \mu_0) p_l = 0. \tag{22.83}$$

After some rearrangement this equation can be put into the form of Bessel's equation of the first kind of order one-third; the linearly independent solutions are

$$p_l^+ = (\mu - \mu_0)^{1/2} J_{1/3} \left[\frac{2}{3} b (\mu - \mu_0)^{3/2} \right] \tag{22.84}$$

and

$$p_l^- = (\mu - \mu_0)^{1/2} J_{-1/3} \left[\frac{2}{3} b (\mu - \mu_0)^{3/2} \right]. \tag{22.85}$$

These Bessel functions are related to the Airy functions $\mathrm{Ai}(x)$ and $\mathrm{Bi}(x)$, as indicated in earlier chapters (e.g., Chapters 5–7). In this section, however, we will continue to work with the Bessel function form.

If we define

$$\beta = \int_{\mu_0}^{\mu} F_\xi^{1/2}(\xi) d\xi, \tag{22.86}$$

then near $\mu = \mu_0$, $\beta \approx (2b/3)(\mu - \mu_0)^{3/2}$, and so in terms of constants A and B, the general solution of (22.81) is, for $\mu > \mu_0$

$$p_l = \beta^{1/2} F_\mu^{-1/4} \left[A J_{1/3}(\beta) + B J_{-1/3}(\beta) \right]. \tag{22.87}$$

As $\mu \to \mu_0$ this reduces to a linear combination of the solutions (22.84) and (22.85). Using the asymptotic form

$$J_n(x) \sim \left(\frac{2}{\pi x} \right)^{1/2} \cos \left(x - \frac{n\pi}{2} - \frac{\pi}{4} \right), \tag{22.88}$$

it follows that

$$p_l \sim \left(\frac{2}{\pi} \right)^{1/2} F_\mu^{-1/4} \left[A \cos \left(\beta - \frac{5\pi}{12} \right) + B \cos \left(\beta - \frac{\pi}{12} \right) \right]$$

$$= A' F_\mu^{-1/4} \exp \left(i \int F^{1/2}(\mu) d\mu \right) + B' F_\mu^{-1/4} \exp \left(-i \int F^{1/2}(\mu) d\mu \right). \tag{22.89}$$

This means that, asymptotically it has the linear combination of solutions as required by (22.76). If $\mu < \mu_0$, then $F_\mu < 0$ and therefore both $F_\mu^{1/2}$ and β are imaginary quantities with branch points at μ_0. We choose the branch specified by $F_\mu^{1/2} = \left| F_\mu^{1/2} \right| \exp(i\pi/2)$ near $\mu = \mu_0$, where $F_\mu^{1/2} = b(\mu - \mu_0)^{1/2}$, so

$$\beta \approx \frac{2b}{3} (\mu - \mu_0)^{3/2} = \frac{2}{3b^2} F_\mu^{3/2} = |\beta| \, e^{3\pi i/2} = -i \, |\beta| . \tag{22.90}$$

Then for $\mu < \mu_0$ the solution (22.87) changes to

$$p_l = |\beta|^{1/2} \, |F_\mu|^{-1/4} \, e^{-i\pi/4} \left[A J_{1/3} \left(|\beta| \, e^{3\pi i/2} \right) + B J_{-1/3} \left(|\beta| \, e^{3\pi i/2} \right) \right] . \tag{22.91}$$

Using the asymptotic form

$$J_n \left(|\beta| \, e^{3\pi i/2} \right) \sim \frac{1}{2} \left(\frac{2}{\pi \, |\beta|} \right)^{1/2} \exp \left[i\pi (n - 1/4) \right]$$

$$\times \left\{ \exp \left[|\beta| + \frac{i\pi}{2} \left(n + \frac{1}{2} \right) \right] + \exp \left[- |\beta| - \frac{i\pi}{2} \left(n + \frac{1}{2} \right) \right] \right\} , \tag{22.92}$$

and after some reduction we find that

$$p_l \to \frac{1}{2} \left(\frac{2}{\pi} \right)^{1/2} |F_\mu|^{-1/4} e^{-3\pi i/4} \left[(B - A) \, e^{|\beta|} + \left(A e^{i\pi/6} + B e^{-i\pi/6} \right) e^{-|\beta|} \right] , \tag{22.93}$$

as $\mu \to -\infty$. For this solution to be bounded as $\mu \to -\infty$ (and hence $u_l(0) = 0$), $A = B$, and then (22.87) has the following asymptotic behavior as $\mu \to +\infty$:

$$p_l \sim \frac{3^{1/2}}{4} \left(\frac{2}{\pi} \right)^{1/2} |F_\mu|^{-1/4} \tilde{A} \cos \left(\beta - \frac{\pi}{4} \right) , \tag{22.94}$$

where $\tilde{A} = A \exp(-3\pi i/4)$. Reverting to the original variable r, we have that

$$F_\mu(\mu) = \left[r^2 \left\{ k^2 - V(r) \right\} - \left(l + \frac{1}{2} \right)^2 \right] = r^2 F(r) . \tag{22.95}$$

As $r \to \infty$,

$$\left[r^2 F(r) \right]^{-1/4} \to (kr)^{-1/2} , \tag{22.96}$$

$$\beta = \int_{\mu_0}^\mu F_\xi^{1/2}(\xi) d\xi = \int_{r_0}^r \left[k^2 - V(\xi) - \frac{(l + 1/2)^2}{\xi^2} \right]^{1/2} d\xi , \tag{22.97}$$

so that (since $\cos(\beta - \pi/4) = \sin(\beta + \pi/4)$)

$$u_l(r) = r^{1/2} p_l(r) \sim B \sin \left\{ \int_{r_0}^r \left[k^2 - V(\xi) - \frac{(l + 1/2)^2}{\xi^2} \right]^{1/2} d\xi + \frac{\pi}{4} \right\} , \tag{22.98}$$

where

$$B = \frac{3^{1/2}}{4} \left(\frac{2}{\pi} \right)^{1/2} k^{-1/2} \tilde{A} . \tag{22.99}$$

22.4.1 The Phase Shift

Recall that, after invoking the Langer transformation in the previous section, r_0 is now the (assumed) unique positive solution of $G_1(r) = 0$ (say), where

$$G_1(r) = k^2 - V(r) - \frac{(l + 1/2)^2}{r^2}. \qquad (22.100)$$

In the absence of a scattering potential the solution (22.98) would take the asymptotic form

$$u_l(r) \sim B' \sin \left\{ \int_{r_0'}^{r} \left[k^2 - \frac{(l + 1/2)^2}{\xi^2} \right]^{1/2} d\xi + \frac{\pi}{4} \right\}, \qquad (22.101)$$

where $r_0' = (l + 1/2)/k$, and $B' \neq B$ in general. The change in phase δ_l induced by the potential $V(r)$ can therefore be written as the difference of integrals:

$$\delta_l = \int_{r_0}^{r} \left[k^2 - V(\xi) - \frac{(l + 1/2)^2}{\xi^2} \right]^{1/2} d\xi - \int_{r_0'}^{r} \left[k^2 - \frac{(l + 1/2)^2}{\xi^2} \right]^{1/2} d\xi. \qquad (22.102)$$

It seems somewhat problematical that δ_l is the difference of two integrals, each of which becomes infinite as $r \to \infty$; this will be partially addressed below.

Note that (22.98) may be rewritten as

$$u_l(r) \sim B \sin \left\{ \int_{r_0}^{r} \left[G_1^{1/2}(\xi) - k \right] d\xi + k(r - r_0) + \frac{\pi}{4} \right\}. \qquad (22.103)$$

If we compare the argument of the sine function in both equations (22.71) and (22.103) we see that an alternative representation of δ_l is

$$\delta_l = \frac{\pi}{2} \left(l + \frac{1}{2} \right) - kr_0 + \int_{r_0}^{\infty} \left[G_1^{1/2}(r) - k \right] dr. \qquad (22.104)$$

If $G_1(r)$ has more than one zero, it is often only necessary to choose r_0 to be the outermost one, but failing that, the contributions to the phase from the inner regions need to be identified.

22.4.2 Some Comments on Convergence

Let us briefly examine some conditions under which the phases δ_l are finite. Since $V(r) \to 0$ as $r \to \infty$, we choose a radial distance R such that, for $r > R$,

$$V(r) \ll \left[k^2 - \frac{(l + 1/2)^2}{r^2} \right]^{1/2}. \qquad (22.105)$$

Therefore

$$\left[k^2 - V(r) - \frac{(l + 1/2)^2}{r^2} \right]^{1/2} \approx \left[k^2 - \frac{(l + 1/2)^2}{r^2} \right]^{1/2}$$

$$- \frac{V(r)}{2 \left[k^2 - r^{-2} (l + 1/2)^2 \right]^{1/2}}, \qquad (22.106)$$

so for $r > R$ equation (22.102) is reduced to

$$\delta_l \approx -\int_R^\infty \frac{V(r)}{2\left[k^2 - r^{-2}(l+1/2)^2\right]^{1/2}}dr. \qquad (22.107)$$

This integral is convergent provided $V(r) \to 0$ faster than r^{-1}, thereby excluding the Coulomb potential, for which $V(r) \sim r^{-1}$. In that important case δ_l becomes logarithmically infinite (see (22.167)). But even if all the δ_l *are* finite, the series (18.53) may still diverge. From equation (22.102) note that the lower limits of the integrals, namely r_0 and r_0', are of $O(l/k)$, so they are large for large l. If $V(r) \to 0$ faster than r^{-1}, the phase would be expected to be small and decrease as l increases. Then instead of the result (22.107), a similar argument yields [66]

$$\delta_l \approx -\int_{R_0}^\infty \frac{V(r)}{2\left[k^2 - r^{-2}(l+1/2)^2\right]^{1/2}}dr, \qquad (22.108)$$

where

$$R_0 = \frac{l+1/2}{k}. \qquad (22.109)$$

In particular, if $V(r) \sim Ar^{-n}$, then

$$\delta_l \approx -\frac{AR_0^{1-n}}{2k}\int_1^\infty \frac{\xi^{1-n}}{\left(\xi^2-1\right)^{1/2}}d\xi = \frac{A'k^{n-2}}{l^{n-1}}, \qquad (22.110)$$

where A' is another constant.

Exercise: Establish result (22.110).

It can be shown further using order-of-magnitude arguments that if $V(r) \approx Ar^{-n}$, then for sufficiently large values of l the terms of the series (18.53) may be approximated by

$$\frac{4\pi}{k^2}(2l+1)\delta_l^2 \propto l^{3-2n}, \qquad (22.111)$$

which converges provided $n > 2$. Thus if $V(r) \to 0$ more rapidly than r^{-2}, the total cross section will be finite. If, however $V(r) \to 0$ more rapidly than r^{-1} but not more rapidly than r^{-2}, the individual phases the scattering amplitude $f(\theta)$ can still be determined (and hence the angular distribution of the scattered particles). Nevertheless, it is possible for the series (18.53) to converge while the scattering amplitude $f(\theta)$ diverges for $\theta \approx 0$. Further details may be found in [66].

22.4.3 The Transition to Classical Scattering

How do the expressions for scattering in the quantum realm transform to their classical counterparts? Simplistically (in view of several discussions in Part I) we might expect that as the wavelength of the incident beam becomes vanishingly small compared to the dimensions of the scattering field, the wave morphs into the particle. Here we continue with the semiclassical approach to this concept. Nonrigorously, we may write the harmonic expansion of the delta "function" as

$$\delta(1 - \cos\theta) = \frac{1}{2} \sum_{l=0}^{\infty} (2l + 1) P_l(\cos\theta). \tag{22.112}$$

This can be verified by multiplying both sides by $\sin\theta\, P_l(\cos\theta)$ and integrating over θ, or established using the result

$$\delta(y - x) = \frac{1}{2} \sum_{l=0}^{\infty} (2l + 1) P_l(y) P_l(x), \tag{22.113}$$

for $x \in [-1, 1]$ and $y \in [-1, 1]$, noting that $P_l(1) = 1$ for all l [114]. The expression (22.112) means that the sum vanishes for all $\theta \neq 0$, so that the -1 term in $\left(e^{2i\delta_l} - 1\right)$ can be omitted when $\theta \neq 0$. Therefore, excluding the case of $\theta = 0$, and recalling the scattering amplitude $f(\theta)$, we have

$$f(\theta) = \frac{1}{2ik} \sum_{l=0}^{\infty} e^{2i\delta_l} (2l + 1) P_l(\cos\theta). \tag{22.114}$$

Because this infinite sum is over l, and classically the wavelength is exceedingly small compared with the range of interaction, many values of l will contribute to $f(\theta)$. In view of this we are justified in focusing on the contributions from large values of l and use the asymptotic form of $P_l(\cos\theta)$ for those values. Thus we have

$$P_l(\cos\theta) \sim \left(\frac{2}{l\pi\,\sin\theta}\right)^{1/2} \sin\left[\left(l + \frac{1}{2}\right)\theta + \frac{\pi}{4}\right], \tag{22.115}$$

and therefore, for $\theta \neq 0$,

$$P_l(\cos\theta) \approx -\frac{1}{k} \sum_l \left(\frac{l}{2\pi\,\sin\theta}\right)^{1/2} (A + B), \tag{22.116}$$

where

$$A = \exp\left(i\left\{2\delta_l + \left(l + \frac{1}{2}\right)\theta + \frac{\pi}{4}\right\}\right); \tag{22.117}$$

$$B = \exp\left(i\left\{2\delta_l - \left(l + \frac{1}{2}\right)\theta - \frac{\pi}{4}\right\}\right). \tag{22.118}$$

Generally, a lot of these large-l terms will cancel because of the extremely rapid oscillations in the exponential terms. However, in the neighborhood of a stationary value of either exponent, constructive interference will occur (i.e., using the stationary phase principle again). For notational convenience we define $\sigma^\pm$ by

$$\sigma^\pm = 2\delta_l \pm \left(l + \frac{1}{2}\right)\theta \pm \frac{\pi}{4}, \tag{22.119}$$

and note that the greatest contributions to the scattering amplitude will be made by those values of l near $l = l_0^\pm$, where $l_0^\pm$ are l values for which $\partial\sigma^\pm/\partial l = 0$, that is, for which

$$\frac{\partial\delta_l}{\partial l} = \mp\frac{\theta}{2}. \tag{22.120}$$

Then from (22.104) we find

$$\int_{r_0}^{\infty} \frac{\partial}{\partial l} \left[k^2 - V(r) - \frac{(l+1/2)^2}{r^2} \right]^{1/2} dr = -\frac{1}{2} \left(\pi \pm \theta \right). \tag{22.121}$$

If we define $\hat{J} = (l + 1/2)$ and $E = k^2$, this condition can be rewritten as

$$\int_{r_0}^{\infty} \frac{\partial}{\partial \hat{J}} \left[E - V(r) - \frac{\hat{J}^2}{r^2} \right]^{1/2} dr = -\frac{1}{2} \left(\pi \pm \theta \right), \tag{22.122}$$

i.e.

$$\int_{r_0}^{\infty} r^{-1} \left[\frac{r^2 E}{\hat{J}^2} \left(1 - \frac{V(r)}{E} \right) - 1 \right]^{-1/2} dr = \frac{1}{2} \left(\pi \pm \theta \right). \tag{22.123}$$

This is exactly the form of the scattering integral found in Chapters 15 and 16 on classical and gravitational scattering. Furthermore,

$$\frac{E}{\hat{J}^2} = \frac{p^2}{\hat{J}^2} = \frac{1}{\hat{b}^2}, \tag{22.124}$$

where p is the momentum of the incident particle and $\hat{b}$ is the impact parameter. Choosing the negative sign in (22.123) (for reasons discussed below), the scattering angle takes the following succinct form

$$\theta = \pi - 2 \int_{r_0}^{\infty} \frac{dr}{r \varpi (r)}, \tag{22.125}$$

where

$$\varpi (r) = \left[\frac{r^2}{\hat{b}^2} \left(1 - \frac{V(r)}{E} \right) - 1 \right]^{1/2}. \tag{22.126}$$

This clarifies the relationship between the most probable angle of scattering (quantum mechanically) and the classical scattering field. So what about the positive sign in (22.123)? As pointed out in [66], the positive sign corresponds to negative values of l so that contributions from the second exponential term in (22.116) never show constructive interference.

22.5 COULOMB SCATTERING: THE ASYMPTOTIC SOLUTION

Prior to examining aspects of the WKB(J) approximation for the Coulomb potential in the next section, we will derive the asymptotic solutions for the wave function and the phase. These two results will then be compared. To study scattering by a Coulomb potential, in particular the total scattering cross section, it is much simpler to express the Schrödinger equation in parabolic cylindrical rather than spherical polar coordinates, though we will examine the latter also. This is because of the strong dependence of the solution on the variable ξ, as demonstrated below.

22.5.1 Parabolic Cylindrical Coordinates (ξ, η, ϕ)

In terms of the spherical coordinate system variable (r, θ, ϕ) the corresponding parabolic cylindrical coordinates are

$$\xi = r - z = r(1 - \cos\theta) = 2r \sin^2\left(\frac{\theta}{2}\right); \qquad (22.127)$$

$$\eta = r + z = r(1 + \cos\theta) = 2r \cos^2\left(\frac{\theta}{2}\right); \quad \phi = \phi, \qquad (22.128)$$

where $0 \leq \xi, \eta < \infty$, and $\phi \in (0, 2\pi)$. The surfaces $\xi = $ constant and $\eta = $ constant are paraboloids of revolution about the z-axis with focus at the origin. It is an orthogonal coordinate system, with surfaces of constant ϕ being planes through the polar axis, just as in the spherical system. The Laplacian takes the form

$$\nabla^2 \equiv \frac{4}{\eta + \xi}\left[\frac{\partial}{\partial\xi}\left(\xi\frac{\partial}{\partial\xi}\right) + \frac{\partial}{\partial\eta}\left(\eta\frac{\partial}{\partial\eta}\right) + \frac{\eta + \xi}{4\eta\xi}\frac{\partial^2}{\partial\phi^2}\right], \qquad (22.129)$$

and so the time-independent Schrödinger equation becomes

$$\left\{\frac{4}{\eta + \xi}\left[\frac{\partial}{\partial\xi}\left(\xi\frac{\partial}{\partial\xi}\right) + \frac{\partial}{\partial\eta}\left(\eta\frac{\partial}{\partial\eta}\right) + \frac{\eta + \xi}{4\eta\xi}\frac{\partial^2}{\partial\phi^2}\right] + k^2 - V(r)\right\}\psi = 0, \qquad (22.130)$$

where now $V(r) = Kr^{-1}$. Cylindrical symmetry and the following plausibility argument implies that the Coulomb wave function (incident plus scattered wave) may be written as $\psi = e^{ikz}g(\xi)$. We know that asymptotically,

$$\psi \sim e^{ikz} + f(\theta)\frac{e^{ikr}}{r} = e^{ikz}\left[1 + \frac{f(\theta)}{r}e^{ik(r-z)}\right]. \qquad (22.131)$$

Furthermore, from classical considerations (see Section 16.3.3) we know that $f(\theta) \propto \csc^2(\theta/2)$. Hence for constant A we have

$$\psi \sim e^{ikz}\left[1 + \frac{A}{r\sin^2(\theta/2)}e^{ik(r-z)}\right] = e^{ikz}\left(1 + \frac{A}{\xi}e^{ik\xi}\right).$$

It therefore seems reasonable to anticipate that $\psi = e^{ikz}g(\xi)$ in what follows. If we substitute this form into equation (22.130), we obtain, after considerable effort,

$$\xi\frac{d^2g}{d\xi^2} + (1 - ik\xi)\frac{dg}{d\xi} - \rho g = 0, \qquad (22.132)$$

where $\rho = K/2$. If this is compared with the standard from of the confluent hypergeometric equation,

$$z\frac{dF^2}{dz^2} + (b - z)\frac{dF}{dz} - aF = 0, \qquad (22.133)$$

with the solution that is regular at the origin (in terms of the confluent hypergeometric function), then

$$F = {}_1F_1(a, b; z) \equiv 1 + \frac{az}{b} + \frac{a(a+1)}{b(b+1)}\frac{z^2}{2!} + \cdots, \qquad (22.134)$$

we can identify $z = ik\xi$, $b = 1$, and $g(\xi) = F(z)$. Then

$$g(\xi) = {}_1F_1(-i\mu, 1; ik\xi),\tag{22.135}$$

where $\mu = K/2k$, whence the solution to the scattering problem (thus far) is

$$\psi(k, z, \xi) = Ce^{ikz} {}_1F_1(-i\mu, 1; ik\xi),\tag{22.136}$$

for C a (possibly complex) constant.

22.5.2 Asymptotic Form of ${}_1F_1(-i\mu, 1; ik\xi)$

Now we need to know the asymptotic form of $\psi(k, z, \xi)$ for large values of $ik\xi$. The asymptotic expansions of ${}_1F_1(a, b; z)$ to leading order for $z = i|z|$ are

$$_1F_1(a, b; z) \sim \frac{\Gamma(b)}{\Gamma(a)} \frac{\exp\left(i\left[|z| - \pi\left(b - a\right)/2\right]\right)}{|z|^{b-a}} + \frac{\Gamma(b)}{\Gamma(b - a)} \frac{\exp\left(i\pi a/2\right)}{|z|^a}.\tag{22.137}$$

After some delicate reduction, the asymptotic form of the wave function is found to be

$$\psi(k, z, \xi) \sim \frac{Ce^{\pi\mu/2}}{\Gamma(1 + i\mu)} e^{ikz} \left[\exp\left(i\mu \ln(k\xi)\right) \right.$$
$$\left. - \frac{\mu\Gamma(1 + i\mu)}{\Gamma(1 - i\mu)} \frac{\exp\left(i\left[k\xi - \mu \ln(k\xi)\right]\right)}{k\xi} \right]\tag{22.138}$$

If we choose C to normalize the amplitude of the radiation, that is, set

$$\frac{Ce^{\pi\mu/2}}{\Gamma(1 + i\mu)} = 1,\tag{22.139}$$

and since $\xi = r - z$, we may rewrite this as

$$\psi(k, z; r) \sim \left[\exp(i[kz + \mu \ln(k(r - z))]) \right.$$
$$\left. - \frac{\mu\Gamma(1 + i\mu)}{\Gamma(1 - i\mu)} \frac{\exp(i[kr - \mu \ln(k(r - z))])}{kr(1 - z/r)} \right].\tag{22.140}$$

We identify the first term as the incident wave and the second as the outgoing scattered wave. On reintroducing the angle θ, we therefore have to leading order,

$$\psi(k, z; r) \sim \exp\left(i\left[kz + \mu \ln\left(2kr \sin^2\left(\theta/2\right)\right)\right]\right)$$
$$- \frac{\mu\Gamma(1 + i\mu)}{\Gamma(1 - i\mu)} \frac{\exp\left(i\left[kr - \mu \ln\left(2kr \sin^2\left(\theta/2\right)\right)\right]\right)}{2kr \sin^2\left(\theta/2\right)}.\tag{22.141}$$

Now

$$\Gamma(1 + i\mu) = |\Gamma(1 + i\mu)| \exp\left[i \arg \Gamma(1 + i\mu)\right] = |\Gamma(1 + i\mu)|e^{i\eta_0},\tag{22.142}$$

and

$$\Gamma(1 - i\mu) = |\Gamma(1 - i\mu)|e^{-i\eta_0},\tag{22.143}$$

since $\Gamma(1 + i\mu) = \overline{\Gamma(1 - i\mu)}$, $\mu \in \mathbb{R}$. Therefore,

$$\frac{\Gamma(1 + i\mu)}{\Gamma(1 - i\mu)} = e^{2i\eta_0},\tag{22.144}$$

so that if we write

$$\psi (k, z; r) \sim \psi_i + \frac{f(\theta)}{r} \exp [i \{kr - \mu \ln (2kr)\}], \tag{22.145}$$

and compare this with the form (22.141), then we can make the identification

$$f(\theta) = \frac{-\mu}{2k \sin^2 (\theta/2)} \exp \left(i \left[\pi + 2\eta_0 - \mu \ln \{\sin^2 (\theta/2)\} \right] \right). \tag{22.146}$$

Since

$$\frac{d\sigma}{d\Omega} = |f(\theta)|^2, \tag{22.147}$$

the differential scattering cross section is found to be

$$\frac{d\sigma}{d\Omega} = \left(\frac{\mu}{2k \sin^2 (\theta/2)} \right)^2 = \left(\frac{K}{4k^2} \right)^2 \csc^4 \left(\frac{\theta}{2} \right), \tag{22.148}$$

which is exactly the Rutherford scattering law. Note that in terms of the confluent hypergeometric function we may write the Coulomb wave function ψ_C as

$$\psi_C = \Gamma(1 + i\mu) e^{-\pi\mu/2} e^{ikz} {}_1F_1 (-i\mu, 1; ik\xi)$$
$$= \Gamma(1 + i\mu) e^{-\pi\mu/2} e^{ikr \cos(\theta)} {}_1F_1 \left(-i\mu, 1; 2ikr \sin^2 \left(\frac{\theta}{2} \right) \right). \tag{22.149}$$

Also, because

$$ {}_1F_1 \left(-i\mu, 1; 2ikr \sin^2 \left(\frac{\theta}{2} \right) \right) = 1 + O(r), \tag{22.150}$$

it follows that the probability distribution at $r = 0$ is

$$|\psi_C (0)|^2 = |C|^2 = |\Gamma(1 + i\mu)|^2 e^{-\mu\pi} = \frac{\pi \mu e^{-\mu\pi}}{\sinh (\pi\mu)} \tag{22.151}$$

$$= \frac{2\pi\mu}{e^{2\pi\mu} - 1} \approx \begin{cases} 2\pi\mu e^{-2\pi\mu} & \text{if } \mu \gg 1 \\ 2\pi|\mu| & \text{if } \mu < 0 \end{cases}. \tag{22.152}$$

Note that $\mu > 0$ corresponds to a repulsive Coulomb potential and $\mu < 0$ to an attractive one. Since $\mu = K/2k$, $\mu \gg 1$ implies a large value of K or a small value of k (energy), or both. Note also from equation (22.141) that both the incident and scattered waves ψ_i and ψ_{sc} are distorted by logarithmic phase factors in contrast to the non-Coulombic case; this is a direct consequence of the long-range nature of the Coulomb potential.

22.5.3 The Spherical Coordinate System Revisited

Let

$$\psi_C (r, \theta; k) = \sum_{l=0}^{\infty} R_l(r) P_l (\cos \theta). \tag{22.153}$$

The radial wave equation is

$$\frac{1}{r^2}\frac{d}{dr}\left(r^2\frac{dR_l}{dr}\right) + \left[k^2 - \frac{K}{r} - \frac{l\,(l+1)}{r^2}\right]R_l = 0.\tag{22.154}$$

Now we write

$$R_l = r^l e^{ikr} f_l(r).\tag{22.155}$$

Collecting terms, equation (22.154) becomes

$$r\frac{d^2 f_l}{dr^2} + [2ikr + 2\,(l+1)]\frac{df}{dr} + [2ik\,(l+1) - K]\,f_l = 0.\tag{22.156}$$

A further change of variable, $z = -2ikr$, results in the standard confluent hypergeometric form again,

$$z\frac{d^2 f_l}{dz^2} + [2\,(l+1) - z]\frac{df_l}{dz} + \left[\frac{K}{2ik} - (l+1)\right]f_l = 0,\tag{22.157}$$

with solution regular at the origin,

$$f_l(r) = C_{l1}F_1\,(l+1+i\mu, 2\,(l+1)\,;\, -2ikr)\,,\tag{22.158}$$

in terms of constants C_l. Returning now to the *spherical* coordinate case, $z = -2ikr$, we have that

$$_1F_1\,(l+1+i\mu, 2(l+1);\, -2ikr) \sim \frac{\Gamma(2l+2)}{\Gamma(l+1-i\mu)}\frac{\exp\left[-i\pi(l+1+i\mu)/2\right]}{(2kr)^{l+1+i\mu}}$$

$$+\frac{\Gamma(2l+2)}{\Gamma(l+1+i\mu)}\left[(2kr)^{(i\mu-(l+1))}\exp\left[(2kr)e^{-i\pi/2}\right]\exp\left[-i\pi(i\mu - (l+1))\right]/2\right]\tag{22.159}$$

$$\sim \frac{\Gamma(2l+2)(2kr)^{-(l+1)}}{\Gamma(l+1-i\mu)}\left\{(2kr)^{-i\mu}\exp\left[\pi\,(\mu - i\,(l+1))\,/2\right]\right\}$$

$$+\frac{\Gamma(2l+2)(2kr)^{-(l+1)}}{\Gamma(l+1-i\mu)}\left\{\frac{\Gamma(l+1-i\mu)}{\Gamma(l+1+i\mu)}(2kr)^{i\mu}\exp\left[-2ikr + \pi(\mu + i(l+1))/2\right]\right\}.\tag{22.160}$$

Let

$$\eta_l = \arg\Gamma(l+1+i\mu);\tag{22.161}$$

with some minor manipulation we obtain,

$$_1F_1\,(l+1+i\mu, 2\,(l+1);\, -2ikr)$$

$$\sim \frac{\Gamma(2l+2)(2kr)^{-(l+1)}}{\Gamma(l+1-i\mu)}e^{\pi\mu/2}i\left\{\exp\left[i\left(-2\eta_l + \mu\ln(2kr) + \frac{\pi l}{2} - 2kr\right)\right]\right\}$$

$$-\frac{\Gamma(2l+2)(2kr)^{-(l+1)}}{\Gamma(l+1-i\mu)}e^{\pi\mu/2}i\left\{\exp\left[-i\left(\mu\ln(2kr) + \frac{\pi l}{2}\right)\right]\right\}.\tag{22.162}$$

By choosing

$$C_l = \frac{e^{-\pi\mu/2}}{2(2l+1)!}\Gamma(l+1-i\mu),\tag{22.163}$$

we find that

$$f_l(r) \sim (2kr)^{-(l+1)} \exp\left[-i\left(\eta_l + kr\right)\right] \sin\left(kr - \frac{\pi l}{2} - \mu \ln(2kr) + \eta_l\right),$$

(22.164)

from which the expression for R_l follows using equation (22.155).

22.6 COULOMB SCATTERING: THE WKB(J) APPROXIMATION

22.6.1 Coulomb Phases

In Section 22.5.3 the asymptotic behavior of the Coulomb wave function was shown to be

$$\psi_l(r) \propto \sin\left(kr - \frac{\pi l}{2} - \mu \ln(2kr) + \eta_l\right),$$

(see (22.164)), where the phase is

$$\eta_l = \arg \Gamma(l + 1 + i\mu).$$

(22.165)

From the WKB(J) approximation, however, we have

$$\psi_l(r) \propto \sin\left(kr - \frac{l\pi}{2} + \delta_l\right),$$

where, as above, δ_l is the difference between two integrals, equation (22.102), rewritten here as

$$\delta_l = \lim_{r \to \infty} \left\{ \int_{r_0}^{r} \left(k^2 - \frac{2\mu k}{\xi} - \frac{\lambda^2}{\xi^2}\right)^{1/2} d\xi - \int_{\lambda/k}^{r} \left(k^2 - \frac{\lambda^2}{\xi^2}\right)^{1/2} d\xi \right\},$$

(22.166)

so (22.166) should be an approximation to the exact form

$$\delta_l = -\mu \ln 2kr + \eta_l = -\mu \ln 2kr + \arg \Gamma(l + 1 + i\mu).$$

(22.167)

We introduce the dimensionless variable $x = kr$ and set

$$R = x^2 - 2\mu x - \lambda^2; \quad x_0 = \mu + \left(\mu^2 + \lambda^2\right)^{1/2}.$$

Then (again in terms of a dummy variable ξ) we have

$$\delta_l = \lim_{x \to \infty} \left\{ \int_{x_0}^{x} \frac{R^{1/2}}{\xi} d\xi - \int_{\lambda}^{x} \frac{\left(\xi^2 - \lambda^2\right)^{1/2}}{\xi} d\xi \right\}.$$

(22.168)

The integrals in (22.168) are standard (but nontrivial) to evaluate analytically; after doing so, the result is found to be

$$\delta_l = \lim_{x \to \infty} \left\{ R^{1/2} - \left(x^2 - \lambda^2\right)^{1/2} - \mu \ln\left[\frac{x - \mu + R^{1/2}}{x\left(\mu^2 + \lambda^2\right)^{1/2}}\right] \right.$$

$$\left. + \lambda\left[\arcsin\left(\frac{\mu x + \lambda^2}{x\left(\mu^2 + \lambda^2\right)^{1/2}}\right) - \arcsin\frac{\lambda}{x}\right]\right\}.$$

(22.169)

Exercise: Verify this result.

Expression (22.169) can be expanded in powers of x^{-1} prior to the evaluation of the limit, which (eventually) leads to the result

$$\delta_l = -\mu + \frac{\mu}{2} \ln \left(\mu^2 + \lambda^2 \right) + \lambda \arcsin \left[\frac{\mu}{\left(\mu^2 + \lambda^2 \right)^{1/2}} \right] - \mu \ln 2x. \quad (22.170)$$

This can be rearranged by defining

$$g(z) = \frac{1}{2} \ln \left(1 + z^2 \right) + z \arcsin \left[\left(1 + z^2 \right)^{-1/2} \right].$$

Then from equations (22.167) and (22.170) we find the WKB(J) approximation to the Coulomb phase to be

$$\eta_l = \mu \left\{ (\ln \mu - 1) + g \left(\frac{\lambda}{\mu} \right) \right\}.$$

Exercise: Verify this result.

How accurate is this approximation? The following table is taken from [41]; the phase shifts were calculated to three decimal places for $\mu = 2$.

l	δ_l (Exact)	δ_l (WKB(J))	
0	0.130	0.110	
1	1.237	1.222	
2	2.022	2.012	(22.171)
3	2.610	2.610	
4	3.074	3.076	

Furthermore, it is shown in particular [41] that for $\lambda \ll \mu$,

$$\delta_0 \, (\text{WKB(J)}) - \delta_0 \, (\text{Exact}) \approx -\frac{1}{24\mu}, \quad (22.172)$$

whereas (more generally) for $\lambda \gg \mu$ and $\lambda \gg 1$,

$$\Delta \delta_{l+1}^{\text{WKB(J)}} = (\delta_{l+1} - \delta_l)_{\text{WKB(J)}} = \frac{\mu}{\lambda} - \frac{\mu}{2\lambda^2} + \frac{\mu}{3\lambda^3} - \frac{\mu^3}{3\lambda^3} \cdots, \quad (22.173)$$

and

$$\Delta \delta_{l+1}^{\text{Exact}} = (\delta_{l+1} - \delta_l)_{\text{Exact}} = \frac{\mu}{\lambda} - \frac{\mu}{2\lambda^2} + \frac{\mu}{4\lambda^3} - \frac{\mu^3}{3\lambda^3} \cdots, \quad (22.174)$$

in other words,

$$\Delta \delta_{l+1}^{\text{WKB(J)}} - \Delta \delta_{l+1}^{\text{Exact}} \approx \frac{\mu}{12\lambda^3}. \quad (22.175)$$

Therefore the WKB(J) phase shift approximation is increasingly accurate as $\mu/\lambda \to 0$.

22.6.2 Formal WKB(J) Solutions for the TIRSE

Consider the time-independent radial Schrödinger equation (TIRSE) in the form

$$y_l''(r) + Q^2(r) y_l = 0, \quad (22.176)$$

where

$$Q^2(r) = k^2 \left[1 - \frac{V(r)}{E} - \frac{l(l+1)}{k^2 r^2} \right]. \tag{22.177}$$

If we seek a normalized solution as

$$y_l(r) = e^{ikS(r)} \tag{22.178}$$

equation (22.176) becomes a form of Riccati's equation for $S'(r)$:

$$\frac{i}{k} S''(r) - \left[S'(r) \right]^2 + \frac{Q^2(r)}{k^2} = 0. \tag{22.179}$$

If a is a characteristic dimension (or radius) for the scattering region, then for short wavelengths, an expansion in terms of the dimensionless parameter $\varepsilon = (ka)^{-1} \ll 1$ is appropriate, or,

$$S(r) = \sum_{j=0}^{\infty} \left(\frac{\varepsilon}{i} \right)^j S_j(r), \tag{22.180}$$

assuming that this converges or is at least asymptotic. The first four orders of ε lead to the equations

$$\left(S_0' \right)^2 = \frac{Q^2}{k^2}; \tag{22.181}$$

$$S_1' S_0' = -\frac{1}{2} a S_0''; \tag{22.182}$$

$$S_2' S_0' = -\frac{1}{2} \left[a S_1'' + \left(S_1' \right)^2 \right]; \tag{22.183}$$

$$S_3' S_0' = -\frac{1}{2} \left[a S_2'' + 2 S_1' S_2' \right], \tag{22.184}$$

and so on. From these a sequence of iterative terms may be generated. Specifically, if $S_j'(r) \equiv b_j(r)$, $j = 0, 1, 2, \dots$, then

$$b_0 = \pm \frac{Q}{k}; \quad b_1 = -\frac{1}{2} a \frac{b_0'}{b_0}; \tag{22.185}$$

$$b_2 = -\frac{a b_1' + b_1^2}{2 b_0}; \quad b_3 = -\frac{a b_2' + 2 b_1 b_2}{2 b_0}, \tag{22.186}$$

and so forth. Then is it possible to write each $b_j(r)$ in terms of $b_0(r)$ and its derivatives; thus

$$b_1 = -\frac{1}{2} a \frac{b_0'}{b_0}; \quad b_2 = \frac{a^2}{4} \left(\frac{b_0''}{b_0'} - \frac{3}{2} \frac{\left(b_0' \right)^2}{b_0^3} \right), \tag{22.187}$$

and so forth.

If

$$b_0(r) = \pm \left[1 - \frac{V(r)}{E} - \frac{l(l+1)}{k^2 r^2} \right]^{1/2} \tag{22.188}$$

is a real quantity, then so are all the subsequent $b_j(r)$, and hence the terms in the expansion (22.180) are alternatingly real and imaginary. This is turn provides corrections to the phase and amplitude respectively. Therefore the solution $y_l(r)$ can be written as the integral

$$y_l(r) = \exp\left\{\int_a^r \frac{1}{\varepsilon}\left[\frac{i}{\varepsilon}b_0 + b_1 + \frac{\varepsilon}{i}b_2 + \dots\right]\right\}. \tag{22.189}$$

As pointed out in [41], if the change $b_0 \to -b_0$ is made, the odd functions b_{2j+1} are unchanged (and so too the amplitude corrections), but the even functions $b_{2j} \to -b_{2j}$ and so provide a complex conjugate solution. Therefore a fundamental set of solutions results from the application of this method.

Exercise: Verify the above results.

22.6.3 The Langer Transformation: Further Justification

In this subsection the Langer transformation is justified by showing that the asymptotic phase of the radial WKB(J) wave function is recovered when the factor $\lambda^2 \equiv l(l+1)$ is replaced by $(l+1/2)^2$.

In the absence of a potential, the WKB(J) wave function for equation (22.176) is

$$y_l''(r) + Q^2(r)y_l(r) = 0, \tag{22.190}$$

where

$$Q^2(r) = 1 - \frac{l(l+1)}{k^2r^2} = 1 - \frac{\lambda^2}{k^2r^2}. \tag{22.191}$$

From equation (22.28) y_l has the form

$$y_l(r) = [Q(r)]^{-1/4} \sin\left\{k\int_{r_0}^r \left(1 - \frac{\lambda^2}{k^2\xi^2}\right)^{1/2} d\xi + \frac{\pi}{4}\right\},$$

that is,

$$y_l(r) = \left(1 - \frac{\lambda^2}{k^2r^2}\right)^{-1/4} \sin\left\{k\int_{r_0}^r \left(1 - \frac{\lambda^2}{k^2\xi^2}\right)^{1/2} d\xi + \frac{\pi}{4}\right\},$$

(see also Chapter 7). The turning point is $r_0 = \lambda/k$. Note that if $x = r_0/r$,

$$\int_{r_0}^r \left(1 - \frac{\lambda^2}{k^2\xi^2}\right)^{1/2} d\xi = r\left[(1 - x^2)^{1/2} - x\,\text{arcsec}\left(\frac{1}{x}\right)\right], \tag{22.192}$$

so that, on expanding this expression for small x, we find

$$\int_{r_0}^r \left(1 - \frac{\lambda^2}{k^2\xi^2}\right)^{1/2} d\xi = r\left(1 - \frac{\pi}{2}x\right) + O(x^2) \approx r - \frac{\pi}{2}r_0. \tag{22.193}$$

To this order the amplitude factor is just unity, so asymptotically

$$y_l(r) \sim \sin\left\{k\left(r - \frac{\pi}{2}r_0\right) + \frac{\pi}{4}\right\} = \sin\left\{kr - \left(\lambda - \frac{1}{2}\right)\frac{\pi}{2}\right\}.$$

The exact solution $\psi(r)$ is proportional to

$$j_l(kr) \sim \sin\left(kr - \frac{l\pi}{2}\right),$$

so these are equivalent if $\lambda = l + 1/2$, whence $(l + 1/2)^2$ replaces the term $l(l + 1)$ in the centrifugal term.

In the case of $V \neq 0$ the asymptotic phase of the radial WKB(J) wave function (noting that now $r_0 \neq \lambda/k$) is conveniently expressed by the difference

$$\delta_l = k \lim_{r \to \infty} \left\{ \int_{r_0}^{r} \left(1 - \frac{V(\xi)}{E} - \frac{\lambda^2}{k^2 \xi^2} \right)^{1/2} d\xi - \int_{\lambda/k}^{r} \left(1 - \frac{\lambda^2}{k^2 \xi^2} \right)^{1/2} d\xi \right\}.$$

(22.194)

(Recall that in quantum mechanical problems, $E = k^2 \hbar^2/(2m)$.) Note that this equation has been encountered earlier in the forms of equation (22.102) and (22.166). Meanwhile we proceed in this limit to obtain

$$\sin \left\{ \int_{r_0}^{r} k \left(1 - \frac{V(\xi)}{E} - \frac{\lambda^2}{k^2 \xi^2} \right)^{1/2} d\xi + \frac{\pi}{4} \right\} \sim \sin \left(kr - \frac{l\pi}{2} + \delta_l \right), \quad (22.195)$$

so that

$$\delta_l = k \lim_{r \to \infty} \left\{ \int_{r_0}^{r} \left[\left(1 - \frac{V(\xi)}{E} - \frac{\lambda^2}{k^2 \xi^2} \right)^{1/2} \right] d\xi - r \right\} + \left(l + \frac{1}{2} \right) \frac{\pi}{2}. \quad (22.196)$$

Equation (22.192) implies that (for $x = \lambda/kr$)

$$\int_{\lambda/k}^{r} \left(1 - \frac{\lambda^2}{k^2 \xi^2} \right)^{1/2} d\xi = \frac{\lambda}{k} \left[\left(\frac{k^2 r^2}{\lambda^2} - 1 \right)^{1/2} - \operatorname{arcsec} \left(\frac{kr}{\lambda} \right) \right] \quad (22.197)$$

$$= \frac{\lambda}{k} \left[\left(\frac{k^2 r^2}{\lambda^2} - 1 \right)^{1/2} - \frac{\pi}{2} + \arcsin \left(\frac{\lambda}{kr} \right) \right]$$

(22.198)

$$\sim r - \frac{\lambda \pi}{2k} = r - \left(l + \frac{1}{2} \right) \frac{\pi}{2k}, \quad (22.199)$$

where $\lambda = l + 1/2$. If this is substituted in expression (22.194), the result (22.196) follows immediately.

The Sabatier Transform $\alpha(r)$

It is sometimes convenient to express the WKB(J) phase δ_l in terms of the variable

$$\alpha(r) \equiv r \left[1 - \frac{V(r)}{E} \right]^{1/2}, \quad (22.200)$$

and a so-called quasi-potential defined by

$$\Omega(\alpha) = 2E \ln \left[\frac{r(\alpha)}{\alpha} \right] = -E \ln \left[1 - \frac{V(r)}{E} \right]. \quad (22.201)$$

To see this, we first recall equation (22.194), rewritten slightly in terms of the impact parameter $p = (l + 1/2)/k$ as

$$\delta_l = k \lim_{r \to \infty} \left\{ \int_{r_0}^{r} \left(1 - \frac{V(r)}{E} - \frac{\lambda^2}{k^2 r^2} \right)^{1/2} dr - \int_{p}^{r} \left(1 - \frac{\lambda^2}{k^2 r^2} \right)^{1/2} dr \right\},$$

(22.202)

r_0 being the largest zero of the integrand $Q(r)$. Using the new variable (22.200), this takes the form

$$\int_{r_0}^{r} \left(1 - \frac{V(r)}{E} - \frac{\lambda^2}{k^2 r^2} \right)^{1/2} dr = \int_{p}^{\alpha} \left(\frac{d \ln r(\alpha)}{d\alpha} \right) (\alpha^2 - p^2)^{1/2} d\alpha.$$

(22.203)

An underlying assumption is that the transformation $\alpha(r)$ may be uniquely inverted to obtain $r(\alpha)$; hence $\alpha(r)$ must be a monotone function. In the second integral α can be a dummy variable, so

$$\delta_l = k \lim_{\alpha \to \infty} \int_{p}^{\alpha} \left(\frac{d \ln r(\alpha)}{d\alpha} - \frac{1}{\alpha} \right) (\alpha^2 - p^2)^{1/2} d\alpha \qquad (22.204)$$

$$= k \lim_{\alpha \to \infty} \int_{p}^{\alpha} \left(\frac{d}{d\alpha} \left[\frac{\ln r(\alpha)}{\alpha} \right] \right) (\alpha^2 - p^2)^{1/2} d\alpha. \qquad (22.205)$$

Noting that $t \sim r$ as $t \to \infty$, an integration by parts provides us with the compact result

$$-\frac{k}{2E} \lim_{\alpha \to \infty} \int_{p}^{\alpha} \frac{\alpha \Omega(\alpha)}{(\alpha^2 - p^2)^{1/2}} d\alpha. \qquad (22.206)$$

From the definition (22.201), the zeros of $\Omega(\alpha)$ coincide with those of $V(r)$; furthermore, if $|V(r)| \ll E$, then $\Omega(\alpha) \approx V(r)$ as a first approximation.

Chapter Twenty-Three

A Sturm-Liouville Equation: The Time-Independent One-Dimensional Schrödinger Equation

This theory [Sturm-Liouville] began with the original work of Sturm from 1829 to 1836 and was then followed by the short but significant joint paper of Sturm and Liouville in 1837, on second order linear ordinary differential equations with an eigenvalue parameter. The details of the early development of Sturm-Liouville theory, from the beginnings about 1830, are given in a historical survey paper of Jesper Lutzen (1984), in which paper a complete set of references may be found to the relevant work of both Sturm and Liouville. The catalogue commences with sections devoted to a brief summary of Sturm-Liouville theory including some details of differential expressions and equations, Hilbert function spaces, differential operators, classification of interval endpoints, boundary condition functions and the Liouville transform. There follows a collection of more than 50 examples of Sturm-Liouville differential equations; many of these examples are connected with well-known special functions, and with problems in mathematical physics and applied mathematics.

[256]

The purpose of this chapter is to provide an anthology of mathematical properties of this very important (and pedagogically valuable) equation. The full Schrödinger equation is of course a partial differential equation in several spatial variables and time, so the time-independent form no longer describes a wave function: with suitable boundary conditions it now reverts to an *eigenvalue* problem. The canonical form for the Sturm-Liouville equation is

$$\frac{d}{dx}\left[p(x)\frac{du}{dx}\right] - q(x)u + \lambda r(x)u \equiv L(u) + \lambda r(x)u = 0, \qquad (23.1)$$

for $u(x) \in [a, b]$, with *homogeneous* boundary conditions

$$A_1 u(a) + B_1 u'(a) = 0; \qquad (23.2)$$

$$A_2 u(b) + B_2 u'(b) = 0. \qquad (23.3)$$

It must be emphasized that in this chapter the notation u, p, λ, and so on has no physical connection with previously used variables (e.g., λ is a parameter, not a wavelength, etc.).

For *regular* (as opposed to *singular*) Sturm-Liouville problems, $p(x) > 0, q(x)$ and $r(x) > 0$ are all real-valued functions with $p(x)$ continuously differentiable; $q(x)$ and $r(x)$ are required at least to be continuous. λ is a parameter which will be identified as a generic eigenvalue in boundary-value problems. In his 1908 dissertation the German mathematician Hermann Weyl generalized the regular Sturm-Liouville theory on a finite closed interval to second-order differential operators with singularities at the endpoints of the interval, possibly semi-infinite or infinite. Unlike the classical case, the spectrum may no longer consist of just a countable set of eigenvalues but may also contain a continuous part. In this case the eigenfunction expansion involves an integral over the continuous part in addition to a sum over the discrete part. Many equations arising mathematical physics can be expressed in one or other of these forms.

23.1 VARIOUS THEOREMS

We consider the one-dimensional Schrödinger equation in the standard dimensionless form (with independent variable x), requiring that $V(x)$ and all its required derivatives are continuous. Clearly this excludes rectangular barrier/well problems; those have already been discussed to some extent, and the topic will be revisited in Part V. In the list of theorems some of the more standard properties of second-order ordinary differential equations have been reproduced, but there are also several less well-known results [59]. In what follows the dependent variable Ψ will be used (as opposed to u) as a reminder of the original context of this equation, though when satisfying appropriate boundary conditions it is an *eigenfunction*, not a wavefunction. Thus we have

$$\Psi'' + [\lambda - V(x)]\Psi = 0, \quad -\infty < x < \infty. \tag{23.4}$$

Clearly this is of the form (23.1) with $p(x) = 1, r(x) = 1$, and $q(x) = V(x)$. However, it is important to note that every equation of the form (23.1) may be written as (23.4); this is called the *Liouville normal form* and will be discussed in Chapter 28.

In most cases, the proofs are summarized, but occasionally they are left as exercises for the reader.

Theorem i: If $\Psi(x_0) \neq 0$ and $\lambda > V(x_0)$, it follows that $\Psi''(x_0)/\Psi(x_0) < 0$, so Ψ is concave toward the x-axis at the point $x = x_0$. Likewise, if $\Psi(x_0) \neq 0$ and $\lambda < V(x_0)$, it follows that $\Psi''(x_0)/\Psi(x_0) > 0$, so Ψ is convex toward the x-axis at the point $x = x_0$.

Theorem ii: If $\Psi(x_0) = 0$, $\Psi'(x_0) \neq 0$, and $\lambda \neq V(x_0)$, then $\Psi''(x_0) = 0$ and $\Psi'''(x_0) \neq 0$, so $\Psi(x)$ has a point of inflection at $(x_0, 0)$.

Theorem iii: If $\Psi(x_0) \neq 0$, $\lambda = V(x_0)$, and $V'(x_0) \neq 0$, then $\Psi''(x_0) = 0$ and $\Psi'''(x_0) \neq 0$, so $\Psi(x)$ has a point of inflection at $(x_0, 0)$.

Theorem iv: Suppose that $\Psi'(x_1) = 0$, $\Psi'(x_2) = 0$, and $\lambda - V(x)$ does not change sign for any $x \in (x_1, x_2)$. Integrating (23.4), we see that

$$\int_{x_1}^{x_2} [\lambda - V(x)] \Psi dx = 0, \tag{23.5}$$

which implies that Ψ vanishes at least once in (x_1, x_2).

Theorem v: Suppose that $\Psi\Psi' = 0$ at x_1 and x_2. By *Rolle's theorem*, $(\Psi\Psi')'$ must vanish at least once in (x_1, x_2), that is,

$$(\Psi\Psi')' = (\Psi')^2 + \Psi\Psi'' = 0 = (\Psi')^2 + [V(x) - \lambda]\Psi^2, \tag{23.6}$$

and this implies that $(\Psi\Psi')' > 0$ if $\lambda < V(x)$. Hence $\Psi\Psi'$ can vanish only once in such a region. This means that the solution $\Psi(x)$ cannot have two zeros, two extrema, or an extremum and a zero in any interval (x_1, x_2) for which $\lambda < V(x)$.

Theorem vi: Any two particular solutions Ψ_1 and Ψ_2 satisfy

$$\Psi_1 \Psi_2' - \Psi_2 \Psi_1' = c, \tag{23.7}$$

where c is a constant. This is just the constant Wronskian $W(\Psi_1, \Psi_2)$, and the result follows from equation (23.4).

Theorem vii: If Ψ_1 and Ψ_2 in theorem vi are linearly dependent, $W(\Psi_1, \Psi_2) = 0$ as is readily shown. Conversely, if $W(\Psi_1, \Psi_2) = 0$, then

$$\frac{\Psi_1'}{\Psi_1} = \frac{\Psi_2'}{\Psi_2} \Rightarrow \ln|\Psi_1| = \ln|\Psi_2| + \text{constant}, \tag{23.8}$$

or $\Psi_1 \propto \Psi_2$. Hence $c = 0$ is a necessary and sufficient condition for Ψ_1 and Ψ_2 to be linearly dependent.

Theorem viii: Let Ψ_1 and Ψ_2 be linearly independent solutions, and x_0 an arbitrary fixed value of x. Then

$$\Psi_2'' + (\lambda - V(x))\Psi_2 = 0. \tag{23.9}$$

Using the method of reduction of order, let $\Psi_2 = \Psi_1 X$, from which it follows that

$$\left[\Psi_1'' + (\lambda - u)\Psi_1\right] X + 2\Psi_1' X' + \Psi_1 X'' = 0, \tag{23.10}$$

leading to

$$\frac{X''}{X'} = -\frac{2\Psi_1'}{\Psi_1},$$

and hence in terms of constant A and B,

$$X = A \int \frac{dx}{\Psi_1^2} + B, \tag{23.11}$$

so the second desired solution is

$$\Psi_2(x) = \Psi_1(x) \int_{x_0}^{x} \frac{d\xi}{\Psi_1^2(\xi)}.$$

Theorem ix: If Ψ_1 and Ψ_2 are two particular solutions satisfying the initial conditions

$$\Psi_1(0) = 1, \quad \Psi_1'(0) = 0, \quad \Psi_2(0) = 0, \quad \Psi_2'(0) = 1, \tag{23.12}$$

and

$$\upsilon(x) = (\Psi_1^2 + \Psi_2^2)^{1/2}, \tag{23.13}$$

then the following conditions hold:

$$\upsilon(0) = 1, \quad \upsilon'(0) = 0, \quad \upsilon(x) > 0. \tag{23.14}$$

These follow directly from the definition of υ, and imply that $\upsilon(x)$ has no zeros. The following is a similar type of result.

Theorem x: If Ψ_1 and Ψ_2 are two particular solutions satisfying the initial conditions in theorem ix and

$$\phi(x) = \arctan(\Psi_2/\Psi_1), \tag{23.15}$$

then

$$\phi(0) = 0 \quad \text{and} \quad \phi'(x) = \upsilon^{-2}, \tag{23.16}$$

so ϕ is an increasing function. This follows from the fact that

$$\phi'(x) = W(\Psi_1, \Psi_2)\upsilon^{-2} = \upsilon^{-2}. \tag{23.17}$$

Theorem xi: The general solution of (23.4) can be written (in terms of constants D and ϕ_0) in the form

$$\Psi(x) = D\upsilon(x) \sin[\phi(x) - \phi_0]. \tag{23.18}$$

Since υ is a monotone function, it is clear that the sinusoidal term alone is responsible for the zeros of Ψ. The proof of this result follows from the above properties; specifically, for arbitrary constants A and B,

$$\Psi(x) = A\Psi_1 + B\Psi_2 = \upsilon(A\cos\phi + B\sin\phi) = D\upsilon \sin[\phi(x) - \phi_0], \tag{23.19}$$

where $D^2 = A^2 + B^2$, and $\phi_0 = \arctan(A/B)$. Furthermore, the following also holds.

Theorem xii: $\upsilon(x)$ is a particular solution of

$$\upsilon'' + [\lambda - V(x)]\upsilon = 0, \quad \upsilon(0) = 1, \quad \upsilon'(0) = 0. \tag{23.20}$$

This is established by direct substitution of equation (23.18) into equation (23.4), noting that

$$\upsilon\phi'' + 2\upsilon'\phi' = 0. \tag{23.21}$$

Theorem xiii: Suppose that $\varepsilon > 0$ is a constant and that $\Phi(x)$ satisfies the equation

$$\Phi'' + [\lambda + \varepsilon - V(x)]\Phi = 0. \tag{23.22}$$

Then the nodes of Ψ and Φ interlace, that is, consecutive zeros of Ψ are separated by zeros of Φ (this is *Sturm's separation theorem*). To prove this we multiply equation (23.4) by Φ and equation (23.22) by Ψ and subtract to obtain

$$\Phi\Psi'' - \Psi\Phi'' = \left(\Phi\Psi' - \Psi\Phi'\right)' = \varepsilon\Phi\Psi. \tag{23.23}$$

Therefore if x_1 and x_2 are successive zeros of Ψ, then

$$\Psi'(x_2)\Phi(x_2) - \Psi'(x_1)\Phi(x_1) = \varepsilon \int_{x_1}^{x_2} \Phi\Psi dx. \tag{23.24}$$

Now if $\Psi > 0$ in (x_1, x_2) and $\Phi > 0$ in $[x_1, x_2]$, we have

$$\Psi'(x_2)\Phi(x_2) - \Psi'(x_1)\Phi(x_1) < 0, \tag{23.25}$$

while

$$\int_{x_1}^{x_2} \Phi\Psi dx > 0. \tag{23.26}$$

Clearly there is a contradiction, and this persists if the sign of Ψ, Φ, or both is changed, and also if the sign of Φ is unchanged in the open interval (x_1, x_2) instead of $[x_1, x_2]$. The only way the contradiction can be resolved is by permitting Φ to have a zero in (x_1, x_2). This theorem can also provide information on what happens to the zeros as λ increases.

23.2 BOUND STATES

Bound states are square-integrable solutions corresponding to classical motion that does not exist outside certain values of x; the motion of the classical particle is bound because $V(x) \to \infty$ as $|x| \to \infty$. Of course, quantum mechanically, the phenomenon of tunneling can occur, and more will be said below about this (and also later in the context of morphology-dependent resonances—see Chapter 27). We define x_1 and x_2 respectively as the least and greatest roots of the equation

$$\lambda - V(x) = 0, \tag{23.27}$$

and denote by I_-, I_i, and I_+ respectively the intervals $(-\infty, x_1)$, (x_1, x_2), and (x_2, ∞). In I_- and I_+ the condition $\lambda < V(x)$ holds, so for bound states the following theorems can be stated

23.2.1 Bound-State Theorems

Theorem I: $\Psi(x)$ has a finite number of zeros. By theorem v, Ψ has at most one zero in the outer regions I_-, I_+, and clearly only a finite number (if any) in the inner interval I_i.

Theorem II: The following asymptotic results concern the function $\phi(x)$ defined by equation (23.15):

$$\lim_{x \to -\infty} \phi(x) = \phi_1; \quad \lim_{x \to \infty} \phi(x) = \phi_2; \quad |\phi_1| < \infty, \quad |\phi_2| < \infty. \tag{23.28}$$

This is seen to be the case by examining the consequences of ϕ decreasing without limit as $x \to -\infty$ or increasing without limit as $x \to \infty$. In each case, the sine term in equation (23.18) would force Ψ to have infinitely many nodes, in contradiction to theorem I.

Theorem III: $\phi'(x) \to 0^+$ as $|x| \to \infty$, otherwise theorem II would be contradicted; recall also that $\phi'(x)$ is never negative.

Theorem IV: Both $\upsilon(x)$ and $\upsilon'(x) \to \infty$ as $|x| \to \infty$. The first result follows from theorem III and the definition of ϕ'. From equation (23.20) it follows that $\upsilon'' \to \infty$ when $|x| \to \infty$ and hence so does $\upsilon'(x)$.

Theorem V: There are four types of solutions for bound states: (a) $|\Psi| \to 0$ as $x \to -\infty$; (b) $|\Psi| \to \infty$ as $x \to -\infty$; (c) $|\Psi| \to 0$ as $x \to \infty$; and (d) $|\Psi| \to \infty$ as $x \to \infty$. To establish these solution behaviors recall the general solution (23.18):

$$\Psi(x) = D_a \upsilon(x) \sin[\phi(x) - \phi_0]. \tag{23.29}$$

The previously arbitrary phase factor ϕ_0 is now ϕ_1, as $x \to -\infty$ by virtue of theorem II, so (23.29) appears to be indeterminate in this limit. We write

$$\Psi(x) = D_a \frac{\sin[\phi(x) - \phi_1]}{\upsilon^{-1}}, \tag{23.30}$$

and apply L'Hôpital's rule in this limit to obtain

$$\lim_{x \to -\infty} \Psi(x) = \lim_{x \to -\infty} -D_a \frac{\phi' \cos[\phi(x) - \phi_1]}{\upsilon^{-2}\upsilon'} = -D_a \frac{\cos[\phi(x) - \phi_1]}{\upsilon'} = 0, \tag{23.31}$$

by theorem IV. The same argument applies to solutions of type (c), that is,

$$\Psi(x) = D_c \upsilon(x) \sin[\phi(x) - \phi_2]. \tag{23.32}$$

Solutions of types (b) and (d) are combinations of the sets

$$\{D_b \upsilon(x) \cos[\phi(x) - \phi_1], D_a \upsilon(x) \sin[\phi(x) - \phi_1]\}, \tag{23.33}$$

and

$$\{D_d \upsilon(x) \cos[\phi(x) - \phi_2], D_c \upsilon(x) \sin[\phi(x) - \phi_2]\}, \tag{23.34}$$

respectively. A careful examination of some graphical results in the light of the Sturm separation theorem viii indicates that as λ increases, the zeros move leftward on the x-axis (all such solutions have a zero in the limit $x \to -\infty$), and after the right-most zero has entered I_i a new zero is created at $x = \infty$.

Theorem VI: If $|\Psi| < \infty$ in the outer intervals I_- or I_+, it tends to zero when $|x|$ increases in these regions. To establish this we consider the following argument. Suppose that $\Psi(x_0) > 0$, where $x_0 \in I_+$. Then by theorem ii, Ψ is convex toward the x-axis for all $x \geq x_0$. If $\Psi'(x_0) \geq 0$, then $\Psi \to \infty$ as $x \to \infty$. If $\Psi'(x_0) < 0$, several possibilities exist: (a) when $x \to \infty$, Ψ remains positive and has a minimum to the right of x_0, then tends to infinity; (b) Ψ approaches the x-axis asymptotically, or (c) Ψ crosses the x-axis to the right of x_0 (having a point of inflection there) and tends to $-\infty$ as $x \to \infty$. Only in case (b) does Ψ remain finite. Similar arguments can be used to establish this result for the remaining cases, that is, $\Psi(x_0) \leq 0$, where $x_0 \in I_+$, and the corresponding two cases when $x_0 \in I_-$.

Theorem VII: Eigenfunctions of (23.4) tend to zero as $|x| \to \infty$. This follows directly from the preceding theorem.

Theorem VIII: The quantity $\Phi = \phi_2 - \phi_1$ is an increasing function of λ, and $\Phi \to \infty$ as $\lambda \to \infty$. This result is established by first differentiating (23.4) partially with respect to λ, multiplying by Ψ, and using (23.4) again to obtain

$$\Psi'' \frac{\partial \Psi}{\partial \lambda} - \Psi \frac{\partial \Psi''}{\partial \lambda} = \frac{d}{dx} \left(\Psi' \frac{\partial \Psi}{\partial \lambda} - \Psi \frac{\partial \Psi'}{\partial \lambda} \right) = \Psi^2. \tag{23.35}$$

Next substitute (23.18), so that (after some algebra),

$$\frac{d}{dx} \left\{ \left(v' \frac{\partial v}{\partial \lambda} - v \frac{\partial v'}{\partial \lambda} \right) \sin^2 (\phi - \phi_0) + \frac{\partial \phi}{\partial \lambda} + v^{-1} \sin 2 (\phi - \phi_0) \right\} = v^2 \sin^2 (\phi - \phi_0) . \tag{23.36}$$

We choose an arbitrary value of x, $x = x_a$ say, set $\phi_0 = \phi(x_a)$, and integrate both sides with respect to x from 0 to x_a. Therefore,

$$\left[\left(v' \frac{\partial v}{\partial \lambda} - v \frac{\partial v'}{\partial \lambda} \right) \sin^2 (\phi - \phi_0) + \frac{\partial \phi}{\partial \lambda} + v^{-1} \sin 2 (\phi - \phi_0) \right]_0^{x_a}$$
$$= \int_0^{x_a} v^2 \sin^2 (\phi - \phi_0) \, dx. \tag{23.37}$$

In view of the initial conditions on v, v', and ϕ' (see above) the only term that remains on the left-hand side is the λ derivative of ϕ, and so we have the result

$$\frac{\partial \phi(x_a)}{\partial \lambda} = \int_0^{x_a} v^2 \sin^2 (\phi - \phi_0) \, dx = \text{sign}(x_a), \tag{23.38}$$

since the integrand is never negative. Therefore it follows that $\phi(\lambda)$ is an increasing function for positive values of x and a decreasing function for negative values of x.

Theorem IX: A necessary and sufficient condition for λ to be an eigenvalue is that

$$\Phi = \phi_2 - \phi_1 = n\pi, \quad n = 1, 2, 3, \ldots . \tag{23.39}$$

To prove sufficiency, it follows from (23.39) that

$$v \sin (\phi - \phi_1) = v \sin (\phi - \phi_2 + n\pi) = \pm v \sin (\phi - \phi_2) , \tag{23.40}$$

and these are the form of the eigenfunctions (23.18), remaining finite as $|x| \to \infty$. To prove necessity, let Ψ be an eigenfunction. Then $\Psi \to 0$ as $|x| \to \infty$ (see theorem VII). But $v(x)$ has no zeros, so both $\phi_1 - \phi_0$ and $\phi_2 - \phi_0$ must be integer multiples of π.

Theorem X: Equation (23.4) has a discrete, nondegenerate eigenvalue spectrum $\{\lambda_n\}$, $n = 1, 2, 3, \ldots$, where

$$\min_x V(x) < \lambda_1 < \lambda_2 < \lambda_3 < \ldots, \tag{23.41}$$

An eigenfunction vanishes when, in particular, $x \to -\infty$ and has the form (23.18). If $\lambda < \min_x V(x)$, Ψ has no zeros other than that at $-\infty$, and hence $\Phi = \phi_2 - \phi_1 < \pi$, and λ is therefore not an eigenvalue. But as λ increases, so does Φ (theorem VIII), and an eigenvalue exists whenever (23.39) occurs.

Theorem XI: An eigenfunction Ψ_n associated with an eigenvalue λ_n has $n + 1$ zeros: one each at $x = \pm\infty$, together with $n - 1$ interior zeros.

Theorem XII: Eigenfunctions are square integrable on $(-\infty, \infty)$; that is, $\Psi_n \in L_2(-\infty, \infty)$.
   Exercise: Prove theorem XII (see also Theorem B in Section 23.2.3).

Theorem XIII: Eigenfunctions belonging to distinct eigenvalues (even if the eigenfunctions are complex valued) are orthogonal.
   Let us prove theorem XIII for the canonical Sturm-Liouville form.

Proof: Let u and v be any two distinct solutions of the Sturm-Liouville problem (23.1) (i.e., not necessarily eigenfunctions). Then

$$\int_a^b \{vL(u) - uL(v)\} dx = \int_a^b \frac{d}{dx}\left[p\left(v\frac{du}{dx} - u\frac{dv}{dx}\right)dx\right]$$

$$= p(b)\left[v(b)u'(b) - u(b)v'(b)\right] \tag{23.42}$$

$$-p(a)\left[v(a)u'(a) - u(a)v'(a)\right] = 0, \tag{23.43}$$

on applying the boundary conditions. If now u and v are two eigenfunctions of the system (ϕ_m and ϕ_n, respectively), corresponding to distinct eigenvalues λ_m and λ_n, then it follows from this result that

$$\int_a^b \{\phi_n L(\phi_m) - \phi_m L(\phi_n)\} dx = (\lambda_n - \lambda_m)\int_a^b r(x)\phi_m\phi_n dx = 0. \tag{23.44}$$

Thus provided $\lambda_m \neq \lambda_n$, the eigenfunctions are orthogonal with respect to the weight function $r(x)$.

Theorem XIV: If $V(x)$ is an even function, then every eigenfunction belonging to the eigenvalue λ_n is an even function if n is odd, and an odd function if n is even. It is easy to see that when $V(x) = V(-x)$, $\Psi_n(-x)$ is an eigenfunction when $\Psi_n(x)$ is (i.e., they each satisfy (23.4) and vanish at $\pm\infty$). They both belong to the same eigenvalue since the spectrum is nondegenerate, implying that $\Psi_n(-x) = c\Psi_n(x)$ for some constant c. But the transformation $x \to -x$ means that $c = \pm 1$

and hence that every eigenfunction is either an even or an odd function when V is even. Furthermore, even functions have an even number of zeros (including zero as an even number), and odd functions have an odd number of zeros, but from theorem XI we note that if n is even, Ψ_n has an odd number of zeros, and hence is an odd function. The result for n odd follows similarly.

Theorem XV: A necessary and sufficient condition for λ to be an eigenvalue is that

$$\int_{-\infty}^{\infty} \frac{dx}{v^2(x)} = n\pi, \quad n = 1, 2, 3, \ldots. \tag{23.45}$$

This follows from the fact (see theorem x) that

$$\phi(x) = \int_0^x \frac{d\xi}{v^2(\xi)}, \tag{23.46}$$

and hence from the definitions of ϕ_1 and ϕ_2,

$$\phi_1 = \int_0^{-\infty} \frac{dx}{v^2(x)}, \quad \text{and} \quad \phi_2 = \int_0^{\infty} \frac{dx}{v^2(x)}. \tag{23.47}$$

Use of these in equation (23.39) yields the desired result.

23.2.2 Complex Eigenvalues: Identities for $\text{Im}(\lambda_n)$ and $\text{Re}(\lambda_n)$

We are given that each member of a set $\{\psi_n\}$ of eigenfunctions satisfies

$$\frac{d}{dx}\left[p(x)\frac{d\psi_n}{dx}\right] + [q(x) + \lambda_n r(x)]\psi_n = 0, \quad x \in [a, b], \tag{23.48}$$

and

$$\left[p\left(\psi_m \frac{d\psi_n}{dx} - \psi_n \frac{d\psi_m}{dx}\right)\right]_a^b = 0. \tag{23.49}$$

Suppose that $p(x)$ and $r(x)$ and the boundary conditions are real, but that q is a complex function of x. We can derive expressions for both $\text{Im}(\lambda_n)$ and $\text{Re}(\lambda_n)$ as follows. The complex conjugate of equation (23.48) is

$$\frac{d}{dx}\left[p(x)\frac{d\bar{\psi}_n}{dx}\right] + [\bar{q}(x) + \bar{\lambda}_n r(x)]\bar{\psi}_n = 0. \tag{23.50}$$

If we multiply equation (23.48) by $\bar{\psi}_n$ and equation (23.50) by ψ_n and subtract the latter from the former, we obtain

$$\left[p(\bar{\psi}_n \psi_n' - \bar{\psi}_n' \psi_n)\right]' + (\lambda_n - \bar{\lambda}_n) r |\psi_n|^2 = (\bar{q} - q)|\psi_n|^2. \tag{23.51}$$

We integrate this expression over $[a, b]$ and set $\psi_m = \bar{\psi}_n$. As a result the integrated terms vanish to yield

$$\text{Im}(\lambda_n) = -\frac{\int_a^b \text{Im}(q(x))|\psi_n|^2 \, dx}{\int_a^b r(x)|\psi_n|^2 \, dx}. \tag{23.52}$$

For the remaining result, we repeat the above manipulations except we add the resulting equations to get

$$\left(p\psi_n'\right)'\bar\psi_n + \left(p\bar\psi_n'\right)\psi_n + \left(\lambda_n + \bar\lambda_n\right)r\,|\psi_n|^2 + (\bar q + q)\,|\psi_n|^2 = 0. \tag{23.53}$$

It is straightforward to show that

$$\left(p\psi_n'\right)'\bar\psi_n + \left(p\bar\psi_n'\right)\psi_n = \left[p\left(|\psi_n|^2\right)'\right]' - 2p\,|\psi_n'|^2, \tag{23.54}$$

so that, on integration,

$$\mathrm{Re}\,(\lambda_n)\int_a^b r(x)\,|\psi_n|^2\,dx = \int_a^b p(x)\,|\psi_n'|^2\,dx - \frac{1}{2}\left[p(x)\left(|\psi_n|^2\right)'\right]_a^b, \tag{23.55}$$

from which the desired result follows. Note that (23.52) and (23.55) are implicit expressions for the eigenvalues, because $\psi_n = \psi_n(x;\lambda)$.

23.2.3 Further Theorems

Lemma 1: In the equation

$$u'' + [\lambda - q(x)]u = 0, \tag{23.56}$$

let q be continuous and satisfy

$$q(x) = \frac{B}{x} + O(x^{-2}),$$

as $x \to \infty$, for some constant B. Then for $\lambda > 0$, every solution has infinitely many zeros and is bounded.

Theorem A: If $q \in C$ satisfies

$$q(x) = \frac{A}{x} + O(x^{-2}), \quad x \to \infty,$$

and

$$q(x) = \frac{B}{x} + O(x^{-2}), \quad x \to -\infty,$$

then the spectrum of (23.56) includes the half-line $\lambda > 0$.

Lemma 2: In equation (23.56) let $q \in C$, where $q(x) \to 0$ as $x \to \infty$, and let $\lambda = -k^2 < 0$. For any ϵ, $0 < \epsilon < k$, for any two solutions $u_1(x)$ and $u_2(x)$ such that for all sufficiently large x,

$$e^{(k-\epsilon)x} \le u_1(x) \le e^{(k+\epsilon)x}$$

and

$$e^{-(k+\epsilon)x} \le u_2(x) \le e^{(-k+\epsilon)x}.$$

Corollary 1: On $(0, \infty)$ let $q(x) \in C$ and satisfy

$$\lim_{x \to \infty} q(x) = q_0.$$

Then every solution of the Schrödinger equation with $\lambda < q_0$ that is bounded on the interval $(0, \infty)$ is in $\mathcal{L}^2(0, \infty)$.

Corollary 2: Let $g(x) \in C$ on the line $(-\infty, \infty)$, and let $g(x)$ tend to limits q_0 and q_1, respectively, as $x \to \pm\infty$. Then every eigenfunction with eigenvalue $\lambda < \min(q_0, q_1)$ is square integrable.

Corollary 3: By using lemma 2 and corollary 2 for $x \to \infty$, and modifying these results suitably for $x \to -\infty$ (i.e., with bounded solution as $x \to -\infty$, and appropriate matching being understood at $x = 0$), the result follows for $\lambda < \min(q_0, q_1)$ (otherwise λ might be less than 0).

Theorem B: Let $q(x)$ be as in theorem A. Then for $\lambda > 0$ the spectrum is continuous. For $\lambda < 0$ the eigenfunctions are square integrable.

Theorem C (Liouville transformation): The substitutions

$$u = [p(x)r(x)]^{-1/4} w, \quad t = \int \left[\frac{r(x)}{p(x)}\right]^{1/2} dx \qquad (23.57)$$

with $p, r \in C^2$ and $q \in C$ transform the Sturm-Liouville equation (23.1) into the "Liouville normal form"

$$\frac{d^2 w}{dt^2} + [\lambda - Q(t)] w = 0, \qquad (23.58)$$

where λ need not refer to an eigenvalue, and

$$Q(t) = \frac{q}{r} + (pr)^{-1/4} \frac{d^2}{dt^2} \left[(pr)^{1/4}\right]. \qquad (23.59)$$

Exercise: Show that Q may be expanded and written in terms of x derivatives as

$$Q(t) = \frac{q}{r} + \frac{p}{4r} \left\{ \left(\frac{p'}{p}\right)' + \left(\frac{r'}{r}\right)' + \frac{3}{4}\left(\frac{p'}{p}\right)^2 + \frac{1}{2}\left(\frac{p'}{p}\right)\left(\frac{r'}{r}\right) - \frac{1}{4}\left(\frac{r'}{r}\right)^2 \right\}.$$

$$(23.60)$$

Corollary 4: The substitutions (23.57) transform regular Sturm-Liouville systems into the form (23.58), and separated and periodic boundary conditions into separated and periodic boundary conditions. The transformed system possesses the same eigenvalues as the original system.

Corollary 5: As can be seen from equation (23.58), functions orthogonal with respect to weight function $r(x)$ transform into orthogonal functions with unit weight. Indeed, suppose that $u(x)$ and $v(x)$ are transformed into functions $U(t)$

and $V(t)$ by the set (23.57). Then

$$\int_c^d U(t)V(t)dt = \int_a^b u(x)v(x)\,[p(x)r(x)]^{1/2}\left[\frac{r(x)}{p(x)}\right]^{1/2}dx = \int_a^b u(x)v(x)r(x)dx.$$
(23.61)

Proofs of all these theorems can be found in [264].

Exercise: Show that the Liouville normal form for Bessel's equation,

$$\frac{d}{dx}\left(x\frac{du}{dx}\right) + \left(x - \frac{n^2}{x}\right)u = 0,$$
(23.62)

is

$$\frac{d^2w}{dx^2} + \left[1 - \frac{n^2 - 1/4}{x^2}\right]w = 0,$$
(23.63)

because $w = x^{1/2}u$ and $x = t$. It has been noted earlier that if $n = 1/2$, the fundamental set of solutions for $w(x)$ is $\{\sin x, \cos x\}$, so the corresponding set for Bessel's equation is $\{x^{-1/2}\sin x, x^{-1/2}\cos x\}$.

Spectral Classification

On the interval $(0, \infty)$, with no singularity except at infinity, there are essentially four different spectral classifications for the differential operator

$$\mathcal{L} = q(x) - \frac{d^2}{dx^2}$$

[153]. These are

1. If $q(x) \to \infty$, there is a purely point spectrum.
2. If $q(x) \to 0$, there is a continuous spectrum in $(0, \infty)$ and a point spectrum (which may be null) in $(-\infty, 0)$.
3. If $q(x) \to -\infty$, and

$$\int^\infty |q(x)|^{-1/2}\,dx \to \infty,$$

   the spectrum is continuous in $(-\infty, \infty)$.
4. If $q(x) \to -\infty$, and

$$\int^\infty |q(x)|^{-1/2}\,dx < \infty,$$

   there is a point spectrum only.

As pointed out by Titchmarsh, in the last three cases additional conditions have to be imposed, so the classification as stated here is incomplete, but most "ordinary examples" come under one of these broad statements [153]. Such examples include, for case (1), $q(x) = x^2$, for which the normalized eigenfunctions can be written in terms of Hermite polynomials $H_n(x)$ of degree n, that is,

$$\frac{\exp\left(-x^2/2\right)}{2^{n/2}\left(n!\right)^{1/2}\pi^{1/4}}H_n(x).\tag{23.64}$$

As an example of (2), if $q(x)$ is of the form

$$q(x) = \frac{A + B\cosh x}{\sinh^2 x},\tag{23.65}$$

then solutions for the operator $-\mathcal{L}+\lambda$ are expressible in terms of hypergeometric functions. An example of case (3) involves solutions formulated in terms of Bessel functions (or modified Bessel functions, depending on the sign of λ) of order one-third (something we have encountered in one form or another several times along the way!). The full analysis of these cases is very technical, and the interested reader is urged to consult [153] and [256] for further details. Nevertheless we can get a flavor of what is involved from the Section 23.3.

23.3 WEYL'S THEOREM: LIMIT POINT AND LIMIT CIRCLE

Given the Sturm-Liouville equation (23.1) where now $\lambda \in \mathbb{C}$, there are two cases to consider. Suppose that $x = b$ is a singular endpoint, either because $b = \infty$ or $p(b)$ vanishes for $b < \infty$. Then:

1. For any λ such that Im $(\lambda) \neq 0$ there is always one solution that is square integrable, that is, $\in \mathcal{L}^2(b; r)$. This is called the *limit-point* case. If Im $(\lambda) = 0$ there is *at most* one square-integrable solution.
2. If for any λ, say λ_0, both solutions are square integrable, then both solutions are square integrable for all λ. This is called the *limit-circle* case.

Following the proof in [206], let $\phi(x, \lambda)$ and $\psi(x, \lambda)$ be two linearly independent solutions of the equation

$$\frac{d}{dx}\left[p(x)\frac{du}{dx}\right] - q(x)u + \lambda r(x)u \equiv L(u) + \lambda r(x)u = 0, \quad a \leq x < b,\tag{23.66}$$

satisfying the following initial conditions at the (regular) point a:

$$\alpha_1\phi(a, \lambda) + \alpha_2 p(a)\phi'(a, \lambda) = 0;\tag{23.67}$$

$$\alpha_2\psi(a, \lambda) - \alpha_1 p(a)\psi'(a, \lambda) = 0,\tag{23.68}$$

where the constants $\alpha_1, \alpha_2 \in \mathbb{R}$ and $|\alpha_1| + |\alpha_2| \neq 0$. Each solution of (23.66) except multiples of ψ can be expressed in terms of the function $m(\lambda)$ as a multiple of

$$u(x, \lambda) = \phi(x, \lambda) + m(\lambda)\psi(x, \lambda), \quad m \in \mathbb{C}.\tag{23.69}$$

$m(\lambda)$ is called the *Weyl-Titchmarsh m-function*. It controls the square integrability of the solution $u(x, \lambda)$.

A similar boundary condition is imposed at a regular point $b_0 < b$, that is,

$$\beta_1 u(b_0, \lambda) + \beta_2 p(b_0) u'(b_0, \lambda) = 0, \tag{23.70}$$

where $\beta_1, \beta_2 \in \mathbb{R}$ and $|\beta_1| + |\beta_2| \neq 0$. The limit $b_0 \to b$ is of interest (where b may be infinite), and following the argument in [206], it can be shown that u satisfies the condition (23.70) if and only if m satisfies the Wronskian relationship

$$p(b_0) W(u, \bar{u}; b_0) = 0,$$

that is,

$$p(b_0) \left\{ W(\phi, \bar{\phi}; b_0) + |m|^2 W(\psi, \bar{\psi}; b_0) + m W(\psi, \bar{\phi}; b_0) + \bar{m} W(\phi, \bar{\psi}; b_0) \right\} = 0.$$

Therefore

$$m = -\frac{h \phi(b_0) + p(b_0) \phi'(b_0)}{h \psi(b_0) + p(b_0) \psi'(b_0)},$$

where $h = \beta_1/\beta_2$. Provided $\operatorname{Im}(\lambda) \neq 0$, as h takes on all real values, m describes a circle in the complex plane with complex center

$$z_0 = -\frac{W(\phi, \bar{\psi}; b_0)}{W(\psi, \bar{\psi}; b_0)},$$

and radius

$$R = \frac{\left(\alpha_1^2 + \alpha_2^2\right)^{1/2}}{2 |\operatorname{Im}(\lambda)| \left[\int_a^{b_0} r(x) |\psi|^2 \, dx\right]}.$$

As b_0 increases with λ fixed, the radius R decreases, and each circle is nested within those for smaller values of b_0. Hence in the limit $b_0 \to b$ (which may be infinite) the limiting circle approaches either a limit-circle $C_b(\lambda)$ or a limit-point $m_b(\lambda)$. In the latter case there is only one solution, $u(x, \lambda) = \phi(x, \lambda) + m_b(\lambda) \psi(x, \lambda)$, satisfying

$$\|u\| \equiv \int_a^b r(x) |u|^2 \, dx < \infty, \tag{23.71}$$

whereas in the former case every solution satisfies this condition. Detailed proofs of Weyl's theorem may be found in several places (e.g., [152], [232]).

Several examples are of interest [206].

Example 1: An "eigensimple" equation!

$$u'' + \lambda u = 0, \quad a < x < \infty, \tag{23.72}$$

where a is finite.

If $\lambda = 0$ the general solution is a linear function, so no solution is in $\mathcal{L}^2(a, \infty)$; this is the limit-point case at infinity. The general solution for $\lambda \neq 0$ is of course

$$u = A \exp\left(i\lambda^{1/2} x\right) + B \exp\left(-i\lambda^{1/2} x\right), \tag{23.73}$$

where the principle value of $\lambda^{1/2}$ is defined by

$$\lambda^{1/2} = |\lambda|^{1/2} e^{i\theta/2}, \quad 0 \le \theta < 2\pi. \tag{23.74}$$

Thus $\lambda^{1/2}$ is well defined, being analytic in the complex plane except along the positive real axis (the origin is a branch point). If $\lambda \notin [0, \infty)$, then $\operatorname{Im} \lambda^{1/2} > 0$ and $\exp\left(i\lambda^{1/2}x\right) \in \mathcal{L}^2(a; \infty)$ but $\exp\left(-i\lambda^{1/2}x\right)$ is not, and if $\lambda \in [0, \infty)$, then *no* solution is in $\mathcal{L}^2(a; \infty)$. In summary then, when $\operatorname{Im} \lambda^{1/2} \ne 0$ there is exactly one solution in $\mathcal{L}^2(a; \infty)$, but when $\operatorname{Im} \lambda^{1/2} = 0$ there is one square-integrable solution if $\lambda < 0$ (because then $\lambda^{1/2} = |\lambda|^{1/2} e^{i\pi/2} = i |\lambda|^{1/2}$) and none if $\lambda \ge 0$. This is the limit-point case, and repeating the argument for the intervals $-\infty < x < b$ and $b < x < \infty$ yields the same results for the singular endpoints.

Example 2: Bessel's equation of order $v \ge 0$,

$$\left(xu'\right)' - \frac{v^2}{x}u + \lambda xu = 0. \tag{23.75}$$

On the interval $0 < x < b < \infty$ the endpoint $x = 0$ is a singular point because $p(0) = 0$ (and if $v \ne 0, q(0) = \infty$), so a solution u is square integrable with respect to the weight function $r(x) = x$, that is, $u \in \mathcal{L}^2(0, b; x)$ or

$$\int_0^b x |u|^2 \, dx < \infty. \tag{23.76}$$

If $\lambda = 0$ and $v \ne 0$, (23.75) reduces to Euler's equation with linearly independent solutions x^v and x^{-v}, both of which are square integrable if $v < 1$, but only $x^v \in \mathcal{L}^2(0, b; x)$ if $v \ge 1$. If both λ and v are zero, the independent solutions 1 and $\ln x$ are both $\in \mathcal{L}^2(0, b; x)$, so for $v < 1$, the origin is in the limit-circle case and in the limit-point case if $v \ge 1$. In fact for $\lambda \ne 0$ and $v \ne 0$, the independent solutions are $J_v(\lambda^{1/2}x)$ and $J_{-v}(\lambda^{1/2}x)$; the former is finite at the origin while the latter $\sim x^{-v}$ there, so for $v < 1$, both solutions are $\in \mathcal{L}^2(0, b; x)$. If $v = 0$, the solutions are $J_0(\lambda^{1/2}x)$ and $Y_0(\lambda^{1/2}x)$, where the J_0 function is finite at the origin and the Y_0 is logarithmically singular there, so again, both are $\in \mathcal{L}^2(0, b; x)$.

In contrast, on the interval $0 < a < x < \infty$, the left end point is regular whereas the right one is singular, so the requirement for square integrability is now simply

$$\int_a^\infty x |u|^2 \, dx < \infty. \tag{23.77}$$

If $\lambda = 0$, neither of the solutions x^v and x^{-v} for $v > 0$ satisfy this requirement and this is also true for $v = 0$, for which the solutions are again 1 and $\ln x$, so the point at infinity is the limit-point case for all values of v. Clearly these results can be combined if both endpoints are singular, that is, $0 < x < \infty$.

Example 3: Hermite's equation

$$u'' + \left(\lambda - x^2\right)u = 0, \quad -\infty < x < \infty. \tag{23.78}$$

This equation arises, for example, in connection with the quantum harmonic oscillator problem. Both endpoints are singular. If $u = z \exp\left(x^2/2\right)$, then $z(x)$ satisfies the equation

$$z'' + 2xz' + (\lambda + 1) z = 0. \tag{23.79}$$

For $\lambda = 1$ the two independent solutions are $z = 1$ and

$$z = \int_0^x e^{-\xi^2} d\xi, \tag{23.80}$$

corresponding to the basis set for u

$$\left\{ e^{x^2/2}, \, e^{x^2/2} \int_0^x e^{-\xi^2} d\xi \right\}. \tag{23.81}$$

Neither of these solutions are square integrable in the intervals $(-\infty, c)$, or (c, ∞), so both endpoints are in the limit-point case.

Example 4: Legendre's equation

$$\left[(1 - x^2) u'\right]' - \lambda u = 0, \quad -1 < x < 1. \tag{23.82}$$

In this equation both endpoints are singular. For $\lambda = 0$ the independent solutions are

$$\left\{ 1, \ln\left(\frac{1+x}{1-x}\right) \right\}, \tag{23.83}$$

and for any $c \in (-1, 1)$ each of the solutions is square integrable in the intervals $(-1, c)$ and $(c, 1)$, so each endpoint is in the limit-circle case.

Exercise: Verify all the details in Examples 1–4.

Details for the construction of Green's functions in both limit-point and limit-circle cases (along with much else) may be found in [206]; further applications to problems in quantum mechanics are discussed in [207].

PART V

Special Topics in Scattering Theory

Chapter Twenty-Four

The S-Matrix and Its Analysis

What is the S-matrix? We have asked this question before, and are now in a position to investigate it further. The following extract provides some useful information.

> Scattering matrix: An infinite-dimensional matrix or operator that expresses the state of a scattering system consisting of waves or particles or both in the far future in terms of its state in the remote past; also called the S-matrix. In the case of electromagnetic (or acoustic) waves, it connects the intensity, phase, and polarization of the outgoing waves in the far field at various angles to the direction and polarization of the beam pointed toward an obstacle. It is used most prominently in the quantum-mechanical description of particle scattering, in which context it was invented in 1937 by J. A. Wheeler to describe nuclear reactions. ... If the potential energy in the Schrödinger equation, or the scattering obstacle, is spherically symmetric, the eigenfunctions of the S-matrix are spherical harmonics and its eigenvalues are of the form $\exp(2i\delta_l)$, where the real number δ_l is the phase shift of angular momentum l. [208]

Throughout this book only the simplest nonrelativistic systems are considered: one-dimensional systems on the real line or spherically symmetric systems on the half-line. We start with the case of a finite symmetric square well in $(-\infty, \infty)$. This has been encountered before, of course, but there are some S-matrix-related features that can usefully be unfolded in this chapter.

24.1 A SQUARE WELL POTENTIAL

We return to the problem of scattering by a square well but locate the well symmetrically with respect to the origin to elucidate more easily some of the properties of the associated scattering matrix. Thus for a square well of depth $V_0 > 0$ and width a centered on the origin, we have

$$V(x) = -V_0, \qquad |x| < \frac{a}{2}; \qquad (24.1)$$

$$= 0, \qquad |x| > \frac{a}{2}, \qquad (24.2)$$

and the incident particle is represented by a wave function

$$\psi(x) = Ae^{ikx}, \qquad x < -\frac{a}{2}, \qquad (24.3)$$

where $k = E^{1/2}$, and a transmitted wave function

$$\psi(x) = Be^{ikx} S(E), \qquad x > \frac{a}{2}. \qquad (24.4)$$

The term $S(E)$ is the natural form of a scattering matrix for this problem (note that we are not interested in the reflected wave here). After some tedious algebra it can be shown that [61]

$$S(E) = \left[\cos Ka - \frac{i}{2} \left(\frac{k}{K} + \frac{K}{k} \right) \sin Ka \right]^{-1}, \tag{24.5}$$

where now $K = (E + V_0)^{1/2}$. Note that if the particle is incident from the right instead of from the left (as here), it is only necessary to change the signs of k and K in this equation. Regardless of direction, however, the transmissivity of the well is

$$T(E) = |S(E)|^2 = \left[1 + \frac{V_0^2 \sin^2 Ka}{4E(E + V_0)} \right]^{-1} \tag{24.6}$$

(see Section 21.5.1 for a barrier and equation (21.70) in particular).

This expression has maxima equal to one whenever

$$\sin Ka = 0, \quad \text{that is,} \quad Ka = n\pi, \quad n = 1, 2, 3, \ldots. \tag{24.7}$$

Equivalently,

$$E = \left(\frac{n\pi}{a} \right)^2 - V_0 > 0. \tag{24.8}$$

These maxima correspond to resonances—perfect transmission—so that the potential is reflectionless, or "transparent." The well contains an integral number of half wavelengths when this condition is satisfied (this is essentially akin to a well-known interference problem in optics).

Returning to (24.5), we examine $S(E)$ as an analytic function of the energy E. For $E > 0$, $0 < T(E) \leq 1$, so poles of $T(E)$ (and $S(E)$) will only occur when $E < 0$, in fact at the bound state energies of the potential well. Crudely speaking, if $S(E)$ is infinite, it means that a transmitted wave can exist with no incident wave, which is exactly the condition for the existence of a bound state. Of course, the transmitted wave does not propagate; it merely decays exponentially outside the well. From (24.5) it is apparent that $S(E)$ has a pole when

$$\cos Ka = \frac{i}{2} \left(\frac{k}{K} + \frac{K}{k} \right) \sin Ka. \tag{24.9}$$

This equation has no solutions if k (and hence K) is real. From the identity

$$2 \cot 2\theta = \cot \theta - \tan \theta, \tag{24.10}$$

the solutions of (24.9) can be recast in terms of the two conditions for odd and even parity bound state solutions to exist, that is,

$$K \cot \frac{Ka}{2} = ik; \tag{24.11}$$

$$K \tan \frac{Ka}{2} = -ik. \tag{24.12}$$

If we choose the branch of $E^{1/2}$ such that if $E = |E| e^{i\theta}$, then $E^{1/2} = |E|^{1/2} e^{i\theta/2}$, and for $E < 0$, we have $k = i |E|^{1/2}$, which gives, as required, the exponential decay of the external wave. Suppose now that a resonance occurs at $E = E_r \equiv k_r^2 > 0$. In the vicinity of such a value of the resonance energy, we may expand the expression

$$\left(\frac{k}{K} + \frac{K}{k} \right) \tan Ka, \tag{24.13}$$

as

$$\left(\frac{k}{K} + \frac{K}{k} \right) \tan Ka = \frac{d}{dE} \left[\left(\frac{k}{K} + \frac{K}{k} \right) \tan Ka \right]_{E_r} (E - E_r) + O(E - E_r)^2. \tag{24.14}$$

To first order in $(E - E_r)$, on simplifying, we find that

$$\left(\frac{k}{K} + \frac{K}{k} \right) \tan Ka \approx a \left[\frac{dK}{dE} \left(\frac{k}{K} + \frac{K}{k} \right) \right]_{E_r} (E - E_r) \equiv \frac{4}{\Gamma} (E - E_r), \tag{24.15}$$

where Γ is an (energy) width. Now we can rewrite the expression (24.5) for $S(E)$ as

$$S(E) \approx \sec Ka \left[1 - \frac{i}{2} \left(\frac{k}{K} + \frac{K}{k} \right) \tan Ka \right]^{-1} \tag{24.16}$$

$$\approx \sec Ka \left[1 - i \frac{2}{\Gamma} (E - E_r) \right]^{-1} \tag{24.17}$$

$$= \sec Ka \left(\frac{i\Gamma/2}{E - E_r + i\Gamma/2} \right) \tag{24.18}$$

$$\approx \pm \left(\frac{i\Gamma/2}{E - E_r + i\Gamma/2} \right), \tag{24.19}$$

since $Ka \approx n\pi$ near resonance. To this order of approximation, therefore, the pole of $S(E)$ lies in the fourth quadrant of the complex E-plane. There is a branch cut along the real axis $E > 0$, since according to our choice of branch, when $\theta = 2\pi^-$, $E^{1/2} = -|E|^{1/2}$. The discontinuity in $E^{1/2}$ implies a discontinuity in $S(E)$ also. The pole occurring at $E = E_r - i\Gamma/2$ is not in the expression defined by (24.5) but rather is in the analytic continuation of $S(E)$ from above to below the positive real axis, and it lies on the second Riemann sheet of $S(E)$. As already noted, the bound states of the well correspond to poles of $S(E)$ located on the negative real energy axis. The closer the resonances are to the real axis, the stronger they become, that is, the more they behave like very long lived bound states.

Finally, a nice connection can be made by introducing the phase $S(E)$, that is, by writing

$$S(E) = |T(E)|^{1/2} e^{i\delta(E)}. \tag{24.20}$$

For notational convenience, we write equation (24.5) as

$$S(E) = [A(E) - i B(E)]^{-1}, \tag{24.21}$$

with obvious choices for A and B. Then it follows that

$$\tan \delta(E) = \frac{B(E)}{A(E)} = \frac{1}{2} \left(\frac{k}{K} + \frac{K}{k} \right) \tan Ka \approx \frac{2}{\Gamma} (E - E_r). \tag{24.22}$$

Hence

$$\delta(E) \approx \arctan \left[\frac{2}{\Gamma} (E - E_r) \right], \tag{24.23}$$

and

$$\frac{d\delta(E)}{dE} \approx \frac{2\Gamma}{\Gamma^2 + 4(E - E_r)^2}; \tag{24.24}$$

this derivative has a maximum value when $E - E_r$, that is, at a resonance, so it varies rapidly there.

24.1.1 The Bound States

Since $V(x) = V(-x)$, the solutions can be expected to exhibit symmetric or anti-symmetric behavior (even or odd parity) for $E < 0$ with $E + V_0 > 0$. Thus $K = (E + V_0)^{1/2}$ and $k = (-E)^{1/2}$. The solutions are, respectively,

$$\psi(x) = A \cos Kx, \qquad |x| < \frac{a}{2}; \tag{24.25}$$

$$= B e^{-k|x|}, \qquad |x| > \frac{a}{2}, \tag{24.26}$$

and

$$\psi(x) = A \sin Kx, \qquad |x| < \frac{a}{2}; \tag{24.27}$$

$$= B e^{-k|x|}, \qquad |x| > \frac{a}{2}. \tag{24.28}$$

Applying the continuity of ψ'/ψ at $|x| = a/2$, we find for even parity modes

$$K \tan \frac{Ka}{2} = k, \tag{24.29}$$

and for odd parity

$$K \cot \frac{Ka}{2} = -k. \tag{24.30}$$

As indicated above, for $E < 0$, $k = i|E|^{1/2}$, so making the transformation $k \to ik$ in equations (24.11) and (24.12) yields the bound state conditions (24.29) and (24.30).

24.1.2 Square Well Resonance: A Heuristic Derivation of the Breit-Wigner Formula

For simplicity let us assume low energy (E) scattering by a square well potential such that $ka \ll 1$ and only s-wave $(l = 0)$ scattering is important. The scattering amplitude $f_0(\theta)$ is given in terms of the phase shift $\delta_0(E)$ by

$$f_0(\delta_0) = \frac{1}{k} e^{i\delta_0} \sin \delta_0 = \frac{1}{k} \left(\frac{\sin \delta_0}{\cos \delta_0 - i \sin \delta_0} \right). \tag{24.31}$$

We now expand this expression about a resonance energy $E = E_r$, recalling that at a resonance, the phase shift increases by $\pi/2$. Then in the neighborhood of the resonance,

$$\sin \delta_0(E) = \sin \delta_0(E_r) + \cos \delta_0(E_r) \left[\frac{d\delta_0}{dE}\right]_{E=E_r} (E - E_r) + O\left((E - E_r)^2\right) \approx 1,$$
(24.32)

because $\delta_0(E) \to \pi/2$ as $E \to E_r$. Similarly,

$$\cos \delta_0(E) = \cos \delta_0(E_r) - \sin \delta_0(E_r) \left[\frac{d\delta_0}{dE}\right]_{E=E_r} (E - E_r)$$
(24.33)

$$+ O((E - E_r)^2)$$

$$\approx -\left[\frac{d\delta_0}{dE}\right]_{E=E_r} (E - E_r) \equiv -\frac{2}{\Gamma} (E - E_r),$$
(24.34)

where

$$\left[\frac{d\delta_0}{dE}\right]_{E=E_r} = \frac{2}{\Gamma}.$$
(24.35)

Therefore we can rewrite the expression (24.31) as

$$f_0(\delta_0) \approx -\frac{1}{k} \frac{\Gamma/2}{(E - E_r) + i\Gamma/2}.$$
(24.36)

The total (low energy) cross section σ_T is therefore approximately given by

$$\sigma_T \approx \frac{4\pi}{k^2} |f_0(\delta_0)|^2 \approx \frac{\pi}{k^2} \frac{\Gamma^2}{(E - E_r)^2 + \Gamma^2/4}.$$
(24.37)

This is the *Breit-Wigner resonance formula*, where Γ is the width of the resonance curve. It describes the absorption of an incident particle to form a metastable state of energy E_r and lifetime proportional to Γ^{-1}.

24.1.3 The Watson Transform and Regge Poles

Although in principle the rainbow problem can be "solved" with enough computer time and resources, numerical solutions by themselves offer little or no insight into the physics of the phenomenon. However, there was a significant mathematical development in the early 20th Century that eventually had a profound impact on the study of scalar and vector scattering: The Watson transform, originally introduced in 1918 by Watson in connection with the diffraction of radio waves around the earth [89] (and subsequently modified by several mathematical physicists in studies of the rainbow problem), is a method for transforming the slowly-converging partial-wave series into a rapidly convergent expression involving an integral in the complex angular-momentum plane. This allows the above transformation to effectively "redistribute" the contributions to the partial wave series into a few points in the complex plane—specifically poles (called Regge poles in elementary particle physics) and saddle points. Such a decomposition means that instead of identifying angular

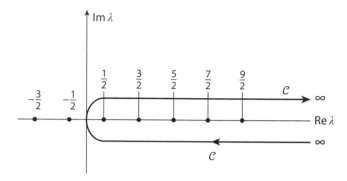

Figure 24.1 Contour for the integral (24.40).

momentum with certain discrete real numbers, it is now permitted to move continuously through complex values. However, despite this modification, the poles and saddle points have profound physical interpretations in the rainbow problem [[26] and references therein]. [149]

Reference was made in Chapters 19 and 20 to the Watson transform (sometimes known also as the *Sommerfeld-Watson transform*). To discuss this, let us broaden the picture, so to speak, from potential barriers and wells to recover an approximation to the abovementioned Breit-Wigner resonance by using this transform. We know that the scattering amplitude $f(\theta)$ can be written variously in terms of the phase shift $\delta_l = \delta_l(k)$ as

$$f(\theta) = \frac{1}{2ik} \sum_{l=0}^{\infty} (2l+1)\left(e^{2i\delta_l} - 1\right) P_l(\cos\theta)$$

$$= \frac{1}{k} \sum_{l=0}^{\infty} (2l+1)\left(e^{i\delta_l} \sin\delta_l\right) P_l(\cos\theta). \tag{24.38}$$

This can be transformed into an integral in the complex l-plane by interpreting each term in (18.13) as the residue at the integer l. This requires

$$f_l = \frac{e^{2i\delta_l} - 1}{2ik} \equiv F(l) \tag{24.39}$$

to be a meromorphic function of l, and the integral can then be decomposed into pole contributions and a so-called background integral taken along a path along the line $\mathrm{Re}\, l = -1/2$. To see this, it is convenient to use the related complex variable $\lambda = l + 1/2$. We shall indicate how to arrive at the important result [41]

$$f(\theta) = \frac{1}{i} \int_C \lambda F\left(\lambda - \frac{1}{2}\right) \frac{P_{\lambda-1/2}(-\cos\theta)}{\cos\lambda\pi} d\lambda. \tag{24.40}$$

If we note that $\cos\lambda\pi = -\sin l\pi$ has zeros at the real points (Figure 24.1)

$$l = 0, \pm 1, \pm 2, \ldots \quad \left(\lambda = \pm\frac{1}{2}, \pm\frac{3}{2}, \pm\frac{5}{2}, \ldots\right), \tag{24.41}$$

with corresponding residues $(-1)^{l+1}/\pi$. If the remaining terms in the integrand of (24.40) are analytic for $\lambda \in (0, \infty)$ (see [63]), the integral over $\mathcal{C}$ is readily decomposed into a sum of *clockwise* circular integrals around the poles, such that

$$f(\theta) = \frac{1}{\pi i} \sum_{l=0}^{\infty} (-2\pi i) (-1)^{l+1} \left(l + \frac{1}{2} \right) F(l) P_l(- \cos \theta), \qquad (24.42)$$

and the Legendre polynomial property that

$$P_l(\cos \theta) = (-1)^l \, P_l (- \cos \theta), \qquad l = 0, 1, 2, \ldots, \qquad (24.43)$$

yields the required result:

$$f(\theta) = \sum_{l=0}^{\infty} (2l + 1) \, F(l) P_l (\cos \theta). \qquad (24.44)$$

To proceed further, the path of integration is deformed into the line $\mathrm{Re}\,\lambda = 0$ and an infinitely large semicircle for $\mathrm{Re}\,\lambda > 0$. If the integrand falls away at least as fast as $|\lambda|^{-(1+\varepsilon)}$, $\varepsilon > 0$, then the contribution from the semicircular contour is zero (though we do not prove that here). The integral along the imaginary λ-axis and the small loop integrals around the poles that the path "passes over" under deformation make up the rest of the contributions. If the finite number of poles are identified by $\lambda = \lambda_n$ and the residues of $F(l)$ at those poles are denoted by β_n, the pole contributions can be represented symbolically by

$$f_{\mathrm{p}}(\theta) = 2 \sum_n \beta_n \lambda_n \frac{P_{\lambda_n - 1/2} (- \cos \theta)}{\cos \lambda_n \pi}. \qquad (24.45)$$

The background integral, for which $\lambda = iy$, $y \in \mathbb{R}$, can be written as

$$f_{\mathrm{b}}(\theta) = -\frac{1}{\pi i} \int_{-\infty}^{\infty} yF \left(iy - \frac{1}{2} \right) \frac{P_{iy-1/2} (- \cos \theta)}{\cosh \pi y}, \qquad (24.46)$$

and therefore,

$$f(\theta) = f_{\mathrm{p}}(\theta) + f_{\mathrm{b}}(\theta) \qquad (24.47)$$

$$= 2 \sum_n \beta_n \lambda_n \frac{P_{\lambda_n - 1/2} (- \cos \theta)}{\cos \lambda_n \pi}$$

$$- \frac{1}{\pi i} \int_{-\infty}^{\infty} yF \left(iy - \frac{1}{2} \right) \frac{P_{iy-1/2} (- \cos \theta)}{\cosh \pi y}. \qquad (24.48)$$

A pole of $F(l)$ in the complex l-plane is referred to as a *Regge pole*, named for the celebrated Italian mathematical physicist Tullio Regge. The path traced out by such a pole as the energy varies is called a *Regge trajectory*.

Example

Suppose that a pole of $F(l)$ lies near a physical value of l, that is, an integer value, $l = N$, say. We shall investigate the behavior of a single pole as the energy varies slightly, and drop the subscript n accordingly. Therefore

$$f_p(\theta) = 2\beta\lambda \frac{P_{\lambda-1/2}(-\cos\theta)}{\cos\lambda\pi}, \tag{24.49}$$

with

$$\lambda = N + \frac{1}{2} + \xi + i\eta; \qquad |\xi| \ll 1, \ |\eta| \ll 1. \tag{24.50}$$

Consequently,

$$\sin\left(\lambda - \frac{1}{2}\right)\pi = (-1)^N \sin\left[(\xi + i\eta)\pi\right]. \tag{24.51}$$

Now for complex index $\lambda - 1/2 = \nu$, the Legendre polynomials can be expressed in terms of those with integer indices. Thus

$$P_\nu(s) = \frac{\sin\nu\pi}{\pi} \sum_{n=0}^{\infty} (2n+1) \frac{P_n(-s)}{(\nu-n)(\nu+n+1)}. \tag{24.52}$$

Therefore we have

$$P_{\lambda-1/2}(-\cos\theta) = \frac{(-1)^N}{\pi} \sin\left[(\xi + i\eta)\pi\right]$$
$$\times \sum_{n=0}^{\infty} \frac{(2n+1) P_n(\cos\theta)}{(N-n+\xi+i\eta)(N+n+1+\xi+i\eta)}, \tag{24.53}$$

and hence

$$f_p(\theta) = \frac{2\beta}{\pi}\left(N + \frac{1}{2} + \xi + i\eta\right) \sum_{n=0}^{\infty} \frac{(2n+1) P_n(\cos\theta)}{(N-n+\xi+i\eta)(N+n+1+\xi+i\eta)}. \tag{24.54}$$

Following [41], we note that the factor $(N - n + \xi + i\eta)$ in the denominator is very small when $n = N$, so this term will in general dominate all the rest. Retaining only this term, we arrive at the approximation

$$f_p(\theta) \approx \frac{\beta}{\pi} (2N+1) \frac{P_n(\cos\theta)}{\xi + i\eta}. \tag{24.55}$$

How does this expression depend on the energy E_N of the particle? Since the physical state of angular momentum $l = N$ lies close to the pole, its real part E_0 will differ only slightly from E_N, and its imaginary part will be small. Writing the form

$$E_N = E_0 - i\frac{\Gamma}{2}, \tag{24.56}$$

so that, from (24.50),

$$\xi(E_N) + i\eta(E_N) = 0. \tag{24.57}$$

If now E is any real energy corresponding to a physical state, close to E_N, then

$$\xi(E) + i\eta(E) \approx \left(\frac{d\lambda}{dE}\right)_{E_N} (E - E_N), \tag{24.58}$$

so from the approximation (24.55), we have

$$f_p(\theta) \approx \frac{\beta}{\pi} \frac{(2N+1)}{(d\lambda/dE)_{E_N}} \frac{P_n(\cos\theta)}{E - E_0 + i\Gamma/2}, \tag{24.59}$$

from which the differential scattering cross section has the familiar Breit-Wigner resonance form

$$f_p(\theta)^2 \propto \frac{1}{4(E - E_0)^2 + \Gamma^2}, \tag{24.60}$$

E_0 now being the resonance energy, and Γ the resonance width. More can be deduced if we denote the values of ξ and η at $E = E_0$ by ξ_0 and η_0, respectively, on the pole trajectory. Then

$$\lambda = N + \frac{1}{2} \approx \lambda(E_0) + \left(\frac{d\lambda}{dE}\right)_{E_0} (E_N - E_0), \tag{24.61}$$

so that

$$\xi_0 + i\eta_0 \approx i\frac{\Gamma}{2}\left(\frac{d\lambda}{dE}\right)_{E_0}. \tag{24.62}$$

Therefore $\xi_0 = 0$ and to this level of approximation,

$$\eta_0 \approx \frac{\Gamma}{2}\left(\frac{d\lambda}{dE}\right)_{E_0} \tag{24.63}$$

is proportional to the resonance line width.

The Watson Transform and the Poisson Summation Formula

As noted earlier, the Watson transform was originally motivated by the desire to understand diffraction of radio waves around the earth (prior to the discovery of the ionosphere). The topic of primary interest was therefore the neighborhood of the earth's surface in the shadow zone of the transmitter. Once the transformation to the complex l-plane had been made, with the corresponding freedom of path deformation, rapidly convergent asymptotic estimates could be made by paths suitably chosen to emphasize a few dominant pole contributions as opposed to dealing with those from many partial waves. In the shadow region the dominant (complex) poles λ_n are now called Regge poles, and their imaginary part (associated with angular damping) grows rapidly with n, so in general only those few poles closest to the real axis need to be considered. This leads to a rapidly convergent residue series, each term of which corresponds to a "creeping wave" generated by tangentially incident rays that are radiation damped as they travel around the surface of the sphere.

In the lit region Watson's method fails: in addition to Regge poles, the now-dominant "background integral" in the l-plane (or equivalently the λ-plane, where $\lambda = l + 1/2$) must be considered. The significant contributions to this integral come from saddle points, which are, in general, complex. When these are real, however, they correspond to the rays of geometrical optics and provide the associated contributions; by taking account of higher-order terms in the saddle-point method, the WKB(J) series can be recovered to any desired accuracy. There is a penumbral region between the lit and shadow regions: this was investigated by Fock [237]; he described the behavior of the wave field in this region in terms of a new function (subsequently named the *Fock function*). It is in this region, near the edge, that the creeping waves are generated. The Fock function interpolates smoothly between the geometrically lit (or WKB(J)) region and the diffracted ray or creeping wave region. While subsequent attempts were made to apply the method to both impenetrable and transparent spheres (see [25] for further references), and significant advances were made, there was no theory going beyond the classic Airy approximation in a comprehensive manner for all scattering angles. In due time, however, the original Watson transform method was extended to the scattering amplitude or the wave field by means of the related Poisson summation formula [236], [239] (and see also Section 28.1).

A succinct formal proof of the latter follows [234]. Suppose that $a(\lambda)$ is a continuous function for $\lambda \in (0, \infty)$ and that

$$a_l = a\left(l + \frac{1}{2}\right). \tag{24.64}$$

From the theory of (complex) Fourier series, with standard Dirichlet conditions [235], we may expand $a(l + x)$ as such a series in the interval $(0, L)$. Thus

$$a(l + x) = \sum_{m=-\infty}^{\infty} A_m e^{-2m\pi i x/L}, \tag{24.65}$$

where l is an integer, with Fourier coefficents defined by

$$A_m = \int_l^{l+1} a(\lambda) e^{2m\pi i \lambda} d\lambda.$$

If we set $x = 1/2$ in equation (24.65), and $L = 1$

$$A_m = (-1)^m \int_l^{l+1} a(\lambda) d\lambda,$$

hence

$$a_l = a\left(l + \frac{1}{2}\right) = \sum_{m=-\infty}^{\infty} (-1)^m \int_l^{l+1} a(\lambda) e^{2m\pi i \lambda} d\lambda.$$

If we now sum over all l, we obtain the Poisson summation formula

$$\sum_{l=0}^{\infty} a_l = \sum_{m=-\infty}^{\infty} (-1)^m \int_0^{\infty} a(\lambda) e^{2m\pi i \lambda} d\lambda, \tag{24.66}$$

with an appropriate choice of deformed path in the complex λ-plane depending on the problem of interest. This choice, which is also different for different angular regions, leads to Regge pole contributions and saddle-point-dominated background integrals. A saddle point on the real axis is also a point of stationary phase in (24.66), and this characterizes an extremal path (i.e., a ray in geometrical optics or a classical orbit in the particle context). The integer m in equation (24.66) has the topological significance of a *winding number*, which is associated with the number of circumlocutions performed by a path around the center of the sphere.

The *Fock functions* are defined as

$$f(z) = \frac{e^{-i\pi/6}}{2\pi} \int_{\Gamma} \frac{e^{iz\xi}}{\text{Ai}(\xi e^{2i\pi/3})} d\xi;$$

$$g(z) = -\frac{e^{i\pi/6}}{2\pi} \int_{\Gamma} \frac{e^{iz\xi}}{\text{Ai}'(\xi e^{2i\pi/3})} d\xi,$$

where Γ is a contour starting at infinity in the angular sector $\pi/3 < \arg \xi < \pi$, passing between the origin and the pole of the integrand nearest the origin and ending at infinity in the angular sector $-\pi/3 < \arg \xi < \pi/3$. They can be generalized to include integer powers of ξ in the integrand. To leading order, their asymptotic behavior is [237]

$$f(z) \sim 2iz \exp(-iz^3/3), \quad z \to -\infty,$$
$$\sim 0, \quad z \to \infty,$$

and

$$g(z) \sim 2 \exp(-iz^3/3), \quad z \to -\infty$$
$$\sim 0, \quad z \to \infty.$$

24.2 MORE DETAILS FOR THE TIRSE

Let us return to the time-independent radial Schrödinger equation (TIRSE) for $k \in \mathbb{R}$,

$$\left[\frac{d^2}{dr^2} - \frac{l(l+1)}{r^2} - V(r) + k^2\right] u(r) = 0, \tag{24.67}$$

and now consider l to be a suitably differentiable real variable. On differentiating (24.67) with respect to l and forming the Wronskian of u and $\partial u/\partial l$, we obtain the result

$$\frac{d}{dr} W\left(u, \frac{\partial u}{\partial l}\right) = (2l+1) \frac{u^2}{r^2}, \tag{24.68}$$

so that

$$\left[W\left(u, \frac{\partial u}{\partial l} \right) \right]_0^\infty = (2l+1) \int_0^\infty \frac{u^2(r)}{r^2} dr > 0. \tag{24.69}$$

Exercise: Derive (24.68).

The Wronskian is zero at the lower limit, and using the known asymptotic form of $u(r)$ and the Jost function result that

$$f_l(k) = |f_l(k)| \exp\left[-i\delta_l(k) \right], \tag{24.70}$$

we can show that

$$k \, |f_l(k)|^2 \left(\frac{\pi}{2} - \frac{\partial \delta_l}{\partial l} \right) > 0 \Rightarrow \frac{\partial \delta_l}{\partial l} < \frac{\pi}{2}. \tag{24.71}$$

This inequality means that no physical phase shift $\delta_{l+1}(k)$ can exceed the preceding lower one by more than $\pi/2$:

$$\delta_{l+1}(k) < \delta_l(k) + \frac{\pi}{2}. \tag{24.72}$$

Exercise: Derive (24.71).

By *Levinson's theorem* [63] (and see below), the number of bound states of angular momentum l is

$$n_l = \frac{\delta_l(0)}{\pi}, \tag{24.73}$$

hence from the inequality (24.72),

$$n_{l+1} < n_l + \frac{1}{2}. \tag{24.74}$$

But the n_l are all integers, so it follows that

$$n_{l+1} \le n_l. \tag{24.75}$$

This means that the number of bound states with angular momentum $l + 1$ can never exceed the number with angular momentum l. If we carry out a similar calculation using (24.67) but with respect to $E = k^2$ rather than l (where now $l = l(E)$), we obtain the equation corresponding to (24.68), namely,

$$\frac{d}{dr} W \left(u, \frac{\partial u}{\partial E} \right) = \left(\frac{2l+1}{r^2} \frac{dl}{dE} - 1 \right) u^2. \tag{24.76}$$

On integration, therefore,

$$(2l+1) \frac{dl}{dE} \int_0^\infty \frac{u^2(r)}{r^2} dr = \int_0^\infty u^2(r) dr \Rightarrow \frac{dl}{dE} > 0 \tag{24.77}$$

[63]–[65].

24.3 LEVINSON'S THEOREM

From the definition

$$S_l(k) = e^{2i\delta_l(k)}, \tag{24.78}$$

it follows that the logarithmic derivative of the scattering matrix $S_l(k)$ is related to the derivative of the phase shift:

$$\frac{S_l'(k)}{S_l(k)} = 2i\delta_l'(k). \tag{24.79}$$

In view of (24.79) it is useful to consider the integral

$$\mathcal{I} = \int_{-\infty}^{\infty} \frac{S_l'(k)}{S_l(k)} dk = 4i \int_0^{\infty} \delta_l'(k)\, dk = 4i\left[\delta_l(\infty) - \delta_l(0)\right], \tag{24.80}$$

since $\delta_l(k)$ is an odd function of its argument and is here regarded as a continuous function of k. We also know that $S_l(k)$ can be expressed in terms of the Jost functions $f_l(k)$ and $f_l(-k)$, namely,

$$S_l(k) = (-1)^l \frac{f_l(k)}{f_l(-k)}, \tag{24.81}$$

so that

$$\mathcal{I} = \int_{-\infty}^{\infty} \frac{S_l'(k)}{S_l(k)} dk = \int_{-\infty}^{\infty} \frac{d}{dk} \left[\ln f_l(k) - \ln f_l(-k)\right] dk \tag{24.82}$$

$$= 2 \int_{-\infty}^{\infty} \frac{d}{dk} \left[\ln f_l(k)\right] dk. \tag{24.83}$$

Exercise: Prove that $S_l(k)$ is an odd function of its argument.

Now the function $\ln f_l(k)$ tends to zero exponentially as $k \to \infty$ along any direction in the lower half k-plane, so the integral $\mathcal{I}$ is closed from below. The *Principle of the Argument* states that if (i) a function $f(z)$ is analytic on and inside a positively oriented simple closed path $\mathcal{C}$, except at a finite number of poles in the interior of $\mathcal{C}$, and (ii) $f(z)$ has no zeros on $\mathcal{C}$, then

$$\frac{1}{2\pi i} \int_{\mathcal{C}} \frac{f'(z)}{f(z)} dz = N_z - N_p, \tag{24.84}$$

where N_z and N_p are the number of zeros and poles of $f(z)$ in $\mathcal{C}$ respectively (where the orders of zeros and poles are counted). A proof of this theorem can be found in most texts on functions of a complex variable. Since $f_l(k)$ is analytic in the lower half-plane, only its zeros on the negative imaginary axis contribute to the integral, so that

$$\mathcal{I} = -4\pi i N_p = -4\pi i N_l, \tag{24.85}$$

because N_l is the number of bound states associated with the zeros of $f_l(k)$. Comparing equation (24.80) with equation (24.85), it follows that

$$\delta_l(0) - \delta_l(\infty) = N_l \pi, \tag{24.86}$$

unless there is a bound state at $k = 0$ (i.e., if $f_l(0) = 0$), in which case this result is modified to

$$\delta_l(0) - \delta_l(\infty) = \left(N_l + \frac{1}{2} \right) \pi \tag{24.87}$$

[65]. If the range of the potential is finite, $\delta_l(\infty) = 0$, and so the zero-energy scattering phase shift is determined by the number of bound states. In particular, for $l = 0$ and $f_0(0) = 0$, we have

$$S_0(0) = \exp[2i\delta_0(0)] = -1, \tag{24.88}$$

and for all $l \neq 0$, we have $S_l(0) = 1$, so the partial scattering amplitude

$$f_l(k) = \frac{1}{2ik} [S_l(k) - 1] \tag{24.89}$$

is finite at $k = 0$ except when $l = 0$.

Chapter Twenty-Five

The Jost Solutions: Technical Details

The nonrelativistic quantum mechanical two-body problem can be described conveniently and nicely in terms of the Jost functions and Jost solutions of the Schrödinger equation. When defined for all complex values of the momentum, these functions contain complete information about the underlying physical system. In contrast to the conventional treatment of the two-body problem, based on the scattering amplitude and physical wave-function, in the Jost function approach the bound, virtual, scattering, and resonance states are treated on an equal footing and simultaneously. Also the Jost function is a more fundamental quantity than the S-function since it is free of ambiguities caused by redundant zeros, while the S-function may have redundant poles.

[257]

25.1 ONCE MORE THE TIRSE

Returning to (24.67) but using the variable $\lambda = l + 1/2$, $\lambda \in \mathbb{C}$, we have

$$\left[\frac{d^2}{dr^2} - \left(\frac{\lambda^2 - 1/4}{r^2}\right) - V(r) + k^2\right] u(\lambda, k, r) = 0. \tag{25.1}$$

Now if $\phi(\lambda, k, r)$ is the solution of (25.1) that is regular at the origin, then we know that

$$\lim_{r \to 0^+} r^{-(\lambda+1/2)} \phi(\lambda, k, r) = 1; \tag{25.2}$$

correspondingly the other linearly independent solution is $\phi(-\lambda, k, r)$, with the behavior

$$\lim_{r \to 0^+} r^{(\lambda-1/2)} \phi(-\lambda, k, r) = 1. \tag{25.3}$$

It is readily seen that

$$W[\phi(\lambda, k, r), \phi(-\lambda, k, r)] = -2\lambda. \tag{25.4}$$

Let the Jost solutions of (25.1) be $f(\pm\lambda, k, r)$, so that, in particular,

$$\lim_{r \to \infty} e^{ikr} f(\lambda, k, r) = 1 \tag{25.5}$$

(see also Section 17.2); note also that

$$W[f(\lambda, k, r), f(-\lambda, k, r)] = 2ik. \tag{25.6}$$

We know that $f(\lambda, k, r)$ is an entire function of λ, because the coefficient of $u(\lambda, k, r)$ in (25.1) is entire and the boundary condition (25.5) is independent of λ, and furthermore, if $\lambda \in \mathbb{R}$, $f(\lambda, k, r)$ is an even function of λ. And if $V(r)$ is a real function,

$$f(\lambda, k, r) = \bar{f}(\bar{\lambda}, \bar{k}, r), \tag{25.7}$$

as is readily demonstrated. Following [65] we can prove that $\phi(\lambda, k, r)$ is an analytic function for Re $\lambda > 0$ by comparing the properties of a new equation with those of (25.1). To that end, let

$$r = r_0 e^{-\rho}, \quad u = w e^{-\rho/2}. \tag{25.8}$$

With these changes of variables equations (25.1) and (25.2) are transformed to the pair

$$\left[\frac{d^2}{d\rho^2} - \lambda^2 - \hat{V}(\rho)\right] w = 0, \tag{25.9}$$

and

$$\lim_{\rho \to \infty} e^{\lambda\rho} w(\rho) = 1, \tag{25.10}$$

where

$$\hat{V}(\rho) = \left[V(e^{-\rho}) - k^2\right] r_0^2 e^{-2\rho}. \tag{25.11}$$

By directly comparing (25.1) and (25.2) with (25.9) and (25.5), we note that, with this new potential,

$$w(\rho) = f\left(\frac{1}{2}, -i\lambda, \rho\right). \tag{25.12}$$

Hence $w(\rho)$ is an analytic function for Im $-i\lambda < 0$, or Re $\lambda > 0$, so that $\phi(\lambda, k, r)$ is also analytic in the same region (i.e., Re $\lambda > 0$.) The *Jost function* $f(\lambda, k)$ is defined as the Wronskian of $f(\lambda, k, r)$ and $\phi(\lambda, k, r)$, that is,

$$f(\lambda, k) = W[f(\lambda, k, r), \phi(\lambda, k, r)]$$
$$= f(\lambda, k, r)\phi'(\lambda, k, r) - f'(\lambda, k, r)\phi(\lambda, k, r) \tag{25.13}$$

(see Chapters 17 and 26).

We write $\partial f(\lambda, k, r)/\partial r$ as $f'(\lambda, k, r)$ in what follows. Because (i) $f(\lambda, k, r)$ and $f'(\lambda, k, r)$ are entire functions of λ, and (ii) $\phi(\lambda, k, r)$ and $\phi'(\lambda, k, r)$ are analytic functions of λ for Re $\lambda > 0$, $f(\lambda, k)$ is analytic in λ Re $\lambda > 0$ and fixed k. But as will be seen below $f(\lambda, k)$ is an analytic function of k for Im $k < 0$ (and fixed λ); hence the Jost solution is analytic in the domain

$$D^-(\lambda, k) \equiv (\text{Re } \lambda > 0) \otimes (\text{Im } k < 0). \tag{25.14}$$

The general solution of (25.1) is a linear combination of any two linearly independent solutions and, in fact, for the regular solution we have

$$\phi(\lambda, k, r) = \frac{1}{2ik}[f(\lambda, k)f(\lambda, -k, r) - f(\lambda, -k)f(\lambda, k, r)], \qquad (25.15)$$

where (25.6) has been used. The asymptotic form of $\phi(\lambda, k, r)$ is

$$\phi(\lambda, k, r) \sim \frac{1}{2ik}[f(\lambda, k)e^{ikr} - f(\lambda, -k)e^{-ikr}], \qquad (25.16)$$

where we have replaced $f(\lambda, \pm k, r)$ with $e^{\pm ikr}$, respectively. If we parametrize the Jost functions in (25.16) as

$$f(\lambda, k) = \tau(\lambda, k)\exp[i\delta(\lambda, k) - (\lambda - 1/2)/2i\pi]; \qquad (25.17)$$

$$f(\lambda, -k) = \tau(\lambda, k)\exp[-i\delta(\lambda, k) + (\lambda - 1/2)/2i\pi], \qquad (25.18)$$

where $\tau(\lambda, k)$ is the complex amplitude, then (25.16) becomes

$$\phi(\lambda, k, r) \sim \frac{1}{k}\tau(\lambda, k)\sin\left[kr + \delta(\lambda, k) - \frac{1}{2}\pi\left(\lambda - \frac{1}{2}\right)\right], \qquad (25.19)$$

which agrees with the standard partial-wave analysis in quantum mechanics texts [60], [62]. The S-matrix is then given by

$$S(\lambda, k) = e^{2i\delta(\lambda, k)} = \left[\frac{f(\lambda, k)}{f(\lambda, -k)}\right]e^{i\pi(\lambda - 1/2)} \equiv (-1)^l\left[\frac{f(\lambda, k)}{f(\lambda, -k)}\right] \text{ if } l \in \mathbb{Z}. \qquad (25.20)$$

(See (26.43)). Therefore $S(\lambda, k)$ is a meromorphic function in the related domain

$$D^+(\lambda, k) \equiv (\operatorname{Re}\lambda > 0) \otimes (\operatorname{Im}k > 0). \qquad (25.21)$$

Note that the S-matrix is proportional to the ratio of the coefficients of the outgoing and incoming waves in (25.16). From (25.7) and (25.20) we find that for real $V(r)$, $S(\lambda, k)$ satisfies the unitarity property

$$\bar{S}(\bar{\lambda}, \bar{k}) = S^{-1}(\lambda, k). \qquad (25.22)$$

But of course, if $k \in \mathbb{R}$ then

$$\bar{S}(\bar{\lambda}, k) = S^{-1}(\lambda, k), \qquad (25.23)$$

so if for real k, $S(\lambda, k)$ has a pole at $\lambda = \lambda_p$, then both $\bar{S}$ and S have a zero at $\lambda = \bar{\lambda}_p$. We already know that if both k and λ are real, then $S(\lambda, k)$ is a phase, usually written, as noted several times earlier, in terms of the real phase shift $\delta(\lambda, k)$ (though if λ or k is complex, so too is δ) as

$$S(\lambda, k) = e^{2i\delta(\lambda, k)}. \qquad (25.24)$$

If we multiply (25.1) by $\bar{u}$ and subtract its complex conjugate, for real values of k the result is

$$\frac{d}{dr}[W(\bar{u}, u)] = 2i\operatorname{Im}(\lambda^2)\frac{|u|^2}{r^2}, \qquad (25.25)$$

so on using (25.15), (25.17), and (25.18) and integrating over $(0, \infty)$, we obtain

$$2i \operatorname{Im}(\lambda^2) \int_0^\infty \frac{|\phi|^2}{r^2} dr = \frac{1}{2k} \left[|f(\lambda, k)|^2 - |f(\lambda, -k)|^2 \right]$$

$$= \frac{1}{2k} |\tau(\lambda, k)|^2 (A - A^{-1}), \qquad (25.26)$$

where

$$A = \exp\left[-2\left(\operatorname{Im}\delta - \frac{\pi}{2}\operatorname{Im}\lambda\right)\right]. \qquad (25.27)$$

Exercise: Derive (25.26).

If $\operatorname{Im}\lambda < 0$ and $\operatorname{Re}\lambda > 0$, the left-hand side of equation (25.26) is negative, implying that $\operatorname{Im}\delta > (\pi/2)\operatorname{Im}\lambda$. $\operatorname{Im}\delta$ is bounded below when these conditions are satisfied, and therefore (25.20) is bounded from above, and hence possesses no poles in the lower half of the λ-plane. Furthermore, by virtue of result (25.23), it follows that $S(\lambda, k)$ has no zeros in the upper half of the λ-plane. From (25.20), poles in the complex λ-plane for real energy (or $k^2 \in \mathbb{R}$) are Regge poles determined by the condition $f(\lambda, -k) = 0$, in other words, where $\lambda = \lambda(k)$, say. For $E = k^2 < 0$ we write $k = i\kappa$, $\kappa > 0$, corresponding to values of E on the so-called physical Riemann sheet. From expression (25.16) we see that the asymptotic behavior of the regular solution is

$$\phi(\lambda, k, r) \sim \frac{1}{2ik} f(\lambda, k) e^{ikr} \sim \phi(\lambda, k, r) \sim -\frac{1}{2\kappa}[f(\lambda(k), i\kappa)e^{-\kappa r}, \quad r \to \infty,$$
$$(25.28)$$

and this, together with the fact that $\phi(\lambda, k, 0) = 0$, provides the result

$$2i \operatorname{Im}(\lambda^2) \int_0^\infty \frac{|\phi|^2}{r^2} dr = 0 \Rightarrow \operatorname{Im}\{l(k)[l(k)+1]\} \int_0^\infty \frac{|\phi|^2}{r^2} dr = 0. \quad (25.29)$$

The integral converges because $l(k) > -1/2$, so we conclude that $l(k) \in \mathbb{R}$ and takes on physical values (i.e., integers).

25.2 THE REGULAR SOLUTION AGAIN

Consider the solutions of the simplified equation (25.1) as functions of k. Depending on the topic of interest, different conditions can be imposed on the potential. For example, to ensure that the potential is less singular than r^{-2} at the origin and decreases faster than r^{-3} at infinity, we require [64]

$$\int_0^\infty r |V(r)| dr = M_0 < \infty, \qquad (25.30)$$

and

$$\int_0^\infty r^2 |V(r)| dr = M_\infty < \infty. \qquad (25.31)$$

For fixed real λ, the regular solution $u = \phi(\lambda, k, r)$ satisfies the boundary conditions at $r = 0$:

$$\phi(\lambda, k, 0) = 0; \quad \phi'(\lambda, k, 0) = 1, \qquad (25.32)$$

where the prime refers to the derivative with respect to r. For real values of k, $\phi(\lambda, k, r) \in \mathbb{R}$ and is an even function of k. If, in contrast $k \in \mathbb{C}$, then since polynomials are entire functions, the coefficient of u in (25.1) is an entire function, of k, and by a theorem of Poincaré, since the boundary conditions are independent of k, the solution $\phi(\lambda, k, r)$ is an entire function of k. We can convert (25.1) into an integral equation using the method of variation of parameters. We obtain

$$\phi(\lambda, k, r) = r^{\lambda+1/2} + \frac{1}{2}\lambda^{-1} \int_0^r [(\xi/r)^\lambda - (r/\xi)^\lambda]$$

$$\times (r\xi)^{1/2}[k^2 - V(\xi)]\phi(\lambda, k, \xi)d\xi, \tag{25.33}$$

with boundary conditions at $r = 0$

$$\lim_{r \to 0} \phi(\lambda, k, r) = 0; \quad \lim_{r \to 0} \phi'(\lambda, k, r) = \lim_{r \to 0}\left(\lambda + \frac{1}{2}\right)r^{\lambda-1/2}. \tag{25.34}$$

Equation (25.33) is a Volterra integral equation, and we write its solution as a perturbation expansion

$$\phi(\lambda, k, r) = \sum_{n=0}^{\infty} \phi_n(\lambda, k, r), \tag{25.35}$$

where

$$\phi_0(\lambda, k, r) = r^{\lambda+1/2},$$

and

$$\phi_{n+1}(\lambda, k, r) = \frac{1}{2}\lambda^{-1}\int_0^r [(\xi/r)^\lambda - (r/\xi)^\lambda](r\xi)^{1/2}[k^2 - V(\xi)]\phi_n(\lambda, k, \xi)d\xi. \tag{25.36}$$

The perturbation expansion for $\phi(\lambda, k, r)$ is bounded term-by-term and is unrestrictedly convergent. In fact, adapting the analysis in [65] for $\lambda = 1/2$, we can show that for real values of λ,

$$|\phi_n(\lambda, k, r)| \leq O\left\{\frac{r}{1 + |kr|}\exp\left[|\mathrm{Im}\,(kr)|\frac{M_0}{n!}\right]\right\}. \tag{25.37}$$

We may therefore conclude, following [64], that the $\phi_n(\lambda, k, r)$ and hence $\phi(\lambda, k, r)$ are analytic for $k \in \mathbb{C}$ and so the physical solution (defined originally for $k \in \mathbb{R}$) can be analytically continued to complex k. A similar approach for the Jost solution yields

$$f(\lambda, k, r) = e^{-ikr} + k^{-1}\int_r^\infty [\sin k(\xi - r)]$$

$$\times \left[V(\xi) + \left(\lambda^2 - \frac{1}{4}\right)\xi^{-2}\right]f(\lambda, k, \xi)d\xi. \tag{25.38}$$

We can write the solution of (25.38) as a perturbation expansion

$$f(\lambda, k, r) = \sum_{n=0}^{\infty} g_n(\lambda, k, r), \tag{25.39}$$

where $g_0 = e^{-ikr}$, and

$$g_{n+1}(\lambda, k, r) = k^{-1} \int_r^{\infty} [\sin k(\xi - r)] \left[V(\xi) + \left(\lambda^2 - \frac{1}{4} \right) \xi^{-2} \right] g_n(\lambda, k, \xi) d\xi.$$

$$(25.40)$$

Since

$$|g_n(\lambda, k, r)| \leq \frac{M_0^n}{n!}, \qquad (25.41)$$

the perturbation expansion for $f(\lambda, k, r)$ is bounded for any λ. In fact, again following [65], for $k \in \mathbb{C}$ and $\lambda = 1/2$ (i.e., $l = 0$) it can be shown, using (25.13), that

$$f\left(\frac{1}{2}, k \right) = 1 + \int_0^{\infty} V(r) e^{-ikr} \phi\left(\frac{1}{2}, k, r \right) dr, \qquad (25.42)$$

and that the following bound is valid:

$$\left| f(\frac{1}{2}, k) - 1 \right| \leq O\left\{ \exp (M_0) \left[\int_0^{\infty} \frac{|V(r)| r}{1 + |kr|} dr \right] \exp (r \, [|\mathrm{Im}(k)| + \mathrm{Im}(k)]) \right\}.$$

$$(25.43)$$

If $\mathrm{Im}\, k < 0$, the exponent vanishes (and the resulting bound then applies for real values of k), and the Jost function is analytic in the lower half of the k-plane. If $\mathrm{Im}\, k > 0$, the integral (25.42) diverges as $r \to \infty$ and more restrictions must be imposed on the behavior of $V(r)$ at infinity to establish analyticity. Note that the integral (25.42) is continuous for $\mathrm{Im}\, k \leq 0$, which of course includes the so-called physical region (i.e., $k \in \mathbb{R}^+$) and so the physically meaningful Jost function is contiguous to where it is analytic in the lower half of the k-plane. For general values of λ the generalization of (25.42) leads to

$$f(\lambda, k) = 1 - i \int_0^{\infty} V(r) h_{\lambda-1/2}^{(1)} \phi(\lambda, k, r) dr, \qquad (25.44)$$

where $h_{\lambda-1/2}^{(1)}$ is a spherical Hankel function of the first kind. In this case also $\phi(\lambda, k, r)$ is an entire function in the complex k-plane, and the Jost function is analytic in the lower half of that plane. From the boundary conditions (25.34) and the condition

$$\lim_{r \to \infty} e^{ikr} f(\lambda, k, r) = 1, \qquad (25.45)$$

it can also be shown that

$$\bar{\phi}(\lambda, \bar{k}, r) = \phi(\lambda, k, r); \qquad \bar{f}(\lambda, -\bar{k}, r) = f(\lambda, k, r). \qquad (25.46)$$

Theorem: The zeros of the Jost function $f(\lambda, k)$ for $\mathrm{Im}\, k < 0$ are associated with the bound states of the system.

To prove this, suppose $f(\lambda, k_n) = 0$ for some $k = k_n$. From (25.15) it follows that

$$\phi(\lambda, k_n, r) = \frac{f(\lambda, k_n, r)}{C_n}, \qquad (25.47)$$

where C_n is in general a complex constant. Now $\phi(\lambda, k_n, 0) = 0$ and for $\operatorname{Im} k_n < 0$, $f(\lambda, k_n, r)$ vanishes exponentially as $r \to \infty$; therefore $\phi(\lambda, k_n, r) \in \mathcal{L}^2(0, \infty)$. We now establish that $k_n^2 \in \mathbb{R}$ by proceeding in a similar manner to the method for obtaining (25.25), but for real λ and complex k to obtain

$$\frac{d}{dr}\left[W(\phi(\lambda, k_n, r), \bar{\phi}(\lambda, k_n, r))\right] = 2i \operatorname{Im}\left(k_n^2\right) |\phi(\lambda, k_n, r)|^2. \tag{25.48}$$

From equations (25.46) and (25.47) it follows that

$$\bar{\phi}(\lambda, k_n, r) = \frac{\bar{f}(\lambda, k_n, r)}{\bar{C}_n} = \frac{f(\lambda, -\bar{k}_n, r)}{\bar{C}_n} \sim \frac{\exp\left[i\bar{k}_n r\right]}{\bar{C}_n}, \tag{25.49}$$

and this decreases exponentially as $r \to \infty$ if $\operatorname{Im} k_n < 0$. Using the analogous property of $\phi(\lambda, k, r)$ and integrating (25.48) over $(0, \infty)$ yields the result

$$\operatorname{Im}\left(k_n^2\right) \int_0^\infty |\phi(\lambda, k_n, r)|^2 \, dr = 0, \tag{25.50}$$

and because $\phi(\lambda, k_n, r) \in \mathcal{L}^2(0, \infty)$, we can conclude that k_n^2 is real. The next step is to show that $k_n^2 < 0$. To establish this we assume that $k_n^2 > 0$ to obtain a contradiction. If $k_n \in \mathbb{R}$, then $f(\lambda, k) = 0$ for some real values of k, and by virtue of (25.46) $f(\lambda, -k) = 0$ also. Then (25.15) implies that $\phi(\lambda, k_n, r) \equiv 0$ in contradiction to the boundary condition (25.2). Hence $f(\lambda, k_n) = 0$ only for $k_n^2 < 0$, so the zeros must lie on the imaginary axis of the lower half of the k-plane (i.e., where $k_n = -i\kappa_n, \kappa_n > 0$). The associated discrete energy levels are given by $E_n = -\kappa_n^2$. The solution (25.47) is square integrable since

$$\phi(\lambda, -i\kappa_n, r) \sim \frac{e^{-\kappa_n r}}{C_n}, \qquad r \to \infty, \tag{25.51}$$

where, from (25.15),

$$C_n = -2\frac{\kappa_n}{f(\lambda, i\kappa_n)}, \tag{25.52}$$

so the functions $\phi(\lambda, -i\kappa_n, r)$ are real. It can be further shown that the zeros of $f(\lambda, k)$ are simple. We sketch the proof here. Differentiate the definition (25.13) with respect to k^2 (or E) and set $k^2 = k_n^2$. Using expression (25.47), it can be shown that

$$\frac{df(\lambda, k_n)}{dk_n^2} = \frac{1}{C_n} W\left[\frac{df(\lambda, k_n, r)}{dk_n^2}, f(\lambda, k_n, r)\right]$$
$$- C_n W\left[\frac{d\phi(\lambda, k_n, r)}{dk_n^2}, \phi(\lambda, k_n, r)\right]. \tag{25.53}$$

Now both $f(\lambda, k_n, r)$ and $\phi(\lambda, k_n, r)$ satisfy equation (25.1), have the same exponential decay at infinity, and $\phi(\lambda, k_n, 0) = 0$, so it can be shown that

$$W[f(\lambda, k_n, r), f(\lambda, k, r)] = (k^2 - k_n^2) \int_r^\infty f(\lambda, k_n, r')f(\lambda, k, r')dr', \tag{25.54}$$

and

$$W[\phi(\lambda, k_n, r), \phi(\lambda, k, r)] = (k_n^2 - k^2) \int_0^r \phi(\lambda, k_n, r') \phi(\lambda, k, r')dr'. \tag{25.55}$$

By differentiating these expressions with respect to k^2 and then setting $k^2 = k_n^2$, we arrive at

$$W\left[\frac{d\phi(\lambda, k_n, r)}{dk_n^2}, \phi(\lambda, k_n, r)\right] = \int_0^r \phi^2(\lambda, k_n, r')dr' \qquad (25.56)$$

and

$$W\left[\frac{df(\lambda, k_n, r)}{dk_n^2}, f(\lambda, k_n, r)\right] = -\int_r^\infty f^2(\lambda, k_n, r')dr'. \qquad (25.57)$$

These may then be substituted in (25.53) to yield the final result

$$\frac{df(\lambda, k_n)}{dk_n} = -2k_n C_n \int_0^\infty \phi^2(\lambda, k_n, r)dr \neq 0, \qquad (25.58)$$

because $C_n \neq 0$ and the functions $\phi(\lambda, k_n, r)$ are real. Hence the zeros of $f(\lambda, k)$ in the lower half of the k-plane are simple.

By virtue of the above theorem we see that the Jost function $f(\lambda, -k)$ is analytic in the upper half of the complex k-plane and has simple zeros on the positive imaginary semi-axis, corresponding to the bound states of the system. However, from (25.20) these are exactly the poles of the S-matrix $S(\lambda, k)$, so that each bound state can be associated with a zero of the S-matrix on the negative imaginary semi-axis, and a symmetrically located pole on the positive semi-axis.

25.3 POLES OF THE S-MATRIX

Reverting to the explicit l dependence of the system, and as we have noted already, the time-*dependent* wave function can be written asymptotically as

$$\psi_l(r, t) \sim \frac{1}{r}\left\{(-1)^l e^{-ikr} - S_l(k)e^{ikr}\right\}e^{-iEt}, \qquad (25.59)$$

where $E = k^2$, and rewriting (25.20),

$$S_l(k) = (-1)^l \left[\frac{f(l, k)}{f(l, -k)}\right]. \qquad (25.60)$$

Both functions $f(l, \pm k)$ are defined for $k \in \mathbb{C}$, and so $S_l(k)$ is similarly defined. From (25.60) the following symmetry property can be noted for $l \in \mathbb{Z}$:

$$S_l(-k) = S_l^{-1}(k), \qquad (25.61)$$

illustrating the above mentioned relationship between zeros and poles of the S-matrix. This also follows from the exchange $k \to -k$ in (25.59). Furthermore, note from (25.20) that

$$S_l(k) = e^{2i\delta_l(k)}, \qquad (25.62)$$

which implies that $\delta_l(k)$ is an odd function of k if k is real,

$$\delta_l(k) = -\delta_l(-k), \qquad \mod(n\pi), \qquad n \in \mathbb{Z}. \qquad (25.63)$$

For arbitrary values of l the relation (25.61) has to be replaced by

$$S_l(-k) = e^{2il\pi} S_l^{-1}(k),$$ (25.64)

from which (25.63) takes the form

$$\delta_l(k) = -\delta_l(-k) + l\pi.$$ (25.65)

Using the result from (25.46) in the Jost function, we find that

$$\bar{S}_l(\bar{k}) = S_l^{-1}(k) = S_l(-k).$$ (25.66)

This means that the S-matrix in the complete k-plane can be determined from knowledge of it in any quadrant. To see this, suppose that, given some point $k = k_0$,

$$S_l(k_0) = S_l^0,$$ (25.67)

so at the remaining three symmetric points it follows that

$$S_l(\bar{k}_0) = \left(\bar{S}_l^0\right)^{-1}; \quad S_l(-\bar{k}_0) = \bar{S}_l^0; \quad S_l(-k_0) = \left(S_l^0\right)^{-1}.$$ (25.68)

These properties indicate that the values of the scattering matrix at points symmetric about the imaginary k-axis are complex conjugates, so $S_l(k)$ is real on the imaginary axis and therefore $\delta_l(k)$ is imaginary there, that is, for $\kappa \in \mathbb{R}$,

$$\delta_l(i\kappa) = -\bar{\delta}_l(i\kappa).$$ (25.69)

Figure (25.1) illustrates a generic distribution of such S-matrix features [117]; crosses correspond to bound-state poles, circles to resonance poles (and squares to their conjugate poles), and triangles to virtual states.

It also follows from property (25.66) that for points symmetric with respect to the real axis, $|S_l(k)| = 1$ on the real axis, and $\delta_l(k)$ is a real quantity there. In fact (25.66) indicates that if $S_l(k)$ has a zero at a point k, then it has poles at the points $-k$ and $\bar{k}$ and a zero at $-\bar{k}$, the pairs of zeros and poles being symmetric with respect to the imaginary axis (i.e., the line $\mathrm{Re}\, k = 0$). Indeed, it may be proved using the probability conservation law for the time-dependent Schrödinger equation that, coupled with the complex energy W associated with the nonstationary states of the system, in the upper half of the complex k-plane the poles of $S_l(k)$ lie only on the line $\mathrm{Re}\, k = 0$. More specifically, if $k = k_\mathrm{r} + ik_\mathrm{i}$, and

$$W = E - \frac{i}{2}\Gamma; \quad E = k_\mathrm{r}^2 - k_\mathrm{i}^2; \quad \Gamma = -4k_\mathrm{r}k_\mathrm{i},$$ (25.70)

then we can use the conservation law

$$\frac{\partial}{\partial t} \int_V |\psi|^2 \, dr = -\int_S i \left(\psi \nabla \bar{\psi} - \bar{\psi} \nabla \psi\right) ds,$$ (25.71)

with the asymptotic form

$$\psi_l(k; r, t) = \frac{1}{r} u_l(r) e^{-iWt} \sim -\frac{1}{r} S_l(k) \exp\left[i(kr - Wt)\right].$$ (25.72)

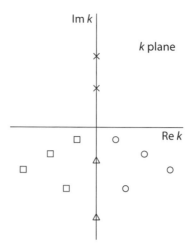

Figure 25.1 Typical pole distribution for the S-matrix.

Note that because of the assumed pole in $S_l(k)$ it is justifiable to neglect the first term in (25.59) by comparison. If we perform the volume integration over a sphere of sufficiently large radius R such that the asymptotic form (25.72) is valid, then it follows that

$$k_r k_i \int_0^R u_l^2(r)\, dr = -Dk_r \,|S_l(k)|^2\, e^{-2k_i R} < 0, \qquad (25.73)$$

D being a real positive constant. This condition is certainly satisfied if $k_r = 0$, in which case the poles lie on the imaginary axis, or if $k_r \neq 0$ and $k_i < 0$, (i.e., the poles lie in the lower half of the complex k-plane), thus proving the assertion. As pointed out by Sitenko [65], the scattering matrix $S_l(k)$ describes various physical processes, the simplest in one sense being that if k is real, the scattering can be described in terms of real phase shifts that in turn determine the scattering cross section, something of great practical importance. But the poles of $S_l(k)$ associated with both pure imaginary and complex k-values are representative of bound, so-called virtual and resonance (or quasi-stationary) states of the system.

Returning to the bound states for which $k_n = -i\kappa_n$, $\kappa_n > 0$, we may write the square-integrability condition for (25.59) as

$$\int_0^\infty \left| (-1)^l\, e^{-\kappa_n r} - S_l(-i\kappa_n) e^{ikr} \right|^2 dr < \infty, \qquad (25.74)$$

which is satisfied provided that $S_l(-i\kappa_n) = 0$. As already noted, a bound state corresponds to a zero of $S_l(k)$ on the negative imaginary semi-axis; it is associated with a pole at the point $i\kappa_n$ on the positive portion of this axis. However, in some situations there are other zeros on the negative imaginary semi-axis that are *not* associated with bound states, but they do not arise if the potential $V(r)$ has finite range.

Poles of $S_l(k)$ lying on the negative imaginary semi-axis are associated with virtual (or "anti-bound") states, and with zeros of $S_l(k)$ on the positive imaginary semi-axis. Thus if a pole exists at $k_p = -i\kappa_p$, $\kappa_p > 0$, then $S_l(i\kappa_p) = 0$. From (25.15) the wave function for a virtual state cannot be normalized, since it behaves as $\exp(\kappa_p r)$. These states correspond to negative energies $E_p = -\kappa_p^2$ but are associated with poles of $S_l(k)$ on the second Riemann sheet of the (complex) energy.

25.3.1 Wavepacket Approach

Suppose that for a potential vanishing for $|x| > a/2$, such as (but not restricted to) the potential well/barrier discussed earlier (see Section 24.1), the incoming wave function is of the form [61]

$$\psi_i(x, t) = \int_0^\infty f(k) e^{ika/2} e^{i(kx - Et)} dk, \tag{25.75}$$

where $f(k)$ is a real function. The integrand is chosen such that the phase $k(x + a/2)$ is zero at $t = 0$ (when the wavepacket reaches the well at $x = -a/2$). For the transmitted packet in the region beyond the well, $x > a/2$,

$$\psi_t(x, t) = \int_0^\infty \left[S(E) e^{-ika/2} \right] f(k) e^{ika/2} e^{i(kx - Et)} dk. \tag{25.76}$$

Following [61], we suppose that $f(k)$ has at least one sharp resonance, so that to a good approximation, $f(k)$ is appreciably different from zero in the neighborhood of the resonant energy E_0, and that $f(k)$ varies slowly in the vicinity $D(E_0)$ of the resonance. Thus we are in a position to write

$$\psi_t(x, t) \approx f(k_0) \int_{D(E_0)} S(E) e^{i[k(x - a/2) - Et]} dk. \tag{25.77}$$

We can also approximate $S(E)$ by the form already derived for the square well, that is,

$$S(E) \approx \frac{i\Gamma/2}{E - E_0 + i\Gamma/2}, \tag{25.78}$$

to within a slowly varying factor. Hence

$$\psi_t(x, t) \approx \frac{i\Gamma}{2} f(k_0) \int_{D(E_0)} \frac{\exp\left(i\left[k(x - a/2) - Et\right]\right)}{E - E_0 + i\Gamma/2} dk. \tag{25.79}$$

Noting that in this formulation, $k^2 = 2E$, and using the Taylor polynomial for $k(E)$ expanded about E_0, ψ_t can be rewritten as

$$\psi_t(x, t) \approx \frac{i\Gamma}{2} \frac{f(k_0)}{k_0} \int_{D(E_0)} \frac{\exp\left(i\left\{(k_0 + [E - E_0]/k_0)(x - a/2) - Et\right\}\right)}{E - E_0 + i\Gamma/2} dE. \tag{25.80}$$

The exponential term can be rearranged into the form

$$\exp\left(i\left[k_0\left(x - \frac{a}{2}\right) - E_0 t\right]\right) \times \exp\left\{-i(E - E_0)\left[t - \frac{(x - a/2)}{k_0}\right]\right\}. \tag{25.81}$$

The first factor can be temporarily ignored to investigate the behavior of the integral in E as $|(E - E_0)|$ becomes increasingly large. As this occurs the denominator in (25.80) increases so that it is justifiable (to the present level of approximation) to extend the limits on the integral from $-\infty$ to ∞ without much loss in precision, so we define

$$I_\psi = \int_{-\infty}^{\infty} \frac{\exp\left\{-i\,(E - E_0)\left[t - \frac{(x-a/2)}{k_0}\right]\right\}}{E - E_0 + i\Gamma/2}\,dE, \qquad (25.82)$$

in order to see how to evaluate it in the complex E-plane. The real part of the exponent is

$$\Xi = \frac{1}{k_0}\,(\operatorname{Im} E)\left(k_0 t + \frac{a}{2} - x\right), \qquad (25.83)$$

so for $x > k_0 t + a/2$, $\Xi \to 0$ as $|(E - E_0)| \to \infty$, provided that $\operatorname{Im} E > 0$, and the complex contour can be closed in the upper half of the complex E-plane. Then it follows that $I_\psi = 0$, because the simple pole at $E = E_0 - i\Gamma/2$ lies in the lower half-plane. For $x < k_0 t + a/2$ we require that $\operatorname{Im} E < 0$ to close the contour in the lower half-plane, and by the residue theorem, we have

$$\psi_t(x, t) \approx \frac{\pi\,\Gamma f(k_0)}{k_0}\,\exp\left(i\left[k_0\left(x - \frac{a}{2}\right) - E_0 t\right]\right)$$
$$\times \exp\left\{-\frac{\Gamma}{2}\left[t - \frac{(x - a/2)}{k_0}\right]\right\}. \qquad (25.84)$$

The first exponential is of course the resonance-related harmonic wave, while the second has the character of a temporal decay term for any given x, but at any time t it increases exponentially in x (for $x > a/2$).

We conclude this chapter with a neat little result.

The Parity Theorem for Symmetric One-Dimensional Potentials

Let

$$u''(x) + \left[k^2 - V(x)\right]u = 0, \qquad (25.85)$$

where $V(x) = V(-x)$, and $|u|, |u'| \to 0$ as $|x| \to \infty$. Then equation (25.85) and its boundary conditions are invariant with respect to the parity transformation $x \to -x$. We can always write $u(x)$ as the sum of even and odd parts, but more generally, let

$$u(x) = Au_e(x) + Bu_o(x), \qquad (25.86)$$

where $u_e(x)$ and $u_o(x)$ are even and odd functions, respectively. Now if $u(x)$ is a solution of (25.85) and the boundary conditions, so is

$$u(-x) = Au_e(x) - Bu_o(x). \qquad (25.87)$$

Provided no degeneracy is involved (i.e., multiple eigenvalues are not associated with a given eigenfunction), then the solutions must be proportional, that is, $u(-x) = \alpha u(x)$, or

$$Au_e(x) - Bu_o(x) = \alpha \left[Au_e(x) + Bu_o(x) \right]. \tag{25.88}$$

But since $u_e(x)$ and $u_o(x)$ are linearly independent, this is true if and only if $A = \alpha A$ and $B = -\alpha B$, that is, either $\alpha = 1$ and $B = 0$ (so u is an even function), or $\alpha = -1$ and $A = 0$ (u is an odd function), and the theorem is proved.

Exercise: Show that $U(x)$ in (25.85) can always be written as the sum of even and odd functions.

Chapter Twenty-Six

One-Dimensional Jost Solutions: The S-Matrix Revisited

For one dimensional Schrödinger operators, high-energy asymptotic expansions of the Jost solutions, Green's function, various scattering quantities, and other related functions have long been treated by both mathematicians and physicists. There is very extensive literature on this subject. ... Most of the earlier work and work done by physicists is either formal or under excessively strong conditions on potentials.

[258]

The most important feature of the transition amplitudes for various physical processes in the energy representation is their analyticity in the upper half of the complex plane. Sometimes one can find out more about singularities in the lower half-plane, investigating dynamical equations specific for the physical problem, like the Schrödinger equation for scattering. The one-dimensional "scattering," i.e., the potential-barrier problem, is somewhat more complicated than the scattering off a force center, since the system has two channels, corresponding to two waves running in the opposite directions from the potential region in the final state. Thus instead of one analytical function, the partial wave scattering amplitude, one deals with two analytical functions, transmission and reflection amplitudes. (The unitarity condition holds in both cases.) The familiar arguments of the scattering theory must be extended properly to the one-dimensional case.

[259]

26.1 TRANSMISSION AND REFLECTION COEFFICIENTS

The left-right transmission and reflections coefficients can be expressed in terms of the elements of the S-matrix for one-dimensional scattering problems on $(-\infty, \infty)$. As noted in Chapter 25 (and Chapter 17) we can express all solutions of the Schrödinger equation as linear combinations of a solution $f_1(x, k)$ that reduces to e^{ikx} as $x \to \infty$, and another solution $f_2(x, k)$ that reduces to e^{-ikx} as $x \to -\infty$, that is, as we have seen,

$$\lim_{x \to \infty} \left[e^{-ikx} f_1(x, k) \right] = 1; \quad \lim_{x \to -\infty} \left[e^{ikx} f_2(x, k) \right] = 1, \qquad (26.1)$$

where $k \in \mathbb{R}$. f_1 and f_2 are called the fundamental or Jost solutions, and we examine their properties in more detail in this chapter. It has already been shown that

$$W(f_1(x, k), \overline{f_1}(x, k)) \equiv W(f_1(x, k), f_1(x, -k)) = -2ik, \quad k \neq 0, \quad (26.2)$$

and

$$W(f_2(x, k), \overline{f_2}(x, k)) \equiv W(f_2(x, k), f_2(x, -k)) = 2ik, \quad k \neq 0. \quad (26.3)$$

Therefore the pairs $(f_1, \overline{f_1})$, $(f_2, \overline{f_2})$ are linearly independent (note that the definition of the Wronskian used here is the standard one, i.e., $W(f, g) = fg' - f'g$; it is the negative of that used in [39]). Since any third solution to a second-order ordinary differential equation can be expressed as a linear combination of the two linearly independent solutions, we may write, on suppressing the x dependence,

$$f_2(k) = c_{11} f_1(k) + c_{12} f_1(-k), \quad (26.4)$$

and

$$f_1(k) = c_{21} f_2(-k) + c_{22} f_2(k). \quad (26.5)$$

Substituting for $f_1(k)$ from equation (26.5) into equation (26.4) and equating coefficients, it transpires that

$$f_2(k) = c_{11}(k)c_{21}(k) f_2(-k) + c_{11}(k)c_{22}(k) f_2(k)$$
$$+ c_{12}(k)c_{21}(-k) f_2(k) + c_{12}(k)c_{22}(-k) f_2(-k). \quad (26.6)$$

For consistency it is necessary to satisfy the following conditions:

$$1 = c_{11}(k)c_{22}(k) + c_{12}(k)c_{21}(-k), \quad (26.7)$$

$$0 = c_{11}(k)c_{21}(k) + c_{12}(k)c_{22}(-k). \quad (26.8)$$

A similar substitution of equation (26.4) into equation (26.5) yields

$$1 = c_{21}(k)c_{12}(-k) + c_{22}(k)c_{11}(k), \quad (26.9)$$

$$0 = c_{21}(k)c_{11}(-k) + c_{22}(k)c_{12}(k). \quad (26.10)$$

Using equations (26.2), (26.3), (26.4), and (26.5) in the various Wronskian relations gives a great deal more information, and after some reduction it can be shown that

$$c_{11}(k) = \frac{W[f_1(-k), f_2(k)]}{2ik}; \quad c_{22}(k) = \frac{W[f_1(k), f_2(-k)]}{2ik}; \quad (26.11)$$

$$c_{12}(k) = c_{21}(k) = \frac{W[f_2(k), f_1(k)]}{2ik}. \quad (26.12)$$

In Chapter 17 we examined oblique incidence of waves on a submarine ridge, and using normal mode theory converted it to an essentially one-dimensional problem.

In this case we have a strict one-dimensional problem and define reflection and transmission coefficients for a wave of unit amplitude incident from the right as

$$R_R(k) = \frac{c_{11}(k)}{c_{12}(k)}; \quad T_R(k) = \frac{1}{c_{12}(k)}; \tag{26.13}$$

and from the left as

$$R_L(k) = \frac{c_{22}(k)}{c_{21}(k)}; \quad T_L(k) = \frac{1}{c_{21}(k)} \tag{26.14}$$

[39] (R_L and R_R correspond to R_2 and R_1 in Chapter 17 respectively). Clearly here $T_L(k) = T_R(k)$ (corresponding to $T_1 = T_2$ in Chapter 17). Furthermore, since $f_1(x, k) = \overline{f_1}(x, -k)$ (with a similar relation for f_2) for real k and potential $V(x)$, it follows that

$$c_{12}(-k) = \overline{c_{12}}(k), c_{11} = -\overline{c_{22}}(k) = -c_{22}(-k). \tag{26.15}$$

On using the consistency conditions (26.7)–(26.10), it follows that

$$|c_{12}(k)|^2 = 1 + |c_{11}(k)|^2 = 1 + |c_{22}(k)|^2, \tag{26.16}$$

that is,

$$1 = |T(k)|^2 + |R_R(k)|^2 = |T(k)|^2 + |R_L(k)|^2. \tag{26.17}$$

Also

$$R_R(k)T(-k) + R_L(-k)T(k) = 0, \tag{26.18}$$

and

$$R_L(k) = \overline{R_L}(-k); \quad R_R(k) = \overline{R_R}(-k). \tag{26.19}$$

Other notations include, e.g.,

$$c_{12}(k) = a(k), R_+ = R_R(k) = \frac{c_{11}(k)}{c_{12}(k)} = \frac{b(k)}{a(k)},$$

and

$$R_- = R_L(k) = c_{22}.$$

26.1.1 Poles of the Transmission Coefficient: Zeros of $c_{12}(k)$

From equation (26.16), which is valid for real k only, c_{12} can never be zero when Im $k = 0$. Thus all the zeros of c_{12} (or c_{21}) must be off the real k-axis. In addition, for a real potential any poles in the upper half of the k-plane must lie on the imaginary axis. This is most readily shown by considering the one-dimensional Schrödinger-type equation for a pole occurring when $k = k_p$, say. Then

$$f_1'' + k_p^2 f_1 = V(x) f_1, \tag{26.20}$$

along with the corresponding complex conjugate equation

$$\overline{f_1''} + \overline{k_p^2 f_1} = V(x)\overline{f_1}. \tag{26.21}$$

Utilizing the usual device of multiplying each equation by the corresponding version of f_1, subtracting, and integrating (this time over the real line), we obtain

$$\left[\overline{f_1} f_1' - f_1 \overline{f_1'}\right]_{-\infty}^{\infty} = \left(\overline{k_p^2} - k_p^2\right) \int_{-\infty}^{\infty} |f_1|^2 \, dx. \tag{26.22}$$

Because of the Wronskian properties (equation (26.2)), the left-hand side of this equation is zero, and hence

$$\left(\operatorname{Re} k_p\right) \left(\operatorname{Im} k_p\right) \int_{-\infty}^{\infty} |f_1|^2 \, dx = 0. \tag{26.23}$$

But it is already known that $\operatorname{Im} k_p \neq 0$, so $\operatorname{Re} k_p = 0$.

26.2 THE JOST FORMULATION ON $[0, \infty)$: THE RADIAL EQUATION REVISITED

We reconsider the radial equation (21.76) in order to reexamine the well-known problem of the square-well potential from the perspective of the Jost solution [67]:

$$\frac{d^2 u_l(r)}{dr^2} + \left[k^2 - \frac{l(l+1)}{r^2} - V(r)\right] u_l(r) = 0. \tag{26.24}$$

In contrast with the requirements on $V(r)$ discussed in connection with equations (25.30) and (25.31), we can impose less restrictive ones here (see [67]); as before, $V(r)$ must be a real function vanishing sufficiently fast at infinity and must be almost everywhere continuous; but we only require that

$$\int_c^{\infty} |V(r)| \, dr = M(c) < \infty; \tag{26.25}$$

$$\int_0^{c'} r |V(r)| \, dr = N(c') < \infty, \tag{26.26}$$

where c and c' are arbitrary constants greater than zero. This conditions are readily satisfied for nonsingular $V(r)$ with compact support.

26.2.1 Jost Boundary Conditions at $r = 0$

For small r,

$$\frac{l(l+1)}{r^2} \gg |k^2 - V(r)|,$$

and we begin our discussion of (26.24) by neglecting the term $[k^2 - V(r)]$. The equation then becomes

$$\frac{d^2 u_l(r)}{dr^2} - \left[\frac{l(l+1)}{r^2}\right] u_l(r) = 0, \tag{26.27}$$

which has a regular point at $r = 0$, and the exact solution is

$$u_l(r) = \alpha r^{l+1} + \beta r^{-l}. \tag{26.28}$$

Using (26.28) as a guide, define two linearly independent solutions of (26.27) with the behavior

$$\phi(r) = r^{l+1}[1 + o(1)], \tag{26.29}$$

$$\phi_1(r) = r^{-l}[1 + o(1)], \tag{26.30}$$

where o is an order symbol (see Appendix A). We also define $\lambda = l + 1/2$, so that (26.30) becomes even in λ:

$$\frac{d^2 u_{\lambda-1/2}(r)}{dr^2} + \left[k^2 - \frac{(\lambda^2 - 1/4)}{r^2} - V(r) \right] u_{\lambda-1/2}(r) = 0. \tag{26.31}$$

Since the Wronskian of the two linearly independent solutions of (26.31) is nonzero and constant, we can evaluate the Wronskian of $\phi(\lambda, k, r)$ and $\phi(-\lambda, k, r)$ with $r^{\lambda+1/2}$. Thus

$$W[\phi(\lambda, k, r), \phi(-\lambda, k, r)]$$
$$= \phi(\lambda, k, r)\phi'(-\lambda, k, r) - \phi'(\lambda, k, r)\phi(-\lambda, k, r) = -2\lambda. \tag{26.32}$$

26.2.2 Jost Boundary Conditions as $r \to \infty$

For large r, we can neglect the term $r^{-2}(\lambda^2 - 1/4) + V(r)$ in (26.31), which therefore becomes [67]

$$\frac{d^2 u_{\lambda-1/2}(r)}{dr^2} + k^2 u_{\lambda-1/2}(r) = 0, \tag{26.33}$$

which possesses the exact solution

$$u_{\lambda-1/2}(r) = \alpha e^{-ikr} + \beta e^{ikr}. \tag{26.34}$$

Therefore, we can construct the Jost solution $f(\lambda, k, r)$ with the asymptotic behavior

$$\lim_{r \to \infty} e^{ikr} f(\lambda, k, r) = 1, \tag{26.35}$$

Note the following relation:

$$W[f(\lambda, k, r), f(\lambda, -k, r)] = 2ik, \tag{26.36}$$

where we have evaluated the Wronskian by substituting for $f(\lambda, \pm k, r)$ its asymptotic behavior, $e^{\pm ikr}$.

26.2.3 The Jost Function and the S-Matrix

At the risk of some repetition from section 25.1, we restate some of its features. The Jost function $f(\lambda, k)$ is defined as

$$f(\lambda, k) = W[f(\lambda, k, r), \phi(\lambda, k, r)]$$
$$= f(\lambda, k, r)\phi'(\lambda, k, r) - f'(\lambda, k, r)\phi(\lambda, k, r). \tag{26.37}$$

The general solution of (26.31) is

$$\phi(\lambda, k, r) = \frac{1}{2ik}[f(\lambda, k)f(\lambda, -k, r) - f(\lambda, -k)f(\lambda, k, r)]. \qquad (26.38)$$

Asymptotically

$$\phi(\lambda, k, r) \sim \frac{1}{2ik}\left[f(\lambda, k)e^{ikr} - f(\lambda, -k)e^{-ikr}\right]. \qquad (26.39)$$

The Jost functions can be written as

$$f(\lambda, k) = \tau(\lambda, k)\exp[i\delta(\lambda, k) - (\lambda - 1/2)/2i\pi]; \qquad (26.40)$$

$$f(\lambda, -k) = \tau(\lambda, k)\exp[-i\delta(\lambda, k) + (\lambda - 1/2)/2i\pi], \qquad (26.41)$$

$\tau(\lambda, k)$ again being the complex amplitude. Then we have

$$\phi(\lambda, k, r) \sim \frac{1}{k}\tau(\lambda, k)\sin\left[kr + \delta(\lambda, k) - \frac{1}{2}\pi\left(\lambda - \frac{1}{2}\right)\right]. \qquad (26.42)$$

The S-matrix is

$$S(\lambda, k) = e^{2i\delta(\lambda, k)} = \left[\frac{f(\lambda, k)}{f(\lambda, -k)}\right]e^{i\pi(\lambda - 1/2)}. \qquad (26.43)$$

Note that the S-matrix is proportional to the ratio of the coefficients of the outgoing and incoming waves in equation (26.39).

26.2.4 Scattering from a Constant Spherical Inhomogeneity

Using these ideas, we consider the problem of scattering from a constant spherical inhomogeneity, with the standard notation used above:

$$V(r) = -V_0, \quad k(r) = K, \quad 0 < r < a;$$

$$V(r) = 0, \quad k(r) = k, \quad r > a.$$

The solutions in the two regions are [67]

$$u_{\lambda-1/2}^{(1)}(k, r) = r\left[Aj_{\lambda-1/2}(Kr) + By_{\lambda-1/2}(Kr)\right], \qquad (26.44)$$

and

$$u_{\lambda-1/2}^{(2)}(k, r) = r\left[Ch_{\lambda-1/2}^{(1)}(kr) + Dh_{\lambda-1/2}^{(2)}(kr)\right], \qquad (26.45)$$

where $j_{\lambda-1/2}(Kr)$, $y_{\lambda-1/2}(k_2r)$, $h_{\lambda-1/2}^{(1)}(kr)$, and $h_{\lambda-1/2}^{(2)}(kr)$ are spherical Bessel, Neumann, and Hankel functions of the first kind and second kind, respectively. Choosing $u_{\lambda-1/2}^{(1)}(Kr)$ to be $\phi(\lambda, k, r)$ and imposing the boundary conditions at $r = 0$, we find that $B = 0$, and

$$\phi(\lambda, k, r) = 2^{\lambda+1/2}\pi^{-1/2}K^{-\lambda+1/2}\Gamma(\lambda + 1)rj_{\lambda-1/2}(Kr), \qquad (26.46)$$

$$\phi'(\lambda, k, r) = 2^{\lambda+1/2}\pi^{-1/2}K^{-\lambda+1/2}\Gamma(\lambda + 1)$$

$$\times [j_{\lambda-1/2}(Kr) + Krj'_{\lambda-1/2}(Kr)], \qquad (26.47)$$

where the prime denotes differentiation with respect to the argument of the function, Γ is the gamma function, and we have used the following series representation for $j_{\lambda-1/2}(Kr)$:

$$j_{\lambda-1/2}(Kr) = \sum_{n=0}^{\infty} \frac{(-1)^n \pi^{1/2}(Kr/2)^{\lambda+2n-1/2}}{2n!\Gamma(\lambda+n+1)}, \quad \lambda - \frac{1}{2} \neq -1, -2, -3, \ldots.$$

(26.48)

Choosing $u_{\lambda-1/2}^{(2)}(kr)$ to be $f(\lambda, k, r)$ and imposing the boundary conditions at $r = \infty$, we find that $C = 0$, $D = k \exp\left[-i\pi(\lambda+1/2)/2\right]$, and

$$f(\lambda, k, r) = k \exp\left[-i\pi(\lambda+1/2)/2\right] r h_{\lambda-1/2}^{(2)}(kr);$$

(26.49)

$$f'(\lambda, k, r) = k \exp\left[-i\pi(\lambda+1/2)/2\right] [h_{\lambda-1/2}^{(2)}(kr) + kr h_{\lambda-1/2}^{(2)\prime}(kr)], \quad (26.50)$$

where the following asymptotic form for $h_{\lambda-1/2}^{(2)}(kr)$ has been used:

$$\lim_{kr \to \infty} h_{\lambda-1/2}^{(2)}(kr) = \frac{1}{kr} \exp\left(-i\left[kr - \frac{\pi}{2}(\lambda+1/2)\right]\right).$$

(26.51)

Since the point $r = a$ is the common domain of $\phi(\lambda, k, r)$ and $f(\lambda, k, r)$, the Jost function can be evaluated at $r = a$ so that

$$f(\lambda, k) = 2^{\lambda+1/2}\pi^{-1/2}\Gamma(\lambda+1)k^{-\lambda+1/2}Kae^{-i(\pi/2)(\lambda+1/2)}(A - B), \quad (26.52)$$

where

$$A = Kj'_{\lambda-1/2}(Ka)h_{\lambda-1/2}^{(2)}(ka); \quad (26.53)$$

$$B = kj_{\lambda-1/2}(Ka)h_{\lambda-1/2}^{\prime(2)}(ka)]. \quad (26.54)$$

Similarly,

$$f(\lambda, -k) = 2^{\lambda+1/2}\pi^{-1/2}\Gamma(\lambda+1)K^{-\lambda+1/2}ke^{-i(\pi/2)(\lambda+1/2)}ae^{i\pi(\lambda-1/2)}(C + D), \quad (26.55)$$

where

$$C = -Kj'_{\lambda-1/2}(Ka)h_{\lambda-1/2}^{(1)}(ka); \quad (26.56)$$

$$D = kj_{\lambda-1/2}(Ka)h_{\lambda-1/2}^{\prime(1)}(ka). \quad (26.57)$$

and the following identities have also been used:

$$h_{\lambda-1/2}^{(2)}(kre^{i\pi}) = h_{\lambda-1/2}^{(2)}(-kr) = (-1)^{\lambda-1/2}h_{\lambda-1/2}^{(1)}(kr) \quad (26.58)$$

$$= e^{i\pi(\lambda-1/2)}h_{\lambda-1/2}^{(1)}(kr), \quad (26.59)$$

$$h_{\lambda-1/2}^{(2)\prime}(-kr) = (-1)^{\lambda+1/2}h_{\lambda-1/2}^{(1)\prime}(kr) = e^{i\pi(\lambda+1/2)}h_{\lambda-1/2}^{(1)\prime}(kr) \quad (26.60)$$

$$= -e^{i\pi(\lambda-1/2)}h_{\lambda-1/2}^{(1)\prime}(kr), \quad (26.61)$$

for $\lambda - 1/2 = 0, 1, 2, \ldots$.

The S-matrix is then given by

$$S(\lambda, k) = -\frac{k j_{\lambda-1/2}(Ka) h_{\lambda-1/2}'^{(2)}(ka) - K j_{\lambda-1/2}'(Ka) h_{\lambda-1/2}^{(2)}(ka)}{k j_{\lambda-1/2}(Ka) h_{\lambda-1/2}'^{(1)}(ka) - K j_{\lambda-1/2}'(Ka) h_{\lambda-1/2}^{(1)}(ka)}. \tag{26.62}$$

We can calculate the Jost function for $\lambda = 1/2$ using equation (26.52) to find, for $l = 0$,

$$f\left(\frac{1}{2}, k\right) = \frac{1}{2} e^{-ika} \left[\left(1 - \frac{k}{K}\right) e^{-iKa} + \left(1 + \frac{k}{K}\right) e^{iKa} \right] \tag{26.63}$$

$$= \left[\cos Ka + i \frac{k}{K} \sin Ka \right] e^{-ika}, \tag{26.64}$$

where the following relations have been used:

$$j_0(Ka) = \sin Ka / (Ka),$$

$$j_0'(Ka) = \cos Ka / (Ka) - [\sin Ka / (Ka)^2],$$

$$h_0^{(2)}(ka) = -e^{-ika} / (ika),$$

$$h_0^{(2)'}(ka) = e^{-ika} [1 + 1/(ika)] / (ka).$$

The bound states are given by the zeros of equation (26.64), that is, by

$$\cos Ka + i \frac{k}{K} \sin Ka = 0. \tag{26.65}$$

Note the similarity between this result and the expression (24.11) corresponding to a one-dimensional potential well located symmetrically about the origin. A similar analysis for a three-region potential can be found in [274].

Chapter Twenty-Seven

Morphology-Dependent Resonances:

The Effective Potential

> The cross-section for scattering of electromagnetic energy by a
> dielectric sphere exhibits a series of sharp peaks as a function of the size
> parameter. These peaks are a manifestation of scattering resonances in
> which electromagnetic energy is temporally trapped inside the particle.
> A physical interpretation is that the electromagnetic wave is trapped by
> almost total internal reflection as it propagates around the inside surface
> of the sphere and that after circumnavigating the sphere the wave
> returns to its starting point in phase. These resonances are now
> generally referred to as morphology-dependent resonances (MDRs).
>
> [97]

> Dielectric microspheres (or similar geometries) are optical structures
> that exhibit resonant properties, meaning they can select very narrow
> segments of the incoming signal's spectrum for further manipulation
> and processing. The optical resonances of a microsphere are frequently
> called the whispering gallery modes (WGMs). In general, microspheres
> belong to the same group of devices as Fabry-Perot interferometers and
> fiber Bragg gratings. The optical resonances in microspheres are a
> function of their morphology, meaning, their geometry, and dielectric
> properties (refractive index). Any perturbation to their morphology
> (shape, size or refractive index) caused by a change in the surrounding
> environment will lead to a shift in the resonances (WGM). By tracking
> these morphology-dependent shifts of WGMs, it is possible to measure
> the change in a given environmental property.
>
> [260]

27.1 SOME FAMILIAR TERRITORY

Throughout this book it has been emphasized that the essential mathematical analysis for a scalar wave equation is independent of context: it can be thought of
in terms of classical mathematical physics (e.g., the scattering of sound waves)
or in quantum mechanical terms (e.g., the nonrelativistic scattering of particles

by a square potential well (or barrier) of radius a and depth (or height) V_0). In either case we can consider a scalar plane wave impinging in the direction $\theta = 0$ on a sphere of radius a. In what follows, a boldface letter refers to a vector quantity, thus here, $\mathbf{r} = \langle |r|, \theta, \phi \rangle$ (or $\langle r, \theta, \phi \rangle$) denotes a position vector in space using a spherical coordinate system. Suppose that we had started with the classical wave equation for a dependent variable $\tilde{\psi}(\mathbf{r}, t) = \psi(\mathbf{r})e^{-i\omega t}$. In the standard scalar electromagnetic problem, the angular frequency ω, wavenumber k, and (constant) refractive index $n > 1$ are related by $\omega = kc/n$, c being the speed of light in vacuo. Then for a penetrable (or "transparent") sphere, the spatial part of the wave function $\psi(r)$ satisfies the scalar Helmholtz equation:

$$\nabla^2 \psi + k^2 n^2 \psi = 0, \quad r < a, \tag{27.1a}$$

$$\nabla^2 \psi + k^2 \psi = 0, \quad r > a. \tag{27.1b}$$

We can expand the wave function $\psi(\mathbf{r})$ as

$$\psi(\mathbf{r}) = \sum_{l=0}^{\infty} B_l(k) u_l(r) r^{-1} Y_l^m(\theta, \phi) \equiv \sum_{l=0}^{\infty} A_l(k) u_l(r) r^{-1} P_l(\cos \theta), \tag{27.2}$$

where $r = |\mathbf{r}|$, as noted above, and the coefficients $A_l(k)$ will be unfolded below (recall that the coefficients A_l and B_l are related by a multiplicative normalization constant, though that need not concern us here). The reason the spherical harmonics $Y_l^m(\theta, \phi)$ reduce to the Legendre polynomials in the above expression is because the cylindrical symmetry imposed on the system by the incident radiation renders it axially symmetric (i.e, independent of the azimuthal angle ϕ). The equation satisfied by $u_l(r)$ for a spherically symmetric potential $V(r)$ is

$$\frac{d^2 u_l(r)}{dr^2} + \left[k^2 - V(r) - \frac{l(l+1)}{r^2} \right] u_l(r) = 0, \tag{27.3}$$

but the potential $V(r)$ is now k dependent, that is,

$$V(r) = \begin{cases} k^2(1 - n^2), & r < a, \\ 0, & r > a. \end{cases} \tag{27.4}$$

Since $n > 1$ in the sphere, this potential corresponds to that of a spherical potential well of wavenumber-dependent depth $V = k^2(n^2 - 1)$. This leads very naturally to a discussion of the effective potential, wherein the potential $V(r)$ is combined with the "centrifugal barrier" term $l(l+1)/r^2$. A rather detailed study of the radial wave equations was carried out in [97], specifically for the Mie solution of electromagnetic theory (discussed in Chapter 20). A crucial part of the analysis in [97] was the use of the effective potential for the tranverse electric (TE) mode of the Mie solution (See Appendix C), but we may still refer to the scalar problem here without any loss of generality. This effective potential is defined as

$$U_l(r) = \begin{cases} V(r) + l(l+1)/r^2 = k^2(1-n^2) + l(l+1)/r^2, & r \le a, \\ l(l+1)/r^2 \approx \lambda^2/r^2, & r > a. \end{cases} \tag{27.5}$$

It should be noted here that λ as defined here is not of course the wavelength of the incident radiation. For large enough values of l, $[l(l+1)]^{1/2} \approx l + 1/2$, which as we have seen is just the Langer correction (for all values of l) in the WKB(J) solution. It is clear that $U_l(r)$ has a discontinuity at $r = a$ because of the addition of the potential well to the centrifugal barrier. This is a tall and thin enhancement corresponding to a barrier surrounding a well and suggests the possible existence of resonances, particularly between the top of the former and bottom of the latter, where there are three turning points (at which the energy $k^2 = U_l(r)$). Such resonances are called "shape resonances" or "morphology-dependent resonances"; they are *quasi-bound states* in the potential well that escape by tunneling through the centrifugal barrier (see Figure 27.1). The widths of these resonances depend on where they are located; the smaller the number of nodes of the radial wave function within the well, the deeper that state lies in the well. This in turn determines the width (and lifetime) of the state, because the tunneling amplitude is exponentially sensitive to the barrier height and width [97]. The lifetime of the resonance (determined by the rate of tunneling through the barrier) is inversely proportional to the width of the resonance, so the deepest states have the longest lifetimes. (To avoid confusion of the node number n with the refractive index, the latter has temporarily been written as N.)

Note that as k^2 is reduced, the bottom B of the potential rises (and for some value of k the energy will coincide with the bottom of the well); however, at the top of the well, $U_l(a) = \lambda^2/a^2$ is independent of k^2, but if k^2 is increased it will eventually coincide with the top of the well (T in the figure). Consider a value of k^2 between the top and the bottom of the well: in this range there will be three radial turning points, the middle one obviously occurring at $r = a$ and the largest at $r = b$, for which $U_l(a) = \lambda^2/a^2$. The smallest of the three ($r_{\min}$) is found by solving the equation

$$k^2 = \frac{\lambda^2}{r_{\min}^2} - (n^2 - 1)k^2, \tag{27.6}$$

to obtain, in terms of the impact parameter $b(\lambda) = \lambda/k$,

$$r_{\min} = \frac{\lambda}{nk} \equiv \frac{b}{n}. \tag{27.7}$$

By applying Snell's law for given b, it is readily shown that the distance of nearest approach of the equivalent ray to the center of the sphere is just $r_{\min}$; indeed, there are in general many nearly total internal reflections (because of internal incidence beyond the critical angle for total internal reflection) in the sphere between $r = b/n$ and $r = a$. This is analogous to orbiting in a ray picture; on returning to its original location after one circumnavigation just below the sphere surface, a ray must do so with constructive interference. The very low leakage of these states allows the resonance amplitude and energy to build up significantly during a large resonance lifetime which in turn can lead to nonlinear optical effects. In acoustics these are called whispering gallery modes.

The energy at the bottom of the well (i.e. $\lim_{r \to a^-} U_l(r)$) corresponding to the turning point at $r = a$ is determined by the impact parameter inequalities

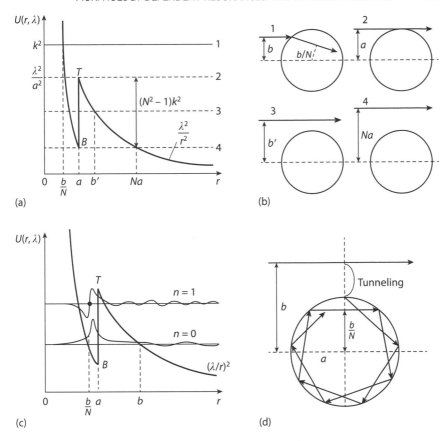

Figure 27.1 (a) The effective potential for a transparent sphere of radius a is illustrated, showing four energy levels: above the top of the potential well, at the top, in the middle, and at the bottom of the well. (b) The corresponding incident rays and impact parameters. Case 2 shows a tangentially incident ray; note that in Case 1 the refracted ray is shown. It passes the center at a distance of $l = b/N$; that this is the case is readily shown from simple geometry; by Snell's law $\sin i = N \sin r = b/a$, and since $l = a \sin r$ the result follows directly. (c) This is similar to panel (a) but with resonant wave functions shown, corresponding to node numbers $n = 0$ and $n = 1$ (the latter possessing a single node). (d) The tunneling phenomenon is illustrated for an impact parameter $b > a$, being multiply reflected after tunneling, between the surface $r = a$ and the caustic surface $r = b/N$ (the inner turning point).

$a < b < na$, or in terms of $\lambda = kb$,

$$U_l(a^-) = \left(\frac{\lambda}{na}\right)^2 < k^2 < \left(\frac{\lambda}{a}\right)^2 = U_l(a^+). \tag{27.8}$$

This is the energy range between the top and bottom of the well (and in which the resonances occur). To cross the forbidden region $a < r < b$ requires tunneling through the centrifugal barrier, and near the resonance energies, the usual

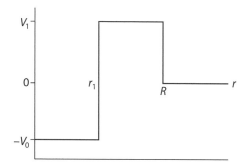

Figure 27.2 The rectangular well-barrier potential.

oscillatory/exponential matching procedures can lead to very large ratios of internal to external amplitudes; these resonances correspond to quasi-bound states of electromagnetic radiation (that would be bound in the limit of zero leakage).

It is important to note that studies of this kind have led to new insights into the phenomenon of diffraction known as tunneling [239]:

> A new approximation to the short-wavelength scattering amplitude from an impenetrable sphere is ... uniform in the scattering angle and it is more accurate than previously known approximations (including Fock's theory of diffraction) by up to several orders of magnitude. It remains valid in the transition to long-wavelength scattering. It leads to a new physical picture of diffraction, as tunneling through an inertial barrier.... The new physical picture of diffraction can be extended to more general curved surfaces. In suitable curvilinear coordinates, diffraction around the obstacle edges results from tunneling through the associated inertial barriers (generalizations of the centrifugal barrier). This picture is reminiscent of the general relativistic theory of light deflection by a gravitational field. However, the gravitational deflection takes place at the level of geometrical optics, whereas diffraction and tunneling are typical wave effects, associated with a much finer scale. [238]

27.1.1 A Toy Model for $l \neq 0$ Resonances: A Particle Analogy

In very simplistic terms, to delay a particle for a long time in in some potential, a barrier is needed to prevent the escape of the particle. To investigate this as simply as possible mathematically, consider the potential shown in Figure 27.2. The one-dimensional potential consists of an attractive rectangular well and a repulsive rectangular barrier. This $l = 0$ rectangular template can be used in principle to mimic the shape of effective potential wells, like those discussed above for $l > 0$. By choosing r_1, R, V_0, and V_1 appropriately, we can obtain an approximate representation of $V_l(r)$ in the case of an effective potential in the presence of a centrifugal barrier [121].

As is well known by now, the time-independent Schrödinger wave equation in physical variables is

$$\frac{d^2\psi(r)}{dr^2} + [E - V(r)]\psi(r) = 0, \tag{27.9}$$

where the potential function is defined as

$$\text{Region 1 } (0 < r < r_1): V(r) = -V_0 < 0,$$
$$\text{Region 2 } (r_1 < r < R): V(r) = V_1 > 0, \tag{27.10}$$
$$\text{Region 3 } (r > R): V(r) = 0.$$

Case 1: $0 < k^2 < V_1$

To find the solution of the Schrödinger wave equation that describes the motion of a particle in the potential, consider first the case that $0 < k^2 < V_1$ for energies below the top of the barrier. Note that in these units the wave number is defined as $k = E^{1/2}$. In Region 1 we have

$$\frac{d^2\psi(r)}{dr^2} + (E + V_0)\psi(r) = 0, \tag{27.11}$$

where $\bar{k}^2 = E + V_0 = k^2 + V_0$, which yields

$$\psi(kr) = A\sin(\bar{k}r), \tag{27.12}$$

with the boundary condition $\psi(0) = 0$. In Region 2,

$$\frac{d^2\psi(r)}{dr^2} + (E - V_1)\psi(r) = 0, \tag{27.13}$$

where $E = k^2$, so let $\kappa^2 = V_1 - k^2$, which yields

$$\psi(kr) = B(\sinh(\kappa r) + \gamma\cosh(\kappa r)), \tag{27.14}$$

with two conditions of continuity of the logarithmic derivative at $r = r_1$ and $r = R$. In Region 3,

$$\frac{d^2\psi(r)}{dr^2} + E\psi(r) = 0. \tag{27.15}$$

We normalize the wavefunction to unit amplitude, with phase shift δ_0, and we then have

$$\psi(kr) = \sin(kr + \delta_0). \tag{27.16}$$

By using all the boundary conditions, we obtain

$$\gamma = \frac{(\kappa/\bar{k})\tan(\bar{k}r_1) - \tanh(\kappa r_1)}{1 - (\kappa/\bar{k})\tan(\bar{k}r_1)\tanh(\kappa r_1)}, \tag{27.17}$$

$$\delta_0 = -kR + \arctan\left[\left(\frac{k}{\kappa}\right)\frac{\gamma + \tanh(\kappa R)}{1 + \gamma\tanh(\kappa R)}\right]. \tag{27.18}$$

Rearranging δ_0 yields

$$\delta_0 = -kR + \arctan\left[\frac{kR}{\kappa R}\frac{(\kappa r_1/\bar{k}r_1)\tan(\bar{k}r_1) + \tanh(\kappa(R - r_1))}{[1 + (\kappa r_1/\bar{k}r_1)\tan(\bar{k}r_1)\tanh(\kappa(R - r_1))]}\right], \tag{27.19}$$

$$B = \frac{\sin(kR + \delta_0)}{\sinh(\kappa R) + \gamma \cosh(\kappa R)}, \tag{27.20}$$

$$A = \frac{B(\sinh(\kappa r_1) + \gamma \cosh(\kappa r_1))}{\sin(\bar{k}r_1)}. \tag{27.21}$$

Since

$$\lim_{\kappa \to 0^+} \gamma = 0, \tag{27.22}$$

(27.19) becomes $\delta_0 = -kR + \arctan(\infty) = -kR + \pi/2$.

Case 2: $k^2 > V_1$

We next consider the case $k^2 > V_1$, for which energies are above the top of the barrier. Then the radial wave function $\psi(r)$ in each region is given as follows. In Region 1, we have

$$\frac{d^2\psi(r)}{dr^2} + (E + V_0)\psi(r) = 0, \tag{27.23}$$

where $\bar{k}^2 = E + V_0 = k^2 + V_0$, which yields

$$\psi(kr) = \tilde{A}\sin(\bar{k}r), \tag{27.24}$$

with the boundary condition $\psi(0) = 0$. In Region 2,

$$\frac{d^2\psi(r)}{dr^2} + (E - V_1)\psi(r) = 0, \tag{27.25}$$

where $\tilde{\kappa}^2 = E - V_1 = k^2 - V_1$, which yields

$$\psi(kr) = B(i\sin(\tilde{\kappa}r) + i\tilde{\gamma}\cos(\tilde{\kappa}r)), \tag{27.26}$$

again with two conditions of continuity of the logarithmic derivative at $r = r_1$ and $r = R$. In Region 3,

$$\frac{d^2\psi(r)}{dr^2} + E\psi(r) = 0. \tag{27.27}$$

By once again normalizing the wavefunction to unity, with phase shift δ_0, we have

$$\psi(kr) = \sin(kr + \delta_0). \tag{27.28}$$

Using the boundary conditions, it follows that

$$\tilde{\gamma} = \frac{(\tilde{\kappa}/\bar{k})\tan(\bar{k}r_1) - \tan(\tilde{\kappa}r_1)}{1 + (\tilde{\kappa}/\bar{k})\tan(\bar{k}r_1)\tan(\tilde{\kappa}r_1)}, \tag{27.29}$$

$$\delta_0 = -kR + \arctan\left[\left(\frac{k}{\tilde{\kappa}}\right)\frac{\tilde{\gamma} + \tan(\tilde{\kappa}R)}{1 - \tilde{\gamma}\tan(\tilde{\kappa}R)}\right]. \tag{27.30}$$

Rearranging δ_0 yields

$$\delta_0 = -kR + \arctan\left[\frac{kR}{\tilde{\kappa}R}\frac{(\tilde{\kappa}r_1/\bar{k}r_1)\tan(\bar{k}r_1) + \tan(\tilde{\kappa}(R - r_1))}{[1 - (\tilde{\kappa}r_1/\bar{k}r_1)\tan(\bar{k}r_1)\tan(\tilde{\kappa}(R - r_1))]}\right], \tag{27.31}$$

where

$$B = \frac{\sin(kR + \delta_0)}{i \sin(\tilde{\kappa} R) + \tilde{\gamma} \cosh(\tilde{\kappa} R)}, \tag{27.32}$$

$$A = \frac{Bi(\sin(\tilde{\kappa} r_1) + \tilde{\gamma} \cosh(\tilde{\kappa} r_1))}{\sin(\tilde{k} r_1)}. \tag{27.33}$$

Consider first the case where $\kappa^2 > 0$. Let $x = kR$ be a dimensionless variable, and $r_1 = \alpha R$, where $0 < \alpha < 1$. Then if $\bar{k} = \sqrt{k^2 + V_0}$,

$$\bar{k} R = \sqrt{k^2 R^2 + V_0 R^2} = \sqrt{x^2 + \rho_0},$$

where $\rho_0 = V_0 R^2$.

If $\kappa = \sqrt{V_1 - k^2}$ ($k^2 \leq V_1$), then similarly,

$$\kappa R = \sqrt{V_1 R^2 - k^2 R^2} = \sqrt{\rho_1 - x^2},$$

where $\rho_1 = V_1 R^2$. Also,

$$\kappa(R - r_1) = \sqrt{\rho_1 - x^2} - \alpha \kappa R$$

$$= \sqrt{\rho_1 - x^2} - \alpha \sqrt{\rho_1 - x^2}$$

$$= (1 - \alpha)\sqrt{\rho_1 - x^2}.$$

Note that

$$\kappa r_1 = \alpha \kappa R = \alpha \sqrt{\rho_1 - x^2},$$

$$\bar{k} r_1 = \bar{k} R \cdot \frac{r_1}{R} = \alpha \sqrt{\rho_0 + x^2},$$

and

$$1 - \alpha = 1 - \frac{r_1}{R} = \frac{R - r_1}{R}.$$

Therefore equation (27.19) can be written in dimensionless form as

$$\delta_0 = -x + \arctan \left\{ \left[\frac{x}{(\rho_1 - x^2)^{1/2}} \right] \right.$$

$$\times \frac{\left[\left(\frac{\rho_1 - x^2}{\rho_0 + x^2}\right)^{1/2} \tan(\alpha \sqrt{\rho_0 + x^2}) + \tanh((1 - \alpha)\sqrt{\rho_1 - x^2}) \right]}{1 + \left(\frac{\rho_1 - x^2}{\rho_0 + x^2}\right)^{1/2} \tan(\alpha \sqrt{\rho_0 + x^2}) \tanh((1 - \alpha)\sqrt{\rho_1 - x^2})} \right\}. \tag{27.34}$$

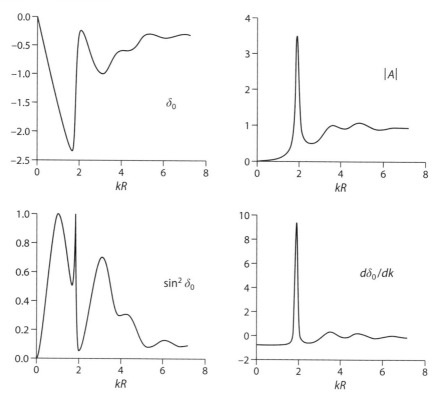

Figure 27.3 Scattering features for phase shift and amplitude.

Similarly, for the case $\kappa^2 < 0$ or $\kappa^2 = -\tilde{\kappa}^2$, equation (27.31) becomes

$$\delta_0 = -x + \arctan\left\{\left[\frac{x}{\left(x^2 - \rho_1\right)^{1/2}}\right]\right.$$

$$\times \left.\frac{\left[\left(\frac{x^2-\rho_1}{x^2+\rho_0}\right)^{1/2}\tan(\alpha\sqrt{\rho_0 + x^2}) + \tan((1-\alpha)\sqrt{x^2 - \rho_1})\right]}{1 - \left(\frac{x^2-\rho_1}{x^2+\rho_0}\right)^{1/2}\tan(\alpha\sqrt{\rho_0 + x^2})\tan((1-\alpha)\sqrt{x^2 - \rho_1})}\right\}. \quad (27.35)$$

Note that

$$\tilde{\kappa} R = \sqrt{k^2 R^2 - V_1 R^2} = \sqrt{x^2 - \rho_1},$$

$$\tilde{\kappa} r_1 = \alpha\sqrt{x^2 - \rho_1},$$

and

$$\tilde{\kappa}(R - r_1) = \tilde{\kappa} R(1 - \alpha) = (1 - \alpha)\sqrt{x^2 - \rho_1}.$$

27.1.2 Resonances

Figure 27.3 plots the relevant physical parameters for scattering for the case $E > V_1$. For the choice $\alpha = 0.5$, $V_0 = 4$, and $V_1 = 20$, a significant feature of this figure is seen to be the resonance that occurs at $kR \simeq 1.8$. At this energy, the results show that the following four characteristics are necessary to define a resonance appropriately (see also a similar analysis in [54]).

(1) The phase shift δ_0 suddenly increases by approximately π and passes through $-\pi/2$. We can distinguish the features of the phase shift plot as resonances. For example, if the phase shift undergoes a large increase, then the resonance interpretation will be very useful. But if the phase shift is gradually increasing and does not change by much, the resonance will vary broadly and merge into the background.

(2) The scattering strength, measured by $\sin^2 \delta_0$, has a sharp maximum and reaches $\sin^2 \delta_0 = 1$. If the scattering strength goes through its maximum value (i.e., $\sin^2 \delta_0 = 1$), then δ must pass through $\pi/2$. However, such a maximum is not a resonance unless δ is increasing with energy. Therefore, this condition is not sufficient for the existence of a resonance. For example, the peak at $kR = 1$ in Figure 27.3 does not correspond to a resonance because the time delay ($\propto d\delta_0/dk$) is negative and there is no resonant state. However, the very narrow peak at $kR \simeq 1.8$ in $\sin^2 \delta_0$ is a sufficient condition for a resonance, because $|d\delta_0/dk|$, is large there.

(3) The amplitude $|A|$ of the wave in the interior region (the well) has a sharp maximum; it signifies that at the resonance, the amplitude of oscillation of the system is large.

(4) The specific time delay $r_1^{-1} d\delta_0/dk$ has a sharp maximum.

Wavepackets Revisited

By using a wavepacket approach [54], the modification of the wavefunction by the potential (namely, by the phase shift $\delta_0(k)$) plays a crucial role in providing information about a time-dependent scattered wavepacket. Let

$$\psi(x, t) = \int_0^x g(k) e^{i(kx - \omega t + \delta_0)} dk. \tag{27.36}$$

The positive spectral function $g(k)$ is taken to have a peak at $k = k_0$, corresponding to a group velocity v_0. This peak is determined by the requirement of a stationary phase at $k = k_0$, or

$$\frac{d}{dk}[kx - \omega t + \delta_0(k)]_{k=k_0} = 0, \tag{27.37}$$

provided

$$\left[x - \frac{d\omega}{dk} t + \frac{d\delta_0(k)}{dk}\right]_{k=k_0} = 0, \tag{27.38}$$

or, at $k = k_0$, with $v_g = d\omega(k)/dk$, we obtain

$$x = v_g(k_0)\left[t - \frac{1}{v_g(k)}\frac{d\delta_0}{dk}\right]_{k=k_0} = v_g(k_0)(t - T(k_0)). \tag{27.39}$$

We can write

$$T = \frac{1}{v_g}\frac{d\delta_0}{dk} \tag{27.40}$$

in terms of the well/barrier width, R (or just the well, r_1) as

$$T = \frac{1}{r_1}\frac{d\delta_0}{dk} \cdot \frac{R}{v_g} \tag{27.41}$$

(at $k = k_0$). So T is just the specific time delay suffered by a wavepacket in units of the transit time for a free particle to cross the distance R.

Chapter Twenty-Eight

Back Where We Started

28.1 A BRIDGE OVER COLORED WATER

It seems appropriate to return, full circle, as it were (pun somewhat intended), to the rainbow directory in the Introduction. I wrote there that the rainbow functions as a sort of template for the topics of this book. And I also noted in the preface that this book is intensely personal, so I would like to share the genesis of my fascination with rainbows (and the first mention of a rainbow in the Bible is (perhaps not surprisingly) in the book of Genesis!). In fact, in 2007 I gave a talk titled "Noah's Arc: Asin in the Sky," referring to the story of Noah and the sign of the rainbow in Genesis chapter 9, and also to the occurrence of that inverse trigonometric function in elementary rainbow theory. The title was probably more creative than the talk. But I digress . . .

In the late 1990s (I cannot recall the exact year) I attended a conference on mathematical biology held at the University of North Carolina, Chapel Hill, where two of my graduate students were presenting papers. Early one evening I was walking back from a delightful visit to a used-book store (with a large package under my arm) and was confronted by an intense and beautiful, almost semicircular rainbow directly ahead of me. In my excitement I pointed this out to several other people who were walking in the opposite direction, and received (as you might well expect) very mixed reactions! So profound was that optical experience that I spent the rest of the evening drafting a table of contents for what would turn out to be, several years later, two books ([122] and [212]). Since that time there have been quite a few more papers, chapters, and talks (and a senior-level class) on this and other topics in the realm of "Nature's mathematics." One of those articles was a book review with a mathematical appendix [148]. The following is an extract (with minor modifications and without the appendix) from my review of *The Rainbow Bridge: Rainbows in Art, Myth, and Science* by Raymond L. Lee, Jr., and Alistair B. Fraser [169] (see http://www.ams.org/notices/200211/fea-adam.pdf for the full review).

> This is a magnificent and scholarly book, exquisitely produced, and definitely not destined only for the coffee table. It is multifaceted in character, addressing "rainbow-relevant" aspects of mythology, religion, the history of art, art criticism, the history of optics, the theory of color, the philosophy of science, and advertising! The quality of the reproductions and photographs is superb. The authors are experts in meteorological optics, but their book draws on many other

sub-disciplines. It is a challenge, therefore, to write a review about a book that contains no equations or explicit mathematical themes, for what is primarily a mathematical audience. However, while the mathematical description of the rainbow may be hidden in this book, it is nonetheless present. Clearly, such a review runs the risk of giving a distorted picture of what the book is about, both by "unfolding" the hidden mathematics and suppressing, to some extent, other important and explicit themes: the connections with mythology, art and science. To a degree this is inevitable, if such a weighted metric is to be used. Lee and Fraser are intent on exploring bridges from the rainbow to all the places listed above; and in the opinion of this reviewer they have succeeded admirably. The serious reader will glean much of value, and mathematicians in particular may benefit from the tantalizing hints of mathematical structure hidden in the photographs and graphics.

In this review I have included several direct quotations to illustrate both the writing style and features of the book that I found most intriguing. Frankly, I found some of the early chapters harder to appreciate than later ones. This is a reflection, no doubt, of my own educational deficiencies in the liberal arts, but upon completing the book, I was moved to suggest to the chair of the Art Department that it might be interesting to offer a team-taught graduate seminar in Art, using this book as a "text". He is now avidly reading my copy of the book. The ten chapters combined trace the rainbow bridge to "the gods"; as a sign and symbol ("emblem and enigma"); to the "growing tension between scientific and artistic images of the rainbow"; through the inconsistencies of the Aristotelian description and beyond, to those of Descartes and Newton, and the latter's theory of color; to claims of a new unity between the scientific and artistic enterprises; the evolution of scientific models of the rainbow to relatively recent times; and the exploitation and commercialization of the rainbow. All these bridges, the authors claim, are united by the human appreciation of the rainbow's compelling natural beauty. And who can disagree? An appendix is provided ("a field guide to the rainbow") comprising nineteen basic questions about the rainbow, with non-technical (but scientifically accurate) answers for the interested observer. This is followed by a set of chapter notes and a bibliography, both of which are very comprehensive. The more technical scientific aspects of the distribution of light within a rainbow are scattered liberally throughout the latter half of the book; indeed, the reader more interested in the scientific aspects of the rainbow might wish to read the last five chapters "in parallel" with the first five (as did I). The technical aspects referred to are explained with great clarity.

But what *is* a rainbow? Towards the end of this review, several mutually nonexclusive but complementary levels of explanation will be noted, reflecting the fact that there is a great deal of physics and mathematics behind one of nature's most awesome spectacles. At a more basic observational level, surely *everyone* can describe the colored arc of light we call a rainbow, we might suggest, certainly as far as the *primary* bow is concerned. In principle, whenever there is a primary rainbow, there is a larger and fainter *secondary* bow. The primary (formed by light being refracted twice and reflected once in raindrops) lies beneath a fainter secondary bow (formed by an additional reflection, and therefore fainter because of light loss), which is not always easily seen. There are several other things to look for: faint pastel fringes just below the top of the primary bow (supernumerary bows, of which more anon); the reversal of colors in the secondary bow; the dark region between the two bows; and the bright region below the primary bow. This observational description, however, is probably not one that is universally known; in some cultures it is considered unwise to even look at a rainbow. Indeed, in many parts of the world, it appears, merely pointing at a rainbow is considered to be a foolhardy act. Lee and Fraser state that "getting jaundice, losing an eye, being struck by lightning, or simply disappearing are among the unsavory aftermaths of rainbow pointing."

The historical descriptions are in places quite breathtaking; we are invited to look over the shoulder, as it were, of Descartes and Newton as they work through their respective accounts of the rainbow's position and colors. While the color theory of Descartes was flawed; his geometric theory was not. Commenting on the latter, the authors point out "Descartes' seventeenth-century analysis of the rainbow bears out Plato's great faith in observations simplified and clarified by the power of mathematics." Newton, on the other hand, eschews Descartes cumbersome ray-tracing technique and "silently invokes his mathematical invention of the 1660s, differential calculus, to specify the minimum deviation rays of the primary and secondary rainbows." Later in the book they remark that Aristotle and later scientists in antiquity "constructed theories that primarily *describe* natural phenomena in mathematical or geometric terms, with little or no concern for physical mechanisms that might *explain* them." This contrast goes to the heart of the difference between "Aristotelian" and mathematical modeling.

I was pleasantly surprised to learn that the English painter John Constable was quite an avid amateur scientist: he was concerned that his paintings of clouds and rainbows should accurately reflect the science of the day, and he took great trouble to acquaint himself with Newton's theory of the rainbow (many other details can be found in pp. 80–87 of the book). In a similar vein, the writer John Ruskin was a detailed observer of nature, his goal being "that of transforming close observation into faithful depiction of a purposeful, divinely shaped nature." The poet John Keats implies in his *Lamia* that Newton's natural philosophy destroys the beauty of the rainbow ("... There was an awful rainbow once in heaven: We know her woof, her texture; she is given in the dull catalog of common things. Philosophy will clip an Angel's wings, ..."). Keats' words reflected a continuing debate: did scientific knowledge facilitate or constrain poetic descriptions of nature? His contemporary William Wordsworth apparently held a different opinion, for he wrote "The beauty in form of a plant or an animal is not made less but more apparent as a whole by more accurate insight into its constituent properties and powers."

In a subsection of chapter 4 entitled *The Inescapable (and Unapproachable) Bow*, the authors address, amongst other things, some common misperceptions about rainbows. Just as occurs when we contemplate our visage in a mirror, what we see is not an *object*, but an *image*; near the end of the chapter this rainbow image is beautifully portrayed as "a mosaic of sunlit rain". Optically, the rainbow is located at infinity, even though the raindrops or droplets sprayed from a garden hose are not. To 'place' the rainbow in the sky, note that the *antisolar point* is 180° from the sun, on a line through the head of the observer. ... rays of sunlight are deviated from this line; the ray of minimum deviation (sometimes called the rainbow or Descartes ray) is deviated by about 138° for the primary bow. Therefore there is a concentration of deviated rays near this angle, and so for the observer, the primary bow is an arc of a circle of radius 180° − 138° = 42° centered on the antisolar point. Thus the rainbow has a fixed angular radius of about 42° despite illusions (and allusions) to the contrary. The corresponding angle for the secondary bow is about 51°, though in each instance the angle varies a little depending on the wavelength of the light. Without this phenomenon of *dispersion*, there would be only a 'whitebow'! It is perhaps an occupational hazard for professional scientists and mathematicians to be a little frustrated by incorrect depictions or explanations of observable phenomena and mathematical concepts in literature, art, the media, etc. From the point of view of an artist, however (Constable notwithstanding), scientific accuracy is not necessarily a prelude to artistic expression; indeed to some it may be considered a hindrance. Nevertheless, we can identify perhaps with the authors, who in commenting on the paintings of Frederic Church (in particular his *Rainy Season in the Tropics*) write "Like Constable, he in places bends the unyielding rule that all shadows must be radii to the bow (that is, they must converge on the rainbow's center)." The same rule applies to sunbeams, should they be observable concurrently with a rainbow.

By the middle of the eighteenth century, the contributions of Descartes and Newton notwithstanding, observations of supernumerary bows were a persistent reminder of the inadequacy of current theories of the rainbow. As Lee and Fraser so pointedly remark, "One common reaction to being confronted with the unexplained is to label it inexplicable." This led to these troubling features being labeled spurious; hence the unfortunate addition of the adjective *supernumerary* to the rainbow phenomenon. Now it is known that such bows "are an integral part of the rainbow, not a vexing corruption of it."; an appropriate, though possibly unintended, mathematical pun. By focusing attention on the light *wavefronts* incident on a spherical drop, rather than the rays normal to them, it is easier to appreciate the self-interference of such a wave as it becomes "folded" onto itself as a result of refraction and reflection within the drop. This is readily seen from figures 8.7 and 8.9 in the book, from which the true extent of the rainbow is revealed: the primary rainbow is in fact the *first interference maximum*, the second and third maxima being the first and second supernumerary bows respectively (and so on). The angular spacing of these bands depends on the size of the droplets producing them. The width of individual bands and the spacing between them decreases as the drops get larger. If drops of many different sizes are present, these supernumerary arcs tend to overlap somewhat and smear out what would have been obvious interference bands for droplets of uniform size. This is why these pale blue or pink or green bands are then most noticeable near the top of the rainbow: it is the near-sphericity of the smaller drops that enable them to contribute to this part of the bow; larger drops are distorted from sphericity by the aerodynamic forces acting upon them. Nearer the horizon a wide range of drop size contributes to the bow, but at the same time it tends to blur the interference bands. In principle, similar interference effects also occur above the secondary rainbow, though they are very rare, for reasons discussed in the penultimate chapter. Lee and Fraser summarize the importance of "spurious" bows with typical metaphorical creativity: "Thus the supernumerary rainbows proved to be the midwife that delivered the wave theory of light to its place of dominance in the nineteenth century."

It is important to recognize that not only were the Cartesian and Newtonian theories unable to account for the presence of supernumerary bows, but also they both predicted an *abrupt* transition between regions of illumination and shadow (as at the edges of Alexander's dark band, when rays only giving rise to the primary and secondary bows are considered). In the wave theory of light such sharp boundaries are softened by *diffraction*, which occurs when the normal interference pattern responsible for rectilinear propagation of light is distorted in some way. Diffraction effects are particularly prevalent in the vicinity of *caustics*. In 1835 Potter showed that the rainbow ray may be interpreted as a caustic, i.e. the envelope of the system of rays constituting the rainbow. The word caustic means "burning", and caustics are associated with regions of high intensity illumination (with geometrical optics predicting an infinite intensity there). Thus the rainbow problem is essentially that of determining the intensity of (scattered) light in the neighborhood of a caustic. This was exactly what Airy attempted to do several years later in 1838. The principle behind Airy's approach was established by Huygens in the 17th century: Huygens' principle regards every point of a wavefront as a secondary source of waves, which in turn defines a new wave-front and hence determines the subsequent propagation of the wave. As pointed out by Nussenzveig [16], Airy reasoned that if one knew the amplitude distribution of the waves along any complete wavefront in a raindrop, the distribution at any other point could be determined by Huygens' principle. Airy chose as his starting point a wavefront surface inside the raindrop, the surface being orthogonal to all the rays that constitute the primary bow; this surface has a point of inflection wherever it intersects the ray of minimum deviation—the "rainbow ray". Using the standard assumptions of diffraction theory, he formulated the local intensity of scattered light in terms of a "rainbow integral", subsequently renamed the *Airy integral* in his honor; it is related to the now familiar Airy function Ai(X), analogous to the Fresnel integrals which also arise in diffraction theory. There are several equivalent representations of the Airy integral in the literature; here it suffices to note that

$$\text{Ai}(X) \propto \int_{-\infty}^{\infty} \exp\left[i\left(\frac{s^3}{3} + Xs\right)\right] ds.$$

While the argument of the Airy function is arbitrary at this point, X refers to the set of control space parameters in the discussion below on diffraction catastrophes. In this case it represents the deviation or scattering angle coordinate. One severe limitation of the Airy theory for the optical rainbow is that the amplitude distribution along the initial wavefront is unknown: based on certain assumptions it has to be guessed. There is a natural and fundamental parameter, the *size parameter*, β, which is useful in determining the domain of validity of the Airy approximation; it is defined as the ratio of the droplet circumference to the wavelength λ of light. In terms of the wavenumber k this is $\beta = 2\pi a/\lambda \equiv ka$, in terms of the droplet radius a. Typically, for sizes ranging from fog droplets to large raindrops, β ranges from about 100 to several thousand. Airy's approximation is a good one only for $\beta \gtrsim 5000$ and angles sufficiently close to that of the rainbow ray. In light of these remarks it is perhaps surprising that an exact solution does exist for the rainbow problem, as indicated below.

In the epilogue to the final chapter of the book, one subsection is entitled *Airy's Rainbow Theory: The Incomplete "Complete" Answer*. This theory did go beyond the models of the day in that it quantified the dependence upon the raindrop size of (i) the rainbow's angular width, (ii) its angular radius, and (iii) the spacing of the supernumeraries. Also, unlike the models of Descartes and Newton, Airy's predicted a non-zero distribution of light intensity in Alexander's dark band, and a finite intensity at the angle of minimum deviation (as noted above, the earlier theories predicted an infinite intensity there). However, spurred on by Maxwell's recognition that light is part of the electromagnetic spectrum, and the subsequent publication of his mathematical treatise on electromagnetic waves, several mathematical physicists sought a more complete theory of scattering, because it had been demonstrated by then that the Airy theory failed to predict precisely the angular position of many laboratory-generated rainbows. Among them were the German physicist Gustav Mie who published a paper in 1908 on the scattering of light by homogeneous spheres in a homogeneous medium, and Peter Debye who independently developed a similar theory for the scattering of electromagnetic waves by spheres. Mie's theory was intended to explain the colors exhibited by colloidally dispersed metal particles, whereas Debye's work, based on his 1908 thesis, dealt with the problem of light pressure on a spherical particle. The resulting body of knowledge is usually referred to as Mie theory (or more accurately, the Mie solution), and typical computations based on it are formidable compared with those based on Airy theory, unless the drop size is sufficiently small. A similar (but non-electromagnetic) formulation arises in the scattering of sound waves by an impenetrable sphere, studied by Lord Rayleigh and others in the 19th century. Mie theory is based on the solution of Maxwell's equations of electromagnetic theory for a monochromatic plane wave from infinity incident upon a homogeneous isotropic sphere of radius a. The surrounding medium is transparent (as the sphere may be), homogeneous and isotropic. The incident wave induces forced oscillations of both free and bound charges in synchrony with the applied field, and this induces a secondary electric and magnetic field, each of which has components inside and outside the sphere. Of crucial importance in the theory are the scattering amplitudes $(S_j(k, \theta), j = 1, 2)$ for the two independent polarizations, θ being the angular variable; these amplitudes can be expressed as an infinite sum called a partial wave expansion. Each term (or "partial wave") in the expansion is defined in terms of combinations of Legendre functions of the first kind, Riccati-Bessel and Riccati-Hankel functions (the latter two being rather simply related to spherical Bessel and Hankel functions respectively).

It is obviously of interest to determine under what conditions such an infinite set of terms can be truncated, and what the resulting error may be by so doing. However, it turns out that the number of terms that must be retained is of the same order of magnitude as the size parameter β, i.e. up to several thousand for the rainbow problem. On the other hand, the "why is the sky blue?" scattering problem —*Rayleigh scattering*—requires only one term because the scatterers are molecules which are much smaller than a wavelength of light, so the simplest truncation—retaining only the first term—is perfectly adequate. Although in principle the rainbow problem can be "solved" with enough computer time and resources, numerical solutions by themselves (as Nussenzveig points out) offer little or no insight into the physics of the phenomenon.

The *Watson transform*, originally introduced by Watson in connection with the diffraction of radio waves around the earth (and subsequently modified by Nussenzveig in his studies of the rainbow problem), is a method for transforming the slowly-converging partial-wave series into a rapidly convergent expression involving an integral in the complex angular-momentum plane. The Watson transform is intimately related to the Poisson summatiom formula

$$\sum_{l=0}^{\infty} a\left(l + \frac{1}{2}, x\right) = \sum_{m=-\infty}^{\infty} e^{-im\pi} \int_0^{\infty} a(\lambda, x) e^{2\pi i m \lambda} d\lambda,$$

for an "interpolating function" $a(\lambda, x)$, where x denotes a set of parameters and $\lambda = l + \frac{1}{2}$ is now considered to be the complex angular momentum variable.

But why *angular momentum*? Although they possess zero rest mass, photons have energy $E = hc/\nu$ and momentum $E/c = h/\nu$ where h is Planck's constant and c is the speed of light in vacuo. Thus for a non-zero impact parameter b_i a photon will carry an angular momentum $b_i h/\nu$ (b_i being the perpendicular distance of the incident ray from the axis of symmetry of the sun-raindrop system). Each of these discrete values can be identified with a term in the partial wave series expansion. Furthermore, as the photon undergoes repeated internal reflections, it can be thought of as orbiting the center of the raindrop. Why *complex* angular momentum? This allows the above transformation to effectively "redistribute" the contributions to the partial wave series into a few points in the complex plane—specifically poles (called *Regge* poles in elementary particle physics) and saddle-points. Such a decomposition means that instead of identifying angular momentum with certain discrete real numbers, it is now permitted to move continuously through complex values. However, despite this modification, the poles and saddle points have profound physical interpretations in the rainbow problem.

In the simplest Cartesian terms, on the illuminated side of the rainbow (in a limiting sense) there are two rays of light emerging in parallel directions: at the rainbow angle they coalesce into the ray of minimum deflection, and on the shadow side, according to geometrical optics, they vanish (this is actually a good definition of a caustic curve or surface). From a study of real and complex "rays" it happens that, mathematically, in the context of the complex angular momentum plane, a rainbow is the *collision of two real saddle points* . But this is not all: this collision does not result in the mutual annihilation of these saddle points; instead a single complex saddle point is born, corresponding to a complex ray on the shadow side of the caustic curve. This is directly associated with the diffracted light in Alexander's dark band.

Lee and Fraser point out that as far as most aspects of the optical rainbow are concerned, Mie theory is esoteric overkill; Airy theory is quite sufficient for describing the outdoor rainbow. I found particularly valuable the following comment by the authors; it has implications for mathematical modeling in general, not just for the optics of the rainbow. In their comparison of the less accurate Airy theory of the rainbow with the more general and powerful Mie theory, they write "Our point here is not that the exact Mie theory describes the natural rainbow inadequately, but rather that the approximate Airy theory can describe it quite well. Thus the supposedly outmoded Airy theory generates a more natural-looking map of real rainbow colors than Mie theory does, even though Airy theory makes substantial errors in describing the scattering of monochromatic light by isolated small drops. *As in many hierarchies of scientific models, the virtues of a simpler theory can, under the right circumstances, outweigh its vices.*" [Italics added] [148]

But as you will see, we are not yet quite done; there are still some more classical ray/scattering connections to be made!

28.2 RAY OPTICS REVISITED: LUNEBERG INVERSION AND GRAVITATIONAL LENSING

28.2.1 Abel's Integral Equation and the Luneberg Lens

Suppose that a ray starting from a point A on an axis of symmetry enters a transparent spherical scatterer at point P_0, traverses the sphere in a clockwise manner (see Figure 28.1) and exits at point P_1, finally reaching the axis again at point B. Outside the sphere the refractive index $n = 1$. Recall from the discussion of Bouguer's law in Chapter 3 (see Figure 3.1) that the angles θ, ϕ, and ψ are clearly related via

$$\psi = \theta + \phi. \tag{28.1}$$

To convert this to a form more appropriate to the discussion below, let us use the supplements of angles θ and ψ, leaving equation (28.1) unchanged of course. Applying Bouguer's law,

$$nr \sin\phi = K, \tag{28.2}$$

to the points A and B in particular, it follows that

$$r_0 \sin\alpha_0 = r_1 \sin\alpha_1 = K. \tag{28.3}$$

Using equation (28.1), the differential form of (28.2) takes the form

$$d\psi = d\theta + \left(\frac{dn}{n} + \frac{dr}{r}\right)\tan\phi, \tag{28.4}$$

but recalling that the change $\theta \to \pi - \theta$ was made, this result simplifies to give the element of "ray bending" magnitude as

$$|d\psi| = \frac{dn}{n}\tan\phi, \tag{28.5}$$

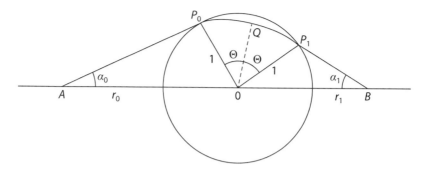

Figure 28.1 The ray path through a transparent sphere.

or using (28.2) again,

$$d\psi = \frac{K\,dn}{n(r)\left[r^2n^2(r) - K^2\right]^{1/2}},\tag{28.6}$$

where $d\psi > 0$ is chosen to conform to ψ increasing in the clockwise sense.

From Figure 28.1 it can be seen that the total bending between P_0 and P_1 is

$$2\Theta = 2K\int_{n(1)}^{n(r_T)}\frac{dn}{n(r)\left[r^2n^2(r) - K^2\right]^{1/2}} = 2K\int_{1}^{n(r_T)}\frac{dn}{n(r)\left[r^2n^2(r) - K^2\right]^{1/2}},\tag{28.7}$$

since $n(1) = 1$ In Chapter 3 the angle Θ was expressed in terms of an integral over radius (not over n) as

$$\Theta = \int_{r_T}^{1}\frac{dr}{r\left[r^2n^2(r) - K^2\right]^{1/2}}.\tag{28.8}$$

There are two basic approaches to this problem, expressing the integrals over r or n, or equivalently, utilizing the angle differences $\theta_1 - \theta_0$ or $\psi_1 - \psi_0$, favored by [240] and by [241], respectively. The two integrals are equivalent, but we adopt the latter approach here (especially since that result is more readily available online). Returning to the form (28.7), note that the upper limit $n(r_T)$ is the refractive index at the mid-point of the arc $P_0 P_1$ and satisfies $n(r_T)r_T = K$. Because the total bending is equal to $\alpha_0 + \alpha_1$ it must be the case that

$$\int_{1}^{n(r_T)}\frac{dn}{n(r)\left[r^2n^2(r) - K^2\right]^{1/2}} = \frac{1}{2K}\left(\arcsin\frac{K}{r_0} + \arcsin\frac{K}{r_1}\right), \quad K \in (0, 1].\tag{28.9}$$

Make the following identifications:

$$\rho = rn; \quad 1 - \rho^2 = \xi(\rho); \quad 1 - K^2 = \eta; \quad f(\xi) = \ln n,$$

and

$$g(\eta) = \frac{1}{2K}\left(\arcsin\frac{K}{r_0} + \arcsin\frac{K}{r_1}\right).$$

The idea is to determine $f(\xi)$ and hence n (where $g(\eta)$ is known if r_0 and r_1 are given). Hence (28.9) is transformed into an Abel integral equation, that is,

$$\int_0^\eta \frac{f'(\xi)}{(\eta - \xi)^{1/2}} d\xi = g(\eta),$$

because $\xi(1) = 0$ and $\xi(\rho_T) = 1 - K^2 = \eta$. Given that in this problem n is continuous at the spherical boundary, $f(0) = 0$ and so the solution is

$$f(\xi) = \frac{1}{\pi} \int_0^\xi \frac{g(\eta)}{(\xi - \eta)^{1/2}} d\eta. \qquad (28.10)$$

Therefore if we define

$$\varpi(\rho, a) = \frac{1}{\pi} \int_\rho^1 \frac{\arcsin(K/a)}{\left(K^2 - \rho^2\right)^{1/2}} dK, \qquad (28.11)$$

it follows from equation (28.10) that

$$\ln n = \frac{1}{\pi} \int_\rho^1 \left(\arcsin \frac{K}{r_0} + \arcsin \frac{K}{r_1} \right) \frac{dK}{\left(K^2 - \rho^2\right)^{1/2}} = \varpi(\rho, r_0) + \varpi(\rho, r_1).$$

Hence

$$n(\rho) = \exp\left[\varpi(\rho, r_0) + \varpi(\rho, r_1)\right],$$

and

$$r(\rho) = \rho \exp(-\left[\varpi(\rho, r_0) + \varpi(\rho, r_1)\right]).$$

When considering the rainbow problem, rays from the sun are approximately parallel, so we can set $\varpi(\rho, \infty) = 0$, obviously simplifying the forms for $n(\rho)$ and $n(\rho)$. An exact solution can then be found for the special case of $a = 1$ in (28.11). This is certainly the case for $r_1 = 1$, and corresponds to a parallel beam of rays being focused at a point on the far surface of the lens. In this instance,

$$\varpi(\rho, 1) = \frac{1}{\pi} \int_\rho^1 \frac{\arcsin K}{\left(K^2 - \rho^2\right)^{1/2}} dK = \frac{1}{2} \ln\left[1 + \left(1 - \rho^2\right)^{1/2}\right]. \qquad (28.12)$$

Exercise: Prove this last result.

Hint: See [6] or [240].

Then we have

$$n(\rho) = \exp\left[\varpi(\rho, r_1)\right] = \left\{\left[1 + \left(1 - \rho^2\right)^{1/2}\right]\right\}^{1/2},$$

and

$$r(\rho) = \rho \exp\left(-\left[\varpi(\rho, r_1)\right]\right) = \frac{\rho}{\left\{\left[1 + \left(1 - \rho^2\right)^{1/2}\right]\right\}^{1/2}},$$

from which it follows that

$$n = \left(2 - r^2\right)^{1/2}, \qquad (28.13)$$

which is the Luneberg refractive index profile [240]. For more recent studies on the Luneberg inverse problem, see [246] (see also [74]).

Exercise: Verify (without using the ϖ function) that the solution (28.13) satisfies the integral equation (28.9) for $r_0 = \infty$ and $r_1 = 1$.

Orbiting

A careful analysis of the integral (28.8) for Θ in the neighborhood of the inverse square-root singularity yields two possibilities depending on whether $\rho(r) = rn(r)$ is a monotone increasing function. They are

(i) *The monotone case.* If $\rho'(r_T) \neq 0$, then in the neighborhood of r_T the integral for Θ has the dominant behavior $(r - r_T)^{1/2}$, which tends to zero as $r \to r_T^+$.

(ii) *The non-monotone case.* If $\rho'(r_T) = 0$ somewhere, then in the neighborhood of r_T the integral for Θ has the dominant behavior $\ln |r - r_T|$, which tends to $-\infty$ as $r \to r_T^+$.

To see this, expand the quantity ρ^2 about $r = r_T$; the radicand then takes the form

$$\rho^2 - K^2 = \rho^2(r_T) - K^2 + (r - r_T)\left[\rho^2\right]_{r_T}'$$

$$+ \frac{1}{2}(r - r_T)^2 \left[\rho^2\right]_{r_T}'' + O\left[(r - r_T)^3\right]. \qquad (28.14)$$

On simplifying this expression (and neglecting extraneous multiplicative and additive constants), we find that if $\left[\rho^2\right]_{r_T}' > 0$, the integral for Θ has the functional form

$$\Theta \propto \int (r - r_T)^{-1/2}\, dr \propto (r - r_T)^{1/2} \to 0, \qquad (28.15)$$

as $r \to r_T^+$. In contrast, if $\left[\rho^2\right]_{r_T}' = 0$, then

$$\Theta \propto \int |r - r_T|\, dr \propto \ln |r - r_T| \to -\infty, \qquad (28.16)$$

as $r \to r_T^+$.

In scattering theory, the logarithmic singularity (ii) is associated with the phenomenon of orbiting (see Chapter 15). For more details of this distinction in an optical context see [73].

28.2.2 Connection with Classical Scattering and Gravitational Lensing

Weak Gravitational Fields

In the study of gravitational lenses, the field is said to be weak if the gravitational potential $V(\mathbf{r})$ satisfies the strong inequality $|V(\mathbf{r})| \ll c^2$, c being the speed of light in vacuo. In such a weak gravitational field, light is bent in accordance with an effective index of refraction given by

$$n_{\mathrm{w}}(\mathbf{r}) \approx 1 - \frac{2V(\mathbf{r})}{c^2} = 1 + \frac{2GM}{c^2 r}, \qquad (28.17)$$

for a spherically symmetric field associated with a mass M, since then $V(\mathbf{r}) = V(r) = -GM/r$. G is the universal gravitational constant. If this approximation is formally substituted in (15.77) of Chapter 15 (or equivalently (3.119) in Chapter 3 for direct transmission through a refracting sphere), then the deflection of a ray by a weak gravitational field is given by

$$\theta \approx \pi - 2b \int_{r_0}^{\infty} \frac{dr}{r^2 \left[1 + (2R_{\mathrm{g}}/r) - (b^2/r^2)\right]^{1/2}} \qquad (28.18)$$

[242], where $b \gg R_{\mathrm{g}}$ is the impact parameter, and $R_{\mathrm{g}} = 2GM/c^2$ is the gravitational radius. Clearly, in this weak-field approximation the gravitational potential corresponds to an attractive Coulomb potential. Therefore the standard results from classical Coulomb scattering apply, in particular,

$$\cot\left(\frac{|\theta|}{2}\right) = \frac{b}{R_{\mathrm{g}}}, \qquad (28.19)$$

whence for small angles,

$$|\theta| \approx \frac{2R_{\mathrm{g}}}{b} = \frac{4GM}{bc^2}, \qquad (28.20)$$

which is a formula used to predict the deflection of light by the sun [243].

Black Hole Lens

The larger refraction effects near a Schwarzschild black hole can be attributed to a refractive index of the form [245]

$$n_{\mathrm{bh}}(r) = \frac{\left(1 + R_{\mathrm{g}}/4r\right)^3}{1 - R_{\mathrm{g}}/4r}. \qquad (28.21)$$

In the weak-field limit, $n_{\mathrm{bh}}(r) \to n_{\mathrm{w}}(r)$, as is readily seen. The following statement (with some minor notational changes) is of interest in this context:

> For external observers probing the region near the origin with electromagnetic waves, a static black hole and a classical refracting lens described by (28.21) are indistinguishable. [A] ray bending angle θ_{w} as a function of impact parameter b is needed for study of the scattering and focusing properties of these lenses. From the appropriate Abel transform,
>
> $$\theta_{\mathrm{w}} = \frac{2 \arccos\left(-R_{\mathrm{g}}/b\right)}{\left(1 - R_{\mathrm{g}}^2/b^2\right)^{1/2}} - \pi, \qquad (28.22)$$
>
> for the weak-field lens. The transform for the black hole lens yields
>
> $$\theta_{\mathrm{bh}} = 2\left(\frac{R_{\mathrm{g}}}{b}\right) + \frac{15\pi}{16}\left(\frac{R_{\mathrm{g}}}{b}\right)^2 + 16\left(\frac{R_{\mathrm{g}}}{b}\right)^3 + \frac{3465\pi}{1024}\left(\frac{R_{\mathrm{g}}}{b}\right)^4 + \cdots \qquad (28.23)$$
>
> for small angles [245].

Note that as $R_{\mathrm{g}}/b \to 0$, θ_{w} and θ_{bh} both tend to the value $2R_{\mathrm{g}}/b$.

Outline of θ_w-Derivation

We utilize the result (28.7) in the form

$$\theta_w = 2b \int_b^\infty \frac{dn}{n\left(\rho^2 - b^2\right)^{1/2}} = 2b \int_b^\infty \frac{1}{n}\frac{dn}{d\rho}\frac{d\rho}{\left(\rho^2 - b^2\right)^{1/2}}, \qquad (28.24)$$

where, as above

$$\rho = rn \equiv r + R_g, \qquad R_g < r. \qquad (28.25)$$

After some manipulation θ_w takes the form

$$\theta_w = -2R_g \int_0^{\pi/2} \frac{\cos\beta}{b - R_g \cos\beta}\,d\beta. \qquad (28.26)$$

Exercise: Show that

$$\int \frac{\cos x}{A + B\cos x}\,dx = \frac{1}{B}\left\{ x - \frac{2A}{\left(A^2 - B^2\right)^{1/2}} \arctan\left[\left(\frac{A - B}{A + B}\right)^{1/2} \tan\frac{x}{2}\right] + C\right\}. \qquad (28.27)$$

Using the form (28.27), it transpires that

$$\theta_w = \left| \pi - \frac{4b}{\left(b^2 - R_g^2\right)^{1/2}} \arctan\left(\frac{b - R_g}{b + R_g}\right)^{1/2}\right|. \qquad (28.28)$$

We are not interested in the sign of the bend angle since a ray tube will have rays that bend on all sides of the gravitational mass, so the results (28.22) and (28.28) can be shown to be equivalent for $R_g \ll b$.

   Exercise: Show the equivalence of equations (28.22) and (28.28) in the weak-field limit.

   And now (apart from the Appendices), we *are* done!

Appendix A

Order Notation: The "Big O," "Little o,"
and "~" Symbols

(1) If $f(z)$ and $g(z)$ are two functions of a complex variable z defined on some domain D, which possess limits as $z \to z_0$ (which may be zero or the point at infinity) in D, then

$f(z) = O(g(z))$ as $z \to z_0$ if $\exists$ constants $K > 0$ and $\delta > 0$ s.t. $|f| \leq K|g|$ whenever $0 < |z - z_0| < \delta$. If $|f| \leq K|g|$ $\forall z \in D$, then $f(z) = O(g(z))$ in D.

(2) If $f(z)$ and $g(z)$ are such that, for any $\varepsilon > 0$, $|f| \leq \varepsilon|g|$ whenever z is in a small δ-neighborhood of z_0, then $f(z) = o(g(z))$ as $z \to z_0$.

Practically speaking, this means that provided $g(z)$ is not zero in a δ-neighborhood of z_0 (except possibly *at* z_0), $f(z) = O(g(z))$ implies that f/g is bounded, so that f is bounded by a multiple of g as $z \to z_0$. Also $f(z) = o(g(z))$ implies that $f/g \to 0$ as $z \to z_0$, that is, f must vanish more rapidly than g as $z \to z_0$.

Now here's an informal little lemma: the only two points that need concern us are zero and the point at infinity! The infomal little proof is: suppose z_0 is neither of these. Then either let $\xi = z - z_0$ or $\xi = (z - z_0)^{-1}$.

(3) $f(z)$ is said to be *asymptotically equivalent* to (or equal to or approaches) $g(z)$ as $z \to z_0$ if

$$\lim_{z \to z_0} \frac{f(z)}{g(z)} = 1,$$

and we write

$$f(z) \sim g(z) \text{ as } z \to z_0.$$

The O, o, and $\sim$ symbols characterize different states of our knowledge about the asymptotic behavior of a function. The $\sim$ symbol tells us what $f(z)$ "looks like" as $z \to z_0$, and in this sense it tells us more about f that the other two symbols do. However, the O symbol is more important than the o in this context because it does provide more specific information. Thus if $f(z) \to 0$ as $z \to 0$, the O-order of $f(z)$ tells us how *rapidly* $f(z) \to 0$, whereas the o-order merely confirms that $f(z) \to 0$. Thus for example, $f(z) = z^2 \ln z + z^3 = O(z^2 \ln z)$ and $o(z)$ as $z \to 0$; it equals $O(z^3)$ and $o(z^n)$, $n > 3$ as $z \to \infty$. The role of the domain is also very important in order relations and asymptotic analysis in general. Hence we conclude that $f(z) = 2z + z \cos z = O(z)$ as $z \to \infty$ if $z \in \mathcal{R}$, since $z \leq f(z) \leq 3z$. However, if $\mathrm{Re}(z) = 0$, or $z = iy$, then $f(z) = O(ye^y)$ as $z \to \infty$ along the imaginary axis. If $f(z) = e^{-z}$ with domain D being $0 < |z| < \infty$, $|\arg z| < \pi/2$, then $f(z) = o(x^n) \forall n$ as $z \to \infty$ in D. If however, the domain D is $0 < |z| < \infty$, $|\arg z| < \pi$, then the gauge function e^{-x} is not terribly useful, since f grows exponentially as $x \to -\infty$. A better description here is $f(z) = O(e^{-z})$. It all depends on the context of the problem [189].

Appendix B

Ray Theory: Exact Solutions

In this appendix (based on [68] and [236]) slightly different notation will be used for the scattering of rays by radially inhomogeneous spheres. Here θ_i will refer to the angle of incidence for the incoming ray, r is the radial distance within a sphere of radius $\tilde{a}$, and $\Theta_k(\theta_i)$ is the total angle of deviation undergone by the ray from its original direction after k internal reflections (the letter a is used in some of the refractive index profiles below). Thus (as throughout Part I of this book) the subscripts 0 and 1 are used to distinguish between the deviations of the exiting ray for direct transmission and the primary bow, respectively. It is a well-known result that the curvature of the ray path is toward regions of higher refractive index n—a consequence of Snell's law of refraction in continuously varying media. This is just the *mirage theorem* discussed in Chapter 3. Thus if $n'(r) < 0$, an incoming ray is concave toward the origin (i.e., it curves around it); if $n'(r) > 0$, it is concave away from it (or curves away from it). If $2\Delta(\theta_i)$ is the angle subtended at the center of the sphere by the ray path inside the sphere, then for *each* internal reflection the ray undergoes an additional deviation of $2\Delta(\theta_i)$. Let us refer to $\Delta(\theta_i)$ as the "basis angle" for the ray path inside the sphere. Hence the total deviation for k internal reflections is

$$\Theta_k(\theta_i) = 2\theta_i - \pi + 2(k+1)\Delta(\theta_i). \tag{B.1}$$

The quantity $\Delta(\theta_i)$ is an improper definite integral to be defined below for a unit sphere (now scaled by the radius $\tilde{a}$). Apart from a few specific $n(r)$ profiles, analytic expressions for $\Delta(\theta_i)$ are difficult to obtain. In this appendix, $\Delta(\theta_i)$ is evaluated for several additional profiles. It is interesting to note that even for simple $n(r)$ profiles (such as the linear profile), the evaluation of the integral is quite challenging. In fact a great deal of elementary (and on occasion, sophisticated) analysis is required to obtain explicit closed-form expressions for $\Delta(\theta_i)$. All these details may be found in [68].

From elementary differential geometry in the plane, the formula for $\Delta(\theta_i)$ is (see equation (3.119))

$$\Delta(\theta_i) = \sin\theta_i \int_{r_c(\theta_i)}^{\tilde{a}} \frac{dr}{r\left[r^2 n^2(r) - \sin^2\theta_i\right]^{1/2}}, \tag{B.2}$$

where the lower limit $r_c(\theta_i)$ is the point at which the integrand is singular, being the largest solution of

$$r_c(\theta_i)n(r_c(\theta_i)) = \sin\theta_i, \tag{B.3}$$

provided $0 < r_c < 1$. The quantity $r_c(\theta_i)$ is the radial point of closest approach to the center of the sphere which is sometimes called the *turning point*. For a kth-order bow to exist for some critical angle of incidence θ_{ic} it is necessary and sufficient that

$$\Theta'_k(\theta_{ic}) = 0. \tag{B.4}$$

Equation (B.2) will be evaluated for ten distinct refractive index profiles. That result can be employed in equation (B.1), and we can the utilize the resulting equation to impose conditions on the refractive index that allow for a zeroth-order bow to exist (see [73]). This technique will be applied only to those refractive index profiles whose derivative of the deviation angle is readily obtainable algebraically (in some cases not without difficulty). It should be noted at the outset that some of the profiles below can singular in the unit sphere, depending on the particular choice of parameters. Nevertheless we will here choose the relevant parameters such that the profiles are not singular in the refracting sphere, though it is not necessary to do so.

B.1 PROFILE 1

The refractive index profile considered is

$$n(r) = \frac{2n_1 r^{1/c-1}}{1 + r^{2/c}}, \qquad n_1 = n(1), \tag{B.5}$$

where c is a positive real constant. This profile is singular at $r = 0$ for $c > 1$. It is also singular at $r = 0$ for $c < 0$. To avoid such singular behavior, it will be assumed that $0 < c \le 1$. When $c = 1$, the refractive index profile in equation (B.5) results in the well-known *Maxwell's fish-eye* profile [12]. If we define the quantity

$$\hat{a} \equiv \frac{\sin\theta_i}{2n_1} < \frac{1}{2},$$

it follows that [68]

$$r_c^{1/c} = \frac{1 \pm \sqrt{1 - 4\hat{a}^2}}{2\hat{a}}.$$

It is readily demonstrated that the $+$ root corresponds to $r_c > 1$, that is, outside the unit sphere, so it follows that

$$r_c(\theta_i) = \left[\frac{1 - \sqrt{1 - 4\hat{a}^2}}{2\hat{a}}\right], \tag{B.6}$$

from which the following remarkably simple result is obtained:

$$\Delta(\theta_i) = \frac{c}{2}\pi. \tag{B.7}$$

Hence, for the refractive index given by profile 1, we find that

$$\Theta_k(\theta_i) = 2\theta_i - \pi + 2\,(k+1)\,\Delta(\theta_i) = 2\theta_i - \pi + (k+1)\,c\pi$$
$$= \pi\,[(k+1)\,c - 1] + 2\theta_i. \tag{B.8}$$

Note that for this profile, $\Theta_k(\theta_i) = 2$ for any value of i. Thus, no bow is possible for any order k, a quite remarkable conclusion given the mathematical complexity of the profile (B.5) compared with a constant refractive index (for which bows do exist).

B.2 PROFILE 2

Consider next the refractive index profile

$$n(r) = n(0) \left(1 - \frac{r^2}{L^2}\right)^{1/2} \equiv n_0 \left(1 - \frac{r^2}{L^2}\right)^{1/2}. \tag{B.9}$$

This is known (not surprisingly) as the parabolic refractive index profile. It has been used in studies of circular optical fibers, where r is the distance from the optical fiber axis. L^2 is a constant that can be determined from knowledge of the core and surface refractive indices. Let $n(1) \equiv n_1 > 1$. Then we find that

$$L^2 = \frac{n_0^2}{n_0^2 - n_1^2}.$$

Let $K = \sin \theta_i$. Using equation (B.3), it is found that

$$r_c^2 = \frac{n_0^2 \pm \left[n_0^4 - 4(n_0^2 - n_1^2)K^2\right]^{1/2}}{2(n_0^2 - n_1^2)}.$$

For the unit sphere, the radial point of closest approach is bounded by

$$0 \le r_c < 1, \quad r_c \in \mathbb{R}.$$

Further restrictions are imposed on $r_c(\theta_i)$ that are dependent on the values of n_0 and n_1. Two cases are summarized here.

Case 1 ($1 < n_1 < n_0$) This corresponds to the refractive index decreasing monotically from the center. It can be shown that

$$r_c = \left\{ \frac{n_0^2 - \left[n_0^4 - 4(n_0^2 - n_1^2)K^2\right]^{1/2}}{2(n_0^2 - n_1^2)} \right\}^{1/2}, \tag{B.10}$$

subject to the constraint

$$n_0^4 > 4(n_0^2 - n_1^2). \tag{B.11}$$

Case 2 ($1 < n_0 < n_1$)

This corresponds to the refractive index increasing monotically from the center. It can be shown that the very same result (B.10) holds in this second case also. Consequently the basis for the interior angle of deviation is

$$\Delta(\theta_i) = \frac{1}{2} \left\{ \arcsin\left[\frac{n_0^2 - 2K^2}{\left[n_0^4 - 4(n_0^2 - n_1^2)K^2\right]^{1/2}}\right] + \frac{\pi}{2} \right\}. \tag{B.12}$$

Therefore, for the refractive index profile 2, we find

$$\Theta_k(\theta_i) = 2\theta_i - \pi + 2(k+1)\Delta(\theta_i)$$

$$= 2\theta_i + (k-1)\frac{\pi}{2} + \arcsin\left[\frac{n_0^2 - 2K^2}{[n_0^4 - 4(n_0^2 - n_1^2)K^2]^{1/2}}\right]. \quad (B.13)$$

As in all these profiles, conditions for the existence of kth-order bows can be found by seeking extrema of $\Theta_k(\theta_i)$.

B.3 PROFILE 3

Consider next the profile

$$n(r) = a - br^2, \quad (B.14)$$

where a and b are real constants. This profile in particular has been used to extend the Airy rainbow theory to nonuniform spheres (see references in [272]). It transpires that the critical radius r_c satisfies the cubic equation

$$br^3 - ar + K = 0, \quad (B.15)$$

so care must be taken in evaluating the three roots (not all of which will be physically interesting). These roots will be denoted by r_1, r_2, and r_c, where $r_c < \min(r_1, r_2)$, if r_1 and r_2 are real. After a great deal of algebra the basis internal angle $\Delta(\theta_i)$ can be expressed succinctly in terms of $\Pi(\phi, \alpha^2, p)$, the *incomplete elliptic integral of the third kind*, where

$$\Pi(\phi, \alpha^2, k) = \int_0^{\sin\phi} \frac{dt}{(1 - \alpha^2 t^2)\sqrt{(1 - t^2)(1 - k^2 t^2)}}, \quad (B.16)$$

and

$$\sin\phi = \left(\frac{r_c^2 - r^2}{r_c^2 - r_2^2}\right)^{1/2}, \quad p = \left(\frac{r_c^2 - r_2^2}{r_c^2 - r_1^2}\right)^{1/2}, \quad \text{and} \quad \alpha^2 = 1 - \frac{r_2^2}{r_c^2}. \quad (B.17)$$

From equation (B.2) we obtain

$$\Delta(\theta_i) = \frac{Ki}{|b|r_c^2\sqrt{r_c^2 - r_1^2}}\Pi(\phi, \alpha^2, p). \quad (B.18)$$

$$\Theta_k(\theta_i) = 2\theta_i - \pi + 2(k+1)\frac{Ki}{|b|r_c^2\sqrt{r_c^2 - r_1^2}}\Pi(\phi, \alpha^2, p), \quad (B.19)$$

where $i = \sqrt{-1}$ should not be confused with the angle of incidence used elsewhere in the book, written here as θ_i. As can be seen from this expression, the calculation of $\Theta_k(\theta_i)$ involves highly complicated calculations that will not be provided here.

B.4 PROFILE 4

Consider the linear profile

$$n(r) = a + br, \tag{B.20}$$

where a and b are constants. The linear refractive index profile has been utilized with respect to absorption measurements of nonlinear optical liquids in the visible and near-infrared spectral region Surprisingly, the algebraic complexity for this seemingly innocuous profile is even worse than that for profile 3. For this reason we state only the final result for $\Delta(\theta_i)$. In what follows let

$$A = \frac{a}{q}, \quad B = \frac{b}{q}, \quad \text{and} \quad q = \sin\theta_i, \quad \theta_i > 0. \tag{B.21}$$

Also define

$$k = -\frac{1}{4B}\left[A^2 \pm i\sqrt{(A^2 + 4B)(4B - A^2)}\right]. \tag{B.22}$$

The positive root is taken if $4B - A^2 > 0$ and the negative root if $4B - A^2 < 0$. Let

$$\sin\phi = \left\{\left[-\frac{(A + \sqrt{A^2 + 4B})}{k(A - \sqrt{A^2 + 4B})}\right]\left[\frac{-\frac{1}{2}(A - \sqrt{A^2 + 4B})r + 1}{\frac{1}{2}(A + \sqrt{A^2 + 4B})r - 1}\right]\right\}^{1/2}, \tag{B.23}$$

and

$$\alpha^2 = \frac{k(A - \sqrt{A^2 + 4B})}{A + \sqrt{A^2 + 4B}}. \tag{B.24}$$

The final result is

$$\Delta(\theta_i) = -\frac{\sqrt{2Bk}}{B}\left[\left(\frac{A + \sqrt{A^2 + 4B}}{2}\right)F(\phi, k) - \sqrt{A^2 + 4B}\,\Pi(\phi, \alpha^2, k)\right], \tag{B.25}$$

where as in equation (B.16), $\Pi(\phi, \alpha^2, k)$ is the incomplete elliptic integral of the third kind, and

$$F(\phi, k) = \int_0^{\sin\phi} \frac{dt}{\sqrt{(1 - t^2)(1 - k^2 t^2)}} \tag{B.26}$$

is the incomplete elliptic integral of the first kind. From equation (B.25) we can compute the deviation angle $\Theta_k(\theta_i)$ given by (B.1). Again, though, the analytic expression for $\Theta_k(\theta_i)$ is extremely complicated and will not be presented here.

B.5 PROFILE 5

Consider next the refractive index profile, For constant b:

$$n(r) = n_1 r^{1/b-1}(2 - r^{2/b})^{1/2}, \quad n_1 = n(1) > 1. \tag{B.27}$$

This profile is singular at $r = 0$ if one of the following conditions is satisfied:

$$b < 0; \qquad b > 1.$$

For a kth-order bow to exist, $b > 2$. Consequently, the second singularity condition $b > 1$ will be met in order to guarantee the existence of a zeroth-order bow. When $b = 1$, this profile corresponds to a *Luneburg lens* [12]. Defining

$$a \equiv \frac{\sin \theta_i}{n_1}, \tag{B.28}$$

it follows that

$$r_c^{2/b} = \frac{2 \pm \sqrt{4 - 4a^2}}{2} = 1 \pm \sqrt{1 - a^2}. \tag{B.29}$$

Because $0 < r_c \leq r \leq 1$, the negative root must prevail, since $a^2 < 1$, so that $r_c \in \mathbb{R}$. Hence

$$r_c(\theta_i) = \left(1 - \sqrt{1 - a^2}\right)^{b/2}. \tag{B.30}$$

Ultimately, the relatively simple result is

$$\Delta(\theta_i) = \frac{b}{2}(\pi - \arcsin a), \tag{B.31}$$

and hence

$$\Theta_k(\theta_i) = 2\theta_i + \pi\,[(k+1)b - 1)] - (k+1)\,b\,\arcsin\left(\frac{\sin \theta_i}{n_1}\right). \tag{B.32}$$

As above, for a kth-order bow to exist for some critical angle of incidence θ_{ic}, it is necessary and sufficient that

$$\Theta_k'(\theta_{ic}) = 0.$$

Thus the condition for extrema is that

$$\cos \theta_{ic} = 2\left[\frac{(n_1^2 - 1)}{(k+1)^2\,b^2 - 4}\right]^{1/2}. \tag{B.33}$$

The consequences of this equation for kth-order bows can be readily inferred. In particular, for a zeroth-order bow ($k = 0$), if we restrict ourselves to the least potentially singular case $b > 0$, then a zeroth-order bow can exist for this profile if $b \geq 2n_1$. Note that a zeroth-order bow cannot exist if $n_1 = 1$.

B.6 PROFILE 6

Consider the "reciprocal linear" as generalized rectangular hyperbolic refractive index profile given by

$$n(r) = (ar + b)^{-1}, \tag{B.34}$$

where again a and b are constants. This profile is singular when $r = -b/a$. It is readily shown that

$$r_c(\theta_i) = \frac{\beta}{1 - \alpha}, \quad \text{where } \alpha = a\sin\theta_i, \quad \beta = \frac{b}{a}\alpha. \tag{B.35}$$

One resulting form of the basis angle $\Delta(\theta_i)$ is:

$$\Delta(\theta_i) = \frac{\alpha}{\sqrt{1 - \alpha^2}} \ln\left[\frac{\sqrt{1 - \alpha^2}\sqrt{1 - (\alpha + \beta)^2} + 1 - \alpha(\alpha + \beta)}{\beta}\right]$$

$$+ \frac{\pi}{2} - \arcsin(\alpha + \beta). \tag{B.36}$$

Using the formula

$$\arccos z = -i\log\left(z + i\sqrt{1 - z^2}\right), \tag{B.37}$$

$\Delta(\theta_i)$ may also be expressed as

$$\Delta(\theta_i) = \frac{\alpha}{\sqrt{\alpha^2 - 1}}\arccos\left[\frac{1 - \alpha(\alpha + \beta)}{\beta}\right] + \arccos(\alpha + \beta),$$

$$= \frac{a\sin\theta_i}{\sqrt{a^2\sin^2\theta_i - 1}}\arccos\left[\frac{1 - a(a + b)\sin^2\theta_i}{b\sin\theta_i}\right] + \arccos[(a + b)\sin\theta_i],$$
$$\tag{B.38}$$

where we have utilized the definitions for α and β. Once more, the expression for $\Theta_k(\theta_i)$ follows quite naturally. More details of this profile may be found in [77].

B.7 PROFILE 7

Next we examine the refractive index profile

$$n(r) = \frac{a}{r\ln(br)}, \tag{B.39}$$

where a and b again are constants. This profile is singular at $r = 0$ and $r = b^{-1}$ and is undefined if $b \leq 0$. If

$$m \equiv \frac{a}{\sin\theta_i},$$

then $r_c(\theta_i)$ is given by

$$r_c(\theta_i) = b^{-1}e^m. \tag{B.40}$$

The basis angle is

$$\Delta(\theta_i) = -\sqrt{a^2\csc^2\theta_i - (\ln b)^2}, \tag{B.41}$$

and therefore, for this refractive index profile, a kth-order bow exists if

$$1 + \frac{(k + 1)a^2\csc^2\theta_{ic}\cot\theta_{ic}}{\sqrt{a^2\csc^2\theta_{ic} - (\ln b)^2}} = 0. \tag{B.42}$$

Since θ_{ic} is in the first quadrant, and the trigonometric functions are all positive in the first quadrant, equation (B.42) cannot be satisfied. Thus a kth-order bow cannot exist for this refractive index profile.

B.8 PROFILE 8

Consider the similar refractive index profile

$$n(r) = \frac{a}{r\sqrt{\ln(br)}}, \tag{B.43}$$

where a and b once more are real constants. This is singular at $r = 0, r = b^{-1}$, and undefined for $b \leq 0$, as was the case for the previous profile (B.39). In addition, it is purely imaginary in the domain $(0, b^{-1})$. Note that now

$$r_c(\theta_i) = b^{-1}e^{m^2}. \tag{B.44}$$

Consequently, after judicious use of inverse trigonometric functions, it transpires that

$$\Delta(\theta_i) = -\left[\sqrt{\ln b}\sqrt{a^2 \csc^2 \theta_i - \ln b} + \frac{a^2}{2} \csc^2 \theta_i \arccos\left(\frac{2\ln b}{a^2} \sin^2 \theta_i - 1\right)\right]. \tag{B.45}$$

Hence

$$\Delta'(\theta_i) = a^2 \csc^2 \theta_i \cot \theta_i \left[\frac{2\sqrt{\ln b}}{\sqrt{a^2 \csc^2 \theta_i - \ln b}} + \arccos\left(\frac{2\ln b}{a^2} \sin^2 \theta_i - 1\right)\right]. \tag{B.46}$$

A kth-order bow exists if

$$1 + (k+1) a^2 \csc^2 \theta_{ic} \cot \theta_{ic} \left[\frac{2\sqrt{\ln b}}{\sqrt{a^2 \csc^2 \theta_{ic} - \ln b}} + \arccos\left(\frac{2\ln b}{a^2} \sin^2 \theta_{ic} - 1\right)\right] = 0. \tag{B.47}$$

The second term in equation (B.47) is positive due to the definition of θ_{ic}. For it to be satisfied, the last term must be negative (i.e., the inverse cosine function must be negative). However, for the set of real numbers, the inverse cosine function is always positive. Therefore, as with the previous profile, a kth-order bow cannot exist for this profile.

B.9 PROFILE 9

Consider the refractive index profile

$$n(r) = \left(ar^{-2} + br^{-1} + c\right)^{1/2}, \tag{B.48}$$

where a, b, and c are real constants. Obviously it is singular at $r = 0$. Let $K = \sin \theta_i$. Then

$$r_c(\theta_i) = \frac{-b \pm \sqrt{b^2 - 4c(a - K^2)}}{2c}. \qquad (B.49)$$

It must be the case that $0 \leq r_c(\theta_i) < 1$ and $r_c(\theta_i) \in \mathbb{R}$. To guarantee that the radial point of closest approach to the center of the sphere, $r_c(\theta_i)$, is a real quantity, it is required that

$$b^2 > 4c(a - K^2).$$

There are numerous cases for this profile where we can study the behavior of the constants that would determine whether we choose the positive or negative sign in equation (B.49). Since we require that $n(r) \in \mathbb{R}$, we will not consider the case where all three constants are negative. Regardless of whether $c > 0$ or $c < 0$, the first condition above would result in the following inequality:

$$a \leq K^2 < a + b + c. \qquad (B.50)$$

In other words it is required that $a + b + c > 1$. Since $0 < K^2 \leq 1$, it follows that

$$a > a - K^2 \geq a - 1. \qquad (B.51)$$

But $a - K^2 < 0$, (and from the equation for $\Delta(\theta_i)$ $K^2 \neq a$) this inequality is satisfied only if $a < 1$. Using $a - K^2 \geq a - 1$ in the condition $b^2 > 4c(a - K^2)$ implies that

$$b^2 > 4c(a - 1). \qquad (B.52)$$

Without loss of generality, we assume that $a < 0$, $b > 0$, and $c > 0$. As a result, we must take the positive root in equation (B.49). Note that a and b must be small and large enough, respectively, so that it is real for all values of r in the domain. Thus we obtain the result

$$\Delta(\theta_i) = \frac{K}{\sqrt{K^2 - a}} \arccos \left[-\frac{2(a - K^2) + b}{\sqrt{b^2 - 4c(a - K^2)}} \right], \qquad (B.53)$$

and hence it follows that

$$\Theta_k(\theta_i) = 2\theta_i - \pi + 2(k + 1)\frac{K}{\sqrt{K^2 - a}} \arccos \left[-\frac{2(a - K^2) + b}{\sqrt{b^2 - 4c(a - K^2)}} \right]. \qquad (B.54)$$

B.10 PROFILE 10

Finally, consider the power-law refractive index profile

$$n(r) = \alpha r^\eta, \qquad (B.55)$$

where η is of either sign, $0 < b \leq r \leq a$, $\alpha > 0$, where a is the sphere's radius, and b is an arbitrary radial distance in the sphere. This profile is obviously singular at $r = 0$ if $\eta < 0$. For a zeroth-order bow to exist, it will be shown that $\eta < 0$. This refractive index profile has been used in connection with a general explanation of both the rainbow ray and the glory ray phenomena in meteorological optics [80]. Using a judicious change of variable, this profile can be transformed into the standard one for a constant refractive index. Hence if $K = \sin\theta_i$, we find that

$$\Delta(\theta_i) = K \int_{r_c(\theta_i)}^{1} \frac{dr}{r\sqrt{\alpha^2 r^{2(\eta+1)} - K^2}} = \int_{r_c(\theta_i)}^{1} \frac{dr}{r\sqrt{Ar^p - 1}}, \tag{B.56}$$

where $p = 2(\eta + 1)$ and $A = (\alpha/K)^2 > 0$. Hence

$$r_c(\theta_i) = \left(\frac{K}{\alpha}\right)^{2/p}. \tag{B.57}$$

To evaluate the integral in equation (B.56), make the change of variables

$$v^2 = Ar^p, \tag{B.58}$$

noting that

$$v(r_c(\theta_i)) = A^{1/2}(r_c(\theta_i))^{p/2} = A^{1/2}\frac{K}{\alpha} = 1, \tag{B.59}$$

with $v(1) = \sqrt{A}$. Using these substitutions, the basis deviation angle becomes

$$\Delta(\theta_i) = \frac{2}{p} \int_{1}^{A^{1/2}} \frac{dv}{v\sqrt{v^2 - 1}} = \frac{1}{\eta+1}\left[\operatorname{arcsec}\left(\frac{\alpha}{K}r^{\eta+1}\right)\right]_{r_c(\theta_i)}^{1}, \tag{B.60}$$

where $\eta \neq -1$. Therefore, if

$$n(r) \equiv n_b, \quad 0 \leq r < b; \tag{B.61}$$

$$= \alpha r^\eta, \quad b \leq r \leq a, \tag{B.62}$$

by continuity at $r = b$,

$$\alpha = n_b b^{-\eta}, \tag{B.63}$$

and so

$$\Delta(\theta_i) = \frac{1}{\eta+1} \operatorname{arcsec}\left(\frac{n_b}{Kb^\eta}\right). \tag{B.64}$$

Hence

$$\Theta_k(\theta_i) = 2\theta_i - \pi + \frac{2(k+1)}{\eta+1} \operatorname{arcsec}\left(\frac{n_b}{Kb^\eta}\right). \tag{B.65}$$

Since

$$\Delta'(\theta_i) = -\frac{1}{\eta+1}\frac{\cos\theta_i}{\sqrt{\alpha^2 - K^2}},$$

a kth-order bow exists if $\Delta'(\theta_i) = -(k+1)^{-1}$. Hence, for the refractive index profile given by the profile (B.55), a kth-order bow exists if

$$\sin\theta_{ic} = \sqrt{\frac{(k+1)^2 - (\eta+1)^2\alpha^2}{(k+1)^2 - (\eta+1)^2}}, \qquad (B.66)$$

where $\eta \neq 0$. The implications of this equation for zeroth-order bows ($k = 0$) in particular are discussed in [68].

Appendix C

Radially Inhomogeneous Spherically Symmetric Scattering: The Governing Equations

C.1 THE TRANVERSE MAGNETIC MODE

Maxwell's equations, in the Gaussian system of units with $\mu = 1$ ([13]) and with $e^{-i\omega t}$ dependence are, from Chapter 19:

$$\nabla \times \mathbf{H} = -k_1 \mathbf{E}; \tag{C.1}$$

$$\nabla \times \mathbf{E} = k_2 \mathbf{H}; \tag{C.2}$$

$$\nabla \cdot (\varepsilon \mathbf{E}) = 0; \tag{C.3}$$

$$\nabla \cdot \mathbf{H} = 0, \tag{C.4}$$

where

$$k_1 = \frac{i\omega}{c}\left(\varepsilon + \frac{4\pi i\sigma}{\omega}\right) = k_1(r), \tag{C.5}$$

$$k_2 = \frac{i\omega}{c}, \tag{C.6}$$

and

$$k^2(r) = -k_1 k_2. \tag{C.7}$$

Now define

$$\mathbf{H} = \nabla \times \mathbf{A}, \tag{C.8}$$

so

$$\nabla \times (\mathbf{E} - k_2 \mathbf{A}) = \mathbf{0}. \tag{C.9}$$

Also, let

$$\mathbf{E} - k_2 \mathbf{A} = -\nabla U. \tag{C.10}$$

Then

$$-\nabla \times \nabla \times \mathbf{A} + k^2(r)\mathbf{A} + k_1(r)\nabla U = \mathbf{0}, \tag{C.11}$$

and with $\mathbf{A} = \hat{\mathbf{r}} A_r(r, \theta, \phi)$ the expansion of $\nabla \times \nabla \times \mathbf{A}$ yields the equation:

$$\frac{\hat{\mathbf{r}}}{r^2 \sin \theta} \left\{ \frac{\partial}{\partial \theta} \left(\sin \theta \frac{\partial A_r}{\partial \theta} \right) + \frac{1}{\sin \theta} \frac{\partial^2 A_r}{\partial \phi^2} \right\}$$

$$+ \hat{\mathbf{r}} k^2(r) A_r + k_1(r) \, \hat{\mathbf{r}} \frac{\partial U}{\partial r} - \frac{\hat{\theta}}{r} \frac{\partial}{\partial \theta} \left\{ \frac{\partial A_r}{\partial r} - k_1(r) U \right\}$$

$$- \frac{\hat{\phi}}{r \sin \theta} \frac{\partial}{\partial \phi} \left\{ \frac{\partial A_r}{\partial r} - k_1(r) U \right\} = 0. \tag{C.12}$$

Clearly, if we set

$$\frac{\partial A_r}{\partial r} = k_1(r) U, \tag{C.13}$$

then we can define the vector $\mathbf{A}$ uniquely through knowledge of $\nabla \times \mathbf{A}$ and $\nabla \cdot \mathbf{A}$ (see Appendix E). Hence

$$U = \frac{1}{k_1(r)} \frac{\partial A_r}{\partial r}, \tag{C.14}$$

and the governing equation becomes the following scalar version:

$$\mathcal{L}_{\theta, \phi} (A_r) + k_1(r) \frac{\partial}{\partial r} \left(\frac{1}{k_1(r)} \frac{\partial A_r}{\partial r} \right) + k^2(r) A_r = 0. \tag{C.15}$$

Here, $\mathcal{L}_{\theta, \phi}$ is the angular part of the Laplacian ∇^2 in spherical polar coordinates:

$$\mathcal{L}_{\theta, \phi} \equiv \frac{1}{r^2 \sin \theta} \frac{\partial}{\partial \theta} \left(\sin \theta \frac{\partial}{\partial \theta} \right) + \frac{1}{r^2 \sin^2 \theta} \frac{\partial^2}{\partial \phi^2}. \tag{C.16}$$

Also

$$\frac{\partial^2 A_r}{\partial r^2} = \frac{1}{r} \frac{\partial}{\partial r} \left(r^2 \frac{\partial}{\partial r} \right) \left(\frac{A_r}{r} \right) = r \mathcal{L}_r \left(\frac{A_r}{r} \right), \tag{C.17}$$

where $\mathcal{L}_r$ is the radial part of the Laplacian, so

$$\mathcal{L}_{\theta, \phi} (A_r) + r \mathcal{L}_r \left(\frac{A_r}{r} \right) + k_1 \left(\frac{1}{k_1} \right)' \frac{\partial A_r}{\partial r} + k^2(r) A_r = 0. \tag{C.18}$$

Therefore, since

$$\frac{1}{r} \mathcal{L}_{\theta, \phi}(A_r) = \mathcal{L}_{\theta, \phi} \left(\frac{A_r}{r} \right), \tag{C.19}$$

we have

$$\left[\nabla^2 + k^2(r) \right] + \frac{k_1}{r} \left(\frac{1}{k_1} \right)' \frac{\partial A_r}{\partial r} = 0. \tag{C.20}$$

C.2 THE TRANVERSE ELECTRIC MODE

Now let

$$\mathbf{E} = \frac{1}{k_1(r)} \nabla \times \mathbf{F}, \tag{C.21}$$

so

$$\nabla \times (\mathbf{H} + \mathbf{F}) = 0;\tag{C.22}$$

therefore if

$$\mathbf{H} + \mathbf{F} = -\nabla\Psi,\tag{C.23}$$

it follows that

$$\nabla \times \mathbf{E} = \nabla \times \left[\frac{1}{k_1(r)}\nabla \times \mathbf{F}\right] = k_2\mathbf{H} = -k_2\mathbf{F} - k_2\nabla\Psi.\tag{C.24}$$

Hence,

$$\nabla \times \left[\frac{1}{k_1(r)}\nabla \times \mathbf{F}\right] + k_2\mathbf{F} + k_2\nabla\Psi = \mathbf{0}.\tag{C.25}$$

In a similar fashion, expanding the $\nabla \times \nabla\times$ term (and introducing a negative sign), we have, with $\mathbf{F} = \hat{\mathbf{r}}F_r(r, \theta, \phi)$,

$$\hat{\mathbf{r}}\left[\frac{1}{k_1(r)}\mathcal{L}_{\theta,\phi}(F_r) - k_2 F_r - k_2\frac{\partial\Psi}{\partial r}\right] - \frac{\hat{\theta}}{r}\frac{\partial}{\partial\theta}\left[\frac{\partial}{\partial r}\left(\frac{F_r}{k_1}\right) + k_2\Psi\right]$$
$$- \frac{\hat{\phi}}{r\sin\theta}\frac{\partial}{\partial\phi}\left[\frac{\partial}{\partial r}\left(\frac{F_r}{k_1}\right) + k_2\Psi\right] = \mathbf{0}.\tag{C.26}$$

Therefore, if

$$\frac{\partial}{\partial r}\left(\frac{F_r}{k_1}\right) + k_2\Psi = 0,\tag{C.27}$$

the governing equation becomes

$$\frac{1}{k_1(r)}\mathcal{L}_{\theta,\phi}(F_r) - k_2 F_r - k_2\left[-\frac{1}{k_2}\frac{\partial^2}{\partial r^2}\left(\frac{F_r}{k_1}\right)\right] = 0,\tag{C.28}$$

since k_2 is constant. Again, since

$$\frac{\partial^2}{\partial r^2}\left(\frac{F_r}{k_1}\right) = r\mathcal{L}_r\left(\frac{F_r}{rk_1}\right),\tag{C.29}$$

we have, on dividing by r

$$\mathcal{L}_{\theta,\phi}\left(\frac{F_r}{k_1 r}\right) + \mathcal{L}_r\left(\frac{F_r}{k_1 r}\right) - \frac{k_2 k_1 F_r}{rk_1} = 0,\tag{C.30}$$

or, since $k_1 k_2 = -k^2(r)$

$$\left[\nabla^2 + k^2(r)\right]\left(\frac{F_r}{k_1 r}\right) = 0.\tag{C.31}$$

Note that from the representations for H and E we may readily construct their components. The notation of [13] and [233] is related to that used in this appendix, for example, $\mathbf{A} = \hat{\mathbf{r}}(r^e\Omega)$ ([233]) $= \hat{\mathbf{r}}(r^e\Pi)$ ([13]) for the TM mode.

Appendix D

Electromagnetic Scattering from a Radially Inhomogeneous Sphere

D.1 A CLASSICAL/QUANTUM CONNECTION FOR TRANSVERSE ELECTRIC AND MAGNETIC MODES

An alternative approach to the problem of electromagnetic scattering from a sphere is readily obtained by directly using vector spherical harmonics. Again, we consider the refractive index $n(r)$ (which may be complex) to be a function of the radial coordinate only, and the sphere has radius a. For $r > a$, $n(r) \equiv 1$. As in much of Appendix C, a time-harmonic dependence of the field quantities, $\exp(-i\omega t)$, is assumed throughout. The governing equation for the electric field $E(r, \theta, \phi)$ is readily found from equations (C.1) and (C.2) to be

$$\nabla \times \nabla \times \mathbf{E} - k^2 n^2(r)\, \mathbf{E} = \mathbf{0}. \tag{D.1}$$

The wavenumber k is $2\pi/\lambda$, λ being the wavelength. As shown in [97], the solution may be found by expanding the electric field in terms of vector spherical harmonics in terms of the so-called transverse electric (TE) and transverse magnetic (TM) modes, respectively:

$$\mathbf{M}_{l,m}(r, \theta, \phi) = \frac{e^{im\phi}}{kr} S_l(r) \mathbf{X}_{l,m}(\theta), \tag{D.2}$$

$$\mathbf{N}_{l,m}(r, \theta, \phi) = \frac{e^{im\phi}}{k^2 n^2(r)} \left[\frac{1}{r} \frac{dT_l(r)}{dr} \mathbf{Y}_{l,m}(\theta) + \frac{T_l(r)}{r^2} \mathbf{Z}_{l,m}(\theta) \right]. \tag{D.3}$$

The vector angular functions in equations (D.2) and (D.3) are defined in a spherical coordinate system as

$$\mathbf{X}_{l,m}(\theta) = \left\langle 0,\, i\pi_{l,m}(\theta),\, -\tau_{l,m}(\theta) \right\rangle, \tag{D.4}$$

$$\mathbf{Y}_{l,m}(\theta) = \left\langle 0,\, \tau_{l,m}(\theta),\, -i\tau_{l,m}(\theta) \right\rangle, \tag{D.5}$$

$$\mathbf{Z}_{l,m}(\theta) = \left\langle l(l+1) P_l^m(\cos\theta),\, 0,\, 0 \right\rangle, \tag{D.6}$$

where $P_l^m(\cos\theta)$ is an associated Legendre polynomial of degree l and order m. The corresponding scalar angular functions are defined as

$$\pi_{l,m}(\theta) = \frac{m}{\sin\theta} P_l^m(\cos\theta), \tag{D.7}$$

$$\tau_{l,m}(\theta) = \frac{d P_l^m(\cos\theta)}{d\theta}. \tag{D.8}$$

The functions $S_l(r)$ and $T_l(r)$ are called the radial Debye potentials, and they respectively satisfy the equations

$$\frac{d^2 S_l(r)}{dr^2} + \left[k^2 n^2(r) - \frac{l(l+1)}{r^2} \right] S_l(r) = 0, \tag{D.9}$$

$$\frac{d^2 T_l(r)}{dr^2} - \left(\frac{2}{n(r)} \frac{dn(r)}{dr} \right) \frac{dT_l(r)}{dr} + \left[k^2 n^2(r) - \frac{l(l+1)}{r^2} \right] T_l(r) = 0. \tag{D.10}$$

Of course, these equations can also be derived directly from equations (C.31) and (C.20), respectively, in Appendix C. What follows is based on [149].

Inpt addition to the appropriate matching conditions at $r = a$, these potentials must also satisfy the boundary conditions $S_l(0) = 0$ and $T_l(0) = 0$. Equation (D.10) may be rewritten in terms of the dependent variable $U_l(r)$, where $T_l(r) = n(r)U_l(r)$, to become

$$\frac{d^2 U_l(r)}{dr^2} + \left[k^2 n^2(r) - n(r)\frac{d^2}{dr^2} \left(\frac{1}{n(r)} \right) - \frac{l(l+1)}{r^2} \right] U_l(r) = 0. \tag{D.11}$$

Provided that $n(0) \neq 0$, $U_l(0) = 0$. Both equations (D.9) and (D.11) may be placed in the form of the canonical time-independent Schrödinger equation, namely,

$$\frac{d^2 S_l(r)}{dr^2} + \left[k^2 - V_S(r) - \frac{l(l+1)}{r^2} \right] S_l(r) = 0, \tag{D.12}$$

$$\frac{d^2 U_l(r)}{dr^2} + \left[k^2 - V_U(r) - \frac{l(l+1)}{r^2} \right] U_l(r) = 0, \tag{D.13}$$

where the k-dependent "scattering potentials" $V_S(r)$ and $V_U(r)$ are defined in $[0, a]$ as

$$V_S(r) = k^2 \left[1 - n^2(r) \right], \tag{D.14}$$

$$V_U(r) = k^2 \left[1 - n^2(r) + \frac{n(r)}{k^2} \frac{d^2}{dr^2} \left(\frac{1}{n(r)} \right) \right]. \tag{D.15}$$

for the TE and TM modes, respectively (the potentials are both identically zero for $r > a$). These potentials are identical for the case of a uniform refractive index. $V_U(r)$ will be regarded as a small perturbation of the potential $V_S(r)$, so we also define

$$\varepsilon(r) \equiv V_U(r) - V_S(r) = n(r)\frac{d^2}{dr^2} \left(\frac{1}{n(r)} \right). \tag{D.16}$$

It is a standard result for potentials vanishing sufficiently fast at infinity [60] that as $r \to \infty$,

$$S_l(r) \sim \sin \left(r - \frac{\pi l}{2} + \delta_l^S(k) \right), \tag{D.17}$$

$$U_l(r) \sim \sin \left(r - \frac{\pi l}{2} + \delta_l^U(k) \right). \tag{D.18}$$

Here $\delta_l^S(k)$ and $\delta_l^U(k)$ are the phase shifts induced by each potential, respectively. Multiplying equations (D.12) and (D.13) by $U_l(r)$ and $S_l(r)$, respectively; subtracting; and integrating, we obtain

$$U_l(r)\frac{dS_l(r)}{dr} - S_l(r)\frac{dU_l(r)}{dr} = -\int_0^r \varepsilon(\eta)\, S_l(\eta)\, U_l(\eta)\, d\eta. \tag{D.19}$$

By utilizing the asymptotic expressions in equations (D.17) and (D.18), we have, in the limit as $r \to \infty$,

$$k \sin\left[\delta_l^U(k) - \delta_l^S(k)\right] = -\int_0^\infty \varepsilon(r)\, S_l(r)\, U_l(r)\, dr = -\int_0^{ka} \varepsilon(r)\, S_l(r)\, U_l(r)\, dr, \tag{D.20}$$

since $n(r)$ is constant for $r > ka$ (or $r > a$). Thus far this equation is exact. If we now consider $\varepsilon(r)$ to be sufficiently small that $U_l(r) \approx S_l(r)$, then $\left|\delta_l^U(k) - \delta_l^S(k)\right| \ll 1$, and we have the relation

$$\delta_l^U(k) \approx \delta_l^S(k) \pm \frac{1}{k}\int_0^{ka} \varepsilon(r)\,[S_l(r)]^2\, dr. \tag{D.21}$$

Whether $\delta_l^U(k) > \delta_l^S(k)$ or not clearly depends on the concavity of $n(r)$. A further approximation can be made if the scattering potential $V_S(r)$ is constant (specifically, $V_S = k^2\left(1 - N^2\right)$ for $n = N, r \leq a$), for then the solution of equation (D.12) can be expressed in terms of a Riccati-Bessel function of the first kind, that is,

$$S_l(r) = \left(\frac{\pi N k r}{2}\right)^{1/2} J_{l+1/2}(Nkr). \tag{D.22}$$

Then we have that

$$\delta_l^U(k) \approx \delta_l^S(k) \pm \frac{\pi N}{2}\int_0^a \left\{n(r)\frac{d^2}{dr^2}\left(\frac{1}{n(r)}\right)\right\}\left[J_{l+1/2}(Nkr)\right]^2 r\, dr$$

$$\equiv \delta_l^S(k) \pm \frac{\pi N}{2}\mathcal{I}(a). \tag{D.23}$$

$\mathcal{I}$ being the integral over $[o, a]$. In the case of a small perturbation about $V_S = 0$, that is, for which $n = N = 1$, the term $\delta_l^S(k)$ in equation (D.23) is zero, and the resulting approximation for $\delta_l^U(k)$ is related to the first Born approximation in quantum scattering theory [62] (see also Section 21.4). In particular, if $\varepsilon(r) = Dr^{-s}$, D being some constant, a closed form solution for $\mathcal{I}$ can be found as $a \to \infty$ [62], namely,

$$\mathcal{I}(\infty) = \int_0^\infty \left[J_{l+1/2}(Nkr)\right]^2 r^{1-s}\, dr = \frac{1}{2}\left(\frac{Nk}{2}\right)^{s-2}\frac{\Gamma(s-1)\Gamma(l - \frac{1}{2}s + \frac{3}{2})}{\left[\Gamma(\frac{1}{2}s)\right]^2 \Gamma\left(l + \frac{1}{2}s + \frac{1}{2}\right)}, \tag{D.24}$$

provided $s > 1$ and $2l > s - 3$.

D.2 A LIOUVILLE TRANSFORMATION

As defined in equations (D.14) and (D.15), the "potentials" $V_S(r)$ and $V_U(r)$ are also k-dependent, which is not the case in potential scattering theory [60], [62]. This has an important consequence: unlike the quantum mechanical case, here pure "bound state" solutions, that is, real square-integrable solutions corresponding to $k^2 < 0$ (Im $k > 0$) do not exist. This can readily be proven [127] for the TE mode (equation (D.12)) that

$$\int_0^\infty \left[\left| \frac{dS_l(r)}{dr} \right|^2 + \frac{l(l+1)}{r^2} |S_l(r)|^2 \right] dr = k^2 \int_0^\infty n^2(r) |S_l(r)|^2 \, dr. \quad (D.25)$$

This cannot be satisfied for $k^2 < 0$ for a real and positive refractive index $n(r)$. In [128] the corresponding result is established from equation (D.11) for $U_l(r)$. Furthermore, a Liouville transformation may be used to define a new k-independent potential [127] (see also Chapter 7). Using the following simultaneous changes of independent and dependent variables in equation (D.9), we have

$$r \to \rho : \rho(r) = \int_0^r n(s) ds, \quad (D.26)$$

$$u_l \to \psi_l : \psi_l(\rho) = (n(r))^{1/2} S_l(r). \quad (D.27)$$

Clearly $n(r)$ must be integrable and nonnegative (in naturally occurring circumstances, $n \geq 1$ and $n(r) = 1$ for $r > a$); also $\rho(0) = 0$. It is easy to establish the following results:

$$\rho(r) = \rho_0 + r - a, \ r \geq a, \quad \text{where } \rho_a = \int_0^a n(s) ds;$$

$$\rho(r) \sim r, \quad r \to \infty;$$

$$r(\rho) = \int_0^\rho \frac{ds}{v(s)}, \quad \text{where } v(\rho) = n(r(\rho))$$

Furthermore, by applying transformations (D.26) and (D.27) to equation (D.12), we find that

$$\left[\frac{d^2}{d\rho^2} - \frac{l(l+1)}{R^2(\rho)} + k^2 \right] \psi_l(\rho) = V(\rho)\psi_l(\rho), \quad (D.28)$$

where

$$R(\rho) = v(\rho)r(\rho) \sim n(0)\rho, \quad \rho \to 0, \quad (D.29)$$

and

$$V(\rho) = [v(\rho)]^{-1/2} \frac{d^2}{d\rho^2} [v(\rho)]^{1/2}. \quad (D.30)$$

Clearly $v(\rho)$ should be at least twice differentiable. Now the new "potential" $V(\rho)$ is independent of the wavenumber k. Note also that $V(\rho) = 0$ for $\rho > \rho_a$.

It is of interest to determine the shape of the potential $V(\rho)$ by inverting $\rho(r)$ for various choices of physical $n(r)$ profiles for $r \in [0, a]$ (with $n(0) = n_0$, $n(a) = n_a$, and $n(r) = 1$ for $r > a$). In what follows only the nonzero potential shapes will be stated (corresponding to $\rho \in [0, \rho_a]$). Thus we have for

$$n(r) = n_a \left[1 - c^2 \left(\frac{r-a}{a} \right)^2 \right]^{-1} ; \quad V(\rho) = \frac{c^2}{n_a^2} > 0, \qquad (D.31)$$

where c is a real constant (i.e., the potential is a spherical barrier). For the profile

$$n(r) = (A + Br)^{-1}, \quad A = n_0^{-1}, \quad B = \frac{n_0 - n_a}{a n_0 n_a}; \quad V(\rho) = \frac{B^2}{4} > 0, \qquad (D.32)$$

which is also a barrier. For the important Maxwell fish-eye profile [12],

$$n(r) = n_0 \left(1 + Br^2 \right)^{-1}, \quad B = \frac{n_0 - n_a}{a^2 n_a}; \quad V(\rho) = -\frac{B}{n_0^2}. \qquad (D.33)$$

In this case, the new potential is a spherical well or barrier as $n_0 > n_a$ or $n_0 < n_a$, respectively. In the latter case the singularity occurring in $n(r)$ is moot since it arises for $r > a$. In many other cases, including $n(r) = n_0 \exp(-\alpha r)$, $n_0 \cos \alpha r$, and $n_0 \cosh \alpha r$, the potentials $V(\rho)$ are rather complicated functions, and there are no significant advantages to using the Liouville transformation in these cases. It is therefore of interest to examine what profiles $n(r)$ give rise to constant potentials $V(\rho)$. Let us define $y(\rho) = [v(\rho)]^{1/2}$ and $V(\rho) = V_0$, where V_0 is a constant of either sign. Then it follows from equation (D.30) that

$$\frac{d^2 y}{d\rho^2} - V_0 y = 0, \qquad (D.34)$$

the general solution being expressible in terms of real or complex exponential functions as $V_0 > 0$ (potential barrier) or $V_0 < 0$ (potential well), respectively. In r-space, $V_0 < 0$ corresponds to a constant refractive index $n = N = (1 + |V_0| k^{-2})^{1/2} > 1$, so we proceed with this physically realistic case. Writing the general solution of (D.34) as

$$y(\rho) = C \cos \left(|V_0|^{1/2} \rho + \eta \right), \qquad (D.35)$$

where C and η, are constants, it follows that

$$r(\rho) = \int_0^\rho \frac{ds}{v(s)} = \left(C^2 |V_0|^{1/2} \right)^{-1} \left[\tan \left(|V_0|^{1/2} \rho + \eta \right) - \tan \eta \right]. \qquad (D.36)$$

This can be inverted to yield

$$\rho(r) = \int_0^r n(s) ds = |V_0|^{-1/2} \left\{ \arctan \left[C^2 |V_0|^{1/2} r + \tan \eta \right] - \eta \right\}. \qquad (D.37)$$

Therefore,

$$n(r) = \rho'(r) = \frac{C}{1 + [Br + \tan \eta]^2}, \qquad (D.38)$$

where $C = n_0 \sec^2 \eta$, and η can be determined from the requirement that $n(a) = n_a$. This is a generalization of the Maxwell fish-eye profile in equation (D.33). The corresponding result for $V_0 > 0$ is

$$n(r) = \frac{C}{1 - [Br + \tanh \eta]^2}. \tag{D.39}$$

Note that in this case a singularity exists for $r > 0$ at $r = B^{-1}(1 - \tanh \eta)$.

Clearly in view of equations (D.31) and (D.32) this approach yields a limited class of possible profiles.

Appendix E

Helmholtz's Theorem

In its most general form this theorem is a combination of three related results valid for vector fields that vanish at the boundaries (or at infinity). We shall prove only the first one here; the second one has been proved in Chapter 19, and the third statements follow from standard vector identities that have been used several times throughout this book.

(1) Any vector field can be written as the sum of a solenoidal (i.e., divergence-free) field and an irrotational (i.e., curl-free) field. This is what is usually referred to as Helmholtz's theorem.

(2) A vector field is uniquely determined if both its divergence and curl are specified.

(3a) An irrotational field can be expressed as the gradient of a scalar field (its scalar potential).

(3b) A solenoidal field can be expressed as the curl of another vector field (its vector potential).

Frequently in physical problems a vector field will turn out to be purely irrotational (as in an electrostatic field), or purely solenoidal (as for magnetic fields or incompressible flow). Statement (1) is very useful in the study of elastic waves in isotropic homogeneous media, a topic addressed in Chapter 10.

E.1 PROOF OF HELMHOLTZ'S THEOREM

Let $\mathbf{V}$ be the vector field of interest. Clearly, for any vector field $\mathbf{Z}$,

$$\nabla \times (\nabla \times \mathbf{Z}) = \nabla (\nabla \cdot \mathbf{Z}) - \nabla^2 \mathbf{Z}. \tag{E.1}$$

If we let

$$\mathbf{V} = -\nabla^2 \mathbf{Z}, \tag{E.2}$$

$$U = \nabla \cdot \mathbf{Z}, \tag{E.3}$$

$$\mathbf{W} = \nabla \times \mathbf{Z}, \tag{E.4}$$

then

$$\mathbf{V} = -\nabla U + \nabla \times \mathbf{W}. \tag{E.5}$$

This is the formal mathematical statement of the result (E.1). We have assumed in (E.2) that $\mathbf{V}$ can be written as the Laplacian of another field $\mathbf{Z}$, but this is just

Poisson's equation, which has the known solution (not proved here)

$$\mathbf{Z}(\mathbf{r}) = \frac{1}{4\pi} \int \frac{\mathbf{V}(\mathbf{r}')}{|\mathbf{r} - \mathbf{r}'|} d\mathbf{r}'. \tag{E.6}$$

Taking the divergence of equation (E.5), we find

$$\nabla \cdot \mathbf{V} = -\nabla^2 U. \tag{E.7}$$

Again, this is Poisson's equation, with solution

$$U(\mathbf{r}) = \frac{1}{4\pi} \int \frac{\nabla' \cdot \mathbf{V}(\mathbf{r}')}{|\mathbf{r} - \mathbf{r}'|} d\mathbf{r}'. \tag{E.8}$$

Now take the curl of equation (E.5) to obtain

$$\nabla \times \mathbf{V} = \nabla \times (\nabla \times \mathbf{W}) = \nabla (\nabla \cdot \mathbf{W}) - \nabla^2 \mathbf{W}. \tag{E.9}$$

But from (E.4),

$$\nabla \cdot \mathbf{W} = 0, \tag{E.10}$$

so we have yet another Poisson's equation, and hence

$$\mathbf{W}(\mathbf{r}) = \frac{1}{4\pi} \int \frac{\nabla' \times \mathbf{V}(\mathbf{r}')}{|\mathbf{r} - \mathbf{r}'|} d\mathbf{r}'. \tag{E.11}$$

This means that we are always able to construct U and **W** in equation (E.5) from knowledge of **V**.

Problem: Prove that the zero vector is the only vector function with zero divergence and curl that vanishes at infinity.

E.2 LAMÉ'S THEOREM

Suppose that a displacement field $\mathbf{u} = \mathbf{u}(\mathbf{x}, t)$ satisfies the vector equation

$$\rho \frac{\partial^2 \mathbf{u}}{\partial t^2} \equiv \rho \ddot{\mathbf{u}} = \mathbf{f} + (\lambda + 2\mu) \nabla (\nabla \cdot \mathbf{u}) - \mu \nabla \times (\nabla \times \mathbf{u}),$$

where the body force **f** and the initial conditions on **u** can be written in terms of Helmholtz potentials as follows:

$$\mathbf{f} = \nabla\phi + \nabla \times \mathbf{\Psi}, \tag{E.12}$$

$$\dot{\mathbf{u}}(\mathbf{x}, 0) = \nabla A + \nabla \times \mathbf{B}, \tag{E.13}$$

$$\mathbf{u}(\mathbf{x}, 0) = \nabla C + \nabla \times \mathbf{D}, \tag{E.14}$$

and $\nabla \cdot \mathbf{\Psi} = 0$, $\nabla \cdot \mathbf{B} = 0$, $\nabla \cdot \mathbf{D} = 0$ (**B** and **D** are general vectors, not the electromagnetic field quantities of Chapter 20). Then there exist potentials ϕ and $\mathbf{\Psi}$ for **u** satisfying the following properties:

$$\mathbf{u} = \nabla\phi + \nabla \times \mathbf{\Psi}, \tag{E.15}$$

$$\nabla \cdot \mathbf{\Psi} = 0, \tag{E.16}$$

$$\ddot{\phi} = \frac{1}{\rho}\phi + \frac{\lambda + 2\mu}{\rho}\nabla^2\phi, \tag{E.17}$$

and

$$\ddot{\boldsymbol{\Psi}} = \frac{1}{\rho}\boldsymbol{\Psi} + \frac{\mu}{\rho}\nabla^2\boldsymbol{\Psi}, \tag{E.18}$$

where $\nabla\phi$ and $\nabla \times \boldsymbol{\Psi}$ are referred to as the P-wave and S-wave components of $\mathbf{u}$, respectively. To prove this theorem we use the fact that for any suitably integrable $f(\mathbf{x}, t)$,

$$\int_0^t \left[\int_0^\tau f\left(\mathbf{x}, \tau'\right)d\tau'\right]d\tau = \int_0^t (t - \tau) f(\mathbf{x}, \tau)\, d\tau,$$

and integrate property (E.17) (recognizing that $\nabla^2\phi = \nabla \cdot \mathbf{u}$) to obtain

$$\phi(\mathbf{x}, t) = \rho^{-1} \int_0^t (t - \tau)\{\phi(\mathbf{x}, \tau) + (\lambda + 2\mu)\nabla \cdot \mathbf{u}(\mathbf{x}, \tau)\}\, d\tau + At + C.$$

In a similar manner,

$$\boldsymbol{\Psi}(\mathbf{x}, t) = \rho^{-1} \int_0^t (t - \tau)\{\boldsymbol{\Psi}(\mathbf{x}, \tau) + \mu\nabla \times \mathbf{u}(\mathbf{x}, \tau)\}\, d\tau + \mathbf{B}t + \mathbf{D}.$$

Exercise: Verify properties (E.15)–(E.18) above using these expressions.

Hint: Integration of the vector equation with respect to time establishes (E.15); (E.16) and (E.17) are trivial, and (E.18) is similar to (E.17) except it involves the use of several vector identities.

Appendix F

Semiclassical Scattering: A Précis
(and a Few More Details)

The classical approximation in principle can be recovered from the exact cross section $\sigma(\theta)$ for angles far from the rainbow angle θ_R by smoothing out the oscillations in the asymptotic form and analytically continuing the result to angles near θ_R; this is basically equivalent to Descartes's theory in the case of the optical rainbow, and as in that case, it diverges at θ_R. The crude semiclassical approach is the unsmoothed asymptotic form of $\sigma(\theta)$ far from θ_R, and it also diverges if continued analytically to θ_R. The *Airy approximation* here means the application to continuous potentials of the previously discussed account of the light intensity in the neighborhood of a caustic, developed by Airy for the optical rainbow problem (see Chapter 5). This gives the form of $\sigma(\theta)$ close to θ_R but rapidly becomes inaccurate as θ deviates from θ_R by more than a degree or two.

Reference has been made earlier to various approximations, including the so-called *uniform approximation*. In [262] the reasons for the existence of these different approximations are carefully explained. Each is valid in a restricted angular domain, and the asymptotic expansion of the scattering amplitude $f(\theta)$ with respect to the (very) small parameter $\hbar$ (when $|\theta - \theta_R|$ is large) changes its form from a series in powers of $\hbar^{1/2}$ to a series in powers of $\hbar^{1/3}$ (when $|\theta - \theta_R| \ll 1$). This phenomenon is familiar for functions defined by ordinary differential equations, which is why the term "crude semiclassical" is associated with the WKB(J) approximation. In contrast, the Airy approximation is called a *transitional* approximation: it heals the wound (as noted in Chapter 6) between one region and another in a smooth, well-defined manner [30]. The uniform approximation, as its name implies, is valid for the whole variable domain [158], [159]. Nussenzveig transformed the eigenfunction expansion of $f(\theta)$ by the Poisson summation formula (first mentioned in Chapter 1) to give a series of integrals [93] (see also [160]); elsewhere [161] the suggestion was made that the Poisson formula is the natural mathematical tool to use when considering an eigenfunction expansion in a region where ray/classical concepts are appropriate. As noted earlier in connection with the Mie series in Chapter 19, such a transformation is exceedingly helpful in practical terms because the standard Faxen-Holtsmark formula (F.2) for $f(\theta)$, while exact, converges extremely slowly for θ in the neighborhood of θ_R; again, many thousand partial waves are often required in this region.

The application of the Poisson summation formula allows $f(\theta)$ to be written in terms of integrals, the integrands of which are rapidly oscillating functions, and it is known that the main contributions to the integrals come from the the neighborhoods

of stationary points on the positive real axis (when they exist) [163]. It suffices here
to note that the points of stationary phase are defined in terms of the deflection angle
$\Theta(l)$; this is just the classical deflection function (mostly denoted elsewhere in this
book by $D(i)$)—rainbow scattering arises whenever $\Theta(l)$ has an extremum. As we
have seen, there are in fact two real stationary (saddle) points if $\theta < \theta_R$; if $\theta > \theta_R$,
they are complex conjugates, having moved away from the real axis, and only one
of these contributes significantly to the amplitude. At $\theta = \theta_R$ they coalesce into a
third-order saddle point, for which the semiclassical approach is not valid (recall
that the direction θ_R defines a caustic direction). The uniform approximation, based
on [164], comes to the rescue (so to speak). Instead of treating the saddle points
separately, as in semiclassical methods, or as essentially coincident, as in the Airy
approximation, this method maps their exact behavior onto that of the stationary
points of the integrand in the Airy function. The resulting equations for the rainbow
cross section σ_R are rather complicated, but in [262] it is shown that on both the lit
and shadow sides of θ_R they reduce in the appropriate limits to that deduced from
the Airy approximation.

In a fundamental paper [88] four mathematical approximations are identified that
can be considered to define the term "semiclassical" approximation. The differential scattering cross section into unit solid angle at θ is

$$\sigma(\theta) = |f(\theta)|^2, \tag{F.1}$$

where $f(\theta)$ is expressed in the form

$$f(\theta) = \frac{1}{2ik} \sum_{l=0}^{\infty} (2l+1)(e^{2i\delta_l} - 1) P_l(\cos\theta), \tag{F.2}$$

δ_l being the phase shift for the lth partial wave. Again, equation (F.2) is known as
the Faxen-Holtsmark formula. The first approximation is:

(i) δ_l is replaced by its WKB(J) approximate value, namely,

$$\delta_l = \frac{\pi}{4} + \frac{\pi l}{2} - r_0 + \int_{r_0}^{\infty} [\kappa(r) - k] \, dr, \tag{F.3}$$

with

$$\kappa(r) = \left[\frac{2m(E-V)}{\hbar^2} - \frac{(l+1/2)^2}{r^2} \right]^{1/2}. \tag{F.4}$$

As noted several times already, r_0 is the turning point of the (classical) motion,
defined by $\kappa(r) = 0$. Physically, this approximation is equivalent to the requirement that the potential V be slowly varying, that is,

$$\left| \frac{1}{kV} \frac{dV}{dr} \right| \ll 1. \tag{F.5}$$

The most important property of the WKB(J) phase shift is its simple relation to the
classical deflection function $\Theta(l)$, namely,

$$\Theta(l) = 2\frac{d\delta_l}{dl}. \tag{F.6}$$

This allows a correspondence of sorts to be made between the quantum and classical results.

(ii) The second approximation concerns the replacement of the Legendre polynomials by their aymptotic forms for large values of l. Thus,

$$(a) \ P_l(\cos\theta) \cong \left[\frac{1}{2}\left(l+\frac{1}{2}\right)\pi \sin\theta\right]^{-1/2} \sin\left[\left(l+\frac{1}{2}\right)\theta+\frac{\pi}{4}\right]; \ \sin\theta \gtrsim \frac{1}{l}, \quad (F.7)$$

and

$$(b) \ P_l(\cos\theta) \cong (\cos\theta)^l J_0\left[\left(l+\frac{1}{2}\right)\theta\right]; \ \sin\theta \lesssim \frac{1}{l}. \quad (F.8)$$

For this approximation to be valid, many l-values must contribute to the scattering at a given angle (and the major contribution to the scattering amplitude comes from values of $l \gg 1$).

(iii) The third approximation is replacement of the sum over l by an integral with respect to the same variable. Again, this means that many partial waves should contribute to the scattering and also that $\delta_l(l)$ should vary slowly and smoothly. The approximations (i) and (ii) render both δ_l and P_l continuously differentiable functions of their arguments (in general), so that (iii) is appropriate.

Approximations (i), (ii)(a), and (iii) (for angles θ not close to 0 or π) yield the semiclassical form of the scattering amplitude,

$$f_{\rm sc} = -k^{-1}(2\pi \sin\theta)^{1/2} \int_0^\infty \left(l+\frac{1}{2}\right)^{1/2} [\exp{(i\phi_+)} - \exp{(i\phi_-)}]dl, \quad (F.9)$$

where the phase functions $\phi_\pm(l,\theta)$ are defined by

$$\phi_\pm(l,\theta) = 2\delta_l \pm \left(l+\frac{1}{2}\right)\theta \pm \frac{\pi}{4}, \quad (F.10)$$

and the result

$$\sum_{l=0}^\infty (2l+1)P_l(\cos\theta) = 0,$$

for $\theta \neq 0$ has been used [62].

(iv) A fourth approximation, not always necessary, is that it may be necessary for the integral in (iii) to be evaluated by the method of stationary phase, or the method of steepest descent. In [62] the integral was examined under circumstances such that there is only one point of stationary phase (corresponding to $\Theta(l)$ varying monotonically between 0 and $\pm\pi$). This is not the case in rainbow or glory scattering, where $\Theta(l)$ is not monotone everywhere in its domain. Because (classically) several incident angular momenta may correspond to the same scattering angle at a given energy, the total cross section will in general be the sum of different contributions from the different branches; and when these are well separated, each will give an independent contribution to the scattering amplitude, which may be evaluated by the method of stationary phase.

The rainbow angle corresponds to the singularity arising in the classical cross section when $d\Theta(l)/dl$ vanishes (the cross section contains a factor $(d\Theta(l)/dl)^{-1}$).

Near a rainbow angle, $\Theta(l)$ may be approximated by the quadratic function

$$\Theta(l) = \theta_R + q(l - l_R)^2, \tag{F.11}$$

where the terms have obvious meanings. On the lit or bright side of the rainbow angle the classical intensity is

$$\sigma_{cl} = \left(\frac{l_R + 1/2}{k^2 \sin \theta_R}\right) |q(\theta - \theta_R)|^{-1/2}, \tag{F.12}$$

and on the dark or shadow side, the classical intensity is zero (if there are no additional contributing branches of the deflection function $\Theta(l)$). This of course is identical with the predictions of geometrical optics for the optical rainbow. In terms of the phase shift δ_l however,

$$\delta_l = \delta_R \pm \frac{\theta_R}{2}(l - l_R) + \frac{q}{6}(l - l_R)^3, \tag{F.13}$$

where the $\pm$ is necessary for the antisymmetry (see equation (F.10) for $\phi_\pm$). For $\theta_R > 0$ the dominant contribution to the integral (F.9) will come from the term containing the factor $\exp(i\phi_-)$; this yields essentially the same functional form for the scattering differential cross section as does the original Airy theory for the intensity of the optical rainbow. Thus

$$f_{sc} = k^{-1} \left[2\pi \frac{(l_R + 1/2)}{\sin \theta}\right]^{1/2} q^{-1/3} e^{i\delta} \, \mathrm{Ai}(x), \tag{F.14}$$

where for $\Theta(l_R) > 0$,

$$x = q^{-1/3}(\theta_R - \theta),$$

is a measure of the deviation from the rainbow angle, and

$$\delta = 2\alpha_R - \frac{\pi}{4} + \left(l_R + \frac{1}{2}\right)(\theta_R - \theta), \tag{F.15}$$

for $\Theta(l_R) > 0$. If $\Theta(l_R) < 0$, then the difference $\theta_R - \theta$ must be replaced by its negative in the expressions for x and δ. The quantity α_R is the intercept of the tangent line to the $(\delta_R, l_R + 1/2)$ curve with the vertical axis:

$$\alpha_R = \left[\delta_l - \left(l + \frac{1}{2}\right)\frac{d\delta_l}{dl}\right]_{l=l_R}, \tag{F.16}$$

(see figure 1 in [262]). The form of the Airy integral used there is [101]

$$\mathrm{Ai}(x) = \frac{1}{2\pi}\int_{-\infty}^{\infty} \exp\left[i\left(xu + \frac{1}{3}u^3\right)\right] du = \frac{1}{\pi}\int_0^{\infty} \cos\left(xu + \frac{1}{3}u^3\right) du. \tag{F.17}$$

The differential cross section near the rainbow angle then has the form

$$\sigma_{sc} = \left[2\pi \frac{(l_R + 1/2)}{k^2 \sin \theta}\right] |q^{-2/3}| \, [\mathrm{Ai}(x)]^2. \tag{F.18}$$

Provided $0 < \Theta(l) < \pi$, or $-\pi < \Theta(l) < 0$, σ_{sc} can be described entirely in terms of the classical cross section, together with interference effects (not discussed here) and rainbow scattering. If however $\Theta(l)$ passes smoothly through 0, $\pm\pi$, and so forth, then the vanishing of $\sin\Theta(l)$ for l, $|d\Theta/dl| < \infty$, leads to a singularity in the cross section for both forward and backward scattering. The optical/meteorological terminology is again borrowed for this situation: it is referred to as a *glory*. In the case, for example, of a backward glory, near $\Theta(l) = \pi$, we may write the approximate form [88]

$$\Theta(l) = \pi + a(l - l_g),\qquad\qquad\text{(F.19)}$$

and then σ_{cl} in the backward direction is just the the sum of two equal contributions from $\Theta < \pi$ and $\Theta > \pi$, that is,

$$\sigma_{cl} = \frac{2l_g}{k^2\,|a|\,(\pi - \theta)}.\qquad\qquad\text{(F.20)}$$

From equation (F.19) the phase shift, for those values of l that contribute most to the glory, may be written as

$$\delta_l = \delta_g + \frac{\pi}{2}(l - l_g) + \frac{a}{4}(l - l_g)^2,\qquad\qquad\text{(F.21)}$$

and after some rearrangements of the integral in the expression for f_{sc}, it can be shown that the glory cross section is

$$\sigma_{sc} = \left(l_g + \frac{1}{2}\right)^2 \left(\frac{2\pi}{k^2\,|a|}\right) J_0^2(l_g\sin\theta),\qquad\qquad\text{(F.22)}$$

when there are no interference effects. For a forward glory the result is similar; there is a different phase term, but this makes no contribution of course to σ_{sc}. Thus the singularity in σ_{cl} is replaced by a finite peak in both forward and backward directions of

$$\sigma_{sc}^{max} = \left(l_g + \frac{1}{2}\right)^2 \left(\frac{2\pi}{k^2\,|a|}\right).\qquad\qquad\text{(F.23)}$$

The Bessel function oscillations may be interpreted as resulting from interference between contributions from the two branches of $\Theta(l)$ near a glory, that is, $\Theta > \pi$ (or 0) and $\Theta < \pi$ (or 0). In [88] it is noted that when the intensity is averaged over several such oscillations, using the result

$$\langle J_0^2(x)\rangle = \frac{1}{\pi x},$$

then σ_{sc} reduces to the classical expression σ_{cl}.

Bibliography

Some of the entries in this bibliography have been annotated with brief comments or descriptions.

[1] M. V. Berry, "Uniform approximation: A new concept in wave theory," *Sci. Prog. Oxf.*, **57** (1969) 43–64.

[2] M. V. Berry and S. Klein, "Colored diffraction catastrophes," *Proc. Natl. Acad. Sci. USA*, **93** (1996), 2614–2619.

[3] G. A. Deschamps, "Ray techniques in electromagnetics," *Proc. IEEE*, **60** (1972) 1022–1035.

[4] C. F. Bohren, "Physics textbook writing: Medieval, monastic mimicry," *Am. J. Phys.*, **77** (2009) 101–103. Craig Bohren is very much a purist; read *anything* he writes — you will be in for a great ride.

[5] G. L. James, *Geometrical Theory of Diffraction for Electromagnetic Waves*, IEE, London and New York, 1980. A lot of interesting material is to be found in this compact volume.

[6] S. Cornbleet, "Geometrical optics reviewed: A new light on an old subject," *Proc. IEEE*, **71** (1983) 471–502. A comprehensive review of the historical development in the field up to 1983.

[7] S. Cornbleet, *Microwave and Optical Ray Geometry*, John Wiley & Sons, Chichester, 1984.

[8] V. M. Tikhomirov, *Stories about Maxima and Minima*, AMS/MAA, 1990.

[9] I. Niven, *Maxima and Minima without Calculus*, MAA, 1981.

[10] D. Pedoe, *A Course in Geometry*, Cambridge University Press, New York, 1970.

[11] H. Helfgott and M. Helfgott, A noncalculus proof that Fermat's principle of least time implies the law of refraction, *Am. J. Phys.*, **70** (2002) 1224–1225.

[12] U. Leonhardt and T. Philbin, *Geometry and Light: The Science of Invisibility*, Dover, New York, 2010. Highly recommended reading for anyone interested in the theory of "cloaking" (so not just for Klingons).

[13] M. Born and E. Wolf, *Principles of Optics*, 7th ed., Pergamon, Oxford, 1999. The *classic* text in optics.

[14] K. Sassen, "Angular scattering and rainbow formation in pendant drops," *J. Opt. Soc. Am.*, **69** (1979) 1083–1089.

[15] R. L. Lee, Jr., "Mie theory, Airy theory, and the natural rainbow," *Appl. Opt.*, **37** (1998), 1506–1519.

[16] H. M. Nussenzveig, "The theory of the rainbow," *Sci. Am.*, **236** (1977) 116–127. A highly readable yet nontrivial article for the rainbow neophyte.

[17] G. P. Können, *Polarized Light in Nature*, Cambridge University Press, Cambridge, 1985. An excellent resource, now freely available from the website http://www.guntherkonnen.com/.

[18] H. E. Edens, "Photographic observation of a natural fifth-order rainbow," *Appl. Opt.*, **54** (2015) B26–B34.

[19] H. E. Edens and G. P. Können, "Probable photographic detection of the natural seventh-order rainbow," *Appl. Opt.*, **54** (2015) B93–B96.

[20] R. Greenler, *Rainbows, Halos and Glories*, Cambridge University Press, Cambridge, 1980. A must-read by a master in the field of meteorological optics.

[21] W. J. Humphreys, *Physics of the Air*, McGrow-Hill, London, 1940. A classic text, now quite hard to find, as is [22].

[22] R. A. R. Tricker, *Introduction to Meteorological Optics*, Elsevier, New York, 1970. See [21]. There is much material here that probably cannot be found elsewhere.

[23] G. B. Airy, "On the intensity of light in the neighbourhood of a caustic," *Trans. Camb. Phil. Soc.*, **6** (1838) 379–403.

[24] J. D. Jackson, "From Alexander of Aphrodisias to Young and Airy," *Phys. Repts.*, **320** (1999) 27–36.

[25] J. A. Adam, "The mathematical physics of rainbows and glories," *Phys. Repts.*, **356** (2002) 229–365.

[26] H. M. Nussenzveig, *Diffraction Effects in Semiclassical Scattering*, Cambridge University Press, Cambridge, 1992. A classic, though it is highly condensed.

[27] R. Haberman, *Elementary Applied Differential Equations*, 3rd ed., Prentice-Hall, Upper Saddle Revir, NJ, 1987. A very comprehensive book, with some quite advanced topics.

[28] M. Abramowitz, I. A. Stegun, eds., *Handbook of Mathematical Functions*, Dover, New York, 1972.

[29] J. F. Nye, *Natural Focusing and Fine Structure of Light: Caustics and Wave Dislocation*, Institute of Physics, Bristol, U.K. 1999.

[30] M. J. Lighthill, *Waves in Fluids*, Cambridge University Press, Cambridge, 1987. Provides the mathematical basis for the principle of stationary phase discussed in Part I.

[31] J. B. Calvert, http://mysite.du.edu/~jcalvert/phys/wkb.htm.

[32] W. D. Lakin and D. A. Sanchez, *Topics in Ordinary Differential Equations*, Dover, New York, 1970.

[33] M. J. Lighthill, "Studies on magneto-hydrodynamic waves and other anisotropic wave motions," *Phil. Trans.*, **252A** (1960) 397–430. One of two seminal papers in my graduate studies. The other is [181]

[34] D. J. Acheson, *Elementary Fluid Dynamics*, Clarendon Press, Oxford, 2005. I have used this book for teaching mathematical fluid dynamics at the senior undergraduate level (see [35]). David was a reviewer for one of the first papers I submitted for publication and (quite appropriately) rejected it!

[35] K. Socha, "Circles in circles: Creating a mathematical model of surface water waves," *Am. Math. Monthly*, **114** (2007) 202–216. Draws on the relevant chapter of [34]; both are excellent resources for the undergraduate student.

[36] J. Billingham and A. C. King, *Wave Motion*, Cambridge University Press, Cambridge, 2006.

[37] H. Lamb, *Hydrodynamics*, Dover, New York, 1945. So much to read, so little time. ...

[38] R. Snieder, *A Guided Tour of Mathematical Methods for the Physical Sciences*, Cambridge University Press, Cambridge, 1981.

[39] G. L. Lamb, *Elements of Soliton Theory*, Cambridge University Press, Cambridge, 1980.

[40] A. Sommerfeld, *Mechanics of Deformable Bodies*, Academic Press, 1964.

[41] S. Flügge, *Practical Quantum Mechanics*, Springer, Berlin, 1999. An excellent source of examples using special functions!

[42] W. M. Ewing and W. S. Jardetzky, *Elastic Waves in Layered Media*, McGraw-Hill, New York, 1957.

[43] C. C. Mei, *The Applied Dynamics of Ocean Surface Waves*, World Scientific, New Jersey, 2003. I found this very useful source of material for an advanced undergraduate course in mathematical fluid dynamics, especially for surface "rays," scattering, and tsunami theory.

[44] G. B. Arfken and H. J. Weber, *Mathematical Methods for Physicists*, 6th edition, Elsevier, Boston, MA, 2005.

[45] http://en.wikipedia.org/wiki/Action_(physics).

[46] A. Small and K. S. Lam, "Simple derivations of the Hamilton-Jacobi equation and the eikonal equation without the use of canonical transformations," *Am. J. Phys.*, **79** (2011) 678–681.

[47] H. Goldstein, *Classical Mechanics*, Addison-Wesley, Reading, MA, 1950. THE classical text on mechanics for those of my generation (at the very least).

[48] A. Orefice, R. Giovanelli, and D. Ditto, "Complete Hamiltonian description of wave-like features in classical and quantum physics," *Found. Phys.*, **39** (2009) 256–272.

[49] H. J. W. Müller-Kirsten, *Introduction to Quantum Mechanics: Schrodinger Equation and Path Integral*, World Scientific, Hackensack, 2006.

[50] A. Messiah, *Quantum Mechanics*, Dover, New York, 1999. At the risk of repetition, another famous and classic text on the subject.

[51] R. G. Newton, *Scattering Theory of Waves and Particles*, Springer-Verlag, New York, 1982. Very comprehensive.

[52] T. W. B. Kibble, *Classical Mechanics*, McGraw-Hill, New York, 1966. Used in my undergraduate physics class, but needs to be supplemented with [47].

[53] C. D. Collinson, *Introductory Mechanics*, Edward Arnold, London, 1980.

[54] W. T. Grandy, Jr., *Scattering of Waves from Large Spheres*, Cambridge University Press, Cambridge, 2000. This is a very detailed book with a good set of mathematical appendices on functions germane to scattering problems.

[55] R. F. Harrington, *Time-Harmonic Electromagnetic Fields*, McGraw-Hill, New York, 1961.

[56] C. F. Bohren and D. R. Huffman, *Absorption and Scattering of Light by Small Particles*, Wiley, New York, 1983. I look forward to seeing the latest edition of this major contribution to the field.

[57] J. A. Stratton, *Electromagnetic Theory*, IEEE Press and John Wiley & Sons, Hoboken, NJ, 2007. Contains a very nice section on the Mie solution of electromagnetic scattering.

[58] C. A. Valagiannopoulos, "An overview of the Watson transformation presented through a simple example," *Prog. Electromag. Res.*, **75** (2007) 137–152.

[59] V. Rojansky, *Introductory Quantum Mechanics*, Prentice-Hall, New York, 1938.

[60] L. I. Schiff, *Quantum Mechanics*, 3rd ed. McGraw-Hill, New York, 1968. I used this book as an undergraduate and still have it, though it is falling apart from much use.

[61] G. Baym, *Lectures on Quantum Mechanics*, W. A. Benjamin, New York, 1969. A treasury of interesting topics not often included in texts of this kind.

[62] N. F. Mott and H. S. W. Massey, *The Theory of Atomic Collisions*, 3rd ed. Clarendon Press, Oxford, 1965.

[63] J. R. Taylor, *Scattering Theory*, John Wiley & Sons, New York, 1972. Quite rigorous and beautiful in its scope.

[64] V. de Alfaro and T. Regge, *Potential Scattering*, North-Holland, Amsterdam, 1965. A joy to own; again like [63], rigorous and beautiful.

[65] A. G. Sitenko, *Scattering Theory*, Springer-Verlag, Berlin, 1991. Very useful for some of the more rigorous proofs.

[66] D. R. Bates ed., *Quantum Theory I. Elements*, Academic Press, New York, 1961. Some classic essays can be found here.

[67] G. V. Frisk and J. A. DeSanto, "Scattering by spherically symmetric inhomogeneities," *J. Acoust. Soc. Am.*, **47** (1970) 172–180.

[68] M. A. Pohrivchak, "Ray- and wave-theoretic approach to electromagnetic scattering from radially inhomogeneous spheres and cylinders," Ph.D. dissertation, Old Dominion University, Norfolk, VA, 2014.

[69] M. J. Adams, *An Introduction to Optical Waveguides*, John Wiley & Sons, New York, 1981.

[70] P. L. E. Uslenghi, "Electromagnetic and optical behavior of two classes of dielectric lenses," *IEEE Trans. Ant. Prop.*, **17** (1969) 235–236.

[71] C. B. Boyer, *The Rainbow, from Myth to Mathematics*, Princeton University Press, Princeton, 1987. Contains fascinating historical details; an important complement to [169].

[72] M. R. Vetrano, J. P. A. J. van Beeck, and M. L. Riethmuller, "Generalization of the rainbow Airy theory to nonuniform spheres," *Opt. Lett.*, **30** (2005) 658–660.

[73] J. A. Adam, "Zero-order bows in radially inhomogeneous spheres: Direct and inverse problems," *Appl. Opt.*, **50** (2011) F50–F59.

[74] J. A. Lock, "Scattering of an electromagnetic plane wave by a Luneburg lens. I. Ray theory," *J. Opt. Soc. Am.*, **A25** (2008) 2971–2979.

[75] J. A. Adam, "Scattering of electromagnetic plane waves in radially inhomogeneous media: Ray theory, exact solutions and connections with potential scattering theory," chapter 3 in A. Kokhanovsky, ed., *Light Scattering Reviews*, **9**, Springer-Verlag, Berlin, 2015.

[76] R. N. Gould and R. Burman, "Some electromagnetic wave functions for propagation in stratified media," *J. Atmos. Terrest. Phys.*, **26** (1964) 335–340.

[77] J. A. Adam and P. Laven, "Rainbows from inhomogeneous transparent spheres: A ray-theoretic approach," *Appl. Opt.*, **46** (2007) 922–929.

[78] B. S. Westcott, "Electromagnetic wave propagation in spherically stratified isotropic media," *Electronic Lett.*, **4** (1968) 572–575.

[79] C. L. Brockman, "High frequency electromagnetic wave backscattering from radially inhomogeneous dielectric spheres," Ph.D. dissertation, University of California, Los Angeles, 1974.

[80] C. L. Brockman and N. G. Alexopoulos, "Geometrical optics of inhomogeneous particles: Glory ray and the rainbow revisited," *Appl. Opt.*, **16** (1977) 166–174.

[81] J. B. Keller and B. R. Levy, "Scattering of Short Waves," in *Interdisciplinary Conference on Electromagnetic Scattering*, August 1962, Milton Kerker, ed. Clarkson College of Technology, Macmillan, New York (1963) 3–24.

[82] H. C. Ohanian and C. G. Ginsburg, "Antibound 'states' and resonances," *Am. J. Phys.*, **42** (1974) 310–315.

[83] H. M. Nussenzveig, "High frequency scattering by a transparent sphere. I. Direct reflection and transmission," *J. Math. Phys.*, **10** (1969) 82–124. This and part II ([84]) are beautifully detailed papers. I just wish I had the capacity to understand them in their entirety!

[84] H. M. Nussenzveig, "High frequency scattering by a transparent sphere. II. Theory of the rainbow and the glory," *J. Math. Phys.*, **10** (1969) 125–176.

[85] H. C. van de Hulst, *Light Scattering by Small Particles*, Dover, New York, 1981. Certainly an important and much-referenced book for workers in meteorological optics (and other fields), but I must confess that I have never found it easy to read.

[86] N. A. Logan, "Survey of some early studies of the scattering of plane waves by a sphere," *Proc. IEEE*, **53** (1965) 773–785. Read this for the *real* history of the Mie solution!

[87] G. Mie, "Beiträge zur Optik trüber Medien, speziell kolloidaler Metallösungen," *Ann. Physik* **25** (1908) 377–445.

[88] K. W. Ford and J. A. Wheeler, "Semiclassical description of scattering," *Ann. Phys.*, **7** (1959) 259–286. A seminal paper on the title topic.

[89] G. N. Watson, "The diffraction of electric waves by the earth," *Proc. Roy. Soc.* (London) **A95** (1918) 83–99.

[90] J. A. Adam, "An initial value problem for magnetoatmospheric waves: I. Theory," *Wave Motion*, **12** (1990) 385–399.

[91] J. A. Adam, "A nonlinear eigenvalue problem in astrophysical magnetohydrodynamics: Some properties of the spectrum," *J. Math. Phys.*, **30** (1989) 744–756.

[92] A. Sommerfeld, *Partial Differential Equations in Physics*, Academic Press, New York, 1964.

[93] H. M. Nussenzveig, "High-frequency scattering by an impenetrable sphere," *Ann. Phys.*, **34** (1965) 23–95. A necessary precursor in the development of [83] and [84].

[94] J. D. Jackson, *Classical Electrodynamics*, 3rd ed., Wiley, New York, 1998. The "Goldstein" of student texts in electromagnetism.

[95] M. V. Berry, "Uniform approximation for potential scattering involving a rainbow," *Proc. Phys. Soc.*, **89** (1966) 479–490.

[96] F. W. J. Olver, *Asymptotics and Special Functions*, Academic Press, New York, 1974.

[97] B. R. Johnson, "Theory of morphology-dependent resonances: Shape resonances and width formulas," *J. Opt. Soc. Am.*, **10** (1993) 343–352. A very informative paper on morphology dependent resonance theory.

[98] H. Jeffreys and B. S. Jeffreys, *Methods of Mathematical Physics*, Cambridge University Press, Cambridge, 1966. One of the first books I ever bought as an undergraduate student. I have never regretted it. It is written from the perspective of classical "British" applied mathematics, which is not surprising, given who the authors are!

[99] M.G.J. Minnaert, *Light and Colour in the Outdoors*, Springer, New York, 1993. A relatively recent printing of the 1954 classic. That is the book that started my interest in meteorological optics, which in turn led ultimately to [122].

[100] M. Minnaert, "On musical air bubbles and the sounds of running water," Phil. Mag., **16** (1933) 235–248.

[101] M. S. Child, *Molecular Collision Theory*, Dover, New York, 1974. Early chapters include an excellent summary of semiclassical theory.

[102] M. V. Berry and C. Upstill, "IV catastrophe optics: Morphologies of caustics and their diffraction patterns," *Progress in Optics*, **18** (1980) 257–346 (ed. E. Wolf) North-Holland, Amsterdam.

[103] D. K. Lynch and W. Livingston, *Color and Light in Nature*, Cambridge University Press, New York, 1995. This is a valuable book, not only for the topics covered but also for the clarity of exposition.

[104] R. Thom, *Structural Stability and Morphogenesis*, Addison-Wesley, New York, 1989.

[105] R. Gilmore, *Catastrophe Theory for Scientists and Engineers*, John Wiley & Sons, New York, 1981.

[106] L. D. Landau and E. M. Lifshitz, *Mechanics*, Pergamon Press, Oxford, 1960.

[107] L. D. Landau and E. M. Lifshitz, *Quantum Mechanics,* Pergamon Press, Oxford 1965. Second only to [60] in terms of how much I have used it.

[108] J. D. Walker, "Multiple rainbows from single drops of water and other liquids," *Am. J. Phys.*, **44** (1976) 421–433.

[109] B. A. Bolt, *Earthquakes: A Primer,* W. H. Freeman, San Francisco, 1978.

[110] M. J. Lighthill, "Group velocity," *J. Inst. Maths. Applics.*, **1** (1965) 1–28. A beautiful exposition by a consummate applied mathematician.

[111] M. J. Lighthill, "Asymptotic behavior of anisotropic wave systems stimulated by oscillating sources," chapter 1 in *Wave Asymptotics* (P. A. Martin and G. R. Wickham, eds.), Cambridge University Press, Cambridge, 1992.

[112] K. Kajuira, "The leading wave of a tsunami," *Bull. Earthquake Res. Inst.*, University of Tokyo, 41 (1963) 525–571.

[113] http://web.ics.purdue.edu/~braile/edumod/slinky/slinky.htm.

[114] National Institute of Standards and Technology, Digital Library of Mathematical Functions. http://dlmf.nist.gov/.

[115] G. B. Whitham, "Group velocity and energy propagation for three-dimensional waves," *Comm. Pure Appl. Math.*, **14** (1961) 675–691.

[116] G. B. Whitham, *Linear and Nonlinear Waves*, John Wiley & Sons, New York, 1974. Much material here for an advanced course on wave motion in many different contexts.

[117] J. A. Adam, "'Rainbows' in homogeneous and radially inhomogeneous spheres: connections with ray, wave, and potential scattering theory," chapter 3 in Bourama, T. ed., *Advances in Interdisciplinary Mathematical Research: Applications to Engineering, Physical and Life Sciences*, **37**, Springer, New York, 2013.

[118] J. A. Adam, "Geometric optics and rainbows: Generalization of a result by Huygens," *Appl. Opt.*, **47** (2008) H11–H13.

[119] R. Aster, "Fundamentals of ray tracing," http://www.ees.nmt.edu/outside/courses/GEOP523/Docs/rays.pdf

[120] N. Marcuvitz, "On field representations in terms of leaky modes or Eigenmodes," *IRE Trans. Antennas Prop.*, **4** (1956) 192–194.

[121] U. Nuntaplook, "Topics in electromagnetic, acoustic and potential scattering theory," Ph.D. dissertation, Old Dominion University, Norfolk, VA, 2013.

[122] J. A. Adam, *Mathematics in Nature: Modeling Patterns in the Natural World*, Princeton University Press, Princeton, NJ, 2003.

[123] E. A. Robinson and D. Clark, "The eikonal equation and the secret Pythagorean theorem," *Leading Edge*, **22** (2003) 749–750.

[124] C. F. Bohren, *Clouds in a Glass of Beer*, John Wiley & Sons, New York, 1987. This book and [125] are very informative, but even more importantly, *fun* to read. Get them for your friends!

[125] C. F. Bohren, *What Light through Yonder Window Breaks?* John Wiley & Sons, New York, 1991.

[126] B. J. Loe and N. Beagley, "The coffee cup caustic for calculus students," *College Math. J.*, **28** (1997) 277–284. http://isites.harvard.edu/fs/docs/icb.topic253321.files/Caustics.pdf.

[127] C. Eftimiu, "Direct and inverse scattering by a sphere of variable index of refraction," *J. Math. Phys.* **23** (1982) 2140–2146.

[128] C. Eftimiu, "Inverse electromagnetic scattering for radially inhomogeneous dielectric spheres," in *Inverse Methods in Electromagnetic Imaging*. Proceedings of the NATO Advanced Research Workshop, Bad Windsheim, West Germany, D. Reidel, Dordrecht (1985) 157–176.

[129] P. A. Martin, "Acoustic scattering by inhomogeneous spheres," *J. Acoust. Soc. Am.*, **111** (2002) 2013–2018.

[130] C. G. Gibson, *Elementary Geometry of Differentiable Curves*, Cambridge University Press, Cambridge, 2001.

[131] K. Maver, "Kelvin ship waves," http://www-f1.ijs.si/~rudi/sola/Kelvin_wave%201.pdf.

[132] A. L. Fetter and J. D. Walecka, *Theoretical Mechanics of Particles and Continua*, Dover, New York, 2003.

[133] R. F. Mudde, "Physics of daily life" (http://s3.amazonaws.com/cramster-resource/45580_vrijeveld06.pdf). An excellent resource for the study of everyday physics.

[134] J. P. McKelvey, "The case of the curious curl," *Am. J. Phys.*, **58** (1990) 306–310.

[135] B. Jeffreys, "Note on the transparency of a potential barrier," *Proc. Camb. Phil. Soc.*, **38** (1942) 401–405.

[136] N. F. Mott and I. N. Sneddon, *Wave Mechanics and Its Applications*, Dover, New York, 1963.

[137] E. Pascuzzi, "The glorious glory," *Phys. Teacher*, **36** (1998) 164–166.

[138] M. J. Buckingham, "On acoustic transmission in ocean-surface waveguides," *Phil. Trans. Roy. Soc. A* **335** (1991) 513–555.

[139] J. W. Blum and D. S. Chen, "Acoustic wave propagation in an underwater sound channel 2. Quantitative theory," *J. Inst. Maths. Applics.*, **8** (1971) 199–220.

[140] J. W. Blum and D. S. Chen, "Acoustic wave propagation in an underwater sound channel 1. Qualitative theory," *J. Inst. Maths. Applics.*, **8** (1971) 186–198.

[141] M. R. Osborne, "On the propagation of sound in a layered medium," *Quart. J. Mech. Appl. Math.*, **13** (1960) 472–486.

[142] M. R. Osborne, "On the equivalence of two methods for solving sound propagation problems in a layered fluid medium," *Quart. J. Mech. Appl. Math.*, **15** (1962) 511–518.

[143] http://www.tsunami.noaa.gov/.

[144] C. O. Hines, "Atmospheric gravity waves: a new toy for the wave theorist," *Radio Sci.*, **69D** (1965) 375–380.

[145] Academic Press, San Diego, 1979.

[146] D. B. Giaiotto and F. Stel, "Waves in the Atmosphere," http://users.ictp.it/~giaiotti/PhD_EFM/l04_waves.pdf.

[147] C. Jargodzki and F. Potter, *Mad about Physics*. Wiley: New York, 2001.

[148] J. A. Adam, "Like a Bridge over Colored Water: A Mathematical Review of The Rainbow Bridge: Rainbows in Art, Myth, and Science," *Notices AMS* **49** (2002) 1360–1371. This is the complete review of [169], some extracts of which have been used in the final chapter of this book (with permission of the AMS).

[149] M. A, Pohrivchak, J. A. Adam and U. Nuntaplook, "Scattering of plane electromagnetic waves by radially inhomogeneous spheres: Asymptotics and special functions," in Bourama Toni, ed., *Mathematical & Statistical Research with Applications to Physical & Life Sciences, Engineering & Technology*, Springer Proceedings in Mathematics and Statistics, **157**, Springer, New York, 2016.

[150] H. Kragh, "Ludvig Lorenz and nineteenth century optical theory: The work of a great Danish scientist," *Appl. Opt.*, **30** (1991) 4688–4695.

[151] M. V. Berry, "Nature's optics and our understanding of light," *Contemp. Phys.*, **56** (2015) 2–16. A superb summary of the nature-physics-mathematics connections in optics.

[152] E. A. Coddington and N. Levinson, *Theory of Ordinary Differential Equations*, McGraw-Hill, New York, 1955.

[153] E. C. Titchmarsh, *Eigenfunction Expansions Associated with Second-Order Differential Equations*, Clarendon Press, Oxford, 1946.

[154] A. E. Lifschitz, *Magnetohydrodynamics and Spectral Theory*, Kluwer Academic, Dordrecht, 1989.

[155] D. Adams, *The Restaurant at the End of the Universe*, Pan Books, London, 1980.

[156] The apostle Paul: 1 Thessalonians 5:21.

[157] T. Pearcey, "The structure of the electromagnetic field in the neighbourhood of a cusp of a caustic," *Phil. Mag.*, **37** (1946) 311–317.

[158] R. E. Langer, "On the connection formulas and the solutions of the wave equation," *Phys. Rev.*, **51** (1937) 669–676.

[159] S. Jorna, "Derivation of Green-type, transitional, and uniform asymptotic expansions from differential equations. I. General theory, and application to modified Bessel functions of large order," *Proc. Roy. Soc. A* (London) **281** (1964) 99–110. Part II (dealing with Whittaker functions) was published immediately following this one (pp. 111–129) and Part III (confluent hypergeometric functions) was published in volume 284 (pp. 531–539).

[160] S. I. Rubinow, "Scattering from a penetrable sphere at short wavelengths," *Ann. Phys.*, **14** (1961) 305–332. In my opinion this paper has not been given the recognition it deserves in the literature on the mathematical theory of rainbows and related optical phenomena.

[161] C. L. Pekeris, "Ray theory vs. normal mode theory in wave propagation problems," *Proc. Symp. Appl. Math.*, **2** (1950) 71–75.

[162] M. Vollmer, *Lichtspiele in der Luft*, Atmosphärische Optik für Einsteiger, Elsevier, Munich, 2006. This compact volume is all about meteorological optics and the physics underlying it; I just wish I had taken more than two years of German in high school!

[163] E. T. Copson, *Asymptotic Expansions*, Cambridge University Press, Cambridge, 1965.

[164] C. Chester, B. Friedman, and F. Ursell, "An extension of the method of steepest descents," *Proc. Camb. Phil. Soc.*, **53** (1957) 599–611.

[165] B. Kinsman, *Wind Waves: Their Generation and Propagation on the Ocean Surface*, Dover, New York, 1984.

[166] M. Reed and B. Simon, *Methods of Modern Mathematical Physics III: Scattering Theory*, Academic Press, San Diego, 1979.

[167] D. B. Pearson, *Quantum Scattering and Spectral Theory*, Academic Press, London, 1988. You want rigor? This is rigorous!

[168] J. A. Lock, "Cooperative effects among partial waves in Mie scattering," *J. Opt. Soc. Am.*, **A5** (1988) 2032–2044.

[169] R. L. Lee, Jr., and A. B. Fraser, *The Rainbow Bridge. Rainbows in Art, Myth and Science*, Pennsylvania State University Press, State Park, 2001.

[170] A. Sommerfeld, *Optics*, Academic Press, New York, 1954.

[171] R. Thom, *Structural Stability and Morphogenesis*, Addison-Wesley, New York, 1989.

[172] T. Poston and I. Stewart, *Catastrophe Theory and Its Applications*, Pitman, Boston, 1978.

[173] V. I. Arnold, "Singularities of smooth mappings," *Russian Math. Sur.*, **23** (1968) 1–43.

[174] Y. A. Kravtsov and Y. I. Orlov, *Caustics, Catastrophes and Wave Fields*, Springer, Berlin, 1999. This book together with [175], [176] contains more than I ever knew existed on geometrical and wave optics! [174] and [175] have many topics in common, however.

[175] Y. A. Kravtsov and Y. I. Orlov, *Geometrical Optics of Inhomogeneous Media*, Springer, Berlin, 1990.

[176] Y. A. Kravtsov, *Geometrical Optics in Engineering Physics*, Alpha Science International, Harrow, U.K., 2005.

[177] D. S. Goodman, "General principles of geometric optics," chapter 1 in M. Bass, ed., *Handbook of Optics*, Volume I, McGraw-Hill, New York, 1995.

[178] B. J. Hoenders, "New ray (characteristic) equations for the equations of physics; Hamilton-Jacobi theory in more-dimensional space," *Pure Appl. Opt.*, **5** (1996) 275–282.

[179] J. A. Shapiro, "Classical Mechanics," 2010, http://www.physics.rutgers. edu/~shapiro/507/book1.pdf.

[180] P. R. Garabedian, *Partial Differential Equations*, John Wiley & Sons, New York, 1964.

[181] D. W. Moore and E. A. Spiegel, "The generation and propagation of waves in a compressible atmosphere," *Astrophys. J.*, **139** (1964) 48–71. One of two seminal papers for my choice of Ph.D. topic. The other one was [33].

[182] J. A. Adam, "Solutions of the inhomogeneous acoustic-gravity wave equation," *J. Phys. A Math. Gen.*, **10** (1977) L169–L173.

[183] J. B. Keller, "Rays, waves and asymptotics," *Bull. Amer. Math. Soc.*, **84** 1978), 727–750. Excellent summary by one of the main contributors to the subject.

[184] C. M. Bender and S. A. Orszag, *Advanced Mathematical Methods for Scientists and Engineers*, McGraw-Hill, Tokyo, 1978. Chock full of excellent topics for graduate classes.

[185] J. Mathews and R. Walker, *Mathematical Methods of Physics*, Benjamin, Reading, MA, 1970.

[186] L. M. Jones, *An Introduction to Mathematical Methods of Physics*, Benjamin Cummings, Menlo Park, CA, 1979.

[187] O. Vallée and M. Soares, *Airy Functions and Applications to Physics,* Imperial College Press, London, 2004.

[188] "NIST digital library of mathematical functions," http://dlmf.nist.gov/.

[189] P. B. Kahn, *Mathematical Methods for Scientists and Engineers*, John Wiley & Sons, New York, 1990.

[190] J. R. Holton, *An Introduction to Dynamic Meteorology*, Elsevier, Burlington, MA, 2004.

[191] A. E. Gill, *Atmosphere-Ocean Dynamics*, Academic Press, New Yotk, 1982.

[192] http://www.met.wau.nl/education/MWS/waves/.

[193] F. Ascani, http://www.soest.hawaii.edu/oceanography/researchers/francois/ RESEARCH/RESEARCH_NOTES/SCIENTIFIC_NOTES/Yanai_Wave. html

[194] C. J. Nappo, *An Introduction to Atmospheric Gravity Waves*, Academic Press, San Diego, 2002.

[195] D. R. Durran, "Mountain waves and downslope winds," in W. Blumen, ed., *Atmospheric Processes over Complex Terrain*, American Meteorological Society, 1990.

[196] R. S. Scorer, *Environmental Aerodynamics*, Ellis Horwood, Chichester, U.K., 1978.

[197] R. S. Scorer, "Theory of waves in the lee of mountains," *Quart. J. Roy. Met. Soc.*, **75** (1949) 41–56.

[198] R. S. Scorer, "The flow over a ridge," *Quart. J. Roy. Met. Soc.*, **79** (1953) 70–83.

[199] C. Eckart, *Hydrodynamics of Oceans and Atmospheres*, Pergamon Press, New York, 1960.

[200] J. A. Adam, "Critical layer singularities and complex eigenvalues in some differential equations of mathematical physics," *Phys. Repts.*, **142** (1986) 263–356.

[201] J. A. Adam, "Asymptotic solutions and spectral theory of linear wave equations," *Phys. Repts.*, **86** (1982) 217–316.

[202] R. Courant and D. Hilbert, *Methods of Mathematical Physics*, volumes I and II, Interscience Publishers, New York, 1966. Encyclopedic classics.

[203] J. A. Adam, *X and the City: Modeling Aspects of Urban Life*, Princeton University Press, Princeton, NJ, 2012.

[204] W. Hergert, *Gustav Mie: From Electromagnetic Scattering to an Electromagnetic View of Matter*, Springer Series in Optical Sciences 169, Springer, New York, 2011.

[205] W. Kauzmann, *Quantum Chemistry: An Introduction*, Academic Press, New York, 1957.

[206] I. Stakgold, *Boundary Value Problems of Mathematical Physics*, Volumes I and II, Macmillan, New York, 1967.

[207] A. M. Perelomov and Y. B. Zel'dovich, *Quantum Mechanics: Selected Topics*, World Scientific, Singapore, 1998.

[208] *McGraw-Hill Dictionary of Scientific & Technical Terms*, 6th ed., McGraw-Hill, Clarinda, IA, 2003.

[209] R. L. Deavenport, "A normal mode theory of an underwater acoustic duct by means of Green's function," *Radio Sci.*, **1** (1966) 709–724.

[210] Australian Government Bureau of Meteorology, "Tsunami facts and information," http://www.bom.gov.au/tsunami/info/.

[211] M. J. Lighthill, *Introduction to Fourier Analysis and Generalised Functions*, Cambridge University Press, Cambridge, 1958.

[212] J. A. Adam, *A Mathematical Nature Walk*, Princeton University Press, Princeton, NJ, 2009.

[213] http://www.pas.rochester.edu/~dmw/phy218/Lectures/Lect_75b.pdf.

[214] O. Bühler, *A Brief Introduction to Classical, Statistical, and Quantum Mechanics*, American Mathematical Society, Providence, RI, 2006.

[215] S. S. Holland, *Applied Analysis by the Hilbert Space Method*, Marcel Dekker, New York, 1990.

[216] http://twistedsifter.com/2012/03/15-incredible-cloud-formations/.

[217] http://glossary.ametsoc.org/wiki/Mountain_wave.

[218] http://glossary.ametsoc.org/wiki/Clear-air_turbulence.

[219] L. N. Howard, "Note on a paper of John W. Miles," *J. Fluid Mech.*, **10** (1961) 509–512. A delightful result: a semicircle theorem.

[220] G. T. Kochar and R. K. Jain, "Note on Howard's semicircle theorem," *J. Fluid Mech.*, **91** (1979) 489–491.

[221] M. B. Banerjee and J. R. Gupta, "On reducing Howard's semicircle for homogeneous shear flows," *J. Math. Anal. Apps.*, **130** (1988) 398–402.

[222] G. Chimonas, "The extension of the Miles-Howard theorem to compressible fluids," *J. Fluid Mech.*, **43** (1970) 833–836.

[223] J. A. Adam, "Magnetohydrodynamic wave energy flux in a stratified compressible atmosphere with shear," *Quart. J. Mech. Appl. Math.*, **31** (1978) 77–98.

[224] J. A. Adam, "Eigenvalue bounds in magnetoatmospheric shear flow," *J. Phys. A. Math. Gen.*, **13** (1980) 3325–3338.

[225] P. S. Cally and J. A. Adam, "Complex eigenvalue bounds in magnetoatmospheric shear flow," *Geophys. Astrophys. Fluid Dyn.*, **23** (1983) 57–67.

[226] S. Chandrasekhar, *Hydrodynamic and Hydromagnetic Stability*, Clarendon Press, Oxford, 1961. Yet another classic!

[227] R. Ehrlich, *The Cosmological Milkshake*, Rutgers University Press, New Brunswick, NJ, 1994. Like Craig Bohren's popular-level books, this one is great fun.

[228] M. Kerker, *The Scattering of Light and Other Electromagnetic Radiation*, Academic Press, New York, 1969. A well-matched competitor to [56].

[229] "Electromagnetic field of an accelerated charge," http://www.tapir. caltech.edu/~teviet/Waves/empulse.html.

[230] C. D. Collinson, "Investigation of planetary orbits using the Lenz-Runge vector," *Bull. IMA.,* **9** (1973) 377–378.

[231] http://physweb.bgu.ac.il/COURSES/PHYSICS3_physics/CLASS_ymeir/ diffraction.pdf.

[232] K. Yosida, *Lectures on Differential and Integral Equations*, Dover, New York, 1991.

[233] P. J. Wyatt, "Scattering of electromagnetic plane waves from inhomogeneous spherically symmetric objects," *Phys. Rev.,* **127** (1962) 1837–1843.

[234] D. M. Brink, *Semiclassical Methods for Nucleus-Nucleus Scattering*, Cambridge University Press, Cambridge,1985.

[235] M. A. Pinsky, *Introduction to Partial Differential Equations with Applications*, McGraw-Hill, New York,1984.

[236] J. A. Adam and M. Pohrivchak, "Evaluation of ray-path integrals in geometrical optics," *Int. J. Appl. Exp. Math.* (2016) 1:108.

[237] V. A. Fock, *Diffraction of Radio Waves around the Earth's Surface*, USSR Academy of Sciences, Moscow, 1946.

[238] H. M. Nussenzveig and W. J. Wiscombe, "Diffraction as tunneling," *Phys. Rev. Lett.,* **59** (1987) 1667–1670.

[239] H. M. Nussenzveig, "Light tunneling," *Prog. Opt.,* **50** (2007) 185–250.

[240] R. K. Luneburg, *Mathematical Theory of Optics*, University of California Press, Berkeley, 1964.

[241] A. Fletcher, T. Murphy, and A. Young, "Solutions of two optical problems," *Proc. Roy. Soc.* (London), **A223** (1954) 216–225.

[242] D. Drosdoff and A. Widom, "Snell's law from an elementary particle viewpoint," *Am. J. Phys.,* **73** (2005) 973–975.

[243] M. Berry, *Principles of Cosmology and Gravitation*, Cambridge University Press, Cambridge, 1976.

[244] G. Fjeldbo, A. J. Kliore, and R. Von Eshleman, "The neutral atmosphere of Venus as studied with the Mariner V radio occultation experiments," *Astron. J.,* **76** (1971) 123–140.

[245] R. Von Eshleman, E. M. Gurrola, and G. F. Lindal, "On the black hole lens and its foci," *Adv. Space Res.,* **9** (1989) 119–122.

[246] M. Sarbort and T. Tyc, "Spherical media and geodesic lenses in geometrical optics," *J. Opt.*, **14** (2012) 1–11.

[247] E. Posmentier, "The tales waves tell," *Cruising World*, (Nov. 2003) 82–85.

[248] "Seismic tomography: IRIS," www.iris.edu/hq/inclass/downloads/optional/269.

[249] J. L. Davis, *Mathematics of Wave Propagation*, Princeton University Press, Princeton, NJ, 2000.

[250] C. B. Officer, *Introduction to the Theory of Sound Transmission with Application to the Ocean*, McGraw-Hill, New York, 1958.

[251] C. B. Officer, *Introduction to Theoretical Geophysics*, Springer-Verlag, New York, 1974.

[252] "Eric Weisstein's World of PHYSICS," http://scienceworld.wolfram.com/physics/GravitationalScattering.html.

[253] Ian Poole, http://www.radio-electronics.com/info/propagation/em_waves/electromagnetic-reflection-refraction-diffraction.php.

[254] A. Doolittle, http://users.ece.gatech.edu/~alan/ECE6451/Lectures/ECE6451L8HarmonicOscilator.pdf.

[255] E. Meissen, "Integral asymptotics and the WKB approximation," http://math.arizona.edu/~meissen/docs/asymptotics.pdf.

[256] W. N. Everitt, "A catalogue of Sturm-Liouville differential equations," http://www.math.niu.edu/SL2/papers/birk0.pdf.

[257] S. A. Rakityansky, S. A. Sofianos, and K. Amos, "A method for calculating the Jost function for analytic potentials," *Il Nuovo Cimento* B, **111** (1996) 363–378.

[258] A. Ryblin, "On a complete analysis of high-energy scattering matrix asymptotics for one dimensional schrödinger operators with integrable potentials," *Proc. Am. Math. Soc.*, **130** (2002) 59–67.

[259] M. S. Marinov and B. Segevdag, "Analytical properties of scattering amplitudes in one-dimensional quantum theory," *J. Phys. A. Math. Gen.*, **29** (1996) 2839–2851.

[260] G. Adamovsky and M. V. Ötügen, "Morphology-dependent resonances and their applications to sensing in aerospace environments," *J. Aerospace Comp. Info. Communi.*, **5** (2008) 409–424.

[261] H.-W. Liu and X.-L. Sun, "Analytical solution for long-wave scattering by a submerged cylinder in an axi-symmetrical pit," *J. Marine Sci. Tech.*, **22** (2014) 542–549.

[262] M. V. Berry and K. E. Mount, "Semiclassical approximations in wave mechanics," *Reps. Prog. Phys.*, **35** (1972) 315–397.

[263] Les Cowley, http://www.atoptics.co.uk/highsky/hgrav.htm. Dr. Cowley's "Atmospheric optics" website (and its sister site, "Optics picture of the day") is a must-view site (as are "Earth science picture of the day" [http://epod.usra.edu/] and "Astronomy picture of the day" [http://apod.nasa.gov/apod/astropix.html]).

[264] G. Birkhoff and G.-C. Rota, *Ordinary Differential Equations*, Wiley, New York (1978).

[265] https://en.wikipedia.org/wiki/Gravity_wave.

[266] T. Young, "The Bakerian Lecture: Experiments and calculations relative to physical optics," *Phil. Trans. Roy. Soc.* London, **94** (1804) 1–16.

[267] P. Laven, Private communication, based on the presentation *Supernumerary arcs and geometrical optics* at the May 2016 Conference on Light and Color in Nature, Granada, Spain.

[268] G. B. Aity, "Supplement to a paper On the intensity of light in the neighbourhood of a caustic", *Trans. Camb. Phil. Soc.*, **8** (1848) 595–599.

[269] S. R. Wilk, *How the Ray Gun Got Its Zap: Odd Excursions into Optics*, Oxford University Press, Oxford, 2013.

[270] J. M. Pernter and F. M. Exner, *Meteorologische Optik*, W. Braumüller, Vienna, 1910.

[271] D. K. Lynch and S. N. Futterman, "Ulloa's observations of the glory, fogbow, and an unidentified phenomenon," *Appl. Opt.*, **30** (1991) 3538–3541.

[272] R. A. R. Tricker, *Bores, Breakers, Waves and Wakes*, Elsevier, New York, 1964.

[273] M. R. Vetrano, J. P. A. J. van Beeck, and M. L. Riethmuller, "Generalization of the rainbow Airy theory to nonuniform spheres," *Opt. Lett.*, **30** (2005) 658–660.

[274] U. Nuntaplook and J. A. Adam, "Scalar wave scattering by two-layer radial inhomogeneities," *Appl. Math. E-Notes*, **14** (2014) 185–192.

Index

Abel's (integral) equation, 531, 533, 535
acoustic cut-off frequency, 61
action, 12, 42, 267, 301–6, 310, 325–26
Airy, Sir George Biddle, 14, 101
Airy approximation, 4, 10, 95, 104, 486, 528, 562–63; integral (*see also* rainbow integral), xx, 14, 103, 130, 132–33, 135, 146, 154, 156, 160, 269, 439, 528, 565; wavefront, 100, 105
Airy diffraction pattern, 409, 411–12
Airy functions, 153*ff*, 435*ff*; related to Bessel functions, 154
angular momentum, 48, 308, 313–14, 316*ff*, 420, 424–25, 484, 488, 530; complex, 3, 10, 349, 384–85, 481, 530
anticorona, 89, 92
Arnold index, 127

barrier/well (*see also* potential barrier/well): centrifugal, 4, 513–16; rectangular, 440–41, 460, 516; spherical, 7, 557; triangular, 434, 438, 440–41
Bauer's expansion, 355, 422, 429
beach, 165, 205, 270, 343; sloping, 342–46
Bessel functions, 351, 362, 410, 423, 443, 471; Hankel functions, 8, 398, 401, 404; modified, 154; Riccati-Bessel/ Ricatti-Hankel, 384, 425, 429, 529; spherical Bessel, 7, 355, 387, 391, 423*ff*, 509; spherical Hankel, 7, 354, 391, 509; spherical Neumann, 509
bifurcation set, 13, 115
billow clouds, 208, 292
black hole lens, 535
bound states, 148, 253, 416, 463–64, 479–80, 488–90, 496, 498, 500, 511
Breit-Wigner formula, 480–82, 485
Brunt-Väisälä frequency, 61, 275, 296
bubbles, 364, 369

cardioid, 120
catastrophes (diffraction), 11–12, 113*ff*, 528; cusp, 14, 114–15; fold, 4–5, 11, 14, 113–15, 120, 126–28
caustics, 3–4, 9, 11–14, 18, 22, 77, 88, 108, 112–17, 119–21, 124, 127, 129–36, 165,

167, 170, 312–13, 417, 515, 527, 530, 562–63; teacup, 116, 119
center-of-mass (CoM) frame of reference, 308
central force, 308, 311, 322, 325*ff*, 424
central potential, 308–9, 325
characteristic equations, 33–34, 39–40, 43, 71–72
characteristics (method of), 34–36, 39–40, 217
confluent hypergeometric equation, 283, 342, 449
connection formulas, 145, 158, 435, 437
coriolis parameter, 63, 274, 279–80
corona, 91–93, 348, 412
cross section (scattering), 84, 87, 308, 311, 500, 512, 562–63, 566; definition 419; differential scattering, 87, 311*ff*, 329–30, 411, 417, 419, 451, 485, 563, 565; partial, 357; Rutherford, 329; total, 311, 328–30, 353–54, 358, 393, 418–19, 430, 446, 448, 481, 564
current density, 307, 378, 418
currents (in water), 162, 170, 345

D'Alembert solution, 55, 58, 359
Darboux (method of), 243, 245, 254
Debye expansion (or series), 4, 9–10, 112; potentials, 554; terms, 4–5
Descartes ray (*see also* rainbow ray), 11, 77, 81, 85, 103, 105, 110, 128–29, 526
diffraction, 1, 4, 7–14, 78, 84, 91–93, 96, 103–5, 122, 124, 264, 332, 348, 372, 377, 384, 397, 401, 409, 411–12, 415–17, 430, 481, 485, 516, 527–28, 530; Airy pattern, 409, 411–12; integral, 13, 124, 406
dispersion, 77, 84, 91, 98, 206, 214, 229, 526
dispersion relations, 54–55, 61–67, 70, 164, 196*ff*, 203*ff*, 239–40, 268, 275*ff*, 344

effective potential, 4, 309, 313, 317–18, 512–16
eikonal equation, 19–20, 39, 41, 45–46, 73–74, 162, 304–5
elliptic integrals, 542–43

energy flux, 52, 72, 165, 287; time-averaged, 352
Euclid, 17–18

Faxen-Holtzmark formula, 7, 429, 562–63
Fermat's principle, 21, 23–27, 32, 42–43, 45, 50, 74, 113, 303
first Born approximation, 426, 555
Fock functions, 487
Fourier transform, 55–57, 65, 148, 158, 238, 243, 257, 265, 271, 288
Fresnel integrals (cosine and sine), 124–25, 528
Froude's law, 227

glory, xix, 54, 76, 89–92, 313, 350, 416–17, 548, 564, 566
gravitational lensing, 531, 534
group speed/velocity, 57, 59–62, 69, 72, 75, 130, 171, 210–12, 214–15, 222, 225, 227, 229, 231, 265–66, 287, 521

Hamiltonian, 2, 34, 39, 41–42, 71, 301–2, 304, 325
Hamilton's principle, 39
Hankel functions (*see* Bessel functions)
Helmholtz equation, 6, 12, 33–34, 53, 238, 242, 305–6, 344, 351, 354, 358, 379–81, 386, 406, 409, 419–20, 513; vector, 379–81
Helmholtz-Kirchhoff integral theorem, 407
Helmholtz representation, 192
Helmholtz's theorem, 559
Hermite's equation, 283, 473
Heron of Alexandria, 18
Heron's problem, 25
Hertzian dipole, 394
Huygens, Christiaan, 83, 527
Huygens' principle, 28, 397, 406, 408, 527–28

impact parameter, 3–4, 7, 53, 79, 84, 97, 112, 128, 309*ff*, 324*ff*, 448, 458, 514–15, 530, 535
instability, 205*ff*, 293–94
inverse problem, 21, 181, 314, 348, 534; Luneberg, 534; Wiechert-Herglotz, 179
inverse scattering, 312
islands, 162, 332, 340

Jost solutions, 336–37, 491, 504; boundary conditions, 507–8; functions, 489, 491, 493, 509

Kelvin (Lord, William Thomson), 208, 216
Kelvin-Helmholtz instability, 208, 293
Kelvin wedge (*see also* wedge angle), 216, 220–21

Kepler problem, 326–7
Kepler's laws (of planetary motion), 309, 321, 326
Kirchhoff-Fresnel diffraction formula, 408
Kirchhoff-Huygens integral, 406

Lagrangian, 37, 71, 301–5, 325; equations, 42; time derivative, 274
Laguerre's equation, 342
Lamé constants, 190, 198
Lamé's theorem, 381, 560
Langer transformation, 442, 445, 456
leaky modes, 240–41. *See also* resonances
Legendre's equation, 140, 421, 474
Lenz-Runge vector, 322
Levinson's theorem, 488–89
limit-circle. *See* Weyl's theorem
limit-point. *See* Weyl's theorem
Liouville normal form, 141, 460, 469–70
Liouville transformation, 141, 335, 469, 556–57
Lorentz-Mie solution. *See* Mie solution
Luneberg lens, 314, 531, 544

Maxwell's equations, 378, 550
Maxwell's fish-eye lens, 540, 557–58
Mie scattering. *See* Mie solution
Mie solution, 2, 7, 10, 94, 349, 371, 377, 384–86, 417, 513
mirage theorem, 19, 48–50, 241, 539
modulus, bulk, 174, 190; shear, 174, 190; of compressibility (*see* modulus, bulk); of rigidity (*see* modulus, shear)
morphology-dependent resonances (MDRs), 463, 512
mountain streams, 364

Navier equations, 189*ff*
nephroid, 117, 120

optical path length, 19, 21, 43, 46, 51, 303
optical theorem, 358
orbiting (scattering), 313, 514, 534
orbits, 44, 313, 317, 319–20, 323
osculating parabolas, 117

parabolic cylindrical coordinates, 449
paraxial approximation, 113, 123
parity theorem, 502
phase plane, 333
phase shift, 22, 111–12, 311, 356, 424, 428–29, 432, 445, 454, 477, 480–82, 488–90, 493, 500, 517–18, 521, 555, 563, 565; Coulomb, 453–54
Plancherel's theorem, 265
Planck's constant, 143, 416, 419, 530; reduced, 143, 427
Poisson summation formula, 8, 485–86, 563

Poisson's ratio, 190
polarization, 112, 371*ff*, 376–77, 384,
    397–98, 405–6, 477, 529; of rainbow,
    85*ff*
potential well (*see also* barrier/well), 6, 148,
    250, 313, 330, 335, 416, 420, 428, 434,
    478, 501, 511, 513–16, 557; barrier, 149,
    416, 426–27, 482, 504, 557
principle of the argument, 489
Ptolemy's inequality, 26–28

quasi-bound states, 514, 516. *See also*
    resonances
quasi-stationary states, 500. *See also*
    resonances

radiation condition, 66–69, 238–38, 247–48,
    288, 360, 362; Sommerfeld, 247,
    358–59, 408
rainbow, 3*ff*, 51, 54, 76*ff*, 95*ff*, 128, 312, 349,
    372, 383*ff*, 415*ff*, 481, 523*ff*, 533, 542,
    548, 562*ff*
rainbow angle, 5–6, 11, 13, 77, 88–89, 95–96,
    110, 112, 312, 315, 415, 530, 562,
    564–65
rainbow integral (*see also* Airy integral), 101,
    103, 154, 528
rainbow ray (*see also* Descartes ray), 3–4, 81,
    97, 103, 105–6, 112, 128, 311, 527–28,
    548
Rayleigh scattering, 93–94, 348–49, 364,
    371, 373, 375, 377, 384, 392, 396, 529
rays, 5, 17*ff*, 33*ff*, 39, 70*ff*, 78*ff*, 121, 162*ff*,
    174*ff*
reflection coefficient, 9, 149, 252, 336, 428,
    504
Regge poles, 3, 8, 385, 481, 485, 494, 530
Regge trajectory, 483
resonances, 8, 247, 416, 463, 478, 512*ff*
Richardson criterion, 294; number, 293,
    295–96
ridge, 164*ff*, 285*ff*, 334, 505
Riemann-Lebesgue lemma, 263

S-matrix, 17, 337*ff*, 356, 477*ff*, 504*ff*; poles
    of, 498*ff*
Sabatier transform, 457
saddle point method, 146
saddle points, 4, 10–11, 147, 385, 415,
    481–82, 485, 530, 563
scattering: amplitude, 4, 7, 9–10, 54, 352*ff*,
    384–85, 411*ff*, 426, 446–47, 480*ff*, 504,
    516, 529, 563*ff*; angle, 3*ff*, 92, 308*ff*,
    324, 409,419, 448; Coulomb, 448*ff*, 535;
    cross section (*see* cross section);
    differential cross section (*see* cross
    section); hard sphere, 328; Lorentz-Mie,
    383; matrix (*see also S*-matrix), 7, 334*ff*,

416, 478, 489, 499, 500; rainbow, 312;
    Rayleigh (*see* Rayleigh scattering),
    Rutherford, 329, 451
Schrödinger equation (time-independent),
    143, 148, 242, 250, 282, 306, 334, 426,
    435, 449, 454, 459*ff*, 487, 491, 499, 504,
    554
Scorer parameter, 286, 291
secret Pythagorean theorem (SPT), 19, 176
seiches, 209
semicircle theorem, 292*ff*
shadow zones, 184
slowness, 20–21, 176, 178
Snellius, Willebrord, 18
Snell's laws (of reflection, refraction), 5, 23*ff*,
    50, 80*ff*, 163, 175, 249, 514–15, 539
Sommerfeld radiation condition. *See*
    Radiation condition
Sommerfeld-Watson transform. *See* Watson
    transform
spherical aberration, 117
spherical harmonics, 8, 379, 421, 477, 513;
    vector form, 553
stationary phase, method of, 64, 67, 130,
    234–36, 260, 271, 404, 564
steepest descent, method of, 434, 564
Sturm-Liouville equation, 141, 459*ff*

Taylor-Goldstein equation, 296
time-independent radial Schrödinger equation
    (TIRSE), 454, 487, 491
transitional approximation, 417, 562
transmission coefficient, 9, 149, 252, 336,
    427, 438–40, 504, 506; poles of, 506
transverse electric mode (TE), 513, 551*ff*, 556
transverse magnetic mode (TM), 550*ff*,
    553*ff*
tsunamis, 165, 201, 229, 255*ff*
tunneling, 4, 311, 313, 416, 437, 463, 514–16
turning point, 146, 150, 152–53, 309, 333,
    437–38, 456, 514–15, 540, 563

uniform approximation, 417, 562–63

virtual states. *See also* resonances, 499

Watson transform, 8, 10, 384, 398, 481–82,
    485–86, 530
wave amplification, 293, 347
wave trapping (by a ridge), 166–67
waveguide, 173, 237*ff*
wavenumber surfaces, 65
wavepacket, 71, 75, 212, 501, 521
waves: acoustic, 57, 61, 277*ff*, 349, 355, 477;
    atmospheric, 61–62, 273*ff*; body, 175,
    191, 193; deep water, 203*ff*, 211, 344;
    edge, 342*ff*; elastic, 189*ff*., 559;
    electromagnetic, 192, 371*ff*, 379, 398*ff*;

waves: (*Continued*) evanescent (*see also* surface waves), 62, 85, 240–41, 246, 277–78, 287*ff*, 415; guided, 239–40; inertia-gravity, 283; internal gravity, 208, 281, 296–97; Kelvin, 280, 284; lee/mountain, 62, 285*ff*; Love, 173, 191, 198; *P*, 174*ff*, 185, 191; plane, 55*ff*, 70, 260, 305, 379; Rayleigh, 173, 191, 197*ff*; Rossby, 63, 279–80, 284; Rossby-gravity, 281, 284; *S*, 174*ff*, 185, 191; seismic, 173*ff*; shallow water, 204*ff*, 212, 229, 268, 332; ship, 214*ff*; surface (*see also* evanescent), 8, 62, 173, 175, 191, 195*ff*, 201, 231*ff*, 240, 263; surface gravity, 48, 57, 164, 171, 200*ff*, 249, 265*ff*, 332*ff*; transient, 255; trapped, 333*ff*; Yanai, 284

wedge angle (ship waves; *see also* Kelvin wedge), 216, 236
Weierstrass approximation theorem, 263
Weyl's theorem, 471–74
Weyl-Titchmarsh *m*-function, 472
whispering gallery modes, 512, 514
wind shear, 292
winding number, 487
WKB(J) approximation, xxi, 2, 4, 137*ff*, 434*ff*, 435*ff*

Young's modulus, 190
Young, Thomas, 76, 111–12

Princeton Series in Applied Mathematics

Chaotic Transitions in Deterministic and Stochastic Dynamical Systems: Applications of Melnikov Processes in Engineering, Physics, and Neuroscience, Emil Simiu

Selfsimilar Processes, Paul Embrechts and Makoto Maejima

Self-Regularity: A New Paradigm for Primal-Dual Interior-Point Algorithms, Jiming Peng, Cornelis Roos, and Tamás Terlaky

Analytic Theory of Global Bifurcation: An Introduction, Boris Buffoni and John Toland

Entropy, Andreas Greven, Gerhard Keller, and Gerald Warnecke, editors

Auxiliary Signal Design for Failure Detection, Stephen L. Campbell and Ramine Nikoukhah

Thermodynamics: A Dynamical Systems Approach, Wassim M. Haddad, VijaySekhar Chellaboina, and Sergey G. Nersesov

Optimization: Insights and Applications, Jan Brinkhuis and Vladimir Tikhomirov

Max Plus at Work, Modeling and Analysis of Synchronized Systems: A Course on Max-Plus Algebra and Its Applications, Bernd Heidergott, Geert Jan Olsder, and Jacob van der Woude

Impulsive and Hybrid Dynamical Systems: Stability, Dissipativity, and Control, Wassim M. Haddad, VijaySekhar Chellaboina, and Sergey G. Nersesov

The Traveling Salesman Problem: A Computational Study, David L. Applegate, Robert E. Bixby, Vasek Chvatal, and William J. Cook

Positive Definite Matrices, Rajendra Bhatia

Genomic Signal Processing, Ilya Shmulevich and Edward R. Dougherty

Wave Scattering by Time-Dependent Perturbations: An Introduction, G. F. Roach

Algebraic Curves over a Finite Field, J.W.P. Hirschfeld, G. Korchmáros, and F. Torres

Distributed Control of Robotic Networks: A Mathematical Approach to Motion Coordination Algorithms, Francesco Bullo, Jorge Cortés, and Sonia Martínez

Robust Optimization, Aharon Ben-Tal, Laurent El Ghaoui, and Arkadi Nemirovski

Control Theoretic Splines: Optimal Control, Statistics, and Path Planning, Magnus Egerstedt and Clyde Martin

Matrices, Moments, and Quadrature with Applications, Gene H. Golub
and Gérard Meurant

Totally Nonnegative Matrices, Shaun M. Fallat and Charles R. Johnson

Matrix Completions, Moments, and Sums of Hermitian Squares,
Mihály Bakonyi and Hugo J. Woerdeman

Modern Anti-windup Synthesis: Control Augmentation for Actuator Saturation,
Luca Zaccarian and Andrew W. Teel

Graph Theoretic Methods in Multiagent Networks, Mehran Mesbahi and
Magnus Egerstedt

*Stability and Control of Large-Scale Dynamical Systems: A Vector
Dissipative Systems Approach*, Wassim M. Haddad and Sergey G. Nersesov

*Mathematical Analysis of Deterministic and Stochastic Problems in Complex
Media Electromagnetics*, G. F. Roach, I. G. Stratis, and A. N. Yannacopoulos

Topics in Quaternion Linear Algebra, Leiba Rodman

Hidden Markov Processes: Theory and Applications to Biology, M. Vidyasagar

Mathematical Methods in Elasticity Imaging, Habib Ammari, Elie Bretin,
Josselin Garnier, Hyeonbae Kang, Hyundae Lee, and Abdul Wahab

Rays, Waves, and Scattering: Topics in Classical Mathematical Physics,
by John A. Adam